SYNCHROTRON RADIATION AND STRUCTURAL PROTEOMICS

Pan Stanford Series on Nanobiotechnology

Series Editor
Claudio Nicolini

Titles in the Series

Published

Vol. 1
Nanobiotechnology and Nanobiosciences
Claudio Nicolini, ed.
2008
978-981-4241-38-0 (Hardcover)
978-981-4241-39-7 (eBook)

Vol. 2
Functional Proteomics and Nanotechnology-Based Microarrays
Claudio Nicolini and Joshua LaBaer, eds.
2010
978-981-4267-76-2 (Hardcover)
978-981-4267-77-9 (eBook)

Vol. 3
Synchrotron Radiation and Structural Proteomics
Eugenia Pechkova and Christian Riekel, eds.
2012
978-981-4267-38-0 (Hardcover)
978-981-4267-93-9 (eBook)

Forthcoming

Vol. 4
Nanobiotechnological Devices and Processes for Energy Compatible with the Environment
Claudio Nicolini and Carlo V. Bruschi, eds.

Vol. 5
Nanobioelectronics: Principles and Applications
Claudio Nicolini and Norbert Hampp, eds.

SYNCHROTRON RADIATION AND STRUCTURAL PROTEOMICS

edited by
Eugenia Pechkova
Christian Riekel

Pan Stanford Publishing

Published by

Pan Stanford Publishing Pte. Ltd.
Penthouse Level, Suntec Tower 3
8 Temasek Boulevard
Singapore 038988

Email: editorial@panstanford.com
Web: www.panstanford.com

British Library Cataloguing-in-Publication Data
A catalogue record for this book is available from the British Library.

Synchrotron Radiation and Structural Proteomics

ISBN 978-981-4267-38-0 (Hardcover)
ISBN 978-981-4267-93-9 (eBook)

Printed in the USA

To my grandmother
Nina Ksandopulo
with love

— E.P.

Contents

Section II
Methods

Preface

The research reported in this volume is dedicated to the recent developments in structural proteomics and synchrotron radiation and constitutes the third volume, edited by Eugenia Pechkova from Italy, and Christian Riekel, of the Pan Stanford Nanobiotechnology Series, edited by Claudio Nicolini.

It represents, with its 15 chapters, the outcome of the very broad international cooperation of numerous countries (Russia, Japan, USA, and Europe) centered around the European Synchrotron Radiation Facility in Grenoble (France) and the Nanoworld Institute in Genova (Italy), formed by the CIRSDNNOB Center at the University of Genova and by the El.B.A. Foundation, at the conclusion of numerous experiments and activities, including the XVIII El.B.A. Nanoforum held at Genova by the Fondazione El.B.A., with headquarters in Rome (Italy).

The volume includes contributions from the advanced research groups operating at different synchrotrons (ESRF, BNL, Spring 8, CHESS) and from structural proteomics groups active all over the world in correlated technologies such as NMR and X-ray crystallography, thereby giving a unique international flavor to the book.

Every chapter, consisting of an updated review article, contains exciting new research results which have never been published so far.

With the major focus on proteomics and synchrotron radiation, several correlated scientific and technological objectives have been carried out and reported in this volume of the Nanobiotechnology series.

We are glad that the coupling of synchrotron radiation and nanotechnology has synergistically expanded our capabilities to understand protein structure and other important events down to the atomic scale, yielding unique complementary information.

Eugenia Pechkova and Christian Riekel
Volume 3 Editors

Claudio Nicolini
Series Editor

Genoa
June 15, 2010

SECTION I
SOURCE AND INSTRUMENTATION

Chapter 1

PROTEIN MICRO- AND NANOCRYSTALLOGRAPHY USING SYNCHROTRON RADIATION

Christian Riekel[a], Manfred Burghammer[a], and Dmitry Popov[b]

[a]*European Synchrotron Radiation Facility, B.P.220, F38043 Grenoble Cedex, France*

[b]*HPCAT, Geophysical Laboratory, Carnegie Institution of Washington, 9700 South Cass Ave., Bldg. 434E, Argonne, IL 60439, USA*

riekel@esrf.fr

We review in this article the current status of protein microcrystallography and the scope for protein nanocrystallography at third-generation synchrotron radiation sources. Practical issues of sample environments, radiation damage, and sample manipulation will be discussed.

1.1 INTRODUCTION

Synchrotron radiation (SR) centers worldwide are making considerable efforts of developing high-throughput protein crystallography for structural proteomics (Arzt *et al.*, 2005). The production of well-diffracting protein crystals remains, however, the limiting step in protein structure analysis (Derewenda, 2004).

Synchrotron Radiation and Structural Proteomics
Edited by Eugenia Pechkova and Christian Riekel

www.panstanford.com

A complementary approach for studying "difficult" protein structures is protein microcrystallography (μPX) (Riekel *et al.*, 2005; Schneider, 2008). "Difficult" implies here proteins which are difficult to crystallize such as membrane proteins. Other reasons for using SR microbeams are the reduction of background scattering from sample environments (Sanishvili *et al.*, 2008) or the study of more perfect domains in a larger crystal. Finally, the mosaic spread can be reduced as the use of microcrystals increases flash-cooling rates and allows using more diluted cryoprotectants (Garman and Mitchel, 1996; Garman and Schneider, 1997; Chinte *et al.*, 2005).

An overview on selected μPX beamlines (BLs) which are operational or in the commissioning phase is given in Table 1.1. The present review will focus principally on μPX developments at the ESRF-ID13 BL.

Table 1.1 Selected parameters of several μPX beamlines which are operational or in the commissioning phase. μPX capabilities are also planned for SOLEIL, Spring-8, Photon Factory, PETRA III, MAX IV, and NSLS-II. (BL: beamline).

SR[web page]	BL[name]	Focus (h×v; μm²)	Flux (ph/s)	Wavelength (nm)	μPX use	Status
ESRF/France[a]	ID13	~1 × 1	~10^{11}	0.095	Partial	Operational
ESRF/France[b]	ID23-2	7 × 4	4 × 10^{11}	0.087	Dedicated	Operational
SLS/Switzerland[c]	X06-SA	15 × 5	2 × 10^{11}	0.217–0.071	Partial	Operational
Diamond/UK[1,d]	I24	5–50	~10^{10} (30 μm)	0.177–0.062	Dedicated	Operational
APS/USA[2,e]	23-ID-B/D	5–20	7 × 10^{10} (5 μm)	0.207–0.062	Partial	Operational
Australia[f]	MX2	37 × 32	4 × 10^{12}	0.225–0.044	Partial	Operational

[1]Evans *et al.*, 2007; [2]Fischetti *et al.*, 2009.

Protein nanocrystallography can be defined as: "using nanotechnology for the production and characterization of protein crystals at the nano- and subnano-scale" (AFM, μPX...) (Pechkova and Nicolini, 2004). The extent to which protein crystallography

[a] www.esrf.eu/UsersAndScience/Experiments/SoftMatter/ID13/

[b] www.esrf.eu/UsersAndScience/Experiments/MX/About_our_beamlines/ID23-2

[c] http://sls.web.psi.ch/view.php/beamlines/px/index.html

[d] www.diamond.ac.uk/Home/Beamlines/MX/I24.html

[e] www.gmca.anl.gov/; μPX capabilities have become available at several APS BLs.

[f] www.synchrotron.org.au/index.php/aussyncbeamlines/macromolecular-crystallography/protein-micro-crystal-and-small-molecule-x-ray-diffraction

with nanometer-sized beams (nanoPX) will find practical applications will depend not only on the scientific interest in studying ultrasmall crystalline domains but also on systematic studies of radiation damage issues, the availability of advanced sample environments including high precision goniometers, sample characterization, and manipulation tools. Pushing the limits to smaller crystals and smaller beam sizes will require the integration of more and more nanotechnology in a μPX BL and annex laboratories, which justifies the term nanocrystallography.

Finally, we mention two further complementary approaches to high-throughput PX which do not require single crystals. Small-angle X-ray scattering (SAXS) allows obtaining low-resolution, averaged solution scattering structures (Svergun, 2007). The use of μSAXS (Riekel *et al.*, 2009) is of interest for microfluidic environments and combinatorial approaches (Toft *et al.*, 2008). Lensless imaging by scanning coherent diffraction imaging (CXDI) (Faulkner *et al.*, 2004; Rodenburg and Faulkner, 2004) using coherent beams could provide in the future complimentary real space information in the <30 nm resolution range (Schroer *et al.*, 2008; Huang *et al.*, 2009; Lima *et al.*, 2009; Nishino *et al.*, 2009).

1.2 SR SOURCES AND INSTRUMENTATION

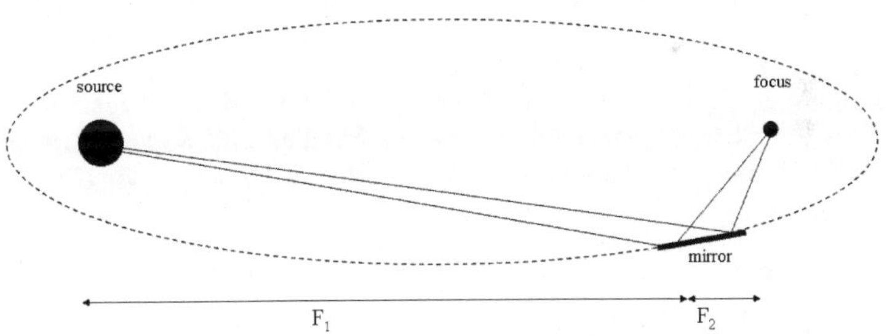

Figure 1.1 Schematic design of mirror focusing system. The horizontal and vertical source size ($s_{h,v}$) is demagnified at the focus by: $s_{h,v} \times (F_2/F_1)$ and the source divergence ($s'_{h,v}$) by: $s'_{h,v} \times (F_1/F_2)$.

Any focusing optical system will reduce the source size and increase the beam divergence as shown schematically in Fig. 1.1 for mirror focusing. A compromise between spot size and divergence at the sample position is therefore required (Riekel *et al.*, 2000).

To obtain enough flux for μPX in a spot of a few micrometers or smaller, a high-brilliance undulator source from a third-generation SR source is required (Riekel *et al.*, 2005) (Fig. 1.2). The photon source parameters of an ESRF low-β undulator, which is optimized for a small source size,[g] are indicated in Table 1.2.

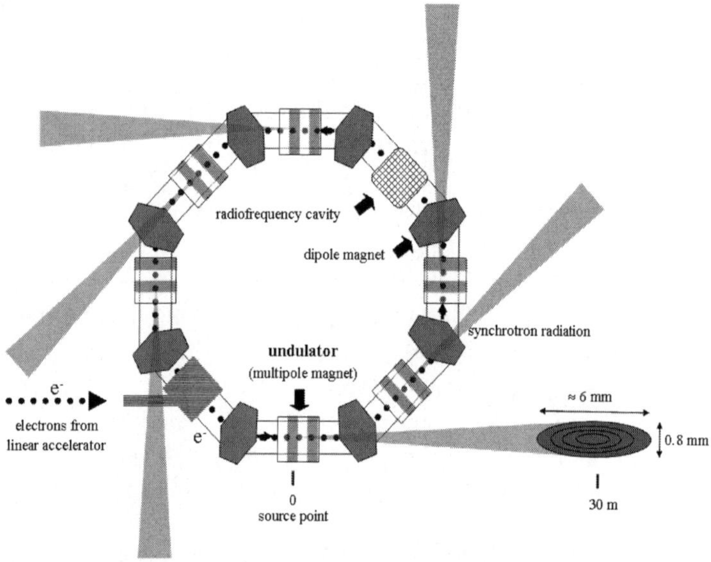

Figure 1.2 Schematic design of a third-generation SR source. The size of the synchrotron radiation beam at 30 m (fwhm) corresponds to an ESRF low-β undulator (Table 1.2). (Reprinted with the permission from Burghammer *et al.* X-ray synchrotron radiation microdiffraction on fibrous biopolymers like cellulose and in particular spider silks. *In Recent Research Developments in Macromolecules*. Vol. 7, pp. 103–125, © 2003, Research Signpost.) See also Color Insert.

Table 1.2 Photon source parameters (full-width-half-maximum (fwhm) values) of ESRF low-β undulator.

Size (μm)	134	25
Divergence (μrad)	0.2	0.01

In practice, a divergence of about 1 mrad for a beam size of about 1 μm can be attained for μPX at an undulator BL. This requires reducing the beam divergence by a collimator and/or the

[g] www.esrf.fr/Accelerators/Performance

limited angular acceptance of the optics. An increase in source brilliance, as expected from future generations of SR sources such as the energy recovery linear accelerator (Bilderback *et al.*, 2005, Bilderback *et al.*, 2010), will provide more flexibility for maintaining a low divergence for a small and intense focal spot.

1.2.1 SR Microfocus Optics and Instrumentation

1.2.1.1 Optics

The principal requirement for an X-ray optical system for μPX is the availability of enough space around the sample to accommodate sample environments, such as a cryoflow system, or various spectroscopic tools.

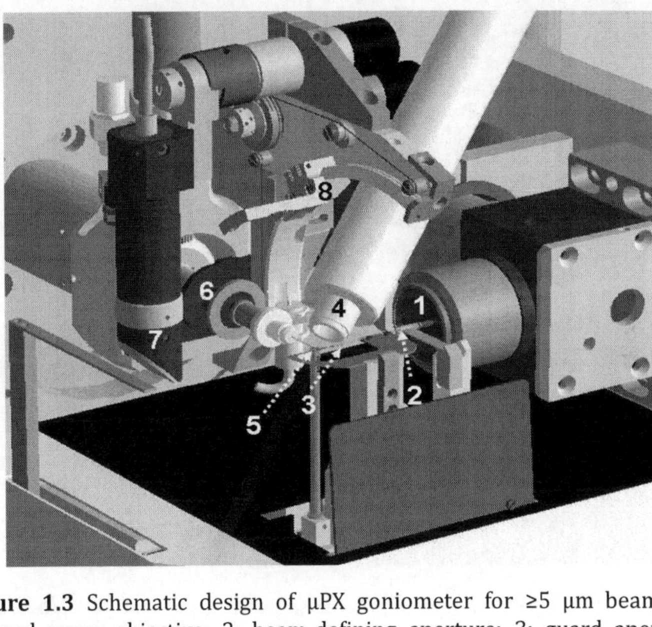

Figure 1.3 Schematic design of μPX goniometer for ≥5 μm beams; 1: pierced zoom objective; 2: beam-defining aperture; 3: guard aperture; 4: cryoflow nozzle; 5: scintillator and photodiode; 6: horizontal rotation axis; 7: illumination; 8: sample transfer system. (Reprinted with the permission from Perrakis *et al.* Protein microcrystals and the design of a diffractometer: current experience and plans at EMBL and ESRF/ID13. Acta Crystallographica D, 55, pp. 1765–1770, © 1999, International Union of Crystallography.) See also Color Insert.

This rules out "near field" X-ray optics such as multiple-bounce tapered glass capillaries (Engström and Riekel, 1996; Riekel, 2000). Mirror optics (Hignette *et al.*, 2001) or single-bounce glass capillaries (Lamb *et al.*, 2007) provide the energy tuning range required for anomalous dispersion experiments and can also be used for large band pass applications, which have not yet been explored for μPX (Riekel *et al.*, 2005). Particularly stable beams down to about 5 μm can be obtained by a beam-defining aperture cutting into a larger beam (Riekel, 2000; Sanishvili *et al.*, 2008; Fischetti *et al.*, 2009). A second (guard) aperture provides a background reduction and is also used in μSAXS pinhole cameras (Riekel *et al.*, 2000).

Figure 1.4 Schematic design of μPX goniometer for 1 μm beam (Volkringer *et al.*, 2007; Moukhametzianov *et al.*, 2008). 1: KB mirror; 2: microionization chamber and aperture; 3: cryoflow system; 4: piezo manipulator; 5: air bearing rotation axis; 6: microscopy objective. The *x/y/z* scanning stage is not visible. (Reprinted with the permission from Volkringer *et al.* A new microdiffraction set-up for nanoporous MOF-type solids. *Nature Materials*, 6, pp. 760–764, © 2007, Nature Publishing Group, division of Macmillan Publishers Limited.) See also Color Insert.

1.2.1.2 Goniometers

A μPX goniometer developed by an EMBL/ESRF collaboration incorporates several advanced features (Perrakis *et al.*, 1999) (Fig. 1.3). Sample observation and alignment is facilitated by an on-axis zoom microscope. The minimum beam size is defined by the integrated aperture to about 5 μm. This prototype has evolved into a commercial system with an integrated sample changer for automatic loading of pre-frozen samples and is installed at several SR BLs such as ESRF-ID23-2 (Table 1.1).[h]

The smallest beam size currently in routine use for μPX is an about 1 × 1 μm^2 beam from a KB mirror system at the ESRF-ID13 BL (Moukhametzianov *et al.*, 2008). The optical system is adapted to a scanning μPX goniometer based on a vertical air bearing rotation axis and an *x*/*y*/*z* scanning unit (Volkringer *et al.*, 2007; Moukhametzianov *et al.*, 2008) (Fig. 1.4). The use of a compact piezo micromanipulator allows keeping the sample in a 1 μm diameter circle of confusion through 360° rotation (Volkringer *et al.*, 2007). This setup does not provide automatic sample transfer features and serves as a R&D system for the use of micron-sized beams.

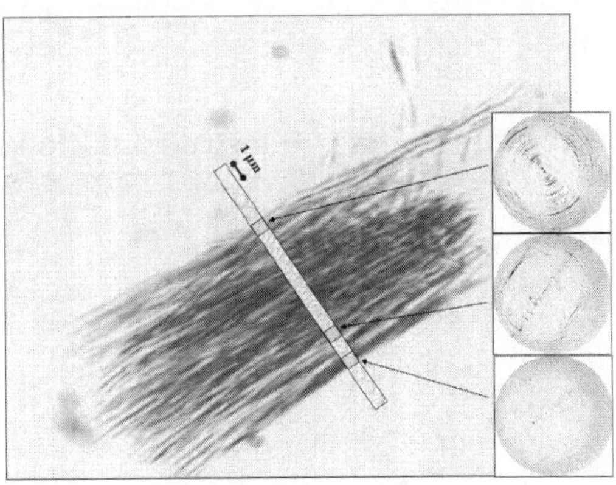

Figure 1.5 Linear diffraction mapping of a bundle of amyloid fibrils through a 0.8 μm beam by a MAR 165 CCD with 1 μm step increment at the ESRF-ID13 BL. The bundle was kept in a borosilicate capillary filled with mother liquor. A zoom of selected patterns is shown (Courtesy D. Eisenberg).

[h] MAATEL, Voreppe, France; web page: www.maatel.com/en/

A scanning goniometer is useful for screening crystals and/or distributing a radiation dose by mapping across extended or multiple crystals. As an example we show in Fig. 1.5 a scan of a bunch of amyloid fibrils of about 1 μm diameter with a 1 μm step increment through a 0.8 μm beam. Selected patterns revealed single crystalline features, which led subsequently to a high-resolution crystal structure determination (Nelson *et al.*, 2005). In the same way one could check the quality of crystals in a cryoloop or directly in a crystal growth environment.

1.2.1.3 Detectors

Pixel detectors such as the PILATUS detector[i] (Brönnimann *et al.*, 2005; Huelsen *et al.*, 2006) are starting to revolutionize protein crystallography with zero readout noise and full frame readout times of a few milliseconds. A 1 KHz readout frame rate has been demonstrated for the Medipix2 chip (Ponchut *et al.*, 2007). The high readout speed allows new types of data collection such as continuous-sample-rotation mode (Huelsen *et al.*, 2006). Angular increments of about 0.02° improve the signal/noise ratio of weak high-resolution reflections. A μPX goniometer should allow in principle any three-dimensional trajectory in continuous scanning mode.

1.2.2 SR Nanofocus Optics and Instrumentation

1.2.2.1 Optics

Third-generation SR sources have resulted in an impressive development of nanofocusing X-ray optical systems, the potential of which is largely unexplored (Yun, 1995; McNulty, 1998, 2001; Noehammer *et al.*, 2005). The demonstrated range of nanometer "hard" X-ray beams (wavelength around 0.1 nm) is currently about 7–140 nm using mirrors (Mimura *et al.*, 2010), Fresnel optics (Noehammer *et al.*, 2005; Kang *et al.*, 2008), capillaries (Bilderback *et al.*, 1994), multilayers, refractive optics (Schroer *et al.*, 2005), and waveguide optics (Pfeiffer *et al.*, 2002). X-ray beam sizes of a few

[i] DECTRIS, Baden, Switzerland, web page: http://dectris.com/

nanometers appear in principle to be possible (Schroer and Lengeler, 2005; Jefimovs *et al.*, 2007).

1.2.2.2 Nanogoniometer

We demonstrate the effort required of making use of nanometer-sized beams by showing the schematic design of an advanced nanogoniometer developed for the ESRF-ID13 BL (Fig. 1.6). This nanogoniometer serves as a test bed for the development of techniques for transfer, alignment, and data collection of ultrasmall protein and biopolymer crystals. It is based on air bearing *x/y* stages developed for microlithography with laser interferometer position encoders, which are integrated in a marble support structure (Newport/Microcontrole).

A vertical air bearing rotation stage (MICOS) provides sub-micrometer eccentricity at the sample position. A microhexapod system is foreseen for sample alignment and *x/y/z* scanning with sub-10 nm resolution. We note the stringent requirements for temperature stability (< ±0.05 °C) and low vibration level for such an equipment requiring a special hutch layout.

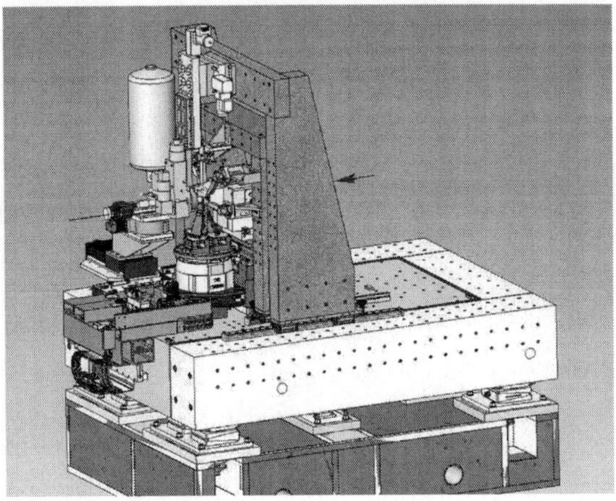

Figure 1.6 Nanogoniometer based on *x/y* air bearing translational stages and a vertical air bearing rotation axis carrying a microhexapod. The mechanical elements requiring the highest stability (e.g., apertures) are located on an inner marble support structure while elements which might introduce vibrations (e.g., cryoflow, fluorescence detector, optical microscope) are installed on a separate outer marble structure. See also Color Insert.

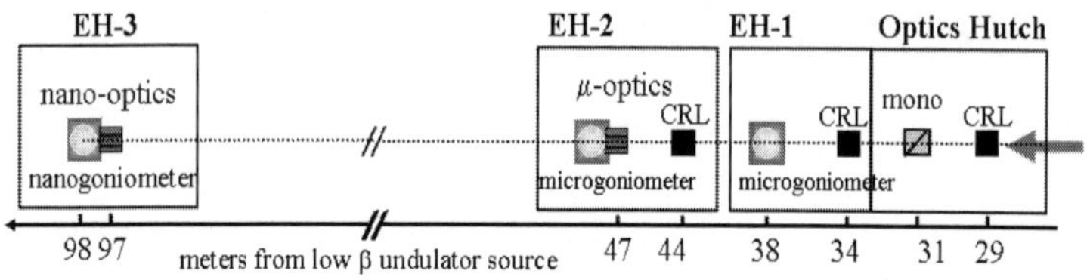

Figure 1.7 The ESRF-ID13 beamline is composed of an optics hutch (OH), two experimental hutches for microfocus applications (EH-1/2), and a hutch for nanofocus applications (EH-3). On-axis focusing parabolic Be compound refractive lenses (CRL) (Schroer *et al.*, 2002) are used for optimizing the SR beam parameters for the optical elements and samples.

This has resulted at the ESRF-ID13 BL in separate experimental hutches for microbeam and nanobeams applications (Fig. 1.7). The installation of X-ray nano-optical systems at nearly 100 m from the source allows maintaining sufficient space for sample environments. The use of on-axis, refractive, parabolic Be-lenses (Lengeler *et al.*, 2001) provides flexibility in adapting the size and divergence of the beam to the acceptance of the optical elements and samples.

1.2.3 Sample Mounts and Manipulation

Sample mounts used for μPX under cryoconditions are often the same as used for high-throughput PX (Fig. 1.8a). The reduced amount of liquid for a Kapton MicroLoop™ (Thorne *et al.*, 2003)[j] can help to further reduce scattering from the surrounding cryocoolant (Sanishvili *et al.*, 2008; Wagner *et al.*, 2009) (Fig. 1.8b).

For samples, which can be easily manipulated in air, glass capillary tips provide a very low scattering background (Fig. 1.8c). The samples stick to the glass surface by a thin viscous film, which avoids local deformations. In μPX, diffraction patterns have to be usually collected at several positions during rotation to distribute the radiation dose. This requires a scanning goniometer as shown in Fig. 1.4. By using metal grids covered by an ultrathin carbon film, multiple crystals can be mounted and successively measured (Riekel *et al.*, 2005) (Fig. 1.8d).

[j] MiTiGen LLC, Ithaca NY; web page: www.mitigen.com

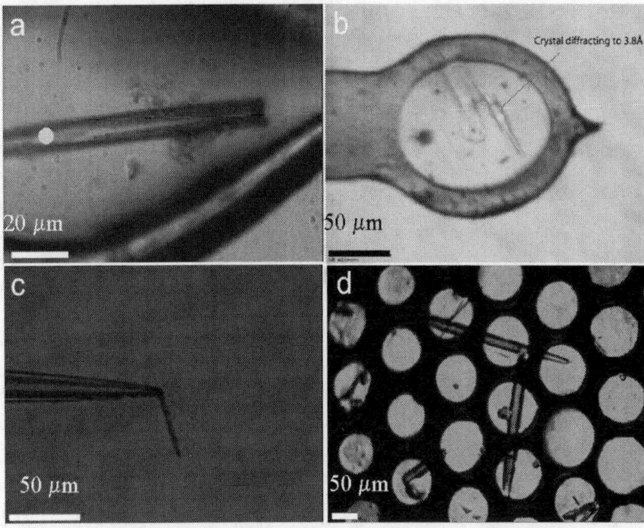

Figure 1.8 (a) Bovine rhodopsin crystal in a cryoloop (Li *et al.*, 2004) (circle corresponds to 5 μm beam). (b) Human β1-adrenergic G-protein-coupled receptor crystals in a Kapton MicroLoopTM (Rasmussen *et al.*, 2007) (circle corresponds to 1 μm beam). (c) Amyloid fibril attached to a glass capillary tip (Nelson *et al.*, 2005). (d) Multiple bovine rhodopsin crystals supported on a metal grid covered with thin carbon foil (Riekel *et al.*, 2005). (Part a: reprinted with permission from Li *et al.* Structure of bovine rhodopsin in a trigonal crystal form. *Journal of Molecular Biology*, 343, pp. 1409–1438, © 2004 Elsevier; Part b: reprinted with permission from Rasmussen *et al.* Crystal structure of the human [bgr]: 2: adrenergic G-protein-coupled receptor. *Nature*, 450, pp. 383–387, © 2007, Nature Publishing Group, division of Macmillan Publishers Limited; Part c: reprinted with permission from Nelson *et al.* Structure of the cross-beta spine of amyloid-like fibrils. *Nature*, 435, pp. 773–778, © 2005, Nature Publishing Group, division of Macmillan Publishers Limited; Part d: reprinted with permission from Riekel *et al.* Protein crystallography microdiffraction. *Current Opinion in Structural Biology*, 15, pp. 556–562, © 2005, Elsevier.) See also Color Insert.

Harvesting of crystals for μPX is usually performed manually. One can expect that this will be at least partially automated as operator-assisted harvesting has been demonstrated for protein crystals down to 10 μm (Viola *et al.*, 2006).

As crystals become smaller, new manipulation techniques from nanotechnology will have to be explored. Some of these techniques could also be integrated into the BL.

Laser microdissection techniques have been used for dissecting polymer and biopolymer samples (Davies *et al.*, 2008b; Seidel *et al.*, 2008). Such techniques could also be used for separating microcrystals from a crystal aggregate (e.g., Fig. 1.9) or for cutting a single domain with higher perfection from a larger crystal showing imperfections (Lovelace *et al.*, 2005).

Figure 1.9 Optical microscopy image of A-amylose single crystal aggregate. (Reprinted with permission from Popov *et al.* Amylose single crystals: unit cell refinement from synchrotron radiation microdiffraction data. *Macromolecules*, 39, pp. 3704–3706, © 2006, American Chemical Society.)

Optical tweezers (Ashkin and Dziedzic, 1971) could replace manual or robotic crystal harvesting techniques. This could, for example, allow placing microcrystals at well-defined places of a grid structure (Fig. 1.8d). A more futuristic vision is to capture microcrystals in multiple traps, created by optical holography. This would allow distributing the radiation dose across multiple copies of a crystal. Such a scheme could well become reality in a not too distant future as the trapping and diffraction from liposomes in a double optical trap has already been demonstrated at the ESRF-ID13 BL (Amenitsch *et al.*, 2007; Cojoc *et al.*, 2007) (Fig. 1.10). Optical tweezers traps with picoNewton forces could be useful for handling very sensitive crystals (e.g., virus crystals). Another possible application of such techniques could be in the manipulation of ultrasmall crystals produced under special growth conditions such as nanoliter volume microdrops (Bodenstaff *et al.*, 2002; Santesson *et al.*, 2003).

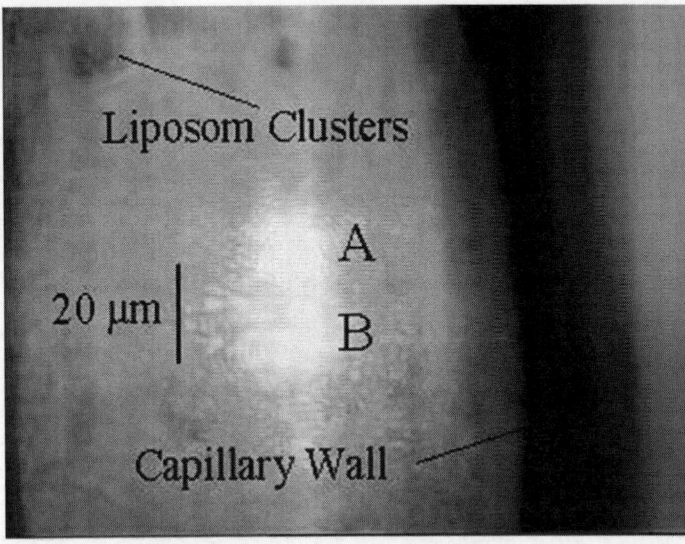

Figure 1.10 Optical microscopy image of two laser-optical trapping spots A, B in a liquid filled capillary, under which two POPE liposome clusters (palmitoyl-oleyl-phosphatidylethanolamine) have been trapped (not seen due to the intense infrared light). A non-trapped cluster is also indicated. (Reprinted with the permission from Cojoc *et al.* Scanning X-ray micro-diffraction of optically manipulated liposomes. *Applied Physics Letters*, 91, art. no. 234107, © 2007, American Institute of Physics.) See also Color Insert.

Inkjet systems (Lee, 2003) have been used for obtaining SAXS solution scattering data from ballistic microdrops of about 80 µm diameter under stroboscopic conditions (Graceffa *et al.*, 2009). In principle one could use microdrop technology for transferring microcrystals into a SR beam.

A modulated free liquid jet can also become the source of a microdrop stream and appears to be less prone to nozzle clogging (Ganan-Calvo, 1998; DePonte *et al.*, 2008).

1.2.4 Software Issues

The data reduction software must fit the requirements introduced by severe radiation damage and restricted scattering power of microcrystals. Due to the intensity loss only limited number of images can be collected on a crystal for a given position (Riekel *et al.*, 2005). Despite these challenges unit cell parameters and intensities of reflections can be successfully calculated from data sets collected on selected crystals of few microns size (Popov *et al.*, 2006).

1.2.4.1 Merging

In most cases the only way to obtain reasonable accuracy of intensity measurements is merging of X-ray data sets collected on many selected crystals or on different parts of the sample which improves the redundancy of the resulting data set. Due to higher multiplicity of a measured reflection its average intensity value is more precise. Effects of the radiation damage strongly reduce completeness of intensity data collected on a crystal, and this problem can be also solved by merging of the data from several crystalline particles or points of a crystal.

The most widely used PX data reduction packages such as XDS (Kabsch, 2001), Denzo/Scalepack (Otwinowski and Minor, 1997) and CCP4 (CCP4, 1994) have features to merge properly μPX data sets. The program BEST (Popov and Bourenkov, 2003), which was developed to plan optimal data collection procedures, can help to overcome the μPX challenges. This software uses a few images collected with small exposure time to estimate experimental parameters which will allow to get data of given highest resolution in shortest period or with lowest absorbed radiation dose.

Application of BEST makes possible obtaining as much data as possible on a selected crystal, hence saves the experimental time. This in turn provides better redundancy of the resulting data set.

1.2.4.2 Multiple crystals

Microcrystals are often arranged into polycrystalline aggregates as shown for A-amylose crystals (Fig. 1.9) or amyloid fibrils (Fig. 1.5). In such a case it is difficult to avoid simultaneously hitting of different crystals by X-ray beam, but reflections from the crystallites can be selected. Useful algorithms to index diffraction patterns collected on multigrain samples are applied in packages ImageD11[k] and CrysAlis.[l]

---

[k] http://fable.wiki.sourceforge.net/imaged11

[l] Oxford Diffraction (2007). Oxford Diffraction Ltd., Xcalibur CCD system, CrysAlisPro Software system.

1.3 RADIATION DAMAGE

The dose limitation in hydrated protein crystals is due to primary and secondary radiation damage (Henderson, 1990; Glaeser *et al.*, 2000; Teng and Moffat, 2002; Ravelli and Garman, 2006; Holton, 2009). Secondary events are due to the generation of photoelectrons which lose their energy while creating radicals.

Cryocooling allows reducing the mobility of the heavier radicals. Simulations show a high photoelectron escape probability for microcrystals of a few micrometers size suggesting that the dose limit due to secondary radiation events has been underestimated (Nave and Hill, 2005; Cowan and Nave, 2008).

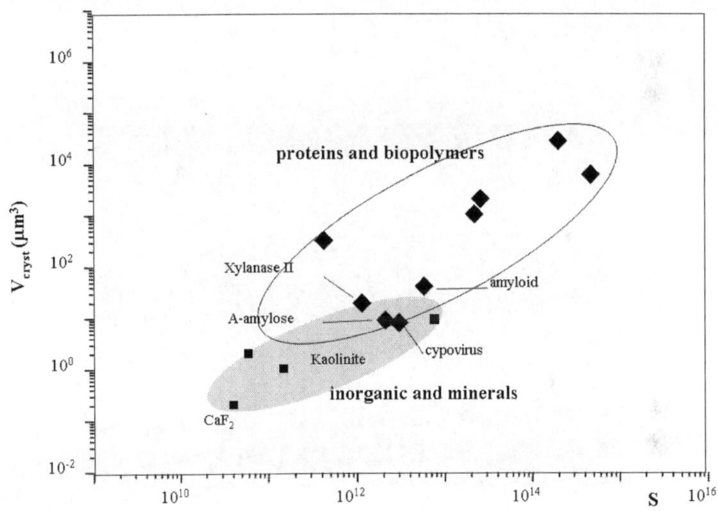

Figure 1.11 Scattering power S as a function of crystal volume for selected protein crystals (Riekel *et al.*, 2005). The scattering power is defined as: $S = (F_{000}/V_{cell})^2 \times \lambda^3 \times V_{cryst}$, where F_{000} is the 0-order structure factor, V_{cell} the unit cell volume, λ the wavelength, and V_{cryst} the crystal volume (Neder *et al.*, 1999). References: Xylanase II: Moukhametzianov *et al.*, 2008; A-amylose: Popov *et al.*, 2006; cypovirus: Coulibaly *et al.*, 2007; Kaolinite: Neder *et al.*, 1996, 1999; CaF_2: Burghammer, 1997; amyloid fibrils: Nelson *et al.*, 2005. (Reprinted with permission from Riekel *et al.* Protein crystallography microdiffraction. *Current Opinion in Structural Biology*, 15, pp. 556–562, © 2005, Elsevier.)

This is corroborated by experimental data on selected protein crystals approaching the 10 μm³ volume range which appear to support a higher dose than previously assumed (Henderson, 1990) (Fig. 1.11). We note the systematically smaller sample volumes and scattering powers of inorganic and mineralogical crystals, which reflect the absence of radiation damage effects.

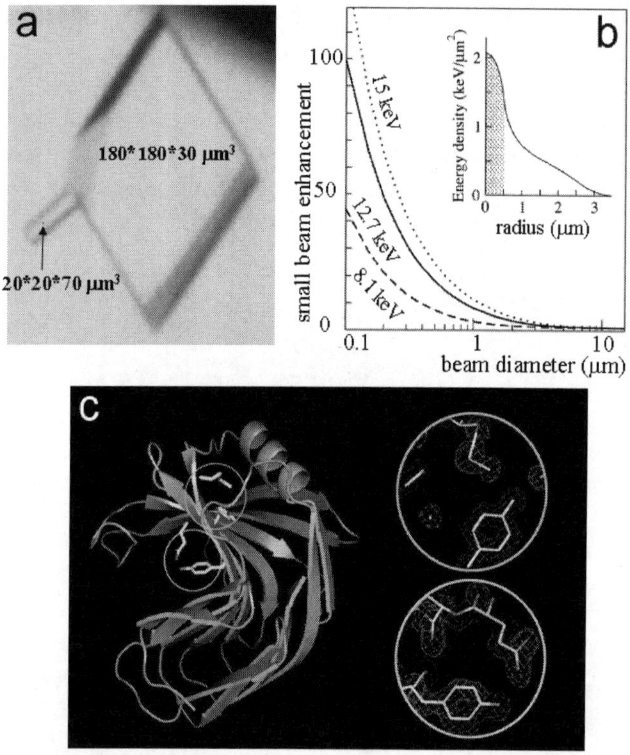

Figure 1.12 (a) The arrow points to the 1 μm beam spot in the smaller part of the Xylanase II crystal. (b) Calculated small beam enhancement (ratio of the total energy deposited in the crystal relative to that actually deposited in the beam volume) as a function of the beam diameter for three energies. Inset: calculated deposited energy density per generated 12.7 keV photoelectron as a function of radius for a 20 μm thick sample illuminated by a 1 μm diameter beam. On average, 12.1 keV is deposited in the whole crystal, but only 1.4 keV of this is contained within the illuminated volume (indicated by the shading). (c) Secondary Xylanase II structure and selected residues shown by electron density map. (Reprinted with the permission from Moukhametzianov *et al.* Protein crystallography with a micron-sized synchrotron radiation beam. *Acta Crystallographica D*, 64, pp. 158–166, © 2008, International Union of Crystallography.) See also Color Insert.

This model is also supported by μPX experiments on a Xylanase II crystal shown in Fig. 1.12a (Moukhametzianov *et al.*, 2008). The smaller part of the crystal shown in Fig. 1.12a was rotated through a 1 μm diameter beam so that fresh material was swept into the beam track at every rotational step. The simulations suggest that the photoelectrons created in the photon beam track deposit most of their energy in the surrounding matrix already at 12.1 keV (Fig. 1.12b).

The refined secondary structure with the weighted electron density map of selected radiation sensitive residues shows no evidence for radiation damage[m] (Fig. 1.12c). The dependence of the small beam enhancement from the beam diameter (Fig. 1.12b) suggests extending radiation damage studies to sub-micrometer beams.

Radiation damage effects are usually spotted via a systematic increase of scale factors (Li *et al.*, 2004; Riekel *et al.*, 2005). Raman spectroscopy is sensitive to specific bonds and intermolecular forces in proteins (Carpentier *et al.*, 2007) and can therefore provide often a more specific probe for radiation damage. In the case of DNA crystals the photon-induced breaking of specific bonds could be monitored by Raman spectroscopy (McGeehan *et al.*, 2007).

The combined use of microRaman and microdiffraction has been demonstrated at the ESRF-ID13 BL for soft condensed matter applications. The microRaman probe head (inVia system by Renishaw) allows focusing a 785 nm laser beam into an about 1 μm focal spot on the sample (Davies *et al.*, 2005, 2009). The SR beam passes through a hole in the focusing mirror (Fig. 1.13). The system has been used for localizing radiation damage in polypropylene films during a scanning diffraction experiment (Davies *et al.*, 2008a). The capability for probing the spatial extent of radiation damage *in situ* by scanning microRaman in a protein crystal would be very useful for μPX. The microRaman system shown in Fig. 1.13 will, however, have to evolve to a somewhat larger focal distance to allow the use of cryoflow systems.

[m] Space group: P21; unit cell: a, b, c: 4.057, 3.896, 5.746 nm; α, β, γ: 90, 110.75, 90°.

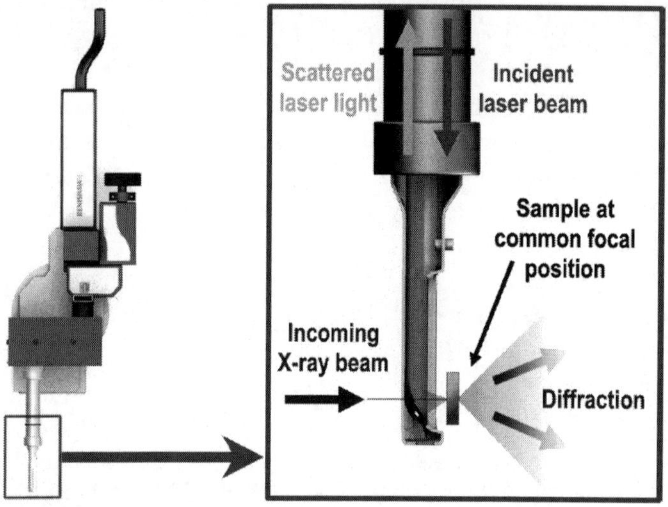

Figure 1.13 MicroRaman probe head. The laser beam is focused on the sample by a parabolic mirror. A hole in the center of the mirror allows the X-ray beam to pass. The common focal position is at 2–3 mm from the exit of the probe head housing.

1.4 CASE STUDY OF μPX CRYSTAL STRUCTURE REFINEMENT

Islet amyloid polypeptide (IAPP) is associated with type II diabetes. Structural studies on IAPP fibrils by electron diffraction, X-ray fiber diffraction, and cryo-electron microscopy have suggested extended "cross-β" strands of amyloid-like fibrils (Summer and Serpell, 2004).

The atomic structure of the IAPP fragment NNFGAIL could be solved using the ESRF-ID13 BL (Wiltzius *et al.*, 2009) (Fig. 1.14). Two to three data sets were collected from a typical crystal shown in Fig. 1.14a for a total of 11 data sets from different points of several crystals. The exposure time was 2–20 seconds for a rotation step of 1°. The data were collected for a 1 μm beam using the μPX goniometer shown in Fig. 1.4. The resulting data set of 0.18 nm highest resolution and redundancy 14 was used to solve the structure with the molecular replacement technique and refine it up to R_{free} = 0.21. The results clearly show atomic details of the cross-β structure (Fig. 1.14b) and can serve as model structure for discussing several proposed fibrillar models.

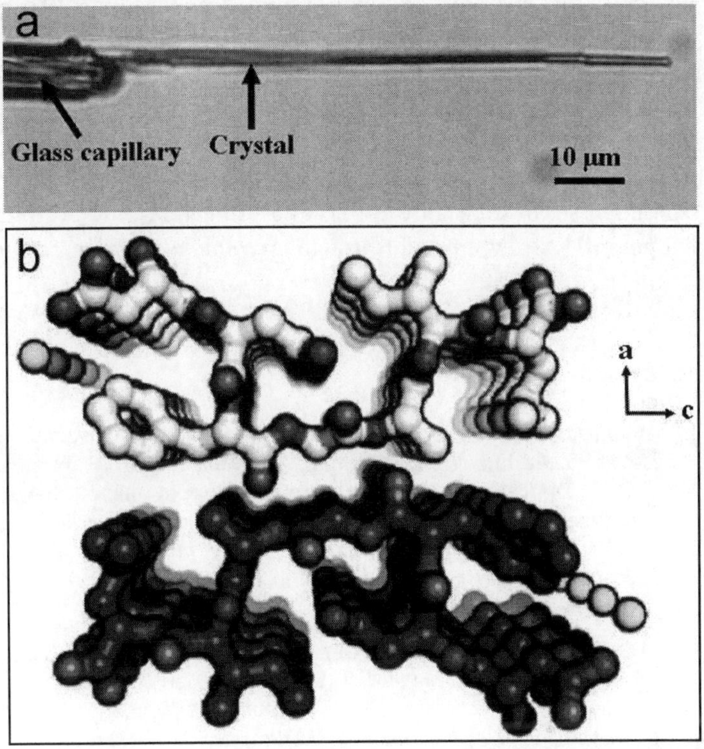

Figure 1.14 (a) Typical crystal of NNFGAIL used to collect intensity data (Wiltzius *et al.*, 2009). (b) The structure of NNFGAIL consists of two close-packed β-sheets, viewed down the sheets. Water molecules are shown as yellow spheres. (Reprinted with the permission from Wiltzius *et al.*, © 2009, The Protein Society.) See also Color Insert.

1.5 CONCLUSIONS AND OUTLOOK

μPX with beam sizes down to a few micrometers is becoming a standard tool at several SR sources. μPX experiments at the 1 μm scale have been demonstrated and smaller beams are technically feasible but largely scientifically unexplored. To make optimum use of current μPX and possibly future nanoPX facilities, the BL infrastructure and laboratory environment should be adapted to advanced nanotechnology instrumentation.

Acknowledgments

Many ideas for the development of μPX are based on a close collaboration with former and present ID13 staff[n] and ID13 users, in particular from EMBL-Grenoble (S. Cusack *et al.*) and MRC Cambridge (G. Schertler *et al.*). We are also grateful to the support of the ID13 technical staff: L. Eybert, M. Nicola, L. Lardiere. We acknowledge contributions to the development of the nanogoniometer and the microRaman setup through the EEC FP6 project *SAXIER*. Software development was supported through a *Human Frontier Science Program* (HFSP) grant.

References

1. Amenitsch, H., Cojoc, D., Rappolt, M., Sartori, B., Laggner, P., Ferrari, E., Garbin, V., Burghammer, M., Riekel, C. and Di Fabrizio, E. (2007). Optical tweezers for sample fixing in micro-diffaction experiments. In *AIP Conference Proceedings: Synchrotron Radiation Instrumentation: Ninth International Conference on Synchrotron Radiation Instrumentation* (eds. Choi, J. Y. and Rah, S.), Vol. 879, pp. 1287–1290, Daegu, Korea, 2006.

2. Arzt, S., Beteva, A., Cipriani, F., Delageniere, S., Felisaz, F., Forstner, G., Gordon, E., Launer, L., Lavault, B., Leonard, G., Mairs, T., McCarthy, A., McCarthy, J., McSweeney, S., Meyer, J., Mitchell, E., Monaco, S., Nurizzo, D., Ravelli, R., Rey, V., Shepard, W., Spruce, D., Svensson, O. and Theveneau. P. (2005). Automation of macromolecular crystallography beamlines. *Progress in Biophysics & Molecular Biology*, **89**, pp. 124–152.

3. Ashkin, A. and Dziedzic, J. M. (1971). Optical levitation by radiation pressure. *Applied Physics Letters*, **19**, pp. 283–285.

4. Bilderback, D. H., Hoffman, S. A. and Thiel, D. J. (1994). Nanometer spatial resolution achieved in hard X-ray imaging and Laue diffraction experiments. *Science*, **263**, pp. 201–203.

5. Bilderback, D. H., Elleaume, P. and Weckert. E. (2005). Energy recovery linac (ERL) coherent hard x-ray sources. *Journal of Physics,* **12**, pp. S773–S797.

6. Bilderback, D.H., Brock, J.D., Dale, D.S., Finkelstein, K.D., Pfeifer, M.A. and Gruner, S.M. (2010). Review of third and next

[n] www.esrf.fr/UsersAndScience/Experiments/SCMatter/ID13/contacts

generation synchrotron light sources. *New Journal of Physics B*, **38**, pp. 1–27.

7. Bodenstaff, E. R., Hoedemaeke, F. J., Kuil, M. E., Vrind, H. P. M. and Abrahams, J. P. (2002). The prospects of protein nanocrystallography. *Acta Crystallographica D*, **58**, pp. 1901–1906.

8. Brönnimann, C., Eikenberry, E. F., Henrich, B., Horisberger, R., Huelsen, G., Pohl, E., Schmitt, B., Schulze-Briese, C., Suzuki, M., Tomizaki, T., Toyokawa, H. and Wagner, A. (2006). The PILATUS 1M detector. *Journal of Synchrotron Radiation*, **13**, pp. 120–130.

9. Burghammer, M. (1997). PhD thesis. Faculty for Geoscience, Ludwig Maximilians University, Munich.

10. Burghammer, M., Müller, M. and Riekel, C. (2003). X-ray synchrotron radiation microdiffraction on fibrous biopolymers like cellulose and in particular spider silks. In *Recent Research Developments in Macromolecules* (ed. Pandalai, S. G.), Vol. **7**, pp. 103–125. Research Signpost, Trivandrum, India.

11. Carpentier, P., Royant, A., Ohana, J. and Bourgeois, D. (2007). Advances in spectroscopic methods for biological crystals. 2. Raman spectroscopy. *Journal of Applied Crystallography*, **40**, pp. 1113–1122.

12. CCP4 (1994). Collaborative Computational Project, Number 4. *Acta Crystallographica D*, **50**, pp. 760–763.

13. Chinte, U., Shah, B., deWitt, K., Kirschbaum, K., Pinkerton, A. A. and Schall, C. (2005). Sample size: an important parameter in flash-cooling macromolecular crystallization solutions. *Journal of Applied Crystallography*, **38**, pp. 412–419.

14. Cojoc, D., Ferrari, E., Garbin, V., Di Fabrizio, E., Amenitsch, H., Rappolt, M., Sartori, B., Laggner, P., Burghammer, M. and Riekel, C. (2007). Scanning X-ray micro-diffraction of optically manipulated liposomes. *Applied Physics Letters*, **91**, art. no. 234107.

15. Coulibaly, F., Chiu, E., Ikeda, K., Gutmann, S., Haebel, P. W., Schulze-Briese, C., Mori, H. and Metcalf, P. (2007). The molecular organization of cypovirus polyhedra. *Nature*, **446**, pp. 97–101.

16. Cowan, J. A. and Nave, C. (2008). The optimum conditions to collect X-ray data from very small samples. Journal of *Synchrotron Radiation*, **15**, pp. 458–462.

17. Davies, R. J., Burghammer, M. and Riekel, C. (2005). Simultaneous microRaman and synchrotron radiation microdiffraction: tools for materials characterization. A*pplied Physics Letters*, **87**, art. no. 264105.

18. Davies, R. J., Burghammer, M. and Riekel, C. (2008a). Micro-Raman spectroscopy as an in situ tool for probing radiation damage during microdiffraction experiments in soft condensed matter. *Macromolecules*, **41**, pp. 7251–7253.

19. Davies, R. J., Koenig, C., Burghammer, M. and Riekel, C. (2008b). On-axis microbeam wide- and small-angle scattering experiments of a sectioned poly(p-phenylene terephthalamide) fiber. *Applied Physics Letters*, **92**, art. no. 101903-1.

20. Davies, R. J., Burghammer, M. and Riekel, C. (2009). A combined microRaman and microdiffraction setup at the European Synchrotron Radiation Facility ID13 beamline. *Journal of Synchrotron Radiation*, **16**, pp. 22–29.

21. DePonte, D. P., Weierstall, U., Schmidt, K., Warner, J., Starodub, D., Spence, J. C. H. and Doak, R. B. (2008). Gas dynamic virtual nozzle for generation of microscopic droplet streams. *Journal of Physics D: Applied Physics*, **41**, art. no. 195505.

22. Derewenda, Z. S. (2004). Rational protein crystallization by mutational surface engineering. *Structure*, **12**, pp. 529–535.

23. Engström, P. and Riekel, C. (1996). Low-angle synchrotron radiation diffraction with glass capillary optics. *Journal of Synchrotron Radiation*, **3**, pp. 97–100.

24. Evans G., Alianelli L., Burt M., Wagner A. and Sawhney K.J.S. (2007) Diamond beamline 124: A flexible instrument for macromolecular micro-crystallography. In: *AIP Conference Proceedings: Synchrotron Radiation Instrumentation: Ninth International Conference on Synchrotron Radiation Instrumentation* (eds. Choi, J. Y. and Rah, S.), Vol. 879, pp. 836–839. Daegu, Korea, 2006.

25. Faulkner, H. M. L. and Rodenburg, H. M. (2004). Movable aperture lensless transmission microscopy: a novel phase retrieval algorithm. *Physical Review Letters*, **93**, art. no. 023903.

26. Fischetti, R. F., Xu, S., Yoder, D. W., Becker, M., Nagarajan, V., Sanishvili, R., Hilgart M. C., Stepanov, S., Makarov, O. and Smith, J. L. (2009). Mini-beam collimator enables microcrystallography experiments on standard beamlines. *Journal of Synchrotron Radiation*, **16**, pp. 217–225.

27. Ganan-Calvo, A. M. (1998). Generation of steady liquid microthreads and micron-sized monodisperse sprays in gas streams. *Physical Review Letters*, **80**, pp. 285–288.

28. Garman, E. and Mitchel, R. P. (1996). Glycerol concentrations required for cryoprotection of 50 typical protein crystallization solutions. *Journal of Applied Crystallography*, **29**, pp. 584–587.

29. Garman, E. F. and Schneider, T. R. (1997). Macromolecular cryocrystallography. *Journal of Applied Crystallography*, **30**, pp. 211–237.

30. Glaeser, R., Facciotti, M., Walian, P., Rouhani, S., Holton, J., MacDowell, A., Celestre, R., Cambie, D. and Padmore, H. (2000). Characterization of conditions required for X-ray diffraction experiments with protein microcrystals. *Biophysical Journal*, **78**, pp. 3178–3185.

31. Graceffa, R., Burghammer, M., Davies, R., Ponchut, C. and Riekel, C. (2009). Studying macromolecular solutions without wall effects by stroboscopic small-angle X-ray scattering. *Applied Physics Letters*, **94**, art. no. 062902.

32. Henderson, R. (1990). Cryo-protection of protein crystals against radiation damage in electron and X-ray diffraction. *Proceedings of the Royal Society of London Series B-Biological Sciences*, **241**, pp. 6–8.

33. Hignette, O., Rostaing, G., Cloetens, P., Rommeveaux, A., Ludwig, W. and Freund A. K. (2001). Submicron focusing of hard x-rays with reflecting surfaces at the ESRF. *In X-Ray Micro- and Nano-Focusing: Applications and Techniques II*. (ed. McNulty, I.), Vol. 4499, pp. 105–116. SPIE, San Diego.

34. Holton, J. M. (2009). A beginner's guide to radiation damage. *Journal of Synchrotron Radiation*, **16**, pp. 133–142.

35. Huang, X., Nelson, J., Kirz, J., Lima, E., Marchesini, S., Miao, H., Neiman, A. M., Shapiro, D., Steinbrener, J., Stewart, A., Turner, J. J. and Jacobsen, C. (2009). Soft X-ray diffraction microscopy of a frozen hydrated yeast cell. *Physical Review Letters*, **103**, art. no. 198101.

36. Huelsen, G., Broennimann, C., Eikenberry, E. F. and Wagner, A. (2006). Protein crystallography with a novel large-area pixel detector. *Journal of Applied Crystallography*, **39**, pp. 550–557.

37. Jefimovs, K., Vila-Comamala, J., Pilvi, T., Raabe, J., Ritala, M. and David, C. (2007). Zone-doubling technique to produce ultrahigh-resolution X-ray optics. *Physical Review Letters*, **99**, art. no. 264801.

38. Kabsch, W. (2001). Integration, scaling, space-group assignment and post refinement. *In International Tables for Crystallography* (eds. Rossmann, M. G. and Arnold, E.), Vol. F, pp. 218–225. Kluwer Academic Publishers, Dordrecht, Netherlands.

39. Kang, H. C., Yan, H., Winarski, R. P., Holt, M. V., Maser, J., Liu, C., Conley, R., Vogt, S., Macrander, A. T. and Stephenson, G. B. (2008). Focusing of hard x-rays to 16 nanometers with a multilayer Laue lens. *Applied Physics Letters*, **92**, art. no. 221114.

40. Lamb, J. S., Cornaby, S., Andresen, K., Kwok, K. L., Park, H. Y., Qiu, X., Smilgies, D. M., Bilderback, D. H. and Pollack, L. (2007). Focusing capillary optics for use in solution scattering small-angle X-ray scattering. *Journal of Applied Crystallography*, **40**, pp. 193–195.

41. Lee, E. R. (2003). *Microdrop Generation* (ed. Lyshevski, S. E.). CRC Press, Boca Raton.

42. Lengeler, B., Schroer, C. G., Benner, B., Gunzler, T. F., Kuhlmann, M., Tummler, J., Simionovici, A. S., Drakapoulos, M., Snigirev, A. and Snigireva, I. (2001). Parabolic refractive X-ray lenses: a breakthrough in X-ray optics. *Nuclear Instruments & Methods A*, **467–468**, pp. 944–950.

43. Li, J., Edwards, P., Burghammer, M., Villa, C. and Schertler, G. F. X. (2004). Structure of bovine rhodopsin in a trigonal crystal form. *Journal of Molecular Biology*, **343**, pp. 1409–1438.

44. Lima, E., Wiegart, L., Pernot, P., Howells, M., Timmins, J., Zontone, F. and Madsen, A. (2009). Cryogenic X-ray diffraction for biological samples. *Physical Review Letters*, **103**, art. no. 198102.

45. Lovelace, J. J., Murphy, C. R., Bellamy, H. D., Brister, K., Pahl, R. and Borgstahl, G. E. O. (2005). Advances in digital topography for characterizing imperfections in protein crystals. *Journal of Applied Crystallography*, **38**, pp. 512–519.

46. McGeehan, J. E., Carpentier, P., Royant, A., Bourgeois, D. and Ravelli, R. B. G. (2007). X-ray radiation-induced damage in DNA monitored by online Raman. *Journal of Synchrotron Radiation*, **14**, pp. 99–108.

47. McNulty, I. (ed.). (1998). *X-ray Microfocusing: Applications and Techniques*, Vol. 3449, pp. 228. SPIE, San Diego.

48. McNulty, I. (ed.). (2001). *X-ray Micro- and Nano-Focusing: Applications and Techniques II*, Vol. 4499, p. 152. SPIE, Argonne.

49. Mimura, H., Handa, S., Kimura, T., Yuomoto, H., Yamakawa, D., Yokoyama, H., Matsuyama, S., Inagaki, K., Yamamura, K., Sano, Y., Tamasaku, K., Nishino, Y., Yabashi, M., Ishikawa, T. and Yamauchi, K. (2010). Breaking the 10 nm barrier in hard-X-ray focusing. *Nature Physics*, **6**, pp. 122–125.

50. Moukhametzianov, R., Burghammer, M., Edwards, P.C., Petitdemange, S., Popov, D., Fransen, M., McMullan, G., Schertler, G. F. X. and Riekel, C. (2008). Protein crystallography with a micron-sized synchrotron radiation beam. *Acta Crystallographica D*, **64**, pp. 158–166.

51. Nave, C. and Hill, M. A. (2005). Will reduced radiation damage occur with very small crystals? *Journal of Synchrotron Radiation*, **12**, pp. 299–303.

52. Neder, R. B., Burghammer, M., Grasl, T., Schulz, H., Bram, A., Fiedler, S. and Riekel, C. (1996). Single crystal diffraction by submicrometer sized kaolinite; observation of Bragg reflections and diffuse scattering. *Zeitschrift fur Kristallographie*, **211**, pp. 763–765.

53. Neder, R. B., Burghammer, M., Grasl, T., Schulz, H., Bram, A. and Fiedler, S. (1999). Refinement of the kaolinite structure from single-crystal synchrotron data. *Clays and Clay Minerals*, **47**, pp. 487–494.

54. Nelson, R., Sawaya, M. R., Balbirnie, M., Madsen, A. Ø., Riekel, C., Grothe, R. and Eisenberg, D. (2005). Structure of the cross-beta spine of amyloid-like fibrils. *Nature*, **435**, pp. 773–778.

55. Nishino, Y., Takahashi, Y., Imamoto, N., Ishikawa, T. and Maeshima, K. (2009). Three-dimensional visualization of a human chromosome using coherent X-ray diffraction. *Physical Review Letters*, **102**, art. no. 018101.

56. Noehammer, B., David, C., Burghammer, M. and Riekel, C. (2005). Coherence-matched microfocusing of hard X-rays. *Applied Physics Letters*, **86**, art. no. 163104.

57. Otwinowski, Z. and Minor, W. (1997). Processing of X-ray diffraction data collected in oscillation mode. *Methods Enzymology*, **276**, pp. 307–326.

58. Pechkova, E. and Nicolini, C. (2004). Protein nanocrystallography: a new approach to structural proteomics. *Trends in Biotechnology*, **22**, pp. 117–122.

59. Perrakis, A., Cipriani, F., Castagna, J. C., Claustre, L., Burghammer, M., Riekel, C. and Cusack, S. (1999). Protein microcrystals and the design of a diffractometer: current experience and plans at EMBL and ESRF/ID13. *Acta Crystallographica D*, **55**, pp. 1765–1770.

60. Pfeiffer, F., David, C., Burghammer, M., Riekel, C. and Salditt, T. (2002). Two-dimensional X-ray waveguides and point sources. *Science*, **297**, pp. 230–234.

61. Ponchut, C., Clement, J., Rigal, J. M., Papillon, E., Vallerga, J., LaMarra, D. and Mikulec, B. (2007). Photon-counting X-ray imaging at kilohertz frame rates. *Nuclear Instruments & Methods A*, **576**, pp. 109–112.

62. Popov, A. N. and Bourenkov, G. P. (2003). Choice of data-collection parameters based on statistic modelling. *Acta Crystallographica D*, **59**, pp. 1145–1153.

63. Popov, D., Burghammer, M., Buleon, A., Montesanti, N., Putaux, J. L. and Riekel, C. (2006). Amylose single crystals: unit cell refinement from synchrotron radiation microdiffraction data. *Macromolecules*, **39**, pp. 3704–3706.

64. Rasmussen, S. G. F., Choi, H. J., Rosenbaum, D. M., Kobilka, T. S., Thian, F. S., Edwards, P. C., Burghammer, M., Ratnala, V. R. P., Sanishvili, R., Fischetti, R. F., Schertler, G. F. X., Weis, W. W. I. and Kobilka, B. K. (2007). Crystal structure of the human [bgr]: 2: adrenergic G-protein-coupled receptor. *Nature*, **450**, pp. 383–387.

65. Ravelli, R and Garman, E. F. (2006). Radiation damage in macromolecular cryocrystallography. *Current Opinion in Structural Biology*, **16**, pp. 624–629.

66. Riekel, C. (2000). New avenues in X-ray microbeam experiments. *Reports on Progress in Physics*, **63**, pp. 233–262.

67. Riekel, C., Burghammer, M. and Müller, M. (2000). Microbeam small-angle scattering experiments and their combination with microdiffraction. *Journal of Applied Crystallography*, **33**, pp. 421–423.

68. Riekel, C., Burghammer, M. and Schertler, G. (2005). Protein crystallography microdiffraction. *Current Opinion in Structural Biology*, **15**, pp. 556–562.

69. Riekel, C., Burghammer, M., Davies, R., Gebhardt, R. and Popov, D. (2009). Fundamentals of non-crystalline diffraction with microfocus techniques. In *Applications of Synchrotron Light to Non-Crystalline Diffraction in Materials and Life Sciences* (eds. García-Gutiérrez, M. C., Nogales, A., Gómez, M. and Ezquerra, T. A.), Vol. 776, pp. 91–104. Springer, Heidelberg.

70. Rodenburg, J. M. and Faulkner, H. M. L. (2004). A phase retrieval algorithm for shifting illumination. *Applied Physics Letters*, **85**, pp. 4795–4797.

71. Sanishvili, R., Nagarajan, V., Yoder, D., Becker, M., Xu, S. L., Corcoran, S., Akey, D. L., Smith, J. L. and Fischetti, R. F. (2008). A 7 μm mini-beam improves diffraction data from small or imperfect crystals of macromolecules. *Acta Crystallographica D*, **64**, pp. 425–435.

72. Santesson, S., Cedergren-Zeppezauer, E. S., Johansson, T., Laurell, T., Nilsson, J. and Nilsson, S. (2003). Screening of nucleation conditions using levitated drops for protein crystallization. *Analytical Chemistry*, **75**, pp. 1733–1740.

73. Schneider, T. R. (2008). Synchrotron radiation: micrometer-sized x-ray beams as fine tools for macromolecular crystallography. *HFSP Journal*, **2**, pp. 302–306.

74. Schroer, C. G., Kuhlmann, M., Lengeler, B., Gunzler, T. F., Kurapova, O., Benner, B., Rau, C., Simionovici, A. S., Snigirev, A. and Snigireva, I. (2002). Beryllium parabolic refractive x-ray lenses. In *Design and Microfabbrication of Novel X-ray Optics* (ed. Mancini, D. C.), Vol. 4783, pp. 10–18. SPIE, Seattle.

75. Schroer, C. G., Kurapova, O., Patommel, J., Boye, P., Feldkamp, J., Lengeler, B., Burghammer, M., Riekel, C., Vincze, L., van der Hart, A. and Kuchler, M. (2005). Hard x-ray nanoprobe based on refractive x-ray lenses. *Applied Physics Letters*, **87**, pp. 24101–24103.

76. Schroer, C. G. and Lengeler, B. (2005). Focusing hard X-rays to nanometer dimensions by adiabatically focusing lenses. *Physical Review Letters*, **94**, art. no. 054802.

77. Schroer, C. G., Boye, R., Feldkamp, J. M., Patommel, J., Schropp, A., Schwab, A., Stephan, S., Burghammer, M., Schoeder, S. and Riekel C. (2008). Coherent X-ray diffraction imaging with nanofocused illumination. *Physical Review Letters*, **101**, art. no. 090901.

78. Seidel, R., Gourrier, A., Burghammer, M., Riekel, C., Jeronimidis, G. and Paris, O. (2008). Mapping fibre orientation in complex-shaped biological systems with micrometre resolution by scanning X-ray microdiffraction. *Micron*, **39**, pp. 198–205.

79. Sumner, M. O. and Serpell, L. C. (2004). Structural characterization of islet amyloid polypeptide fibrils. *Journal of Molecular Biology*, **335**, pp. 1279–1288.

80. Svergun, D. I. (2007). Small-angle scattering studies of macromolecular solutions. *Journal of Applied Crystallography*, **40**, pp. s10–s17.

81. Teng, T. Y. and Moffat, K. (2002). Radiation damage of protein crystals at cryogenic temperatures between 40° K and 150° K. *Journal of Synchrotron Radiation*, **9**, pp. 198–201.

82. Thorne, R. E., Stum, Z., Kmetko, J., O'Neill, J. and Gillilan, R. (2003). Microfabricated mounts for high-throughput macromolecular cryocrystallography. *Journal of Applied Crystallography*, **36**, pp. 1455–1460.

83. Toft, K. N., Vestergaard, B., Nielsen, S. S., Snakenborg, D., Jeppesen, M. G., Jacobsen, J. K., Arleth, L. and Kutter, J. P. (2008). High-throughput small angle x-ray scattering from proteins in solution using a microfluidic front-end. *Analytical Chemistry*, **80**, pp. 3648–3654.

84. Viola, R., Carman, P., Walsh, J., Miller, E., Benning, M., Frankel, D., McPherson, A., Cudney, B. and Rupp, B. (2006). Operator-assisted harvesting of protein crystals using a universal micromanipulation robot. *Journal of Applied Crystallography*, **40**, pp. 539–545.

85. Volkringer, C., Popov, D., Loiseau, T., Guillou, N., Ferey, G., Haouas, M., Taulelle, F., Mellot-Draznieks, C., Burghammer, M. and Riekel C. (2007). A new microdiffraction set-up for nanoporous MOF-type solids. *Nature Materials*, **6**, pp. 760–764.

86. Wagner, A., Diez, J., Schulze-Briese, C. and Schluckebier, G. (2009). Crystal structure of Ultralente-A microcrystalline insulin suspension. *Proteins: Structure, Function, and Bioinformatics*, **74**, pp. 1018–1027.

87. Wiltzius, J. J., Sievers, S. A., Sawaya, M. R., Cascio, D., Popov, D., Riekel, C. and Eisenberg, D. (2008). Atomic structure of the cross-β spine of islet amyloid polypeptide (amylin). *Protein Science*, **17**, pp. 1467–1474.

88. Yun, W. (ed.). (1995). *X-ray Microbeam Technology and Applications*, Vol. 2516, p. 250. SPIE, San Diego.

Chapter 2

SINGLE-BOUNCE MONOCAPILLARY X-RAY OPTICS: DESIGN AND BIOLOGICAL APPLICATIONS

Richard E. Gillilan[a] and Donald H. Bilderback[b,c]

[a]*MacCHESS (Macromolecular Diffraction Facility at CHESS), Cornell University, Ithaca, NY, USA*

[b]*CHESS (Cornell High Energy Synchrotron Source), Cornell University, Ithaca, NY, USA*

[c]*School of Applied Engineering Physics, Cornell University, Ithaca, NY, USA*

reg8@cornell.edu

The explosive growth of genomic information in recent decades has stimulated demand for structures at the molecular scale and driven many technical developments in protein crystallography at synchrotron sources worldwide. Despite improving protein expression and purification techniques, many important targets of study, such as membrane proteins and large multicomponent complexes, can only be isolated in small quantities and only yield tiny crystals, if any. A number of X-ray-focusing schemes have been developed to deliver more photons to smaller samples, thus boosting observable diffraction signals. Single-bounce monocapillary optics are glass tubes having elliptical or parabolic internal surface profiles which, at grazing incidence angles, focus X-rays via total external reflection. Profiles optimized for particular source characteristics and application requirements can be readily produced through a precisely controlled glass drawing process. Focusing optics of this

Synchrotron Radiation and Structural Proteomics
Edited by Eugenia Pechkova and Christian Riekel

www.panstanford.com

kind have been used with success in protein crystallography and small-angle solution scattering, but their use also extends to other areas of growing interest to biologists such as fluorescence mapping scans for metals in tissue, X-ray microscopy, and grazing incidence small-angle scattering.

2.1 INTRODUCTION

Since the introduction of synchrotrons as X-ray sources in the 1960s, continuing advances in accelerator components, beamline optics, and insertion devices have pushed available photon flux to remarkable levels. At the same time, cryogenic sample preparation techniques (Teng, 1990) dramatically increased the amount of data that could be extracted from individual samples, and advancing detector technology improved both sensitivity and throughput. The use of these X-ray sources has escalated dramatically, with the number of reported protein crystal structures, for example, increasing roughly exponentially every year.[a] The first few years of the 21st century have seen much of the process of X-ray structure solution automated as both industry and academics seek to capitalize on vast lists of structural genomics and drug design targets. Unfortunately, some of the most desirable targets are not easily produced in quantity and do not yield crystals of adequate size or quality.

Small crystals may be simply a result of growth and nucleation conditions (Judge *et al.*, 1999), or of small quantities of material to start with. Current trends in automated screening for crystallization conditions have pushed sample sizes into the picoliter regime (Cherezov and Caffrey, 2006). But in many cases, achievable crystal size may be more a reflection of underlying properties of the molecule than a result of crystallization conditions alone. Over 70% of recent drug targets, for example, are membrane-bound proteins, a class that is famously difficult to isolate and crystallize due to the necessary presence of lipids (Lundstrom, 2006).

---

[a]Current statistics can be found at www.rcsb.org

Much work in molecular biology currently focuses on multicomponent complexes and networks of interactions. It is not uncommon for a given protein to interact with multiple partners or to have more than one oligomeric state. Weak interactions and large unit cells can be problematic to crystallography (Xiao *et al.*, 2007).

Conformational flexibility and disorder also pose challenges to protein crystallography by disrupting putative crystal contacts. While it is often expedient to engineer away flexible protein regions to improve crystal formation (Gao *et al.*, 2005), it is the flexible regions that are sometimes of greatest interest in function. Some proteins, by nature of their function, prefer to pack as fibers or aggregates rather than crystals, another factor that may contribute to making crystals small or difficult to form (Sawaya *et al.*, 2007). The problem of small and imperfect crystals is thus unlikely to grow any less important with time. In fact, since the pioneering work of Riekel *et al.* developing protein microcrystallography at ESRF ID13 (Riekel, 2004; Riekel *et al.*, 2005), numerous beamlines around the world have responded to the increasing demand by adapting to handle such samples.[b]

A variety of strategies have been employed to produce microfocused X-ray beams at synchrotrons. Kirkpatrick and Baez (KB) mirrors are now very widely used for this purpose in combination with apertures (Kirkpatrick and Baez, 1948; Riekel *et al.*, 2000; Fischetti *et al.*, 2009). Fresnel zone plates (Yun *et al.*, 1992) and compound refractive lenses (Snigirev *et al.*, 1996) are gaining popularity. A number of emerging technologies are also being investigated such as waveguides (Feng *et al.*, 1995; Lagomarsino *et al.*, 1996; Müller *et al.*, 2000; Pfeiffer *et al.*, 2002), and Laue and kinoform lenses (Aristov *et al.*, 2000; Evans-Lutterodt *et al.*, 2003; Kang *et al.*, 2006).

The use of tapered capillaries to focus X-rays was introduced in the late 1980s (Stern *et al.*, 1988) and has been further

[b]At present, protein microcrystallography is supported at the following facilities and beamlines worldwide: APS (24ID-C, 24ID-E, 23ID-B, 23ID-D), CHESS (F1, F2, A1), ESRF (ID13, ID23-2), SLS (X06SA), Photon Factory (BL17A). Stations are also under construction at APS (21ID-D, 21ID-E), SBC (19ID-2), ESRF (ID29), SSRL (12-2), Soleil (Proxima2), Diamond (I24), Spring8 (BL32XU). This list may not be exhaustive and does not reflect stations that support other X-ray techniques utilizing microfocus optics for biology such as BioSAXS, GISAXS, scanning XRF, and microscopy.

refined and put into use by several groups (Balaic *et al.*, 1996; Huang and Bilderback, 2006; Bjeoumikhov *et al.*, 2008). Glass capillary optics possess a number of useful characteristics: they are potentially inexpensive to produce, mechanically robust, radiation resistant, and achromatic. Without any special surface treatment or coating, glass is capable of greater than 95% reflectivity at grazing incidence angles. A single optic can function over a broad range of energies and bandwidths (from soft X-rays up to as high as 80 keV) without change in spot position, spot size, or beam divergence. Capillaries with short focal lengths can be designed, resulting in very high magnification. By their nature, they are compact, do not require a cumbersome vacuum environment, and combine both horizontal and vertical focusing in a single element.

These kinds of optics do have some limitations. They focus the beam, but are not appropriate for imaging unless used in a double-bounce Wolter mirror configuration (Wolter, 1952). Single capillaries also have a small numerical aperture so their ability to collect light is limited. Finally, current fabrication methods have not reached the same level of precision as those for conventional mirrors.

Capillaries can be used in single-bounce or multiple-bounce configurations, and as single capillaries or as polycapillary bundles. While all these types have proven useful in certain applications, single-bounce monocapillaries provide larger working distances, better control over divergence, and greater overall reflective efficiency. These properties make them more appropriate for the synchrotron source applications we wish to consider here.

We begin by reviewing the basic principles of capillary optics, efforts to model precise optical behavior, and how the essential design parameters are constrained by various application requirements. We then introduce the physics behind the fabrication technique used at CHESS, including current limitations. Alternative methods reported in the literature are also briefly described. A practical discussion of mounting, stability, and alignment issues follows. We finish with a summary of current biological applications that have specifically used single-bounce monocapillary optics. The focus is primarily upon macromolecular crystallography, but small-angle solution scattering and several other methods of biological interest are covered.

2.2 DESIGN CONSIDERATIONS

Single-bounce monocapillary optics are glass tubes in which the internal surface has been tapered in such a way as to act as a focusing mirror for X-rays. Even simple untapered capillaries have proven surprisingly useful as X-ray "guide tubes" or "light pipes" in generating beams as small as 10 μm for such applications as scanning X-ray microfluorescence (Mosher and Stephanikis, 1976; Vetterling and Pound, 1976; Chung and Pantell, 1977; Fukumoto *et al.*, 1992). Stern *et al.* have worked out and tested the basic theory for design of linearly tapered (conical) capillary optics which produce an overall gain in intensity (Stern *et al.*, 1988). Balaic and Nugent developed a formalism for determining optimal design parameters of tapered capillaries and compared the results to three-dimensional ray-tracing simulations for both linear and parabolic profiles (Balaic and Nugent, 1995). Cornaby has given an excellent account (Cornaby, 2008) of the basic principles of capillary optics.[c]

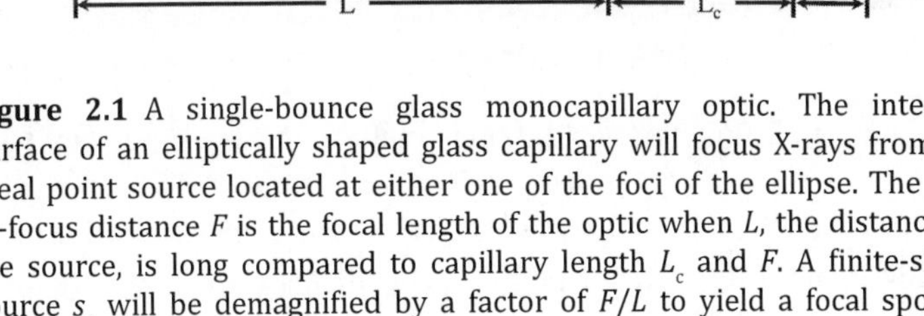

Figure 2.1 A single-bounce glass monocapillary optic. The internal surface of an elliptically shaped glass capillary will focus X-rays from an ideal point source located at either one of the foci of the ellipse. The tip-to-focus distance F is the focal length of the optic when L, the distance to the source, is long compared to capillary length L_c and F. A finite-sized source s_o will be demagnified by a factor of F/L to yield a focal spot of size s_i. The maximum divergence of the focused rays is given by the angle θ_{div} at the focus subtended by the diameter of the capillary exit aperture. A second divergence associated with the optic, θ_m, is always small compared to θ_{div}.

[c]The Handbook of X-ray Single-Bounce Monocapillary Optics, Including Optical Design and Synchrotron Applications by S.W. Cornaby is available free for download at glasscalc.chess.cornell.edu/Cornaby_Handbook_2008.pdf

When reflection angles are constrained to approximately 0.2° or less, the shallow (grazing incidence) configuration of a capillary optic allows nearly complete (>95%) reflection for 10 keV X-rays. This value decreases with increasing energy. For an ideal point source located at finite distance from the capillary, an elliptically shaped internal surface will focus the rays to a point (Fig. 2.1). In practice, a synchrotron source is typically many meters away and the resulting shape is highly eccentric.

Real X-ray sources have finite size. The magnification M relates the finite focal spot size, s_i, to the actual source size, s_0, and can be estimated from the ratio of the distance of the optic from the source (L) and the distance of the optic to the focus (F):

$$M = \frac{s_i}{s_0} = \frac{F}{L}. \tag{2.1}$$

In fact, the focal length and the two distances F and L obey the classical "thin-lens" equation of geometric optics. For most synchrotron applications, L is comparatively long and capillary length L_c is short, consequently F equals the focal length. The maximum divergence, θ_{div}, produced by a capillary is determined by the angle at the focal point subtended by the inside diameter of the tip (Fig. 2.1). This kind of optic generally has a small numerical aperture ($\approx\theta_{div}/2$) and consequently limited light-gathering ability. Total photon flux, f, is proportional to the difference between the inside areas of the base and tip of the capillary:

$$f \propto ID^2_{base} - ID^2_{tip}. \tag{2.2}$$

The maximum expected intensity gain for a very short capillary segment is at most f/s_i^2. More precise calculations of optical performance have been done by a variety of groups using detailed ray-tracing techniques (Furuta *et al.*, 1993; Chen *et al.*, 1994; Wang *et al.*, 1996; Thiel, 1998; Vincze *et al.*, 1998; Vincze and Riekel, 2003). Huang and Bilderback have formulated a practical approach for routine capillary design using an analytical method that requires relatively few input parameters (Huang and Bilderback, 2001, 2006). The formulas derived by Huang and Bilderback have been implemented as a *matlab* script, which is available through an online server.[d]

[d] glasscalc.chess.cornell.edu/ImageProf.shtml

For a simple unfocused X-ray source, horizontal and vertical source size (full width half maximum) and distance to capillary must be known. Focused sources require some additional consideration: if the primary focal point is upstream of the capillary optic, it is effectively the new source. If the primary focal point is downstream, the effective new source can be calculated as the diameter of the beam at any convenient fixed distance upstream (Cornaby, 2008). In these focused cases, the base aperture of the optic may not be large enough to admit the full beam; consequently, the effective source size needs to be reduced appropriately. Nonetheless, the design calculation begins with effective source dimensions and distance.

Four additional parameters are necessary to specify a design: maximum allowed divergence, focal length, capillary length, and expected slope error. Input capillary design parameters are constrained by the specific application requirements that we now consider.

Divergence leads to overlap of Bragg spots on diffraction patterns and places upper limits on the crystallographic unit cell dimensions that can be examined. Macromolecular crystallography typically tolerates no more than 2–4 mrad of divergence. A rough conservative estimate of unit cell limits can be made for typical protein crystallography experiments (12 keV) by assuming 1 mrad is sufficient to resolve a 450 Å unit cell. Thus, a 2 mrad beam should be able to resolve spots from a 225 Å unit cell without difficulty. The actual upper limit achieved will depend on resolution, beam diameter, crystal mosaicity, and other more complex factors.

Small-angle solution scattering applications also generally require low divergence (Riekel *et al.*, 2000). For XRF applications, however, divergence is much less important than flux. The maximum beam divergence produced by a given design is determined by the grazing incidence angle for X-rays right at the exit aperture. This value is limited by the requirement of total external reflection in glass. Geometric considerations therefore restrict the upper limit of divergence to $\theta_{div}=4\theta_{grazing}$ (Cornaby, 2008). But there is another divergence value associated with a capillary. Rays reflecting off opposite sides of the tube converge with angle θ_{div} while rays reflecting from the same side converge with angle $\theta_{m}<1/2\theta_{div}$ (Fig. 2.1). By limiting how much of the internal capillary surface is illuminated, the maximum divergence of the focused beam can be varied from θ_{div} down to θ_m or lower. The ability to control divergence is an important advantage of single-bounce monocapillary optics that we will touch upon later when discussing bifocal optics in the Section 2.5.3.

Single-bounce capillaries permit extremely high magnification ratios with correspondingly short focal lengths. For XRF type applications this can be desirable since the goal is small spot size and high flux. Macromolecular applications often require significant working distance due to the presence of cryogenic sample cooling hardware, scatter guard apertures, and the need for manual sample mounting. For these reasons, focal lengths on the order of 20 mm have been found most practical in macromolecular applications at CHESS. Increasing the focal length increases focal spot size, but also increases overall flux while maintaining constant gain.

Early capillaries tended to be long (>200 mm). Longer optics do indeed collect more photons from a given source, but the intensity profile at the focal spot becomes less sharp with length. It is important to remember that capillaries are not imaging optics and photons arriving at the focal spot have undergone a range of effective magnifications determined by the finite length of the optic (Howell and Horowitz, 1975). Using analytic arguments, Huang and Bilderback have shown that the best compromise between flux and spot size is reached when the optic is approximately twice as long as the focal length (Huang and Bilderback, 2006). Nonetheless, the use of capillary length in customizing beam profile may prove useful in some situations.

The final design parameter, slope error, is an unavoidable consequence of the fabrication process. While every attempt is made to minimize the quantity (currently on the order of 0.03 mrad at CHESS), the effect of slope error must be modeled to determine if design goals can actually be met. Slope error increases spot size and reduces gain. Larger diameter X-ray sources are less affected by slope error; consequently, the present capillary technology produces results that are essentially optically perfect on second-generation synchrotron sources like CHESS.

The formulas derived by Huang and Bilderback are currently limited to relatively distant sources ($L >> L_c$) and correspondingly eccentric elliptical profiles. They do not take into account diffraction effects as would be encountered in approaching nanometer-sized focal spots. Nonetheless, this straightforward analytical approach to capillary design is valid for a wide range of contemporary synchrotron applications.

2.3 GLASS DRAWING TECHNIQUES AND PRINCIPLES

Glass has the advantage that it can be easily softened and drawn, so this is the predominant method used to produce capillaries. Optics manufactured at CHESS use a unique moving furnace design. Thick-walled borosilicate tubing is heated in a short electric furnace under tension (Fig. 2.2). The mass of material entering the heat zone is proportional to $d_o^2v_o$, in which d_o is the initial diameter and v_o is the initial velocity. The amount leaving is proportional to $d_f^2v_f$, where d_f is the final diameter and v_f is the final velocity. If mass is to neither accumulate nor disappear from the heating zone (a statement of conservation of mass), these two values must be equal. Hence $d_f = d_o\text{sqrt}(v_o/v_f)$.[e] This is the open loop control equation of the puller. By controlling the velocities of glass into and out of the furnace, we make the desired diameter versus length profile. If the furnace is moved quickly in response to a small amount of glass extension, then the diameter is reduced just a little. However, if the furnace moves very little in response to a large amount of glass extension, then the diameter is reduced a lot. As the glass yields, the furnace is moved under computer control in such a way as to produce the desired profile. The piece is also constantly rotated about the center bore to help maintain uniformity. The current furnace temperature limit of 900 °C is adequate for borosilicate, soda lime, and other low-melting glasses. Laser-based metrology is employed after the fabrication process to assess profile quality and slope errors.

For elliptically shaped optics approximately 100 mm in length, the figure errors are of order 1 micron with slope errors in the range of 30 to 50 μrad.We believe that these values can be further improved by refining the pulling process.

[e] This result represents steady-state drawing conditions (constant velocity). The full formula in the puller coordinate system has an additional correction term for velocity change that can be found in Cornaby (2008).

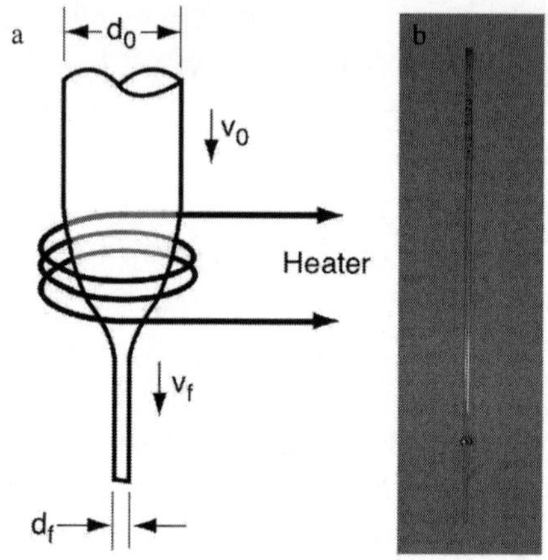

Figure 2.2 (a) Schematic of the glass drawing process for producing capillary optics. A glass tube of initial diameter d_0 is fed into the hot zone of a furnace with a velocity v_0. As glass is drawn out the bottom, the final diameter d_f is determined by the exit velocity v_f of the tube due to conservation of volume. In practice, a program of controlled tension on the tube is combined with furnace movement to produce a precisely controlled capillary profile. (b) Actual 50 mm long glass capillary optic "SF202" used at MacCHESS beamline F1.

These tolerances place a practical limit of 1 μm on the focal spot size for a typical capillary design (Huang and Bilderback, 2006). As slope errors tend to manifest themselves as macroscopic ripples in an otherwise atomically smooth surface (Anderson *et al.*, 1997), it is possible to find small surface patches that can exceed the predicted limits. Snigirev *et al.* used a pre-focused microbeam to illuminate one such patch on a capillary and were able to achieve a 250 nm focal spot (Snigirev *et al.*, 2007). Present capillary drawing technology is thus more than adequate for focal spot sizes commonly used by protein crystallographers (5–20 μm).

Two alternative methods of manufacture have been described in the literature. Solid glass that contains large bubbles will naturally produce elliptical hollow cavities when stretched (Knöchel *et al.*, 1998). This technique has been used with success in a number of protein crystallography applications (Hrmova *et al.*, 2001; Varghese *et al.*, 2002). An alternative fabrication method has also been developed and successfully applied to metal. Slowly withdrawing a metal wire or glass fiber from an etching bath at a controlled speed produces a high-precision mandrel.

The mandrel is then coated with high-Z material using vacuum evaporation, sputtering, or other chemical processes. Once coated, the mandrel can be withdrawn or otherwise chemically etched away to produce a capillary (Hirsch, 1996). Mandrel-based methods have the advantage of being applicable to a wide variety of materials. They also offer good control over profile and straightness of the part. The use of high-Z materials for fabricating capillaries in the future is attractive since it allows for larger grazing incidence angles and consequently greater light-gathering ability.

2.4 MOUNTING, STABILITY, AND ALIGNMENT

The overall quality of a focused X-ray beam depends upon a number of factors. Stability of the X-ray source itself can be an issue of concern when the source size is very small (Igarashi *et al.*, 2008). Imperfections in the optical surface introduced during fabrication have been discussed, but distortions of the ideal shape may also occur as a result of mounting.

Due to the flexibility of glass and the typically very high tolerances needed for good focusing, deformation of the optic under its own weight is a potential concern. With larger X-ray source sizes such as those at CHESS, profile deviations on the order of 5 μm along a 10 cm optic can destroy the focusing. Smaller 1 μm deviations are tolerable but can potentially degrade focal spot quality. The degree of deformation that an optic experiences depends upon mounting, wall thickness, and overall length. According to the theory of flexure for simple beams, degree of deformation increases quadratically with length (Eshbach, 1975). Predicted deformations for short, thick-walled capillaries used in the applications here fall well below a micron and can be reduced through proper support design (Gillilan *et al.*, 2010). Capillaries in use by MacCHESS are supported in a small barrel-like housing having approximately the same form factor as a typical X-ray collimator.

Cornaby has published a detailed design for a small portable optical bench customized for capillary alignment and testing (Cornaby, 2008). For routine use at protein crystallography stations alignment is accomplished by moving the optical table itself. Unlike aperture-based collimators, which can be optimized automatically based on ion-chamber readings, capillary optics generally require specialized alignment techniques.

The procedure involves direct beam visualization using a fluorescent screen. Far-field images of the direct beam are annular in nature (Fig. 2.3). The central spot is due to undeflected beam passing through the center of the optic while the focused radiation appears as a less intense halo. Proper alignment of an optic is indicated when the orientation gives rise to a fully symmetrical far-field pattern.

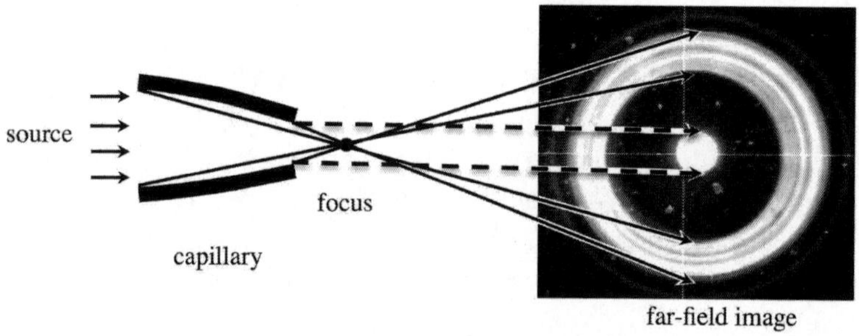

Figure 2.3 Extreme far-field image of a capillary-focused X-ray beam. Unfocused rays (dashed lines) pass through the center of the optic to produce a central spot. Focused rays (solid lines) diverge from the focal spot to form a halo.

2.5 APPLICATIONS

Microfocus optics have been used in a wide variety of biological applications. Among those, protein crystallography is perhaps the most widely utilized. Biological small-angle X-ray solution scattering (BioSAXS), however, has seen a recent surge in interest due to its ability to give information about protein association in solution and low-resolution shape without the need for crystals. In this final section on applications, we focus mainly on these two areas, restricting our discussion to examples using monocapillary optics. Also on the horizon are several techniques that are relatively new to biology: scanning X-ray microfluorescence, grazing incidence small and wide angle scattering, and X-ray microscopy. While uses in these areas are still experimental, monocapillary optics are likely to play an important role here in the near future.

2.5.1 Protein Crystallography

As protein crystals get smaller, it is necessary to illuminate them with higher X-ray flux density to maintain an observable signal. Setting aside the complex issue of radiation damage (Cowan and Nave, 2008), Darwin's formula states that the number of photons per diffraction spot is proportional to the volume of the crystal (neglecting absorption), thus the number of counts in the diffraction pattern deceases as the cube of the diameter of the crystal (Darwin, 1914; Woolfson, 1970). If this were the only problem, it could be remedied by using longer exposures. In practice, crystals are usually suspended in a matrix of amorphous ice mixed with cryoprotectant and any X-rays that fail to hit the crystal produce background scatter that reduces the signal-to-noise ratio. The presence of scatter due to air in the path of the direct beam also degrades the signal, but to a lesser extent. Microfocus optics have thus proven valuable in delivering photons only where they are needed.

The earliest microfocus X-ray optic applied to protein crystallography was actually a multi-bounce glass capillary producing a 5.6 μm diameter beam having a divergence of 2.6 mrad (Bilderback *et al.*, 1994). The experiment produced a Laue diffraction pattern from a 100 μm lysozyme crystal. Several studies have used the Laue method on microcrystals, though not in conjunction with microbeam optics (Hedman *et al.*, 1985).

While Laue experiments have proven extremely advantageous in intervening years for time-resolved protein diffraction problems, the rotation method has remained the most practical and popular technique for routine crystallography. Riekel *et al.* suggest that the Laue method's particularly efficient way of surveying reciprocal space may yet prove useful in microdiffraction (Riekel *et al.*, 2005). Indeed, interest in using the method has been stimulated recently by the possibility that complete diffraction datasets may be assembled from single exposures of multiple microcrystals in the extreme limit of radiation damage (Cornaby, 2008).

Balaic *et al.* were the first to use a single-bounce capillary with a parabolic profile to obtain monochromatic protein crystal diffraction patterns (Balaic *et al.*, 1995). The method was subsequently used on lab sources to solve several unknown structures (Hrmova *et al.*, 2001; Varghese *et al.*, 2002). Bundles of multi-bounce capillaries (polycapillaries) have proven advantageous on laboratory sources due to their ability to collect X-rays from wider angles than monocapillaries (Li and Bi, 1998; Bjeoumikhova, 2008). The properties of single-bounce monocapillaries, however, are better suited to synchrotron sources due to their small spot size, longer focal distances, and more controllable divergence.

The advantages of using microbeam in protein crystallography are well documented in the literature and examples of the use of smaller beams on smaller samples continue to appear (Rey 2007; Sawaya *et al.*, 2007). Recently, X-ray microbeams have become more widely available to the protein crystallography community due in large part to development of the technology at ESRF ID13 and subsequent commercialization (Riekel, 2004). Most facilities today combine KB mirrors with apertures and scatter guards achieving beam sizes down to 5 μm in diameter. We confine our review here purely to applications of monocapillary optics. As discussed previously, capillary optics possess some unique properties, though the basic advantages and strategies of using microbeam in protein crystallography are generally applicable to cases where beam is produced by any of the other methods.

While microbeam has most often been heralded as a means of collecting data on crystals too small for ordinary stations, it should not be overlooked as a means of expanding the capabilities of weaker stations. A case in point is the zinc transport protein CzrB from *Thermis thermophilus* (Cherezov *et al.*, 2008). A complete monochromatic dataset had been obtained for the zinc-free form of the protein but crystals of the zinc form remained elusive and no phases were available for solving the structure. BioSAXS data were collected in the hopes of obtaining a low-resolution envelope and observing a conformational change induced by zinc binding. Remarkably, during the course of the SAXS experiment, showers of small crystals formed in the zinc-containing protein preparation (50 × 50 × 250 μm at the largest). Customarily, a 150 μm diameter beam is used on CHESS F2 station, but such an arrangement would waste some 60% of the beam in this situation. A single-bounce monocapillary optic allowed photons to be concentrated onto an approximately 20 μm spot of the crystal. MAD data collection on the zinc anomalous signal yielded phases for the structure and also resulted in the solution of the zinc-free form via molecular replacement.

The case of CzrB underscores the value of using microbeam in situations where crystallization conditions are serendipitous and not easily reproduced. While it is standard practice to refine initial conditions to yield larger, better formed crystals, this is not always possible in practice. Another example is serinocyclin from *Metarhizium anisopliae* (Krasnoff *et al.*, 2007). Two 60 × 60 × 3 μm thick crystalline flakes were discovered unexpectedly in the bottom of a sample tube that had dried out. A light coating of applied mineral oil allowed the flakes to adhere to a mount without excess solvent. Capillary microbeam was used to scan the flat surface of the flakes to evaluate diffraction quality. Surprisingly, one 20 μm spot on one of the recovered flakes yielded solvable data (Fig. 2.4a,b). A sufficiently complete, high-resolution (1.0 Å) dataset was retrieved from that spot to solve the structure via *ab initio* methods.

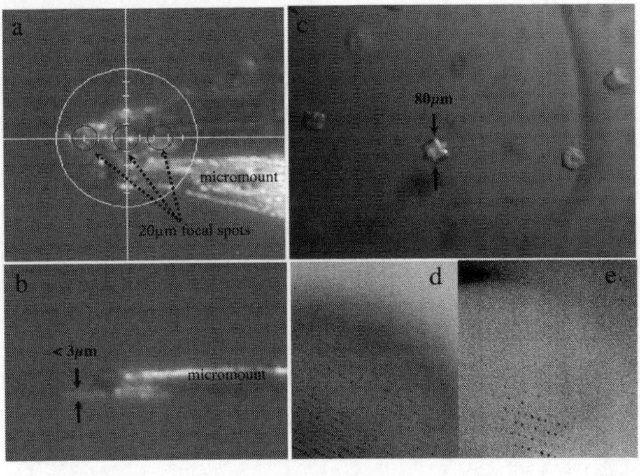

Figure 2.4 Recent uses of capillary microbeam in protein crystallography. Top view (a) and side view (b) of a single serinocylin crystal discovered in the bottom of a sample tube. The thin (<3 μm) plate was exposed at three points (small circles), the central spot yielded high-resolution data. The specimen is shown mounted on a thicker plastic micromount (MiTeGen LLC, NY) with a thin layer of oil. Twinned crystals of the protein CD31 (c), which diffracted with a large P622 unit cell (pattern d), were crushed to yield 20 μm untwined fragments (not shown) with a smaller I222 unit cell (pattern e). See also Color Insert.

Possibly one of the most important applications of microfocused X-rays is in finding regions of good diffraction quality within larger imperfect specimens. The most common problem appears to be the formation of needle-shaped crystals in which multiple lattices or other defects are present on one end but not the other. In addition to the benefit of using a beam close to the diameter of the needle so that photons are not wasted on empty solvent patches, it becomes possible to scan the beam along the length of the needle. Several groups have used this kind of strategy effectively with capillary optics (Lu *et al.*, 2008; Schmiedel *et al.*, 2008; Schmitz *et al.*, 2008).

When protein crystals tend to grow as clusters, microbeam can sometimes allow individual domains to be singled out for exposure. Alternatively, clusters can be mechanically separated as microcrystalline fragments. Liu and Kriksunov recently encountered an extreme example in which larger twinned crystals of the cell adhesion protein CD31 in the space group P622 yielded single lattice fragments of smaller unit cell in space group I222 when physically crushed (Fig. 2.4c–e) (Liu Q. private communication).

2.5.2 BioSAXS

BioSAXS has rapidly gained popularity in recent years as computing and beamline advances have made the technology accessible and more routine (Svergun and Koch, 2003). The extraction of structural information from solutions without the need for crystals is particularly appealing in an era when large complexes of proteins and changes of conformational and oligomeric state are central to molecular biological investigations. Low-resolution molecular envelope reconstructions have been used successfully to build models of complexes from known subdomain structures and fragments (Nagar and Kuriyan, 2005).

In SAXS, the observation of large objects requires that scattering be measured to very small angles. Narrow, intense, highly collimated beams are desirable in this regard since the absence of Bragg reflections results in an inherently weak signal. SAXS using microfocus optics has been quite successful on solid, mineralized, and fibrous biological materials (Paris, 2008). First use of capillary optics for SAXS and WAXS (wide-angle X-ray scattering) was reported by Riekel *et al.* (1998). The capillary optics provided an approximately 3 μm diameter beam, but with limited small-angle resolution in comparison to 10 μm pinhole collimation.

Unlike protein crystallography, successful cryogenic sample preparation protocols in BioSAXS have yet to be fully developed (Iwamoto, 2009). As a consequence, use of BioSAXS on extremely small-volume liquid samples remains limited by radiation damage. A widely used method of reducing the effects of radiation damage is to flow fresh sample past the beam using microfluidic technology (Lipfert, 2006), or alternatively to move the beam to a fresh portion of the sample (Hong and Hao, 2009). In either case, larger quantities of sample are consumed to offset the loss due to damage.

The use of microfluidics in SAXS, however, has another more important dimension: the observation of time-dependent phenomena. In continuous-flow microfluidic systems, a time-dependent phenomenon is initiated at a fixed position along a fluid flow. This can be done either by mixing reactive components or by optical activation. As products flow down the microfluidic channel, the course of the reaction can be probed by examining the small-angle X-ray scatter at points downstream. In this way, a time coordinate is transformed into a position coordinate. High-flux, well-collimated microbeams are ideal for these kinds of applications not only because they minimize sample consumption, but because the size of the beam is directly related to time resolution. Small beams can probe small time differences. Otten *et al.* used a Fresnel zone plate to focus X-rays from ESRF ID10B to a 6 μm spot on a microfluidic device observing intercalation of DNA into multilamellar membranes (Otten *et al.*, 2005). Lamb *et al.* demonstrated that capillary optics could be used to obtain SAXS patterns from cytochrome c within a 2 mm thin-walled polyester (PET) tube of the type used in microfluidic investigations (Lamb *et al.*, 2007). The experiment performed on CHESS D1 station produced a beam approximately 15 μm in diameter with 2 mrad of divergence. In a later application using the same X-ray optics, Lamb *et al.* examined the dimerization of Light Oxygen Voltage sensing protein induced by light activation (Lamb *et al.*, 2008). Photoactivation of the protein was accomplished by placing a laser at one end of the PET tube perpendicular to the X-ray beam and direction of protein flow.

2.5.3 Other Techniques

One of the strengths of optics based on total external reflection is that focal length, divergence, spot position, and size do not change with energy. The maximum energy at which a glass capillary retains its reflectivity varies with design, but can range from 15 keV to as high as 80 keV. This makes capillary optics a good choice for spectroscopic applications and other techniques that require changing energy with sample in place. Barrea *et al.* have recently compared KB mirror focusing optics to single-bounce capillary optics for use in scanning X-ray fluorescence experiments (Barrea *et al.*, 2009). A thin section of mouse prostate tissue was scanned to create images for Cu, Fe, and Zn content using a 10 μm diameter beam.

Another unique advantage of capillary optics is that both vertical and horizontal focusing take place within the same optical element. For any point on the capillary surface there is elliptical curvature in one direction (parallel to the capillary axis) and circular curvature in the other (transverse to the capillary axis). In the language of grazing incidence optics, the short radius focus is *sagittal*, while the longer elliptical focus is *meridional* (Heald and Hasting, 1981). The sagittal focal distance though need not be the same as the meridional focal distance. While fully illuminated capillaries suffer from off-axis aberrations that would be problematic for bifocal designs (Howell and Horowitz, 1975), such difficulties can be avoided by illuminating only a portion of the internal capillary surface. A prototype bifocal capillary optic used in this mode has recently been reported (Cornaby *et al.*, 2010). Anisotropic samples such as fibers and crystalline needles may benefit from optics customized in this way since X-rays can be focused on the sample in one dimension and on the detector in the other dimension (Pechkova *et al.*, 2009). Similar needs can occur in grazing incidence small-angle X-ray scattering (GISAXS) where the large swath of sample illuminated by the grazing beam leads to reduced resolution on the detector. Biological applications of

GISAXS using conventional microbeam are beginning to appear in the literature and will likely be more common in the future (Pechkova *et al.*, 2005; Nicolini and Pechkova, 2006; Pechkova and Nicolini, 2006). To date however, no GISAXS studies of biological systems have been reported using capillary microbeam.

There is one other future application area of capillary optics that deserves mention. X-ray microscopy provides a spatial resolution between visible light microscopy and electron microscopy (15–50 nm) and has been used most notably in three-dimensional tomography of fully hydrated, cryofixed cells (Parkinson *et al.*, 2008). While single capillaries by themselves are not well suited for imaging applications due to aberrations, they have recently been shown to be very effective as a condensing element for X-ray microscopy (Zeng *et al.*, 2008). Capillaries are axially symmetric, have high reflectivity, are achromatic, and have acceptable numerical aperture; consequently, the uniform sample illumination they can provide is potentially superior to present zone plates for this application. Biological applications using this combination of optics have yet to appear in the literature, but the outlook is promising.

Finally, it should be mentioned that single-bounce monocapillary optics have also been used on a wide range of nonbiological samples: high pressure powder diffraction, micro-XANES, confocal X-ray fluorescence, time-resolved powder diffraction, micro-X-ray standing waves, and others areas (Cornaby *et al.*, 2007).

2.6 CONCLUSIONS AND FUTURE DEVELOPMENTS

The ability to illuminate very small samples with well-focused X-ray microbeams continues to be a technology of growing importance in biology, particularly in macromolecular protein crystallography. The field of X-ray optics continues to evolve, presenting users with a variety of choices each having distinct advantages and limitations. The single-bounce capillary technique presently works well for many biological applications to produce beam sizes in the micron range. In this review, we have highlighted the unique capabilities and shown how capillary fabrication parameters can be optimized for particular biological applications.

Single-bounce glass monocapillary optics would work even better if a way could be found to reduce the internal slope errors to near zero in value, i.e., approach a "perfect optic." The smallest beam size that we can presently envision from a single-bounce capillary is 10 nm, a value determined by the penetration of the evanescent wave during total reflection into the bulk of the glass capillary (Bergemann *et al.*, 2003). If such a perfect optic could be made by some new fabrication method, then for a fourth-generation synchrotron source such as the proposed Cornell Energy Recovery Linac source of hard X-rays, we have estimated that a capillary could make an intensity gain of over 80 million times (Huang and Bilderback, 2006). Most likely, though, new fabrication principles will have to be developed along with new metrology methods, to even begin to approach the possibilities.

Acknowledgments

This work was supported by the National Science Foundation and NIH-NIGMS via NSF grant DMR-0225180. MacCHESS is supported by NIH/NCRR award RR-01646. We acknowledge the help of Tom Szebenyi and Sterling Cornaby, who have been extremely helpful in developing the capillary drawing process and providing support in employing capillaries for the various X-ray applications. We also thank Ulrich Englich for valuable feedback on the manuscript.

References

1. Anderson, O., Daalderop, G. H. O. and Bange, K. (1997). X-ray reflectivity investigations of glass surfaces produced by float and draw techniques. *Mikrochimica Acta,* **125**, pp. 63–67.
2. Aristov, V., Grigoriev, M., Kuznetsov, S., Shabelnikov, L., Yunkin, V., Weitkamp, T., Rau, C., Snigireva, I., Snigirev, A., Hoffmann, M. and Voges, E. (2000). X-ray refractive planar lens with minimized absorption. *Applied Physics Letters*, **77**, pp. 4058–4060.
3. Balaic, D. X., Barnea, Z., Nugent, K. A., Garrett, R. F., Varghese, J. N. and Wilkins, S. W. (1996). Protein crystal diffraction patterns using

a capillary-focused synchrotron X-ray beam. *Journal of Synchrotron Radiation*, **3**, pp. 289–295.

4. Balaic, D. X. and Nugent, K. A. (1995). X-ray optics of tapered capillaries. *Applied Optics*, **34**, pp. 7263–7272.

5. Balaic, D. X., Nugent, K. A., Barnea, Z., Garrett, R. and Wilkins, S. W. (1995). Focusing of X-rays by total external reflection from a paraboloidally tapered glass-capillary. *Journal of Synchrotron Radiation*, **2**, pp. 296–299.

6. Barrea, R. A., Huang, R., Cornaby, S., Bilderback, D. H. and Irving, T. C. (2009). High-flux hard X-ray microbeam using a single-bounce capillary with doubly focused undulator beam. *Journal of Synchrotron Radiation*, **16**, pp. 76–82.

7. Bergemann, C., Keymeulen, H. and van der Veen, J. F. (2003). Focusing X-ray beams to nanometer dimension. *Physical Review Letters*, **91**, art. no. 204801.

8. Bilderback, D. H., Hoffman, S. A. and Thiel, D. J. (1994). Nanometer spatial-resolution achieved in hard X-ray imaging and Laue diffraction experiments. *Science*, **263**, pp. 201–203.

9. Bjeoumikhov, A., Erko, M., Bjeoumikhova, S., Erko, A., Snigireva, I., Snigirev, A., Wolff, T., Mantouvalou, I., Malzer, W. and Kanngiesser, B. (2008). Capillary microfocus x-ray lenses with parabolic and elliptic profile. *Nuclear Instruments and Methods in Physics Research Section A*, **587**, pp. 458–463.

10. Bjeoumikhova, A. B. A. S. (2008). Capillary Optics for X-rays. In *Modern Developements in X-Ray and Neutron Optics*, Vol. 137 (eds. A. Erko, Idir, M., Krist, T., Michette, A. G.), p. 287. Springer, New York.

11. Chen, G. J., Cerrina, F., Voss, K. F., Kim, K. H. and Brown, F. C. (1994). The experimental determination of the fraction of neutrons below 1.5 MeV for the Chinese-made am-Be neutron source. *Nuclear Instruments and Methods in Physics Research Section A*, **583**, pp. 407–411.

12. Cherezov, V. and Caffrey, M. (2006). Picolitre-scale crystallization of membrane proteins. *Journal of Applied Crystallography*, **39**, pp. 604–606.

13. Cherezov, V., Hofer, N., Szebeny, D. M. E., Kolaj, O., Wall, J. G., Gillilan, R., Srinivasan, V., Jaroniec, C. P. and Caffrey, M. (2008). Insights into the mode of action of a putative zinc transporter CzrB in Thermus thermophilus. *Structure*, **16**, pp. 1378–1388.

14. Chung, P. S. and Pantell, R. H. (1977). Properties of X-ray guides. *Electronics Letters*, **13**, pp. 527–529.

15. Cornaby, S. (2008). The handbook of X-ray single-bounce monocapillary optics, including optical design and synchrotron applications (Ph.D. dissertation in Physics, Cornell University).

16. Cornaby, S., Smilgies, D. M. and Bilderback, D. (2010). Bifocal miniature toroidal shaped X-ray mirrors. *Advances in X-Ray Analysis*, **53**, pp. 1–6

17. Cornaby, S., Szebeny, T., Haung, R. and Bilderback, D. (2007). Design of single-bounce monocapillary optics. *Advances in X-Ray Analysis*, **50**, pp. 194–200.

18. Cowan, J. A. and Nave, C. (2008). The optimum conditions to collect X-rat data from very small sample. *Journal of Synchrotron Radiation*, **15**, pp. 458–462.

19. Darwin, C. G. (1914). The theory of X-ray reflexion, Part II. *Philosphical Magazine*, **27**, pp. 157–690.

20. Eshbach, O. W. (1975). *Handbook of Engineering Fundamentals*. Prepared by a staff of specialists under the editorship of the late Ovid W. Eshbach and Mott Souders. Wiley, New York.

21. Evans-Lutterodt, K., Ablett, J. M., Stein, A., Kao, C. C., Tennant, D. M., Klemens, F., Taylor, A., Jacobsen, C., Gammel, P. L., Huggins, H., Ustin, S., Bogart, G. and Ocola, L. (2003). Single element ellipitical hard X-ray micro-optics. *Optics Express*, **11**, pp. 919–926.

22. Feng, Y. P., Sinha, S. K., Fullerton, E. E., Siddons, D. P. and Hastings, J. B. (1995). X-ray Fraunhofer diffraction patterns from a thin-film waveguide. *Applied Physics Letters*, **67**, pp. 3647–3649.

23. Fischetti, R. F., Xu, S. L., Yoder, D. W., Becker, M. Nagarajan, V., Sanishvili, R., Hilgart, M. C., Stepanov, S., Makarov, O. and Smith, J. L. (2009). Mini-beam collimator enables microcrystallography experiments on standard beamlines. *Journal of Synchrotron Radiation*, **16**, pp. 217–225.

24. Fukumoto, N., Kobayashi, Y. and Kurahashi, M. (1992) Development of a high-spatial-resolution X-ray-fluorescence element mapping spectrometer for nondestructive analysis of biological samples. *Bunseki Kagaku*, **41**, pp. 545–550.

25. Furuta, K., Nakayama, Y., Shoji, M., Kaigawa, R., Hanamoto, K., Nakano, H. and Hosokawa, Y. (1993). Theoretical consideration of intensity of an X-ray microbeam formed by a hollow glass pipe. *Review of Scientific Instruments*, **64**, pp. 135–142.

26. Gao, X., Bain, K., Bonanno, J., Buchanan, M., Henderson, D., Lorimer, D., Marsh, C., Reynes, J., Sauder, J., Schwinn, K., Thai, C. and Burley, S. (2005). High-throughput limited proteolysis/mass spectrometry for protein domain. *Journal of Structural and Functional Genomics*, **6**, pp. 129–134.

27. Gillilan, R. E., Cook, M. J., Cornaby, S. W. and Bilderback D. H. (2010). Microcrystallography using single-bounce monocapillary optics. *Journal of Synchrotron Radiation*, **17**, pp. 227–236.

28. Heald, S. M. and Hastings, J. B. (1981). Grazing-incidence optics for synchrotron radiation X-ray beamlines *Nuclear Instruments and Methods*, **187**, pp. 553–561.

29. Hedman, B., Hodgson, K. O., Helliwell, J. R., Liddington, R. and Papiz, M. Z. (1985). Protein microcrystal diffraction and the effects of radiation damage with ultra-high-flux synchrotron radiation. *Proceedings of the National Academy of Sciences USA*, **82**, pp. 7604–7607.

30. Hirsch, G. (1996). *Tapered capillary optics*. U.S. Patent 5772903.

31. Hong, X. G. and Hao, Q. (2009). Measurements of accurate X-ray scattering data of protein solutions using small stationary sample cells. *Review of Scientific Instruments*, **80**, pp. 14303–14303.

32. Howell, J. A. and Horowitz, P. (1975). Ellipsoidal and bent cylindrical condensing mirrors for synchrotron radiation. *Nuclear Instruments & Methods*, **125**, pp. 225–230.

33. Hrmova, M., Varghese, J. N., De Gori, R., Smith, B. J., Driguez, H. and Fincher, G. B. (2001). Catalytic mechanisms and reaction intermediates along the hydrolytic pathway of a plant beta-D-glucan glucohydrolase. *Structure*, **9**, pp. 1005–1016.

34. Huang, R. and Bilderback, D. H. (2001). Simulation of microfocused image size from a one-bonce glass capillary. *Nuclear Instruments & Methods in Physics Research Section A*, **467**, pp. 978–981.

35. Huang, R. and Bilderback, D. H. (2006). Single-bounce monocapillaries for focusing synchrotron radiation: modeling, measurements and theoretical limits. *Journal of Synchrotron Radiation*, **13**, pp. 74–84.

36. Igarashi, N., Ikuta, K., Miyoshi, T., Matsugaki, N., Yamada, Y., Yousef, M. S. and Wakatsuki, S. (2008). X-ray beam stabilization at BL-17A, the protein microcrystallography beamline of the Photon factory. *Journal of Synchrotron Radiation*, **15**, pp. 292–295.

37. Iwamoto, H. (2009). Quality evaluation of quick-frozen biological specimens by simultaneous microbeam SAXS/WAXS recordings. *Journal of Synchrotron Radiation*, **16**, pp. 336–345.

38. Judge, R. A., Jacobs, R. S., Frazier, T., Snell, E. H. and Pusey, M. L. (1999). The effect of temperature and solution pH on the nucleation of tetragonal lysozyme crystals. *Biophysical Journal*, **77**, pp. 1585–1593.

39. Kang, H. C., Maser, J., Stephenson, G. B., Liu, C., Conley, R., Macrander, A. T. and Vogt, S. (2006). Nanometer linear focusing of hard X-rays by a multilayer Laue lens. *Physical Review Letters*, **9612**, p. 7401.

40. Kirkpatrick, P. and Baez, A. V. (1948). Formation of optical images by X-rays. *Journal of the Optical Society of America*, **38**, pp. 766–773.

41. Knöchel, A., Gaul, G. and Lechtenberg, F. (1998). *Verfahren zur herstellung von glaskapillaren und deren verwendung*. (Germany). Patent EP0796229 26.11.1998.

42. Krasnoff, S. B., Keresztes, I., Gillilan, R. E., Szebenyi, D. M. E., Donzelli, B. G. G., Churchill, A. C. L. and Gibson, D. M. (2007). Serinocyclins A and B, cycli heptapeptides from Metarhizium anisopliae. *Journal of Natural Produtcs*, **70**, pp. 1919–1924.

43. Lagomarsino, S., Jark, W., DiFonzo, S., Cedola, A., Mueller, B., Engstrom, P. and Riekel, C. (1996). Submicrometer X-ray beam production by a thin film waveguide. *Journal of Applied Physics*, **79**, pp. 4471–4473.

44. Lamb, J. S., Cornaby, S., Andresen, K., Kwok, L., Park, H. Y., Qiu, X., Smilgies, D.-M., Bilderback, D. H. and Pollack, L. (2007). Focusing capillary optics for use in solution small-angle X-ray scattering. *Journal of Applied Crystallography*, **40**, pp. 193–195.

45. Lamb, J. S., Zoltowski, B. D., Pabit, S. A., Crane, B. R. and Pollack, L. (2008).Time-resolved dimerization of a PAS-LOV protein measured with photocoupled smal angle X-ray scattering.*Journal of the American Chemical Society*, **130**, pp. 12226–12227.

46. Li, P. W. and Bi, R. C. (1998). Applications of polycapillary x-ray optics in protein crystallography. *Journal of Applied Crystallography*, pp. 806–811.

47. Lipfert, J., Millett, I. S., Seifert, S. and Doniach, S. (2006). Sample holder for small-angle x-ray scattering static and flow cell measurements. *Review of Scientific Instruments*, **77**, art. no. 46108.

48. Lu, C., Smith, A. M., Fuchs, R. T., Ding, F., Rajashankar, K., Henkin, T. M. and Ke, A. (2008). Crystal structures of the SMA-III/S-MK riboswitch reveal the SAM-dependent translation inhibition mechanism. *Nature Structural & Molecular Biology*, **15**, pp. 1076–1083.

49. Lundstrom, K. (2006). Structural genomics for membrane proteins. *Cellular and Molecualr Life Sciences*, **63**, pp. 2597–2607.

50. Mosher, D. and Stephanakis, S. J. (1976). X-ray “light-pipes.” *Applied Physics Letters*, **29**, pp.. 105–107.

51. Müller, M., Burghammer, M., Flot, D., Riekel, C., Morawe, C., Murphy, B. and Cedola, A. (2000). Microcrystallography with an X-ray waveguide. *Journal of Applied Crystallography*, **33**, pp. 1231–1240.

52. Nagar, B. and Kuriyan, J. (2005). SAXS and the working protein. *Structure*, **13**, pp. 169–170.

53. Nicolini, C. and Pechkova, E. (2006). Structure and growth of ultrasmall protein microcrystals by synchrotron radiation: I. microGISAXS and microdiffraction of P450scc. *Journal of Cellular Biochemistry*, **97**, pp. 544–552.

54. Otten, A., Koster, S., Struth, B., Snigirev, A. and Pfohl, T. (2005). Microfluidics of oft matter investigated by small-angle X-ray scattering. *Journal of Synchrotron Radiation*, **12**, pp. 745–750.

55. Paris, O. (2008). From diffraction to imaging: new avenues in studying hierarchical biological tissues with x-ray microbeams. *Biointerphases*, **3**, pp. FB16–FB26.

56. Parkinson, D. Y., McDermott, G., Etkin, L. D., Le Gros, M. A. and Larabell, C. A. (2008). Quantitative 3D imaging of eukaryotic cells using soft X-ray tomography. *Journal of Structural Biology*, **162**, pp. 380–386.

57. Pechkova, E. and Nicolini, C. (2006). Structure and growth of ultrasmall protein microcrystallography by synchrotron radiation: II microGISAX and microscopy of lysozyme. *Journal of Cellular Biochemistry*, **97**, pp. 553–560.

58. Pechkova, E., Roth, S. V., Burghammer, M., Fontani, D., Riekel, C. and Nicolini, C. (2005). MicroGISAXS and protein nanotemplate crystallization: methods and instrumentation. *Journal of Synchrotron Radiation*, **12**, pp. 713–716.

59. Pechkova, E., Tripathi, S. and Nicolini, C. (2009). MicroGISAXS of Langmuir-Blodgett protein films: effect of temperature on long-range order. *Journal of Synchrotron Radiation*, **16**, pp. 330–335.

60. Pfeiffer, F., David, C., Burghammer, M., Riekel, C. and Salditt, T. (2002). Two-dimensional X-ray waveguides and point source. *Science*, **297**, pp. 230–234.

61. Rey, F. A. (2007). Virology — holed up in a natural crystal. *Nature*, **446**, pp. 35–37.

62. Riekel, C. (2004). Recent developments in microdiffraction on protein crystals. *Journal of Synchrotron Radiation*, **11**, pp. 4–6. Riekel, C., Burghammer, M. and Muller, M. (2000). Microbeam small-angle scattering experiments and their combinatin with microdiffraction. *Journal of Applied Crystallography*, **33**, pp. 421–423.

63. Riekel, C., Burghammer, M. and Schertler, G. (2005). Protein crystallography microdiffraction. *Current Opinion in Structural Biology*, **15**, pp. 556–562.

64. Riekel, C., Engstrom, P. and Martin, C. (1998). Micro X-ray small-angle scattering with synchrotron radiation. *Journal of Macromolecular Sciences Physics*, **B37**, pp. 587–599.

65. Sawaya, M. R., Sambashivan, S., Nelson, R., Ivanova, M. I., Sievers, S. A., Apostol, M. I., Thompson, M. J., Balbirnie, M., Wiltzius, J. J. W., McFarlane, H. T., Madsen, A. O., Riekel, C. and Eisenberg, D. (2007). Atomic structures of amyloid cross-beta spines reveal varied steric zippers. *Nature*, **447**, pp. 453–457.

66. Schmiedel, J., Blaukat, A., Li, S. Q., Knochel, T. and Ferguson, K. M. (2008). Matuzumab binding to EGFR prevents the conformational rearrangement required for dimerization. *Cancer Cell*, **13**, pp. 365–373.

67. Schmitz, K. R., Liu, J. X., Li, S. L., Setty, T. G., Wood, C. S., Burd, C. G. and Ferguson, K. M. (2008). Golgi localization of glycosyltransferases requires a Vps74p oligomet. *Development Cell*, **14**, pp. 523–534.

68. Snigirev, A., Bjeoumikhov, A., Erko, A., Snigireva, I., Grigoriev, M., Yunkin, V., Erko, M. and Bjeoumikhova, S. (2007). Submicrometer hard X-ray focusing using a single-bounce ellipsoidal capillary combined with a Fresnel zone plate. *Journal of Synchrotron Radiation*, **14**, pp. 227–228.

69. Snigirev, A., Kohn, V., Snigireva, I. and Lengeler, B. (1996). A compund refractive lens for focusing high-energy X-rays. *Nature*, **384**, pp. 49–51.

70. Stern, E. A., Kalman, Z., Lewis, A. and Lieberman, K. (1988). Simple method for focusing X-rays using tapered capillaries. *Applied Optics*, **27**, pp. 5135–5139.

71. Svergun, D. I. and Koch, M. H. J. (2003). Small-angle scattering studies of biological macromolecules in solution. *Reports on Progress in Physics*, **66**, pp. 1735–1782.

72. Teng, T. Y. (1990). Mounting of crystals for macromolecular crystallography in a free-standing thin film. *Journal of Applied Crystallography*, **23**, pp. 387–391.

73. Thiel, D. J. (1998). Protein crystallization using a multilayer monochromator. *Journal of Synchrotron Radiation*, **5**, pp. 820–822.

74. Varghese, J. N., Moritz, R. L., Lou, M. Z., van Donkelaar, A., Ji, H., Ivancic, N., Branson, K. M., Hall, N. E. and Simpson, R. J. (2002). Structure of the extracellular domains of the human interleuking-6 receptor alpha-chain. *Proceedings of the National Academy of Sciences USA*, **99**, pp. 15959–15964.

75. Vetterling, W. T. and Pound, R. V. (1976). Measurements on an X-ray light pipe at 5.9 and 14.4 keV. *Journal of the Optical Society of America*, **66**, pp. 1048–1049.

76. Vincze, L., Janssens, K., Adams, F., Rindby, A. and Engstrom, P. (1998). Interpretation of capillary generated spatial and angular distributions of x rays: theoretical modeling and experimental verification using

the European Synchrotron Radiation Facility Optical beam line. *Review of Scientific Instruments*, **69**, pp. 3494–3503.

77. Vincze, L. and Riekel, C. (2003). Status and perspectives of capillary optics at a third-generation synchrotron radiation source. *X-Ray Spectrometry*, **32**, pp. 08–214.

78. Wang, L., Rath, B. K., Gibson, W. M., Kimball, J. C. and MacDonald, C. A. (1996). Performance study of polycapillary optics for hard X-rays. *Journal of Applied Physics*, **80**, pp. 3628–3638.

79. Wolter, H. (1952). Spiegelsysteme streifenden Einfalls als abbildende Optiken für Röntgenstrahlen. *Annalen der Physik*, **445**, pp. 94–114.

80. Woolfson, M. M. (1970). *An Introduction to X-ray Crystallography*. Cambridge University Press, New York.

81. Xiao B., Tarricone, C., Lin, K., Kelly, G. and Justin, N. (2007). Optimizing protien complexes for crystal growth. *Crystal Growth & Design*, **7**, pp. 2213–2218.

82. Yun, W. B., Viccaro, P. J., Lai, B. and Chrzas, J. (1992). Coherent hard X-ray focusing optics and applications. *Review of Scientific Instruments*, **63**, pp. 582–585.

83. Zeng, X. H., Duewer, F., Feser, M., Huang, C., Lyon, A., Tkachuk, A. and Yun, W. B. (2008). Ellipsoidal and parabolic glass capillaries as condensers for X-ray microscopes. *Applied Optics*, **47**, pp. 2376–2381.

Chapter 3

HIGH FIELD SUPERCONDUCTING MAGNETS FOR GENERATION OF SYNCHROTRON RADIATION

Nikolay Mezentsev
Budker Institute of Nuclear Physics of Russian Academy of Science, Lavrentyeva Ave. 11, Novosibirsk, Russia
mezentsev@inp.nsk.su

The first superconducting (SC) insertion devices were designed, fabricated, and installed on electron storage ring more than 30 years ago and used for generation of synchrotron radiation (SR) in Budker INP. Since then, a wide experience of manufacturing and use of such SC insertion devices as SC wave length shifters, multipole wigglers, and superbend is accumulated. A review of various types of SC insertion devices for light sources is given in the report. Their basic characteristics as SR sources are discussed.

3.1 INTRODUCTION

Since the 1980s there has been a vigorous construction of dedicated synchrotron radiation (SR) sources in the developed countries, and by about the 1990s, construction of SR sources had started

Synchrotron Radiation and Structural Proteomics
Edited by Eugenia Pechkova and Christian Riekel

www.panstanford.com

in the developing countries. SR sources began to play a powerful connective role in uniting the scientific researchers of different specialties, working in direct contact, in deciding the solution of the same problems with application of SR.

A large number of the techniques based on the phenomena of diffraction, interference, dispersion, absorption of SR have been developed at interaction with investigated substance. A tendency to use harder SR spectrum for material studies had appeared. Especially, this question was arisen for storage rings with low and middle electron energy. SR spectral characteristics (ε_c — critical photon energy) of bending magnet are defined by two parameters: electron energy E and magnetic field B, $\varepsilon_c \sim E^2B$, and, hence, there are two ways how to make the spectrum harder — to increase electron energy or to increase magnetic field at point of radiation.

To expand opportunities for carrying out of experimental works and thus to prolong a life, expensive installations as electron storage rings superconducting insertion devices (SC ID) may be installed into straight sections of the storage rings to change spectral, angular, and polarizing properties of SR. High field SC wave length shifters (WLS) and multipole wigglers play the role of special magnets shifting SR spectrum to more hard X-ray range, increasing photon flux and brightness of the source.

These devices, as a rule, have zero first and second magnetic field integrals along electron orbit, and, therefore, they are not basic elements of the electron storage rings, and their status does not affect working reliability of all ring. However, such devices may disturb dynamic motion of electron beam in a storage ring, creating linear and nonlinear effects, which should be taken into consideration while designing ID. Replacement of one or several normal conducting bending magnets on SC magnets (superbends) is one more opportunity to increase the rigidity of SR spectrum, but in this case, requirements to reliability of these elements should be much more strong, as they are basic elements of the storage ring.

3.2 SC WLS

3.2.1 Main Requirements on WLS Properties

SC WLS is cheap and effective SR generator for low and middle energy storage rings to make SR spectrum harder. Standard WLS consists of three dipole magnets with alternating sign along an orbit (Fedurin *et al.*, 2000).

The central magnet has the maximal field and is used as a source of radiation. Side magnets are used as corrector magnets for compensation of the central magnet field to minimize orbit distortion in the storage ring.

For this purpose, it is required to satisfy a condition of equality to zero of the first and second integrals of WLS field along an orbit:

$$\phi = \frac{1}{B\rho}\int_{-L/2}^{L/2} B_z(s)ds, \quad \text{first field integral} \tag{3.1}$$

$$\delta = \int_{-L/2}^{L/2} ds' \int_{-L/2}^{s'} \frac{B_z(s'')}{B\rho} ds'', \text{ second field integral,} \tag{3.2}$$

where L is the straight section length; B_z, vertical component of WLS magnetic field; and $B\rho$, rigidity of electron beam. If magnetic field $B_z = 0$ for distance $|s| \geq L/2$, the second field integral may be written as follows:

$$\delta = \phi \cdot L - \int_{-L/2}^{L/2} \frac{s'' B_z(s'')}{B\rho} ds''. \tag{3.3}$$

WLS magnetic system does not disturb orbit in a storage ring if these both integrals are equal to zero. If WLS magnetic system is symmetrical relative $s = 0$ and first field integral $\phi = 0$, the second field integral δ is equal to zero automatically.

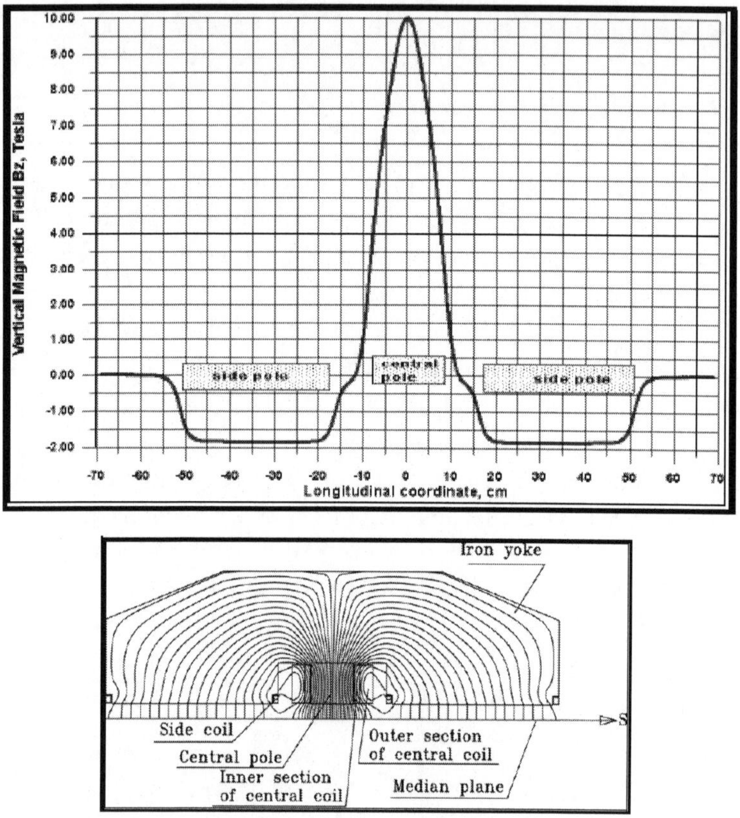

Figure 3.1 The magnetic force lines behavior inside the ½ superconducting WLS magnet iron yoke (right) and longitudinal field distribution in 10 T WLS.

Figure 3.1 shows magnetic force lines behavior inside half of SC WLS and longitudinal field distribution in standard WLS.

3.2.2 Magnetic Field Distribution in WLS

Standard three-pole WLS is similar to two-dimensional magnetic systems (with infinite lateral pole dimensions), where magnetic flux created by central pole is closing through side poles without any additional side coils and first field integral is equal to zero automatically (Fig. 3.1) (Ando *et al.*, 1998; Date *et al.*, 1998; Fedurin *et al.*, 2000).

In reality, lateral dimensions of magnet poles are finite and for correction of first field integral some small side coils are necessary. Disadvantage of standard WLS is orbit distortion inside WLS, so that the radiation point from the maximal field is as much as possible displaced from straight section axis and its position depends on field level. The displacement may be large enough that SR users should readjust beamline optics if they want to work with different field levels. Example of orbit behavior inside standard WLS is shown in Fig. 3.2.

One more disadvantage of standard WLS is presence of two radiation sources visible under the same angle. One radiation source is visible from central pole, and second from side pole, and they are distanced in horizontal coordinate and are a handicap for carrying out of many experiments. For suppression of the second source in a hard range of SR spectrum magnetic field of side poles should be made minimal, but thus the length of side poles and orbit bump will increase.

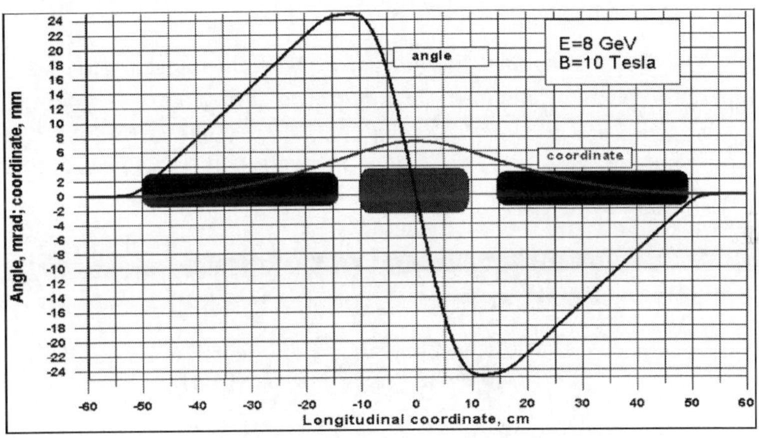

Figure 3.2 Angle deviation and orbit distortion inside the 10 T SC WLS for Spring-8. See also Color Insert.

At least one of the disadvantages may be overcome in case of so-called SC WLS with fixed point of radiation (Borovikov *et al.*, 1998). It consists of two parts: SC magnet similar to standard WLS and two normal corrector magnets outside of cryostat. It requires more space for installation compare with standard WLS, but in this case radiation point from central pole is on straight section axis at any field level.

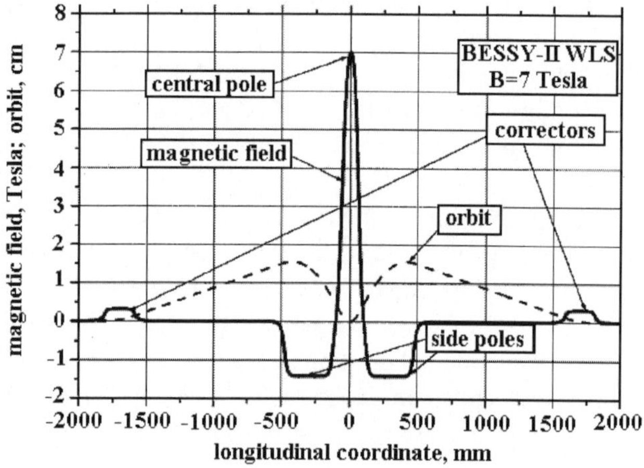

Figure 3.3 Longitudinal magnetic field distribution orbit behavior inside WLS with fixed radiation point.

The SC magnet part has no zero first field integral anymore, but with corrector magnet the integral is zero. Figure 3.3 shows field distribution and orbit distortion inside straight section of WLS with fixed radiation points installed on CAMD LSU (USA) and BESSY (Germany) storage rings (Fig. 3.4).

Figure 3.4 Photo of the 7 T WLS with fixed radiation point for BESSY-2. See also Color Insert.

3.2.3 Spectral SR Properties of WLS

As was mentioned above, installation of high field SC WLS on low energy electron storage rings (CAMD LSU — 1.35 GeV, BESSY-2 — 1.7–1.9 GeV) gives a possibility to carry out experiments in hard X-ray region. As seen in Fig. 3.5, SR spectrum from central pole of SC WLS installed on BESSY-2 storage ring is much wider compare SR from bending magnets (magnetic field of storage ring bending magnets is similar to maximum fields in side poles of WLS). One WLS may provide radiation up to four independent beamlines working with hard X-ray with experiments like X-ray diffraction, X-ray absorption, material study, etc.

There is a unique possibility to achieve very hard X-ray (gamma ray) by installing 10 T WLS on 8 GeV Spring-8 storage ring (Soutome *et al.*, 2003). Critical photon energy is equal to 430 keV of SR radiated from central pole (Fig. 3.6). The spectrum contains a high flux of photons with energy up to 10 MeV. The spectral shape of the WLS radiation is similar to that a blackbody radiation at billions of Kelvin. This means that the radiation from WLS is an ideal photon source for studying nucleosynthesis in nuclear astrophysics. Besides that there is a possibility to use this very hard and intense spectrum as positron and neutron sources with high brightness.

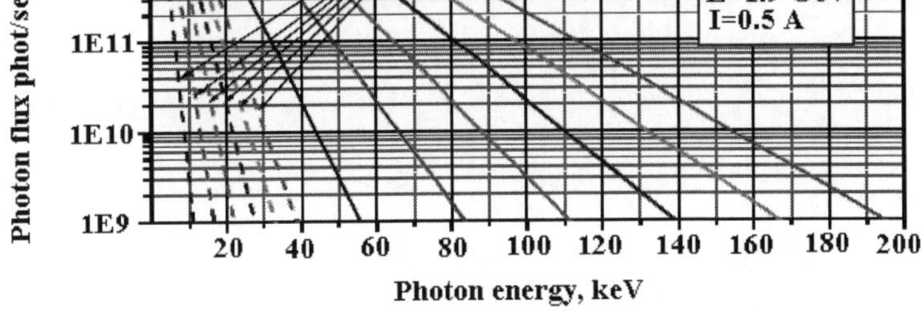

Figure 3.5 Spectral property of SR for some field levels from SC WLS installed on BESSY-2. Spectrum shown from central and side poles for demonstration of difference. See also Color Insert.

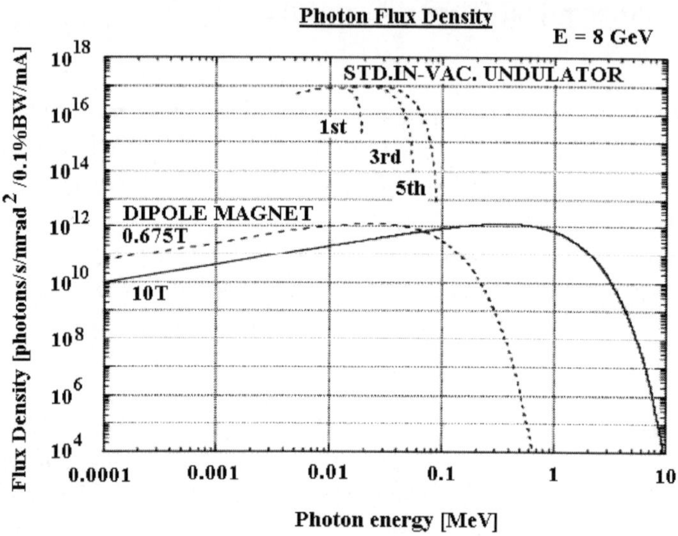

Figure 3.6 Photon flux density from magnetic elements of Spring-8. Solid curve is for 10 T WLS.

Six SC WLS were fabricated by Budker INP for different synchrotron light sources. List of parameters of these WLS is presented in Table 3.1.

Table 3.1 List of parameters of SC WLSs

	Year	Magnetic field, T_{Max}/normal	Magnetic length, mm	Vertical beam aperture, mm	Liquid helium consumption LHe L/hour
WLS for Siberia-1 (Moscow)	1985	5.8 (4.5)	350	22	2–2.5
WLS for PLS (Korea)	1995	7.68 (7.5)	800	26	1.5–2
WLS for LSU-CAMD (USA)	1998	7.55 (7.0)	972	32	1.2–1.6 <0.03*
WLS for Spring-8 (Japan)	2000	10.3 (10.0)	1042	20	0.4–0.6
WLS for (BESSY-II, Germany)	2000	7.5 (7.0)	972	32	0.4–0.5
PSF-WLS (BESSY-II, Germany)	2001	7.5 (7.0)	972	32	0.4–0.5

*After upgrading of cryogenic system in 2009.

3.2.4 Design of SC WLS

The design of WLS SC magnets presented in Table 3.1 is based on three SC dipoles (see Fig. 3.8). Most of these magnets are still working and show good enough reliability. Starting from 2000, the cryogenic systems of the WLS are based on cryocoolers of Leybold and SUMITOMO HI. Nb–Ti SC wire was used for fabrication of central and side poles of WLS with field 7–7.5 T. SC wire parameters are listed in Table 3.2 and Fig. 3.7.

Table 3.2 SC wire properties

SC wire diameter, mm	0.85 (0.92 with insulation)
Ratio NbTi:Cu	0.43
Critical current, A	380 (7 T)

Central pole is fabricated as two sections coils using the same SC wire. Two power supplies are used for feeding of SC part of the WLS. For poles optimization the inner coils and outer coils are feeding by different currents in accordance with critical curve of SC wire. The inner coils and side coils are feeding by one power supply, but outer coils of central pole are feeding by sum of currents of both power supplies. Such way of power supplies feeding enables to keep zero value of the first field integral for any field level by means of redistribution of currents between power supplies.

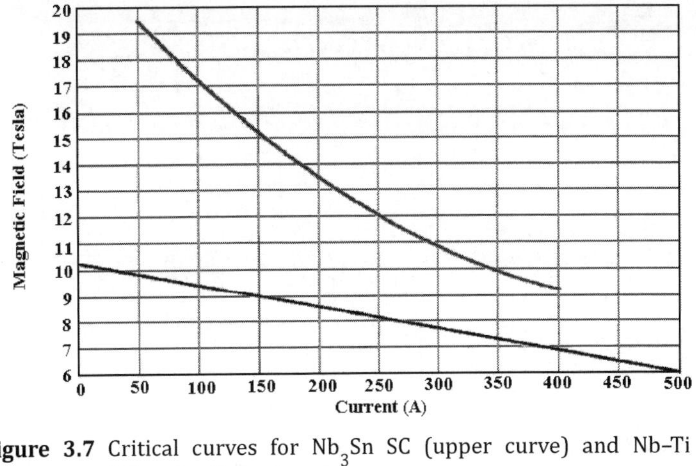

Figure 3.7 Critical curves for Nb_3Sn SC (upper curve) and Nb–Ti SC (lower curve). See also Color Insert.

The WLS design includes a special consideration of a joint action of several forces in the central WLS coils. At a low temperature of 4.2 K the coils are strained due to the difference in the thermal expansion coefficients of the different materials used in the SC WLS. The magnetic field creates a high pressure inside the WLS and expands the coils. To avoid any movement of turns inside coils and a quench of superconductivity, the bandage fixes the position of the coils.

The stainless steel wire bandage is used as a mechanical restrain to the motion of the SC wire inside the coils. Four layers of the stainless steel strained wire of 1 mm in diameter are wound over the SC layers. This bandage is effective mostly for side parts of the central coils where radius is small. The invar wedges are used as a mechanical restrain between the large radius part of the central coils and the iron yoke. The invar thermal expansion coefficient is very small in comparison with the iron one. This allows pressing this part of the coils during cooling down. The ARMCO-iron yoke is used to return the magnetic flux and to support the coils. The upper and the lower WLSs parts are supported by the non-magnetic stainless steel slab located between the halves. The additional iron plates between the upper and lower halves are used to close the stray magnetic flux.

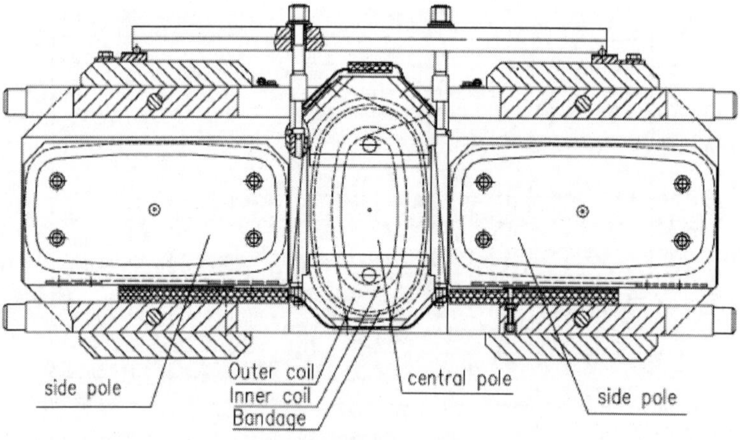

Figure 3.8 Half of SC WLS (median plane view).

Central poles of 10 T SC WLS were divided into three sections. Niobium–tin/Cu SC rectangular wire 1.45 × 0.85 mm^2 in size was used for manufacturing of the inner section of the central pole. Other two sections were manufactured with use of Nb–Ti SC wire. Currents in sections were optimized in accordance with critical curves of superconductors (Fig. 3.7). First section was feeding with first power supply, second section with second power supply, and third section by sum of currents of both power supplies. Iron yoke and side poles of 10 T WLS is very similar to WLS shown in Fig. 3.8.

To provide zero first field integral for any field level the ratio between currents of power supplies should be satisfied to requirements shown in Fig. 3.9.

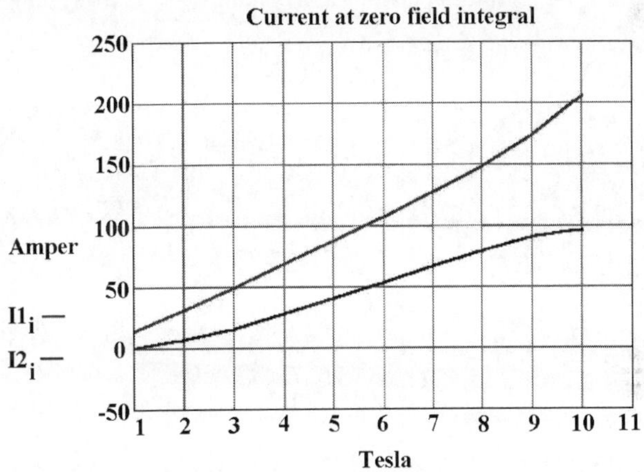

Figure 3.9 Currents in first and second sections depend on maximum magnetic field in median plane of WLS. See also Color Insert.

The first sections of central pole were made by method “wind and react.” After dry winding, the sections were baked out in accordance with temperature schedule shown in Fig. 3.10 and were impregnated with epoxy filled with aluminum oxide. Other Nb–Ti sections of poles were manufactured by “wet reeling” technique with a hot-hardening epoxy resin. Photos of 10 T WLS are presented in Fig. 3.11.

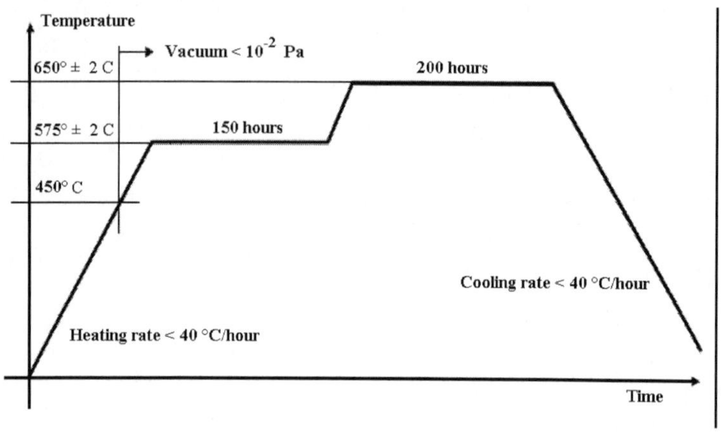

Figure 3.10 Temperature schedule for baking out of Nb–Sn windings in vacuum oven.

Figure 3.11 Photos of magnet disassembled into two parts (left) and assembled magnet of 10 T WLS. See also Color Insert.

3.2.5 Influence of Magnetic Field of SC ID on Beam Parameters

Magnetic field of SC ID may significantly influence the beam dynamics in storage rings. This influence may be divided into three parts:

- orbit distortion, tune shifts, and optics change;
- nonlinear field components may excite some nonlinear resonance and reduce dynamic aperture;

- radiation effects which may change beam emittance, energy spread, etc.

3.2.5.1 SC ID field expansion into multipole components

Magnetic field measurements of WLS are usually carrying out in Cartesian coordinates which will have designations χ, z, σ, thus the axis σ coincides with a longitudinal axis of a WLS, χ and z are horizontal and vertical directions correspondingly. Planes $z = 0$, $\chi = 0$, $\sigma = 0$ are corresponding planes of symmetry of WLS magnetic systems; therefore, the basic members of magnetic field expansion into multipole components in Cartesian coordinates, following K. Steffen (1965), may be written down in the forms:

$$
\begin{aligned}
B_z &= a + \frac{b}{2}\cdot\left(\chi^2 - z^2\right) - \frac{a''}{2}\cdot z^2 + \frac{c}{24}\cdot\left(\chi^4 - 6\chi^2 z^2 + z^2\right) \\
&\quad - \frac{b''}{4}\cdot\chi^2 z^2 + \frac{b''}{12} z^4 + \frac{a''''}{24} z^4 + \ldots \\
B_\chi &= b\cdot\chi z + \frac{c}{6}\cdot\left(z\chi^3 - z^3\chi\right) - \frac{b''}{6}\cdot z^3\chi + \ldots \\
B_\sigma &= a'\cdot z + \frac{b'}{2}\cdot z\left(\chi^2 - \frac{z^2}{3}\right) - \frac{a'''}{6}\cdot z^2 + \ldots
\end{aligned}
\tag{3.3}
$$

The primes in formulas above mean a derivative on longitudinal coordinate σ. Orbit distortion inside WLS may be large enough. We shall designate $x_0(\sigma)\cdot x_0'(\sigma)$ as coordinate and angle of beam orbit deviation to a WLS axis accordingly. In displaced on coordinate $x_0(\sigma)$ and rotated on angle $x_0'(\sigma)$ the basic field multipole components (field, gradient, and sextupole) may be expressed as:

$$
\begin{aligned}
B_z &= a + \frac{b}{2} \cdot x_0^2 + \frac{c}{24} \cdot x_0^4 + \ldots \\
G &= b \cdot x_0 + \frac{c}{6} x_0^3 - x_0' \left(a' + \frac{b'}{2} x_0^2 + \frac{c'}{24} x_0^4 \right) + \ldots \\
S &= b + \frac{c}{2} x_0^2 - 2b' x_0 x_0' - \frac{c}{3} x_0^3 x_0' + a'' x_0'^2 + \frac{b''}{2} x_0^2 x_0'^2 + \ldots
\end{aligned} \tag{3.4}
$$

If magnetic system is homogeneous enough so that orbit deviation $x_0(\sigma)$ is much less than characteristic size of field decrease, the formulas (3.4) may be simplified:

$$
\begin{aligned}
B_z &= a + \frac{b}{2} \cdot x_0^2 \\
G &= b \cdot x_0 - x_0' a' \\
S &= b \cdot 2b' x_0 x_0' + a'' x_0'
\end{aligned} \tag{3.5}
$$

Magnetic field of multipole wigglers may be approximated by formulas:

$$
\begin{aligned}
B_z &= B_0 \cos(k_0 \sigma) \cos(k_x \chi) \cosh(k_z z) \\
B_\chi &= -\frac{k_x}{k_z} B_0 \cos(k_0 \sigma) \sin(k_x \chi) \sinh(k_z z) \\
B_\sigma &= -\frac{k_0}{k_z} B_0 \sin(k_0 \sigma) \cos(k_x \chi) \sinh(k_z z)
\end{aligned} \tag{3.6}
$$

where $k_0 = 2\pi/\lambda_0$, $k_x = 2\pi/\lambda_x$, $k_z = 2\pi/\lambda_z$; λ_0, λ_x, λ_z are wiggler characteristics magnetic field periods in σ-, χ-, z-directions and $k_z^2 = k_x^2 + k_0^2$ are following from Maxwell equations.

Applying these equations (3.6) for field multipole expansion (3.3) one may obtain expansion coefficients:

$$
\begin{aligned}
a(\sigma) &= B_0 \cos(k_0 \sigma) \\
b(\sigma) &= -k_x^2 B_0 \cos(k_0 \sigma) \\
c(\sigma) &= -k_x^4 B_0 \cos(k_0 \sigma)
\end{aligned} \tag{3.7}
$$

Field, gradient, and sextupole components combining (3.4) and (3.7) in approximation $k_x x_0 < 1$, $x_0' < 1$ may be expressed as:

$$
\begin{aligned}
B_z &= B_0 \cos(k_0 s)\left(1 - \frac{1}{2}k_x^2 k_0^2 + \ldots\right) \\
G &= B_0 \left(\cos(k_0 s)(-k_x^2 x_0 + \ldots) + \sin(k_0 s)(x_0' k_0 + \ldots)\right) \\
S &= B_0 \left(\cos(k_0 s)(-k_x^2 - k_0^2 x_0'^2 + \ldots) + \sin(k_0 s)(-2k_x^2 k_0 x_0 x_0' + \ldots)\right)
\end{aligned} \tag{3.8}
$$

3.2.5.2 Orbit distortion by SC ID

Only z-component of magnetic field exists in median plane of considered symmetric magnetic systems so that in an ideal case orbit distortion of particle moving in median plane of these systems leads to orbit distortion only in χ-direction which is described by the following formulas:

$$x_0'(\sigma) = \frac{1}{B\rho}\int_{-L/2}^{\sigma} ds' B_z(s') \tag{3.9}$$

$$x_0(\sigma) = \frac{1}{B\rho}\int_{-L/2}^{\sigma} ds' \int_{-L/2}^{s'} ds'' B_z(s'') \tag{3.10}$$

where L is straight section length of WLS installation, magnetic field is equal to zero, and $B\rho = E/qc$ is rigidity of electrons in storage ring. If $s = L/2$, the formulas (3.9) and (3.10) give field integrals (3.1) and (3.2).

Obviously, the requirement of equality to zero of the first and second integrals can be executed with some accuracy. For the coordination of these requirements with the valid parameters of the storage ring such as emittance, betatron tunes, Twiss parameters, etc., it is possible to use the following expressions:

$$
\begin{aligned}
\phi &< 2\sin(\pi Q_x)\sqrt{\frac{\varepsilon_x}{\beta_x}} \\
\delta &< 2\sin(\pi Q_x)\sqrt{\frac{\varepsilon_x}{\gamma_x}}
\end{aligned} \tag{3.11}
$$

where πQ_x, ε_x is horizontal betatron tune and horizontal emittance correspondingly, and β_x, γ_x is Twiss parameters for horizontal motion.

If the conditions written above are satisfied, it means that orbit distortion in a storage ring due to this SC ID is less than beam size.

The parameter ϕ specifies an accuracy of compensation angle orbit deviation after passing the SC ID and this is criterion of accuracy of set currents in main and correcting coils. The parameter δ is criterion of accuracy of ID manufacturing, a degree of symmetry of B_z magnetic field component in longitudinal direction. Expression (3.3) contains under integral a product of even and odd functions in symmetric limits that automatically leads to zero result. In case of infringement of symmetry of magnetic field distribution as a result of manufacturing imperfection, the parameter δ does not become zero. Usually ID are installing into a straight section with zero dispersion function, but it does not mean that the dispersion is not present there. Dispersion function is created by ID according to the following equation:

$$\eta_x'' = -\frac{1}{B\rho} B_z. \qquad (3.12)$$

As a result, the dispersion function inside the ID may be expressed as:

$$\eta_x(\sigma) = (\eta_{x0}(\sigma) - x_0(\sigma)), \qquad (3.13)$$

where $\eta_{x0}(\sigma)$ is the dispersion function inside the ID if field of ID is switched off.

Orbit distortion inside ID changes the orbit length as:

$$\Delta L = \int_L \sqrt{1 + x_0'^2}\, ds - L, \qquad (3.14)$$

where L is the length of the straight section. If this value is large enough to avoid orbit distortion on other part of storage ring it is necessary to decrease RF frequency.

3.2.5.3 Focusing effect by SC ID

High field SC ID is a focusing element in magnetic structure of the storage ring and creates betatron tune shifts and structural functions changes. Horizontal and vertical betatron motion inside SC ID may be described by the following equations:

$$x'' + K_x \cdot x = 0$$
$$z'' + K_z \cdot z = 0 \tag{3.15}$$

where coefficients of magnetic strength K_x, K_z are equal to:

$$K_x = \frac{B_z^2}{(B\rho)^2} + \frac{1}{B\rho}\left(x'\frac{\partial B_z}{\partial s} - \frac{\partial B_z}{\partial x}\right),\; K_z = -\frac{1}{B\rho}\left(x'\frac{\partial B_z}{\partial s} - \frac{\partial B_z}{\partial x}\right) \tag{3.16}$$

Edge focusing completely compensates focusing by magnetic field in horizontal direction and the main focusing action of SC ID with symmetric magnetic system occurs in vertical direction. The integral values for horizontal and vertical motions may be expressed in view:

$$\int_{-L/2}^{L/2} K_x ds = -\frac{\partial}{\partial x}\int_{-L/2}^{L/2} \frac{B_z ds}{B\rho} = \bar{K}_x L, \tag{3.17}$$

$$\int_{-L/2}^{L/2} K_z ds = \int_{-L/2}^{L/2} \frac{B_z^2}{B\rho^2} ds - \int_{-L/2}^{L/2} K_x ds = \bar{K}_x L, \tag{3.18}$$

where $\bar{K}_x$, $\bar{K}_z$ is average values of ID magnetic strength along ID straight section.

The main results of focusing SC ID are shifts of betatron tunes, possible broadening of half-integer, and integer resonances of betatron motion and beating beta-functions. In approximation for $\sqrt{\bar{K}_{x,z}}L$ values which are small enough the betatron tune shifts may be estimated as:

$$\Delta\nu_{x,z} = \frac{\bar{K}_{x.z} L \beta_{x,z}}{4\pi}\left(1 + \frac{L^2}{12\beta_{x,z}^2}\right), \tag{3.19}$$

The standard behavior of betatron tune shifts versus SC WLS field, installed on BESSY storage ring, is shown in Fig. 3.12. Vertical tune shift is a quadratic dependence on level of magnetic field in

WLS. From Fig. 3.12, one can see that dependence on field level of horizontal tune shift is very weak, that corresponds to two-dimensional field distribution with homogeneous enough lateral field distribution and focusing by magnetic field is fully compensated by edge focusing.

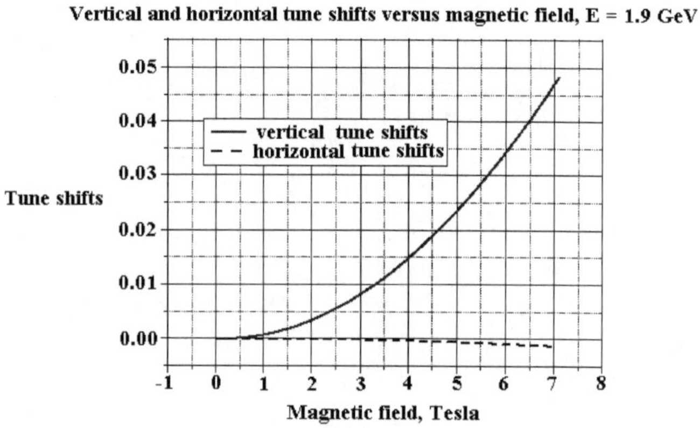

Figure 3.12 Vertical and horizontal betatron tune shifts for BESSY SC 7 T WLS versus magnetic field level. See also Color Insert.

3.2.5.4 Radiation (structural) integrals

Installation of SC high field ID on a storage ring may influent on beam parameters by changing of radiation integrals. The primary goal at wiggler installation on a storage ring is to consider all possible unpleasant changes in structural functions of the storage ring and to provide correct indemnification of all effects which arise with magnetic field of ID. Expressions for changes of radiation integrals for symmetric magnetic systems are as follows:

$$\Delta I_1 = \int_L \frac{(\eta_{x0} - x(s))B_z(s)}{B\rho} ds, \tag{3.20}$$

$$\Delta I_2 = \int_L \frac{B_z^2(s)}{B\rho^2} ds, \tag{3.21}$$

$$\Delta I_3 = \int_L \frac{|B_z(s)|^3}{B\rho^3} ds, \tag{3.22}$$

$$\Delta I_4 = \int_L \left(\frac{B_z^3(s)}{B\rho^3} - \frac{2K_x}{B\rho} \right) (\eta_{x0} - x(s)) ds, \tag{3.23}$$

$$\Delta I_5 = \int_L \frac{|B_z(s)|^3}{B\rho^3} \Big(\gamma_x (\eta_{x0} - x(s))^2 + 2\alpha_x (\eta_{x0} - x(s)) (\eta'_{x0} - x'(s)) + \beta_x (\eta'_{x0}(s))^2 \Big) ds, \tag{3.24}$$

where $\alpha_x, \beta_x, \gamma_x$ are Twiss parameters.

Energy spread and horizontal emittance are the most important beam parameters which may be changed by action of SC ID magnetic field. Ratio of energy spread $(\sigma'E/\sigma E)$ and ratio of emittances $\varepsilon'_x/\varepsilon_x$ with/without magnetic field of ID are described by the following formulas:

$$\left(\frac{\sigma'_E}{\sigma_E} \right)^2 = \frac{1 + \frac{\Delta I_3}{I_3^0}}{1 + \frac{2\Delta I_2 + \Delta I_4}{2I_2^0 + I_4^0}} \approx 1 + \frac{\Delta I_3}{I_3^0} - \frac{\Delta I_{2_4}}{I_2^0} \tag{3.25}$$

$$\frac{\varepsilon'_x}{\varepsilon_x} = \frac{1 + \frac{\Delta I_5}{I_5^0}}{1 + \frac{\Delta I_2 - \Delta I_4}{I_2^0 - I_4^0}} \approx 1 + \frac{\Delta I_5}{I_5^0} - \frac{\Delta I_2}{I_2^0}, \tag{3.26}$$

where I_2^0, I_3^0, I_4^0, I_5^0 are the radiation integrals of the storage ring without SC ID magnetic field.

Energy spread (Fig. 3.14) and emittance (Fig. 3.13) are decreased down to some minimal levels with field increasing, but further field increasing leads to increasing their values.

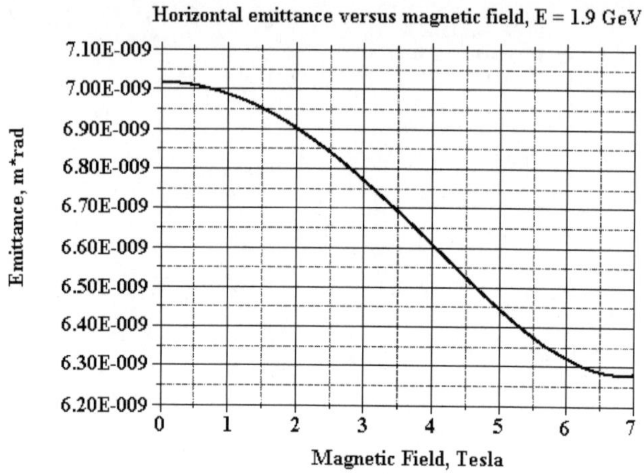

Figure 3.13 Horizontal emittance BESSY storage ring versus magnetic field level in SC 7 T WLS.

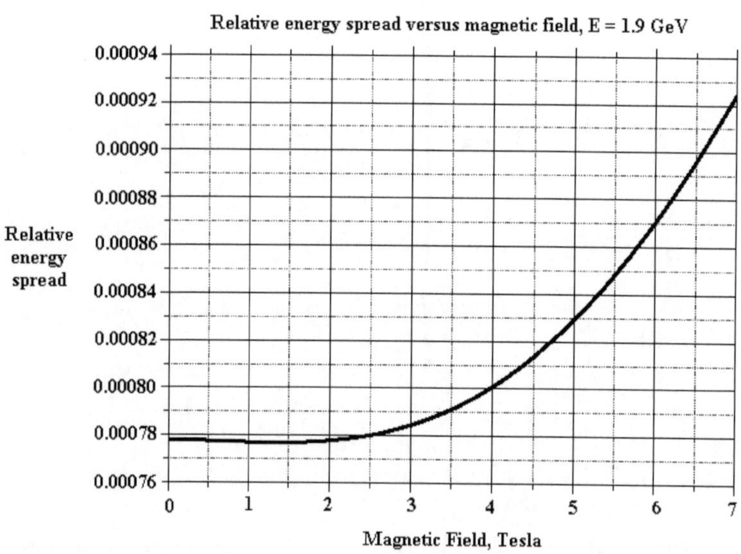

Figure 3.14 Energy spread in BESSY storage ring versus magnetic field level in SC 7 T WLS.

3.3 SC BENDING MAGNETS (SUPERBENDS)

Replacement of conventional bending magnets in existing storage rings used as SR sources on SC high field bending magnets (superbends) is another example of significant improvement of consumer parameters of a SR source. A vivid example of such upgrading is Advanced Light Source (ALS) storage ring where in 2001 three conventional bending magnets with field of 1.3 T have been replaced by superbends with field of 5 T (Robin *et al.*, 2001). The improvement of this upgrade is clear from Fig. 3.15. Long and reliable life of this modernization is a proof that this possibility have right to exist.

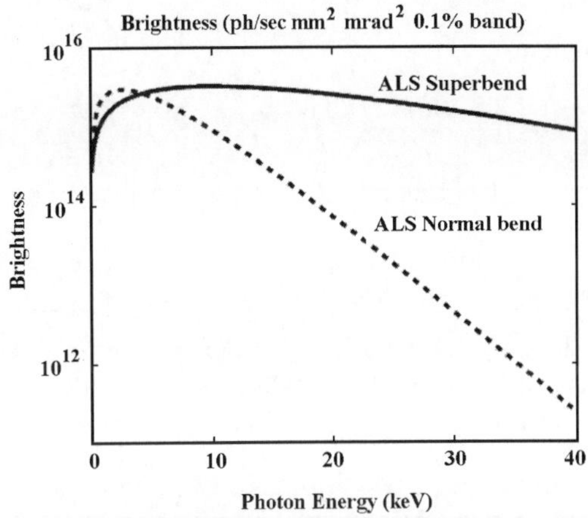

Figure 3.15 Brightness of superbend (solid line) versus the normal conducting bend (dashed line).

Use of superbend as SR source has clear advantages in comparison with WLS. It does not have disadvantages like dependence of orbit distortion on field level and existence of the second source, but it should be not less reliable as conventional magnets, in spite its much more technical complication.

Inspired by this idea, Budker INP in collaboration with BESSY has made the prototype of superbend magnet with nominal field of 8.5 T (Batrakov *et al.*, 2005). The SuperBend (SB) magnet system is dipole consisting of two SC coils assembled above and below the vacuum chamber. The main parameters of the SB magnet are given in Table 3.3.

Table 3.3 The main parameters of the SB prototype magnet for BESSY

Vertical/horizontal aperture, mm	30/75
Pole gap, mm	46
Operating magnetic field, T	3.3–8.5
Maximum magnetic field, T	9.6
Coil material	Nb_3Sn, NbTi
Edge angle, degree	1.3
Current in coil for 8.5 T, A	264
Eff. magnetic length along beam, m	0.1777
Bending angle, degree	11.25
Bending radius, m	0.905
Stored energy for 8.5 T, kJ	180
Cold mass, kg	~2000

The SB magnet coil consists of four sections. The first and second sections are winded from Nb_3Sn wire. The third section is winded from NbTi wire. The fourth section is winded from NbTi wire and used for correction. Special bandaging system is used to prevent winding wire motion under action of pondermotive forces.

Figure 3.16 Photo of assembled pole. See also Color Insert.

The bandaging system consists of bandaging section of the coil and stainless steel ring with 150 bolts. This bolts press on the bandaging section via four special form stainless steel plates. The sketch and photo of SB magnet coils with bandaging system are given in Figs. 3.17 and 3.16. The ARMCO-iron yoke is used to return the magnetic flux and to support the coils. The photo of the assembled SC bending magnet is given in Fig. 3.18.

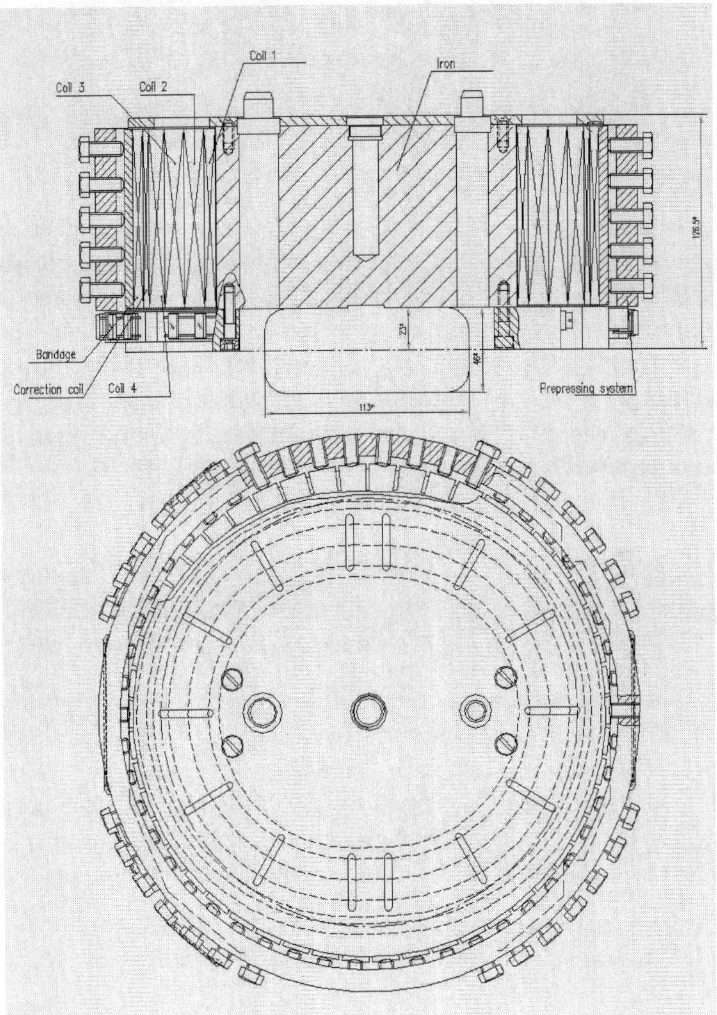

Figure 3.17 Sketch of assembled coil.

Figure 3.18 Photo of assembled magnet. See also Color Insert.

Unfortunately, a decision to replace conventional magnets by superbends on BESSY-2 is still not accepted in spite of successful tests of the prototype superbend and the inspiring experience of ALS already many years successfully working with superbends. But it gives us the basis on designing of new SR sources in which the part of magnets are SC. As a result of such combination electron energy of storage ring may be reduced without deterioration of spectral characteristics.

3.4 SUPERCONDUCTING MULTIPOLE WIGGLERS

SC multipole wiggler represents sign-variable magnetic structure with many poles and with magnetic field defined by critical curve of a SC wire. Beam orbit inside a multipole wiggler moves by a wiggling trajectory and SR is concentrating from all poles in some horizontal angle which equals to one pole orbit angle and that leads to increase of photon flux density by N times where N number of poles in the wiggler. Essentially, multipole wigglers and undulators with period λ_0 and field amplitude B_0 are differing by value of K-parameter:

$$K = 0.934 \cdot \lambda_0[\text{cm}] B_0[T]. \tag{3.27}$$

For $K \sim 1$ spectral properties of radiation (undulator radiation) differ from SR because formation of radiation occurs on all length of magnetic structure. That leads to concentration of radiation due to interference near some selective photon wavelengths λ_n defined by formula below:

$$\lambda_n = \frac{\lambda_0}{n \cdot 2 \cdot \gamma^2}\left(1 + \frac{K^2}{2} + (\gamma\theta)^2\right). \tag{3.28}$$

where n is harmonic number; γ, Lorentz factor; and θ, observation angle. Undulator radiation spectrum turns into SR spectrum if $K > 1$

3.4.1 History

The first SC multipole wiggler was designed and fabricated in Budker INP in 1979 (Fig. 3.19) (Barkov *et al.*, 1978; Artamonov *et al.*, 1980). It had been installed on 2 GeV storage ring VEPP-3 to increase photon flux density with higher energy. Main parameters of the SC wiggler are presented in Table 3.4.

Table 3.4 Main parameters of the first superconducting wiggler

Pole number	20
Pole gap, mm	15
Period, mm	90
Magnetic field amplitude, T	3.5
Vertical beam aperture, mm	7.8

The SC wiggler installation scheme should satisfy several requirements:

- energy injection of electron beam is 350 MeV when wiggler field was switched off and vertical aperture was about 30 mm;
- ramp beam energy up to 2 GeV was performed with zero field in SC wiggler;
- at 2 GeV beam orbit was bumped with use of four special steering magnets to move electron beam into small vertical aperture vacuum chamber of SC wiggler;
- ramp up magnetic field in SC wiggler.

To perform these requirements, the cross section of the vacuum chamber of the SC wiggler was like keyhole — wide area was used for injection, and narrow area was used for creation of magnetic field by the wiggler. These two areas were connected by narrow 5.5 mm slit for electron beam moving from injection area to area with high magnetic field.

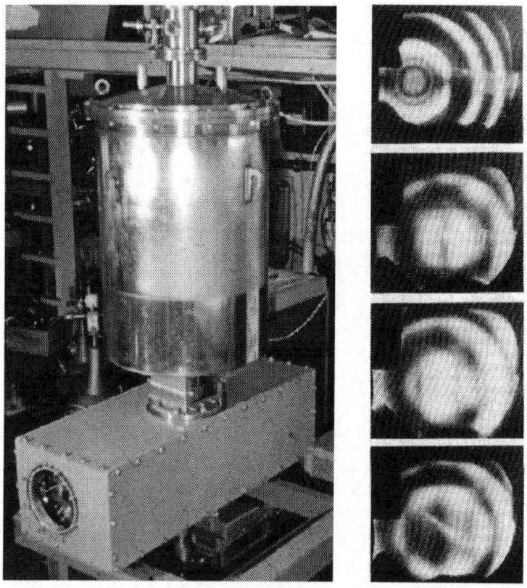

(a) (b)

Figure 3.19 (a) Photo of the first assembled SC wiggler before installing on the ring. (b) Photographs of light intensity from the wiggler through red color filter on a screen for different field level: first, second, third, and fourth harmonics of undulator radiation are visible. See also Color Insert.

Main effect of the wiggler installation on VEPP-3 storage ring was the increase of photon flux by 200 times at photon energy 15–20 keV.

3.4.2 Recent Progress in SC Multipole Wigglers

Table 3.5 shows an array of SC multipole wigglers designed and fabricated in Budker INP for last several years. These SC wigglers may be divided into three groups. High field group consists of wigglers with field of 7–7.5 T and period length is about

150–200 mm (Berger *et al.*, 2001). These wigglers create a large horizontal fan angle, which is possible to divide into several beamlines. Medium field group consists of wigglers with field of 3.5–4.2 T and period length is about 48–60 mm (Zagrando *et al.*, 2002). Low field group consists of short period wigglers (30–34 mm) with field 2–2.2 T (Mezentsev and Perevedentsev, 2005). Short period wigglers have K-value of about 6–7, and its radiation spectrum is similar to undulator spectrum for low photon energy part of spectrum.

Table 3.5 Main parameters of recent SC wigglers

	Year	Magnetic field, T_{Max} normal	Poles number (main + side)	Pole gap, mm	Period mm	Vertical aperture, mm
7 T wiggler (BESSY-II, Germany)	2002	(7.67) 7	13 + 4	19	148	13
3.5 T wiggler ELETTRA (Italy)	2002	(3.7) 3.5	45 + 4	16.5	64	11
2 T wiggler CLS (Canada)	2005	(2.2) 2	61 + 2	13.5	34	9.5
3.5 T wiggler DLS (England)	2006	(3.75) 3.5	45 + 4	16.5	60	11
7.5 T wiggler SIBERIA-2 (Russia)	2007	(7.7) 7.5	19 + 2	19	164	14
4.2 T wiggler CLS (Canada)	2007	(4.34) 4.2	25 + 2	14.5	48	10
4.2 T wiggler DLS (England)	2009	(4.25) 4.2	45 + 4	13.8	48	10
4.1 T wiggler LNLS (Brazil)	2009	(4.19) 4.1	31 + 4	18.4	60	14
2.1 T wiggler ALBA-CELLS (Spain)	2009	2.1	117 + 2	12.6	30.15	8.5

Main wiggler parameters are as follows: B_0 — magnetic field λ_0 — period, and g — pole gap. The parameters may be linked by the following formula (3.29) with some approximation:

$$B_0 = \frac{B_p}{\cosh\left(\pi \dfrac{g}{\lambda_0}\right)} \approx B_p e^{-\pi \frac{g}{\lambda_0}}, \tag{3.29}$$

where B_p is the field parameter that depends on type of SC wire, configuration of pole, etc. Figure 3.20 shows the behavior of B_0/B_p ratio versus g/λ_0. The field exponentially falls down at increase of pole gap and if pole gap equals to the period the B_0/B_p ratio is less than 0.1.

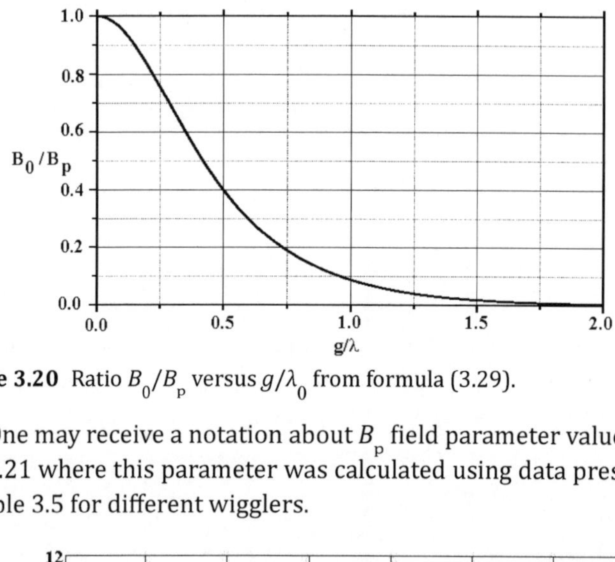

Figure 3.20 Ratio B_0/B_p versus g/λ_0 from formula (3.29).

One may receive a notation about B_p field parameter value from Fig. 3.21 where this parameter was calculated using data presented in Table 3.5 for different wigglers.

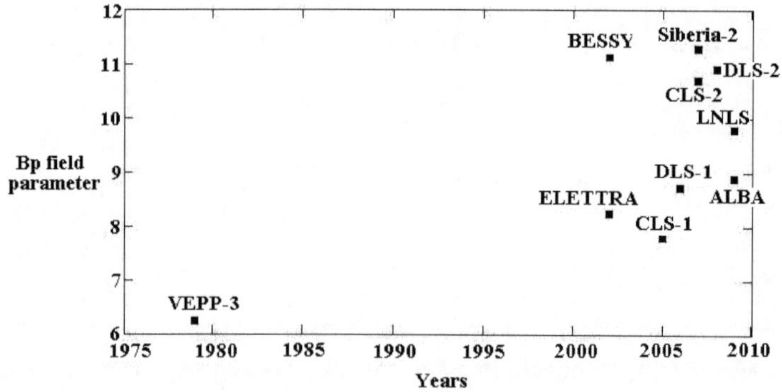

Figure 3.21 B_p field parameter distribution for various fabricated SC wigglers presented in Table 3.5.

The pole gap g value is defined by external requirements for vertical aperture proceeding from reasons of good beam lifetime, injection conditions, etc., while the period λ_0 at the fixed full wiggler length is defined from reasons of the maximal photon flux in required photon energy range. Using given storage ring parameters (energy E, current I, available length L) of wiggler straight section and required photon energy range for a dedicated experiment it is necessary to optimize wiggler parameters on a maximum of photon flux in required spectral range.

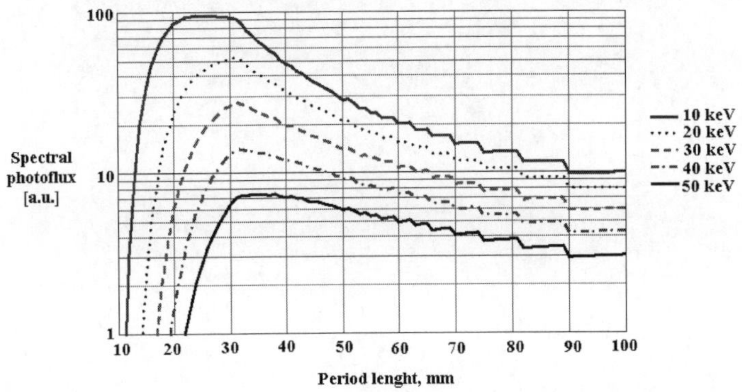

Figure 3.22 Spectral photon flux versus wiggler period length. See also Color Insert.

Figure 3.22 shows example of optimization of wiggler parameters for maximum photon flux in photon energy range 10–50 keV at following conditions: total magnet length should be less than or equal to 2 m, pole gap is 12.4 mm, and total radiated power should be less than or equal to 20 kWatt for beam parameters: E = 3 GeV, I = 0.4 A (ALBA-CELLS photon source). Optimal period length of the wiggler was defined as 31 mm and magnetic field 2.1 T using NbTi SC wire.

3.4.3 Wiggler Magnetic System Design

3.4.3.1 Magnetic structure

The design of windings in all three wiggler groups is approximately identical. The coil represents an iron pole of racetrack form on which the SC round wire is reeled up (Fig. 3.23).

The "wet" winding is used with epoxy filled with GOS (Gd_2O_2S) filler which has anomalous high heat capacity at 4 K (Fig. 3.24), for alignment of temperature contraction coefficient with SC wire and for essential increase in heat capacity of the coil (Alekseev *et al.*, 2007; Keilin *et al.*, 2008).

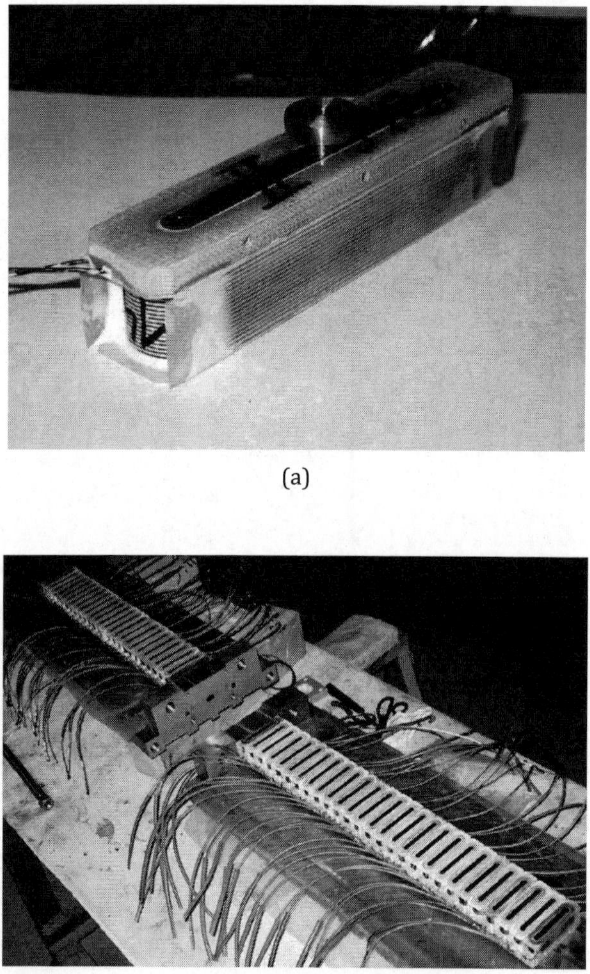

(a)

(b)

Figure 3.23 (a) 4.2 T half pole, and (b) two pole arrays of 4.2 T wiggler before commutation and assembling. See also Color Insert.

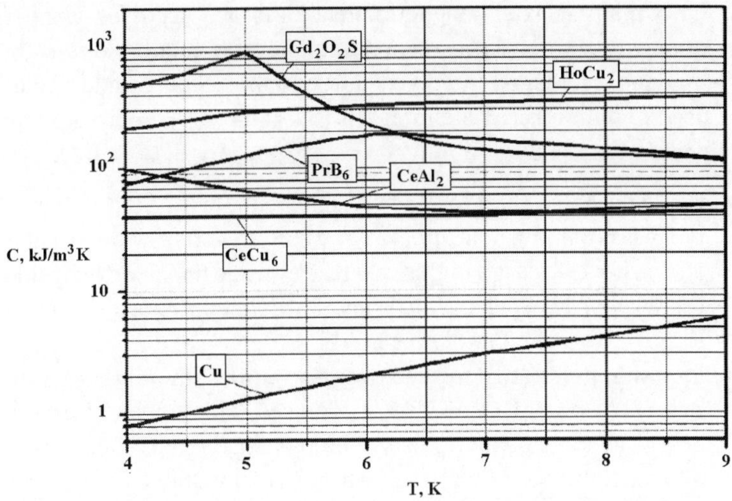

Figure 3.24 Volumetric heat capacities of Cu and different substances with extremely large heat capacity at low temperature in absence of magnetic field.

Wiggler coils of high and medium field groups consist of two sections. Both sections fabricated with use the same SC wire, but sections are feeding with different currents in accordance with critical curve of SC wire (Fig. 3.25).

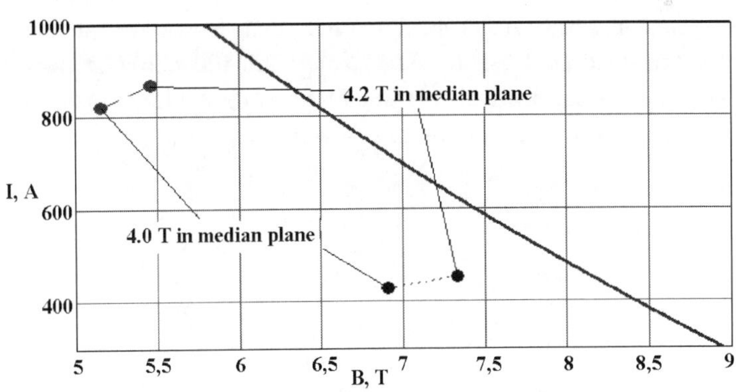

Figure 3.25 Critical curve of SC wire used for medium field wigglers. Blue and magenta points correspond to load lines of first and second coil sections correspondingly relative critical curve at different field level in median plane. See also Color Insert.

Two main wiggler magnetic structures are used for wiggler design: sign-symmetric, corresponding odd pole number and sign-asymmetric, corresponding even pole number. Side wiggler poles are used to obtain zero field integrals (3.1) and (3.2). There are two main variants:

- two side poles (one for each wiggler end) with field integrals twice less of main pole integral. The field structure of the wiggler is ½, −1, +1, −1 ... .½. In this case, the average line of beam inside the wiggler is wiggler off axis and both field integrals are equal to zero if the number of pole is odd.
- four side poles (two for each wiggler end) with field structure 1/4, −3/4, +1, −1, ... −3/4, 1/4 are used. This structure allows having average beam trajectory on the wiggler axis. Due to that it is possible to achieve equality to zero of both field integrals for even and odd pole number.

3.4.3.2 Yoke

The ARMCO-iron yoke is used to return the magnetic flux and to support the coils. The yoke includes two parts which are placed symmetrically above and below of median plane of the wiggler (Fig. 3.26). The upper and the lower wiggler parts are supported by the non-magnetic stainless steel slab located symmetrically between the coils. The additional iron plates between the upper and lower halves are used to close the stray magnetic flux. Coil arrays are pressed by special bronze rods to avoid any movements inside the coils during switching on of magnetic field.

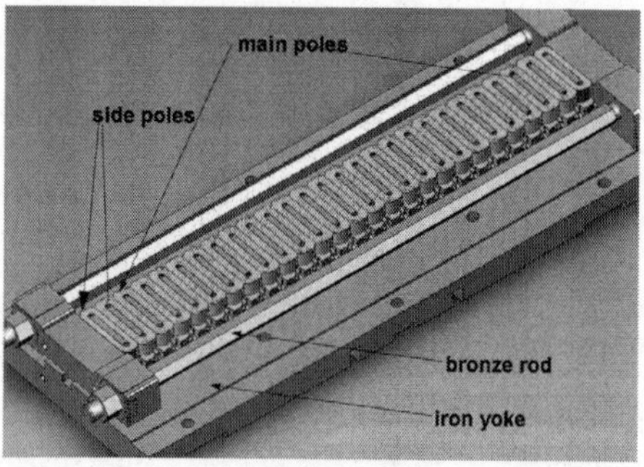

Figure 3.26 Half of assembled wiggler magnet. See also Color Insert.

3.4.3.3 Field distribution and orbit distortion inside SC wigglers

Magnetic measurements of a SC wiggler are making twice: in a bath cryostat before its installation into an own cryostat for revealing manufacturing errors of the magnet and then in the own cryostat. The magnetic measurements in the bath cryostat are carried out by using five LakeShore Hall probes array 6 mm spaced which are placed inside liquid helium at 4.2 K. The Hall probes were mounted on the rotatable disk with 90° rotating angle. To increase the field quality measurements, the relative calibration procedure is provided first. For calibration procedure Hall probes array is rotating in such a way as the probes are passing by one way one by one for relative calibration of the probes. The next measurements are carrying out when the probes are situated perpendicularly to the moving direction. In this case during the probes motion the magnetic map in the one plane is measured. Example of measurement results is shown in Fig. 3.27 of longitudinal magnetic field distribution of Diamond Light Source SC wiggler at set field 4.1 T.

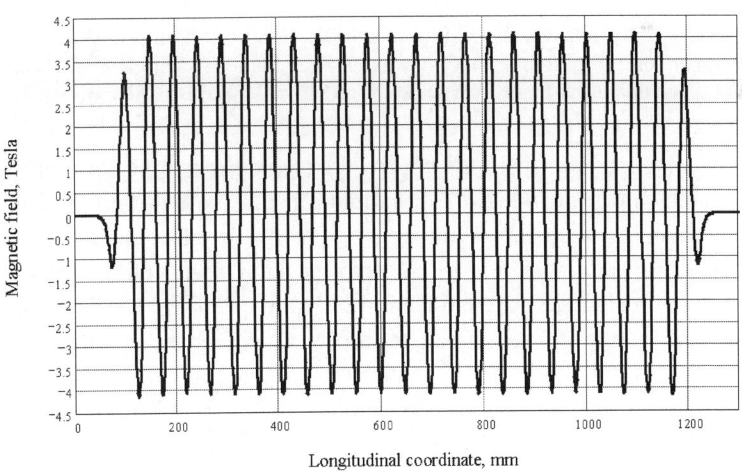

Figure 3.27 Measured longitudinal magnetic field distribution at field level 4.1 T of DLS SC wiggler.

It is important to know residual magnetic field distribution in a SC wiggler if wiggler currents are zero.

Figure 3.28 shows residual field distribution at set zero currents in normal operation mode and after a quench. For normal operation mode the residual field is defined by persistent currents excited in the coils.

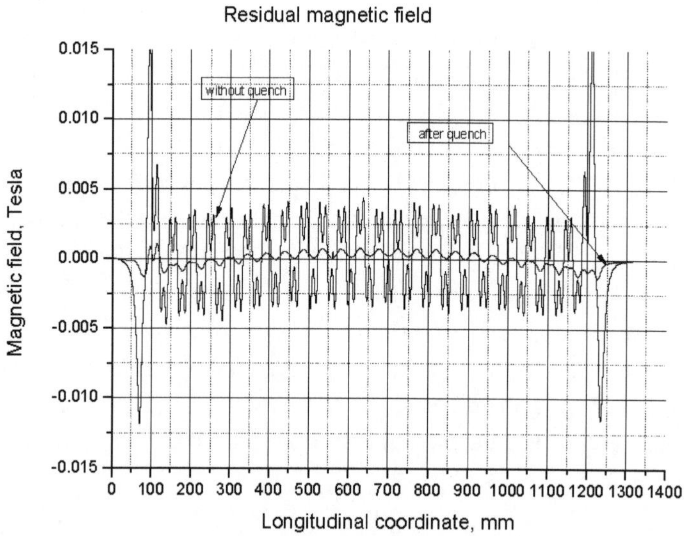

Figure 3.28 Measured residual magnetic field distribution in DLS SC wiggler for zero currents in the wiggler after field ramping down without quench (blue curve) and after quench (red curve). See also Color Insert.

An electron moving through wiggler magnetic field undergoes angular and coordinate displacements inside the wiggler. To avoid any orbit distortions in a storage ring the orbit angle and coordinate at the wiggler exit should be equal to the orbit angle and coordinate at the wiggler entry (conditions (3.1) and (3.2) for the first and second field integrals). As an example, Figs. 3.29 and 3.30 show a behavior of angle and displacement of the orbit inside the wiggler with field shown in Fig. 3.27 under condition of zero field integrals.

The SC coils are feeding by two power supplies which are controlling magnetic field value and zero first field integral. During field ramping up from zero up to nominal value there is a saturation of iron of poles, and it occurs at different currents for the main and side poles. As a result there is a nonlinear field dependence on currents which should be taking into account to prevent orbit distortion during field ramping up and down.

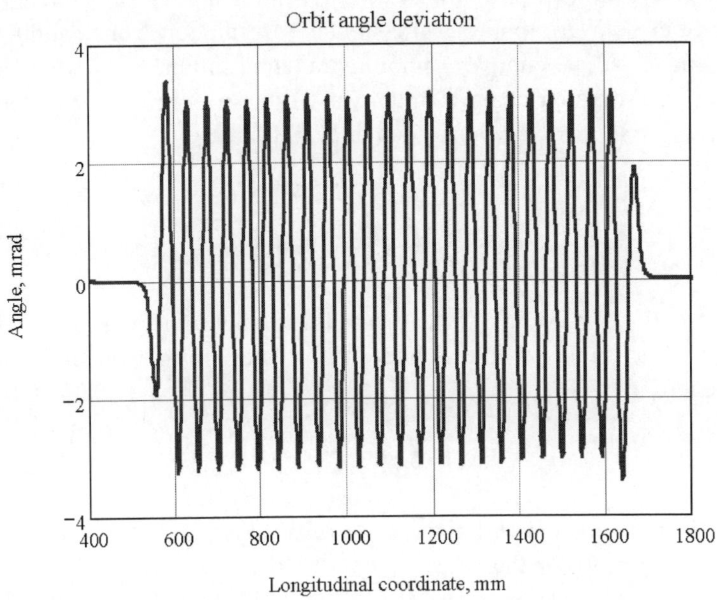

Figure 3.29 Angle orbit deviation at field setting 4.2 T, E = 3 GeV.

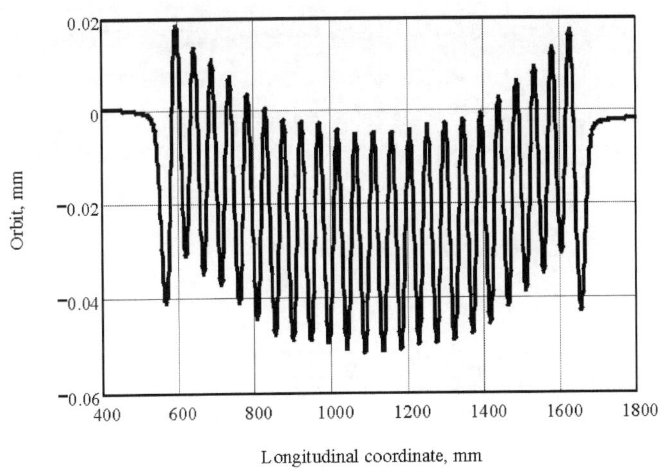

Figure 3.30 Orbit distortion at field setting 4.2 T, E = 3 GeV.

3.4.3.4 Spectral-angular characteristics of radiation from SC wigglers

Spectral characteristics of radiation from wigglers of high and medium field groups are very similar to radiation from bending magnets. Critical energy of photons radiated in horizontal angle θ may be estimated using formula (3.30) due to strong correlation between radiated angle and magnetic field value:

$$\varepsilon_c(\theta) = \varepsilon_{c\,\max}\sqrt{1-\left(\frac{\theta}{\theta_0}\right)^2}, \tag{3.30}$$

where θ_0 is maximal angle orbit deviation in the wiggler, and ε_{c0}, critical energy of the photons radiated from wiggler field maximum:

$$\varepsilon_{c\max}[\text{keV}] = 0.665 \cdot E^2[\text{GeV}] \cdot B_0[\text{T}]. \tag{3.31}$$

Example of spectral-angular characteristics of SR from DLS SC wiggler is shown in Fig. 3.31.

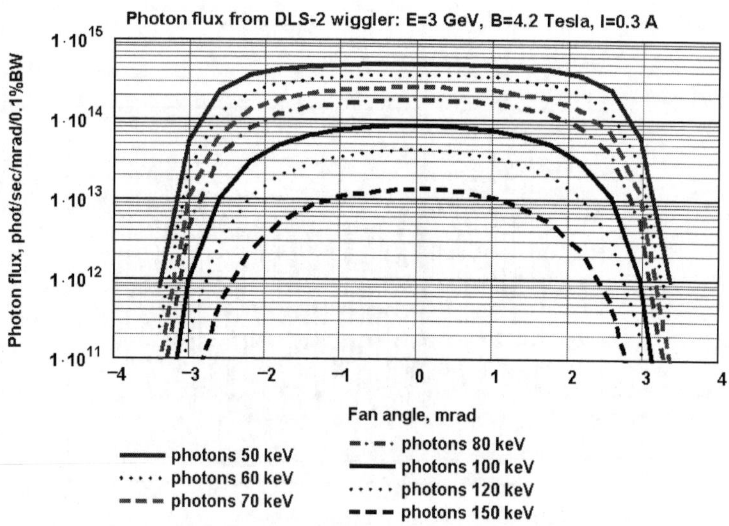

Figure 3.31 Photon flux angle distribution from the wiggler: B = 4.2 T, E = 3 GeV, I = 0.3 A. See also Color Insert.

Spectral characteristics of radiation from low field wigglers consist in that the soft spectrum part has discrete spectrum owing to an interference of radiation from different wiggler poles and it is similar to undulator spectrum with peaks at wavelengths defined by the formula (3.28). The maximum of radiation corresponds to $3/8K^3$ harmonic number. For example, SC wiggler with parameters: $B_0 = 2$ T, $\lambda_0 = 3$ cm, period number = 60, maximum of radiation corresponds to harmonic number near 80. Spectrum discreteness disappears with increasing harmonics number due to several reasons like energy and angle spread in a beam, final number of the wiggler periods. A provisional kind of spectral distribution for above mentioned wiggler is presented in Fig. 3.32.

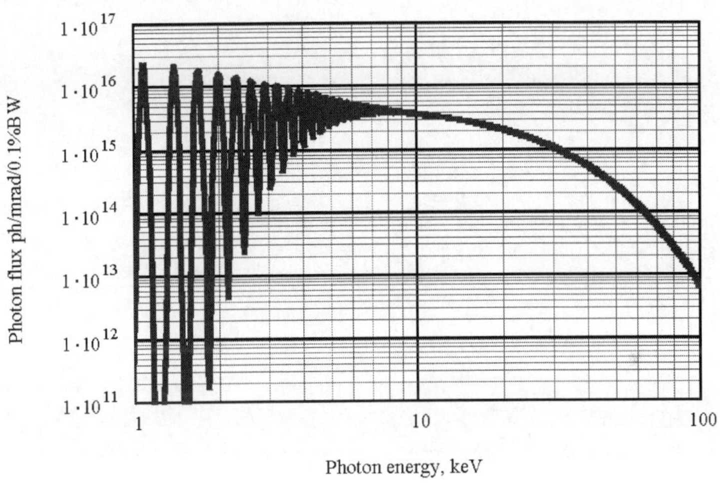

Figure 3.32 Radiation spectrum of SC wiggler with parameters: $B_0 = 2.1$ T, $\lambda_0 = 3$ cm, $N_{pole} = 120$.

3.4.4 Wiggler Cryogenic System

3.4.4.1 Cryostat design

The basic cryostat function is maintenance of SC wiggler magnets at temperature of liquid helium of 4.2 K. The wiggler magnet is placed into a bath with liquid helium and all heat emission inside the magnet and heat in-leak outside lead to liquid helium evaporation process. The main view of a cryostat is shown in Figs. 3.33 and 3.34. The cryostat consists of external vacuum housing, 60 K and 20 K shield screens, liquid helium vessel with a SC multipole magnet inside, throat, vacuum chamber (beam duct) with copper liner inside, upper flange, filling tube, two 2-stage coolers with stage temperature 4.2 K/50 K, and two 2-stage cryocoolers with stages temperature 20 K/50 K for shield screen cooling.

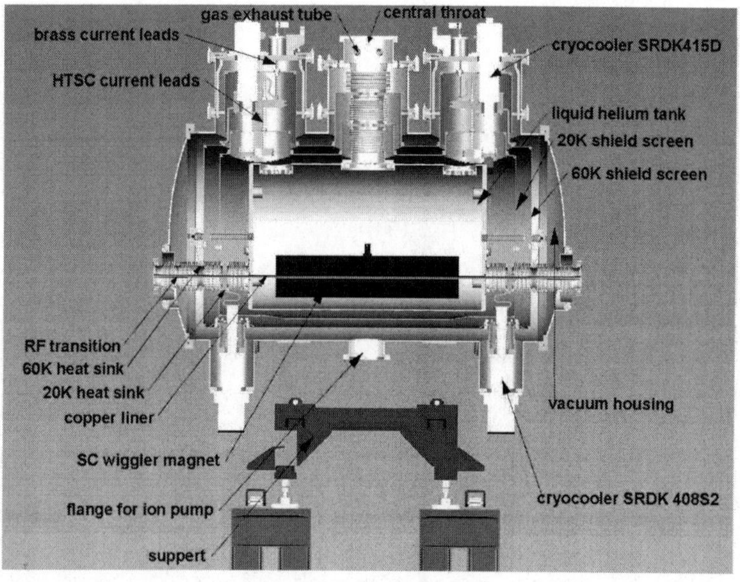

Figure 3.33 Pattern of a cryostat cross section. See also Color Insert.

The inner liquid helium vessel is surrounded by two shield screens to reduce the irradiation heat flux from outside. The temperature on the outside shield screen is about 60 K, and on the inner one the temperature is about 20 K. There is vacuum insulation between the helium vessel and external warm stainless steel vessel to reduce the residual gas heat flux. The SC wiggler magnet is supported by the walls of the helium vessel face flanges with special projections of the SC wiggler body. The helium vessel is suspended with four vertical and four horizontal kevlar straps connected to the external cryostat vessel. These straps pass throughout the both shield screens and attach to bolts on the external housing walls and are used for precise alignment of the vertical magnet position.

Figure 3.34 Photo of SC wiggler installed on CLS storage ring. See also Color Insert.

Cryocoolers allow to get long-term continuous work of a SC wiggler with the average consumption of liquid helium less than 1 L per day. Period between liquid helium refilling is more than 0.5 year.

3.4.4.2 Cold bore vacuum chamber

As it was already mentioned above, the formula (3.29) for multipole wigglers forms a bond of key wiggler parameters: field, period, and pole gap. That strongly limits freedom of a choice. Low limit of beam vertical aperture of modern storage rings is about 6–7 mm by requirements of beam dynamics in storage rings.

The further reduction of SC wiggler period keeping magnetic field still high enough demands reduction of pole gap. In the present cryostat design, it is supposed that inside of pole gap, a vacuum chamber and 20 K shield screen should be installed. The vacuum chamber is a part of liquid helium vessel and has temperature of 4.2 K. Shield screen with temperature 20 K (copper liner) absorbs heat which is raised by electron beam moving inside the copper liner (SR, heating by image currents, etc.). The vacuum between the liner and the vacuum chamber and inside of the liner is the same. In this design of cryostat heat removal is possible only at liner ends where 20 K coolers are connected to liner by means of heat sinks.

Liner material should have heat conductivity as large as possible in temperatures range of 4–40 K. One of the best and cheap materials for liner manufacturing is oxygen-free copper with RRR more than 100. Unfortunately, a copper is soft and a fluid material and consequently between the liner and 4 K the vacuum chamber it is necessary to set spacers with low heat conduction to prevent any contact of the liner to the vacuum chamber.

3.5 CONCLUSIONS AND OUTLOOK

During last 12 years, the technologies of fabrication SC shifters and multipole SC wigglers were developed. Several SC WLS with magnetic field 7–10 T have worked reliably for more than 10 years. Multipole SC wigglers with a high field 7–7.5 T with the period of 150–200 mm, an average field 3.5–4.2 T with the period of 48–60 mm, and a field 1.8–2.2 T with the period of 30–33 mm are fulfilled and reliably working in different SR centers.

The manufacturing techniques of cryostats with magnet immersing in liquid helium with liquid helium consumption less than 1 L per day are created.

Horizontal racetrack design of wiggler poles has practically reached the minimal limit of the wiggler period. For the further reduction of the period, it is necessary to use vertical racetrack arrangement of coils.

The prospect of the further reduction of wiggler pole gap consists in association of functions of a vacuum chamber and a thermal screen. But to combine the functions, it is necessary to depart from the concept of liquid helium bath cryostat to cryostat based on indirect cooling of magnet. In such cryostat the magnet should be placed into isolating vacuum and its cooling should be provided by heat conduction or circulating liquid helium inside small volume tubes.

The vacuum chamber combined with function of the thermal screen should be inserted between magnet poles with some vacuum gap. Such design allows intercepting a heat not only from the ends of the vacuum chamber but also from side surfaces. This strongly weakens requirements of high heat conductivity of the vacuum chamber and gives a possibility to use other materials like aluminum alloys.

Acknowledgments

The presented results have been obtained by large collective of Budker Institute of Nuclear Physics during long-term, persistent, and intense work and first of all by scientists, engineers, and technicians of laboratory 8-2. I am grateful to them for fruitful work. I am grateful to many foreign colleagues from different SR centers for their genuine interest in the work. First of all, I am grateful to Dieter Kraemer and Jos Schouten, who had taken a keen interest in my work and effectively supported it.

References

1. Alekseev, P. A, Keilin, V. E., Kovalev, I. A., Kruglov, S. L., Lazukov, V. N., Medvedev, M. I. and Shutova, D. I. (2007). Investigation of considerable stability increase of composite superconductors doped with extremely large heat capacity substance. *Superconductor Science & Technology*, **20**, pp. 71–76.

2. Ando, A., Date, S., Fedurin, M. G., Hara, M., Kamitsubo, H., V., Kiselev, A., Kulipanov, G. N., Kumagai, N., Mezentsev, N. A., Miyahara, Y., Nakamura, T., Ohkuma, H., Shkaruba, V. A., Skrinsky, A. N., Soutome, K., Takao, M. and Tanaka, H. (1998). Proposal of high-field superconducting wiggler for slow positron source at SPring-8. *Journal of Synchrotron Radiation*, **5**, pp. 360–362.

3. Artamonov, A. S., Barkov, L. M., Baryshev, V. B., Bashtovoy, N. S., Vinokurov, N. A., Gluskin, E. S., Korniukhin, G. A., Kochubei, V. A., Kulipanov, G. N., Mezentsev, N. A., Pindiurin, V. F., Skrinsky, A. N. and Khorev, V. M. (1980). First results on the work with a superconducting "snake" at the VEPP-3 storage ring. *Nuclear Instruments and Methods*, **177**, pp. 239–246.

4. Barkov, L. M., Baryshev, V. B., Kulipanov, G. N., Pindyurin, V. F., Skrinsky, A. N. and Khorev, V. M. A. (1978). A proposal to install a superconductive wiggler magnet on the storage ring VEPP-3 for

generation of the synchrotron radiation. *Nuclear Instruments and Methods*, **152**, pp. 23–29.

5. Batrakov, A. M., Khruschev, S. V., Kraemer, D., Kulipanov, G. N., Lev, V. H., Mezentsev, N. A., Miginsky, E. G., Shkaruba, V. A., Syrovatin, V. M., Tsukanov, V. M., Zjurba, V. K. and Zolotarev, K. V. (2005). Nine tesla superconducting bending magnet for BESSY-II. Proceedings of the XV International Synchrotron Radiation Conference. *Nuclear Instruments and Methods in Physics Research Section A: Accelerators, Spectrometers, Detectors and Associated Equipment*, **543**, pp. 35–41.
6. Berger, D., Fedurin, M., Mezentsev, N., Mhaskar, S., Shkaruba, V., Schaefers, F., Scheer, M. and Weihreter, E.. (2001). A superconducting 7 T multipole wiggler for the BESSY-II synchrotron radiation source. *Proceedings of the 2001 Particle Accelerator Conference*, Chicago, June 18–22, 2001, pp. 2450–2452.
7. Borovikov, V. M., Craft, B., Fedurin, M. G., Jurba, V., Khlestov, V., Kulipanov, G. N., Li, O., Mezentsev, N. A., Saile, V. and Shkaruba, V. A. (1998). Superconducting 7 Tesla wiggler for LSU-CAMD. *Journal of Synchrotron Radiation*, **5**, pp. 440–442.
8. Date, S., Fedurin, M., Hara, M., Kamitsubo, H., Kiselev, A., Kulipanov, G., Kumagai, N., Mezentsev, N., Miyahara, Y., Nakamura, T., Ohkuma, H., Panchenko, V., Plokhoi, V., Shkaruba, V., Skrinsky, A., Soutome, K., Takaoand M. and Tanaka, H. (1998). Slow positron project at Spring-8. *Proceedings of the SRSMS-2 conference*, Kobe, Japan.
9. Fedurin, M., Kulipanov, G., Mezentsev, N. and Shkaruba, V. (2000). Superconducting high-field three-pole wigglers in Budker INP. *Nuclear Instruments and Methods in Physics Research Section A: Accelerators, Spectrometers, Detectors and Associated Equipment*, **448**, pp. 51–58
10. Keilin, V. E., Kovalev, I. A., Kruglov, S. L., Lupanov, D. E. and Shcherbakov, V. I. (2008). Increasing thermomagnetic stability of composite superconductors with additives of extremely large heat capacity substances, *Technical Physics Letters*, **34**, pp. 418–420.
11. Mezentsev, N. and Perevedentsev, E. (2005). Survey of superconducting insertion devices for light sources. *Proceedings of 2005 Particle Accelerator Conference*, Knoxville, Tennessee, USA, May 16–29, 2005, pp. 256–260.

12. Robin, D., Biocca, A., Brown, B., Byrne, W., Calais, D., Chin, M., Colomb, D., Corradi, C., De Vries, J., Fahmie, M., Geyer, A., Harkins, J., Jacobson, S., Krupnick, J., Marks, S., McDonald, J., Mueller, R., Ottens, F., Paterson, A., Pipersky, P., Ritchie, A., Schlueter, R., Spring, J., Steier, C., Taylor, C., Timossi, C., Thur, W., Wandesforde, A., Zbasnik, J. (LBNL), Chen, J., Wang, B., Wang, N. N. M. R. and Livermore, C. A. (Wang NMR) (2001). Superbend Project at the Advanced Light Source. *Proceedings of the 2001 Particle Accelerator Conference*, Chicago, pp. 2632–2634.

13. Soutome, K., Dat, S., Ego, H., Fukui, T., Hara, M., Hosoda, N., Kawashima, Y., Kodera, M., Kumagai, N., Magome, T., Masaki, M., Masuda, T., Matsui, S., Nakamura, T., Noda, T., Ohashi, Y., Ohshima, T., Oishi, M., Saeki, H., Sasaki, S., Shoji, M., Takano, S., Takao, M., Takashima, T., Takebe, H., Tamura, K., Tanaka, H., Taniuchi, Y., Tsumaki, K., Yorita, T., Zhang, C. and Yonehara, H. (2003). Generation of High-Energy Synchrotron Radiation with a 10-T Superconducting Wiggler Installed in the SPring-8 Storage Ring. *Proceedings of the 2003 Particle Accelerator Conference*, Portland, May 12–16, pp. 250–252.

14. Steffen, K. G. (1965). *High Energy Beam Optics*. Wiley.

SECTION II
METHODS

Chapter 4

COHERENT X-RAY DIFFRACTION FOR HIGH-CONTRAST BIOIMAGING

Yoshinori Nishino
Research Institute for Electronic Science, Hokkaido University
Kita 21 Nishi 10, Kita-ku, Sapporo, Hokkaido 001-0021, Japan
yoshinori.nishino@es.hokudai.ac.jp

X-ray diffraction microscopy is high-contrast X-ray imaging utilizing coherent X-ray diffraction. It is especially useful for observing internal structures of thick objects, such as cells or cell organelles, owing to the high penetrating power of X-rays. This chapter describes the basics of X-ray diffraction microscopy and its biological applications.

4.1 INTRODUCTION

X-ray diffraction microscopy (XDM) is an X-ray imaging technique to reconstruct sample structures from the coherent X-ray diffraction data by extending X-ray crystallography to allow imaging of non-crystalline samples (Sayre, 1952; Miao *et al.*, 1999; Robinson *et al.*, 2001). The method is also known as coherent diffractive imaging.

Owing to the high penetrating power of X-rays, XDM allows non-invasive imaging of internal structures of thick samples, which are

Synchrotron Radiation and Structural Proteomics
Edited by Eugenia Pechkova and Christian Riekel

www.panstanford.com

difficult to access with other microscopic methods. For example, X-rays can offer a superior performance compared to electrons for biological samples thicker than ~500 nm because of the limited mean free path of the electron (Henderson, 1995). In life science, non-invasive imaging is of extreme importance to allow the observation of samples closer to their natural condition.

XDM requires no lenses and can overcome various limitations, e.g., the spatial resolution and the sample thickness (the depth of field), which cause problems in conventional lens-based X-ray microscopy. Atomic resolution can be achieved in principle.

Furthermore, XDM has enormous potential for high-contrast imaging of transparent samples, e.g., biomolecular complexes and cell organelles, because it is ideal X-ray phase contrast imaging without contrast degradation due to lenses. Unlike fluorescent light or other microscopic methods, XDM requires no labeling or staining to enhance image contrast. The structural information provided by XDM is the electron density map, which is exactly same as X-ray crystallography does. Therefore, once a high-resolution structure is obtained in XDM, it can be seamlessly combined with the atomic-resolution structure determined by X-ray crystallography to help us understand cell structures and functions from the atomic level.

Figure 4.1 illustrates XDM. In XDM measurements, coherent X-rays illuminate a sample, and the coherent X-ray diffraction patterns are recorded with a two-dimensional X-ray detector, e.g., an X-ray CCD. Here, the coherent diffraction patterns should be sampled at intervals short enough to satisfy the oversampling condition (Miao *et al.*, 1998) to be described later. The sample structures are reconstructed from the oversampled coherent X-ray diffraction data by using an iterative phase retrieval method (Gerchberg and Saxton, 1971; Fienup, 1982), using computers instead of objective lenses. Three-dimensional structure can also be reconstructed from the coherent X-ray diffraction data sets measured at various incident angles by rotating the sample (Miao *et al.*, 2002; Williams *et al.*, 2003; Chapman *et al.*, 2006a; Pfeifer *et al.*, 2006).

The history of XDM dates back to 1952 when Sayre pointed out the possibility of structure determination from the oversampled X-ray diffraction intensities (Sayre, 1952). Along with theoretical studies, Sayre and his collaborators performed coherent X-ray diffraction experiments for non-crystalline samples, e.g., cells, since 1980s. And finally, in 1999, they succeeded in the first image reconstruction for a test pattern sample composed of gold dots (Miao *et al.*, 1999). Since then, the method has been attracting growing interest of scientists due to its high potential, stimulated also by the projects of X-ray free-electron lasers (XFELs). Application works using synchrotron radiation sources become gradually increasing, in parallel with basic research for the improvement of the method and for the use of XFELs.

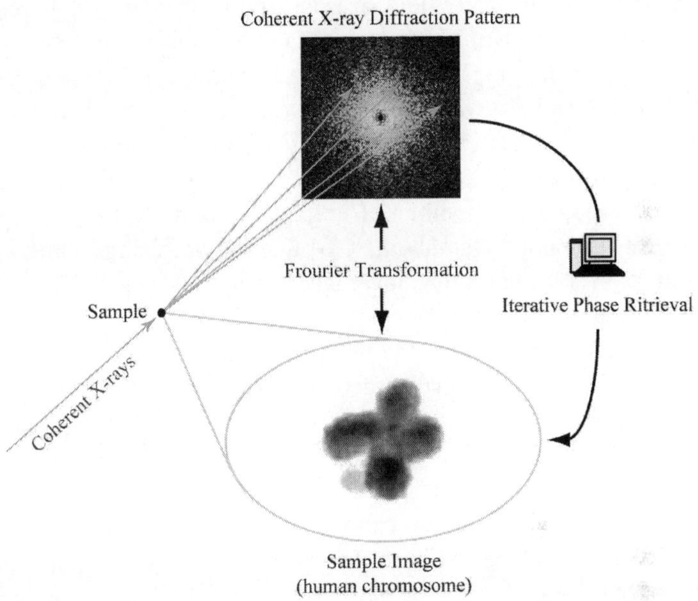

Figure 4.1 Schematic of X-ray diffraction microscopy (XDM). In XDM, a sample is illuminated with coherent X-rays, and the coherent diffraction data are recorded in the far field. The sample structure is reconstructed from the coherent diffraction data by using an iterative phase retrieval algorithm. See also Color Insert.

4.2 COHERENT X-RAY DIFFRACTION

X-ray diffraction occurs when X-rays are elastically scattered from structures with a comparable size to the wavelength of the illuminating X-rays. For crystalline samples, X-ray diffraction intensities concentrate at discrete points satisfying the Bragg conditions. By contrast, when non-crystalline samples are coherently illuminated, diffraction patterns are composed of continuously distributed speckles. Speckles are well known in the scattering of laser light from rough surfaces and generally occur when coherent waves interfere with random phases. Speckled coherent X-ray diffraction patterns are formed by the interference of X-rays scattered from electrons in the sample and are quite sensitive to the sample structure. XDM utilizes this sensitivity of the coherent X-ray diffraction to the sample structure.

It is interesting to note that the illuminating X-rays in X-ray crystallographic measurement are usually not perfectly coherent. Therefore, X-rays scattered from remote electrons do not interfere. In X-ray crystallography, however, it is sufficient for structure determination to have a coherent area much larger than the size of the unit cell, because the sample itself has a periodic structure. Accordingly, the structure obtained in X-ray crystallography is an average over a huge number of unit cells. Similarly, an averaged structure is obtained in small-angle X-ray scattering using incoherent illumination. By contrast, unaveraged structures are obtained in XDM by utilizing coherent X-ray diffraction.

Coherent X-rays, required for XDM measurements, become available with the advancing technology of brilliant synchrotron radiation sources. However, even with the state-of-the-art undulator hard X-ray radiation of storage rings, the coherence area is only 0.1% of the total beam size. With partially coherent undulator radiation, XDM measurement can be performed, e.g., for an isolated particle sample whose size is smaller than the coherence area of the illuminating X-rays. In this way, the sample particle is coherently illuminated. For spatially extended objects, one can use a small aperture (pinhole) or a focusing optics in front of the sample for scanning a small illumination area to cover a larger area and can adopt techniques such as ptychography (Rodenburg *et al.*, 2007; de la Cuesta *et al.*, 2009), keyhole coherent diffractive imaging (Williams *et al.*, 2006; Abbey *et al.*, 2008), or scanning X-ray diffraction microscopy (Thibault *et al.*, 2008) for image reconstruction. With the recent advent of XFELs, X-rays with nearly perfect spatial coherence also became accessible.

The temporal coherence length should be longer than the maximum path difference of the diffracted X-rays, and it becomes relevant for high-resolution imaging. XDM experiments can be performed even with multimode light sources, e.g., spontaneous undulator radiation and XFELs based on self-amplified spontaneous emission, though new scientific possibilities will open with future single-mode XFELs.

XDM reconstructs the sample structure from the diffraction data. In a simple case for small samples, where multiple scattering can be neglected, the complex amplitude $U^{(s)}(\mathbf{r})$ of the scattered field in the far field $\mathbf{r}$ is expressed as

$$U^{(s)}(\mathbf{r}) = -r_e f(\mathbf{K}) \frac{e^{ikr}}{r}, \tag{4.1}$$

for normalized plane wave illumination. Here, r_e denotes the classical electron radius. The complex scattering amplitude $f(\boldsymbol{k})$ is given by the Fourier transformation of the electron density $\rho^{(\mathbf{r}')}$,

$$f(\mathbf{K}) = \int d\mathbf{r}' \rho(\mathbf{r}') e^{-i\mathbf{K}\cdot\mathbf{r}'}, \tag{4.2}$$

according to the kinematical theory of diffraction (the first-order Born approximation). In Eqs. (4.1) and (4.2), $\boldsymbol{K}$ is the scattering vector, $\mathbf{K} = k\mathbf{r}/r - \mathbf{k}_0$, and $\mathbf{K}_0$ denotes the wave vector of the illuminating plane waves. The size of the scattering vector is given by $|\mathbf{K}| = (4\pi/\lambda)$ $\sin(\Theta/2)$ in terms of the wavelength λ and the scattering angle Θ.

As is readily understood from Eq. (4.2), once the complex scattering amplitude $f(\mathbf{K})$ is known, the sample structure can be obtained by applying inverse Fourier transformation. However, the phase of the diffracted wave cannot be measured directly, while the absolute value $|f(\mathbf{K})|$ can be determined from the photon number recorded in each pixel of the two-dimensional detector. To reconstruct the sample structure, the phase has to be retrieved in some way. This is the well-known phase problem in X-ray diffraction. In XDM, the phase problem is solved by using iterative phase retrieval methods to be described later.

A single coherent X-ray diffraction measurement covers a part of the Ewald sphere in reciprocal space (Fourier space). When the measured diffraction pattern covers only a small range of the solid angle, the part of the Ewald sphere can be regarded as a flat plane. In this approximation, a projection image of the sample is obtained by phasing a single coherent diffraction data set:

$$f\left(\frac{2\pi}{\lambda L}x', \frac{2\pi}{\lambda L}y', 0\right) \approx \int dx' \int dy' \left[\int dz' \rho(\mathbf{r}')\right] e^{-i\frac{2\pi}{\lambda L}(xx'+yy')}, \quad (4.3)$$

where L is the distance between the sample plane and the detector plane.

When the projection and the Born approximations are applicable, the coherent X-ray diffraction pattern becomes centrosymmetric for non-absorbing samples. This is because the Fourier transformation of any real-valued function is centrosymmetric. For absorbing samples, the electron density $\rho(\mathbf{r})$ becomes effectively complex, and consequently the diffraction pattern becomes non-centrosymmetric. Here, the electron density $\rho(\mathbf{r})$ is related to the complex refractive index $n(\mathbf{r})$ by

$$\rho(\mathbf{r}) = -\frac{k^2}{4\pi r_e}[n^2(\mathbf{r}) - 1] \approx \frac{k^2}{2\pi r_e}[\delta(\mathbf{r}) - i\beta(\mathbf{r})]. \quad (4.4)$$

In the last approximate equality of Eq. (4.4), we decomposed the complex refractive index as $n(\mathbf{r})=1-\delta(\mathbf{r})+i\beta(\mathbf{r})$ by using the real part $\delta(\mathbf{r})$ corresponding to the phase shift (refraction) and the imaginary part $\beta(\mathbf{r})$ corresponding to the absorption; and we used the fact that the deviation of $n(\mathbf{r})$ from unity is quite small for X-rays.

When the curvature of the Ewald sphere cannot be neglected, defocusing due to the wave propagation inside the sample should be taken into consideration in the two-dimensional reconstruction from a single diffraction data set (Spence *et al.*, 2002; Chapman *et al.*, 2006a). Furthermore, the diffraction pattern looses the centrosymmetry even for non-absorbing samples, because a pair of centrosymmetric points cannot be on a single Ewald sphere.

For thicker samples, Eq. (4.1) is not valid, because the multiplex scattering is not negligible. However, when the measured diffraction pattern covers only a small range of the solid angle, the projection approximation can be applied. For a sample with the thickness t, the projection approximation is valid, when the maximum scattering angle is smaller than sqrt($\lambda/(2t)$). In this approximation, the complex amplitude $U(x, y)$ in the detector plane is given by the Fourier transform of the sample exit wave $U_0(x', y')t(x', y')$:

$$U(x,y)=\frac{e^{iKL}e^{i\frac{k}{2L}(x^2+y^2)}}{i\lambda L}\int dx'\int dy' U_0(x',y')t(x',y')e^{-i\frac{2\pi}{\lambda L}(xx'+yy')} \quad (4.5)$$

Here, $U_0(x', y')$ is the complex amplitude of the sample-illuminating wave, and $t(x', y')$ the transmission function of the sample,

$$t(x',y')\approx \exp\left(-ik\int dz'(1-n(x',y',z')\right). \quad (4.6)$$

For optically thin samples, $\left|k\int dz'\left(1-n(x',y',z')\right)\right|<1$, Eq. (4.5) essentially reduces to Eq. (4.1).

Coherent X-ray diffraction measurements at various incident angles can cover a three-dimensional volume in reciprocal space (Fig. 4.2). From the three-dimensional diffraction data, the three-dimensional structure of the sample can be reconstructed by using three-dimensional Fourier transformation in the iterative phase retrieval method.

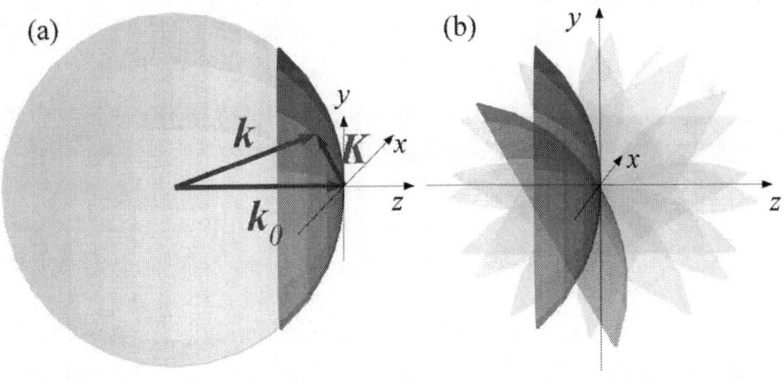

Figure 4.2 Regions in reciprocal space covered by coherent X-ray diffraction experiments. (a) A single diffraction measurement provides diffraction intensities in a near-forward region on the Ewald sphere. In measurements within a sufficiently small range of the solid angle, the measured part of the sphere may be approximated by a plane. (b) Diffraction data sets at a large number of different incident angles can cover nearly an isotropic region around the reciprocal space origin. See also Color Insert.

The above formulation describes coherent X-ray diffraction in the transmission geometry, which provides the electron density map or the transmission function. Coherent X-ray diffraction experiments can also be performed in the reflection geometry by utilizing speckled diffraction patterns near Bragg peaks of a small crystalline sample (Robinson *et al.*, 2001; Williams *et al.*, 2003; Pfeifer *et al.*, 2006). Coherent X-ray diffraction in the reflection geometry can also provide the distribution of crystal deformation (Pfeifer *et al.*, 2006), which is important in materials science.

4.3 OVERSAMPLING

Phase retrieval in XDM is related to the sampling theorem in the information theory. The sampling theorem describes how to perfectly recover a continuous signal from a discrete data set sampled at short and regular intervals. The theorem requires that the sampling interval should be shorter than $1/(2B)$, where B is the highest frequency in the signal. The sampling at intervals shorter than $1/(2B)$ is called the oversampling.

Now, let us apply the sampling theorem to the complex scattering amplitude $f(\mathbf{k})$. Eq. (4.2) shows that $f(\mathbf{k})$ is composed of various frequency components $\exp(-i\mathbf{k} \cdot \mathbf{r})$, and the highest frequency in the k_i direction is given by $a_i/(4\pi)$. Here, a_i is the size of the sample illumination area (the sample size), and the subscript i denotes a component (x, y, z) of the three-dimensional Cartesian coordinate. To recover $f(\mathbf{k})$ by using the sampling theorem, it is required to sample $f(\mathbf{k})$ at intervals shorter than $2\pi/a_i$.

The complex scattering amplitude $f(\mathbf{k})$ is, however, not direct measurable, but the absolute value $|f(\mathbf{k})|$ is. Sampling $|f(\mathbf{k})|$ at intervals of $2\pi/a_i$ provides only a half of information necessary to recover $f(\mathbf{k})$. To make up for the deficit of information, $|f(\mathbf{k})|$ should be sampled at shorter intervals to make the number of the sampling points more than twice. This condition on the sampling interval is called the oversampling condition, and it is a necessary condition for phase retrieval in XDM. Figure 4.3 illustrates the oversampling condition. In discrete Fourier transformation, oversampling $|f(\mathbf{k})|$ in reciprocal space corresponds to adding a zero electron density area around the sample in real space.

The oversampling condition is concisely expressed by using the oversampling ratio σ_i (Miao *et al.*, 1998). The oversampling ratio σ_i is defined as $\sigma_i = 2\pi/(a_i\Delta_i)$, where Δ_i is the sampling interval in the K_i direction. The oversampling condition is then given by $\sigma_x\sigma_y \geq 2$ for two-dimensional reconstruction, and $\sigma_x\sigma_y\sigma_z \geq 2$ for three-dimensional reconstruction.

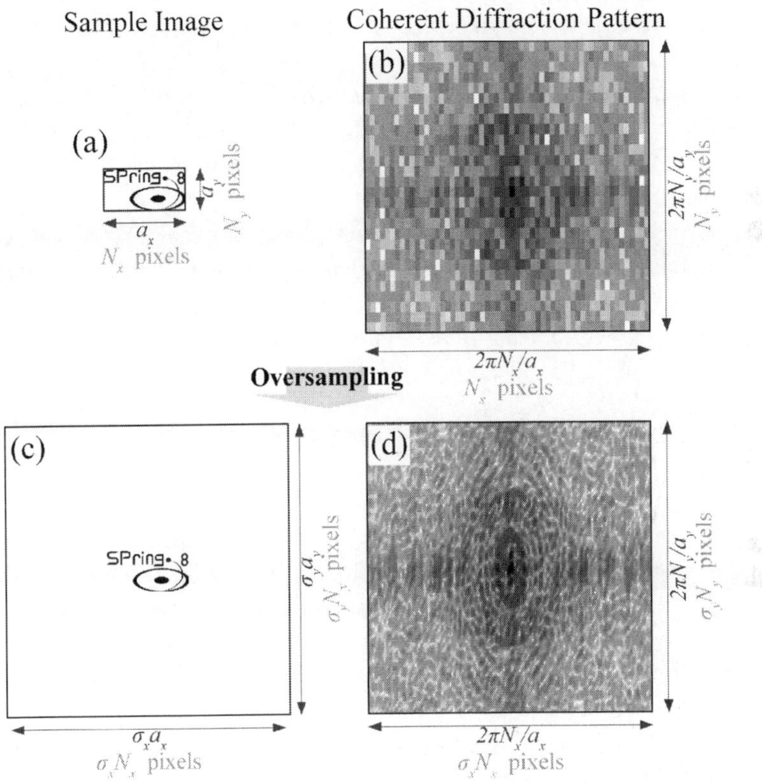

Figure 4.3 Computer simulation of diffraction patterns with different sampling intervals, and the corresponding real-space images. (a) The original sample image and (b) the calculated diffraction pattern are shown at the top. By oversampling the diffraction pattern as in (d), a region with no electron density appears around the sample in real-space image as in (c) in discrete Fourier transformation.

A typical size of a single speckle in coherent X-ray diffraction patterns is $2\pi/a_i$ (angular size $\sim\lambda/a_i$) in the K_i direction. Recalling that the sampling interval in reciprocal space is $\Delta_i=2\pi/(a_i\sigma_i)$, the number of sampling points contained in a single speckle is $\sigma_x\sigma_y$. Therefore, the oversampling condition for two-dimensional reconstruction can be reworded as the following: the size of a single speckle on a two-dimensional detector should be more than twice larger than the pixel size of the detector.

Let us consider here a typical case of using a two-dimensional X-ray detector with a pixel size of a few ten micrometer located at a few meter from the sample. Then, the angular resolution in recording coherent diffraction patterns is a few ten microrad. In this case, micrometer-sized samples can satisfy the oversampling condition for ~1 Å X-rays by producing a speckle pattern with an angular size of ~100 μrad for each speckle.

Although the oversampling condition is a necessary condition for phase retrieval, it is not a sufficient condition. This is because the sampling theorem cannot be used directly for phase retrieval. Instead, in XDM, the phase information is retrieved by using iterative methods.

4.4 ITERATIVE PHASE RETRIEVAL METHODS

XDM utilizes the iterative phase retrieval algorithm originally proposed by Gerchberg and Saxton (1971). Figure 4.4 illustrates the iterative phase retrieval algorithm. The algorithm starts with a random phase set of the complex scattering amplitude $f(\mathbf{K})$ in reciprocal space (or a random electron density $\rho(\mathbf{r})$ in real space). At each iteration, known information is supplied to retrieve the phase as described below.

First, in reciprocal space, the experimentally determined absolute value $|f(\mathbf{K})|$ of the complex scattering amplitude is supplied, while keeping the calculated phase. The complex scattering amplitude after applying the reciprocal space constraints is expressed as

$$g_n'(\mathbf{K}) = \frac{|f(\mathbf{K})|}{|g_n(\mathbf{K})|} g_n(\mathbf{K}), \tag{4.7}$$

where $g_n(\mathbf{K})$ is the calculated complex scattering amplitude at the n-th iteration. Then, the electron density $\rho'_n(\mathbf{r})$ satisfying the reciprocal space constraint is obtained by applying the inverse Fourier transformation to $g'_n(\mathbf{K})$.

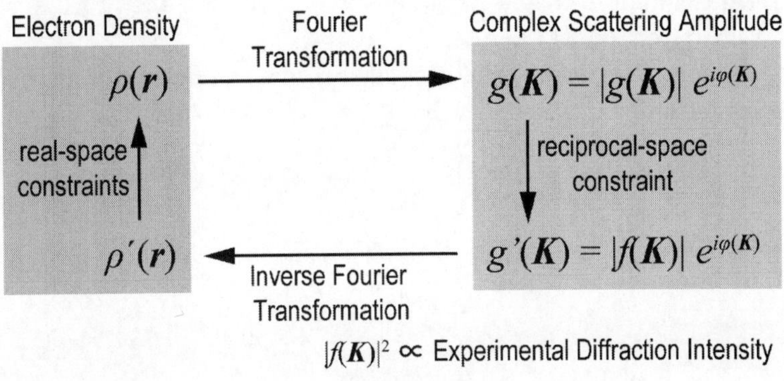

Figure 4.4 Schematic of the iterative phase retrieval method. The phase of the diffracted wave is recovered by iteratively applying the Fourier transformation and the inverse Fourier transformation with constraints in both real and reciprocal spaces at each iteration.

Second, in real space, variety of constraints are available, e.g., that there is no electron density outside the support and that the electron density is physically non-negative. Here, the support denotes the region where the sample is expected to exist. The support size can be determined roughly from the speckle size, because the speckle size is inversely proportional to the sample size. Among various algorithms adopting different real-space constraints, the HIO (hybrid input–output) algorithm proposed by Fienup is widely used. In the HIO algorithm, the electron density $\rho_{n+1}(\mathbf{r})$ at the $(n+1)$-th iteration is obtained from "the input" $\rho_n(\mathbf{r})$ and "the output" $\rho'_n(\mathbf{r})$ of the previous iteration:

$$\rho_{n+1}(\mathbf{r}) = \begin{cases} \rho'_n(\mathbf{r}), & (\mathbf{r} \in S) \wedge (\rho'_n(\mathbf{r}) \geq 0), \\ \rho_n(\mathbf{r}) - \beta\rho'_n(\mathbf{r}), & \text{otherwize} \end{cases} \tag{4.8}$$

where S stands for the support region. In the HIO algorithm, unphysical electron densities are forced to decrease gradually. The parameter β controls how gradually the unphysical electron densities should be decreased, and a good convergence is obtained empirically for β close to unity. The complex scattering amplitude $g_{n+1}(\mathbf{K})$ at the $(n + 1)$-th iteration is obtained by applying the Fourier transformation to the electron density $\rho_{n+1}(\mathbf{r})$. Then, the process is repeated from Eq. (4.7).

The convergence of the iterative process can be monitored by using the index E_n,

$$E_n = \left[\sum_{\mathbf{r} \notin S} |\rho_n'(\mathbf{r})|^2 / \sum_{\mathbf{r} \in S} |\rho_n'(\mathbf{r})|^2 \right]^{1/2}. \tag{4.9}$$

For true solution $\rho'_n(\mathbf{r})$, E_n becomes zero. Mathematical studies show that the solution is unique for two or higher space dimensions except for pathologically rare cases (Barakat and Newsam, 1984). There are trivial non-unique solutions, i.e., spatially shifted images and centrosymmetric images. The trivial non-unique solutions produce exactly the same coherent diffraction pattern for plane wave illumination, and therefore cannot be distinguished.

In an early stage of XDM studies, it was not possible to reconstruct a sample image only from the coherent diffraction data, because experimentally there used to be a large region of missing data behind the beamstop near the diffraction center. To make up for the missing data, a low-resolution real-space image of the sample had been required for image reconstruction. The missing data problem was later solved by the improvements of microscope instruments to make the missing data size smaller and by the improvements of the phase retrieval algorithm, e.g., the iterative improvement of the support (shrink wrap) (Marchesini *et al.*, 2003a) and the iterative normalization (Nishino *et al.*, 2003). These improvements have made it possible to reconstruct a sample image only from the coherent diffraction data.

4.5 BIOLOGICAL APPLICATIONS

In high-resolution bioimaging, cryo-electron microscopy is becoming increasingly powerful. However, for samples thicker than sub-micrometer, X-ray microscopy has the intrinsic advantage due to the high penetrating power. Although micron-sized bio-samples are transparent to hard X-rays, XDM can achieve high image-contrast by utilizing coherent X-ray diffraction.

The first biological application of XDM was for *Escherichia coli* bacteria (Miao *et al.*, 2003). Figure 4.5 shows the coherent diffraction pattern and the reconstructed image of *E. coli* bacteria. The distribution of the proteins stained with manganese oxide was clearly imaged in two dimensions.

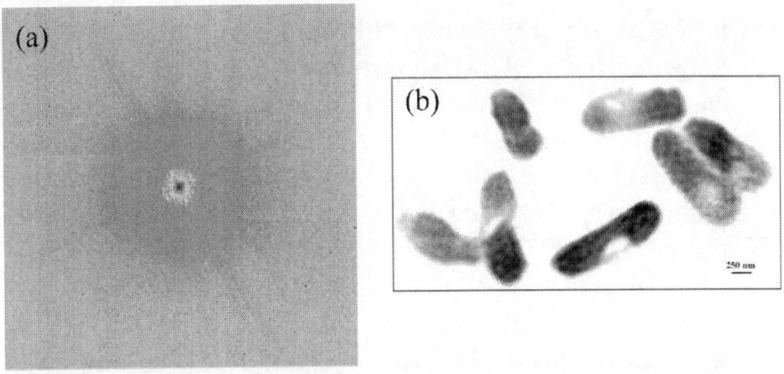

Figure 4.5 Observation of *Escherichia coli* bacteria in XDM. (a) The coherent diffraction pattern and (b) the reconstructed two-dimensional projection image are shown. The higher density regions inside the bacteria are likely the distribution of proteins labeled with $KMnO_4$. (Adapted from Miao *et al.* Imaging whole *E. coli* bacteria by using single-particle x-ray diffraction. *Proceedings of the National Academy of Sciences USA*, 100, pp. 110–112, © 2003, National Academy of Sciences, U.S.A.) See also Color Insert.

The observation of unstained biological samples was first reported for a freeze-dried yeast cell (Shapiro *et al.*, 2005). The complex-valued exit waves were reconstructed with a high quality, and fine features at the cell and nuclear membrane regions were reproduced in the two-dimensional projection images reconstructed from the coherent diffraction data recorded at several different tilt angles. Concerning the two-dimensional observations of unstained biological samples, successful reports followed for different samples, which include intramuscular fish bones at different stages of mineralization (Jiang *et al.*, 2008), blood cells infected with malaria parasites (Williams *et al.*, 2008), and a murine herpesvirus (MHV-68) (Song *et al.*, 2008). The spatial resolution achieved in these examples was a few ten nanometer.

Although the above examples are limited to two-dimensional observation, three-dimensional imaging is obviously important to obtain detail structural information. Recently, the first successful three-dimensional imaging of a cell organelle was performed for an unstained human chromosome (Nishino *et al.*, 2009). Figure 4.6 shows the coherent diffraction pattern and the reconstructed images. The observed images reveal an axial structure with a high electron density, which other microscopic methods have been unable to visualize in the unstained condition, demonstrating the high imaging ability of XDM. Such mesoscopic-scale structures have been difficult to be observed, although molecular structures of their components, e.g., DNA and histone proteins, can be studied by X-ray crystallography or electron microscopy. The resolution in the two-dimensional imaging was 38 nm, and 120 nm for the three-dimensional imaging. The degradation of the resolution in the three-dimensional imaging is caused primarily by radiation damage.

In biological imaging, the observation of frozen-hydrated samples is also highly important: a hydrated state is desirable for observing samples closer to their natural condition; in addition, radiation damage, which often puts limits on the achievable resolution in biological imaging, can be mitigated by cryogenically cooling the sample. Empirical studies show that 10 nm resolution in three-dimensional XDM is attainable for frozen-hydrated biological samples (Marchesini *et al.*, 2003b; Howells *et al.*, 2009). Recently, two different experimental methodologies were proposed for the observation of frozen-hydrated biological samples. The first uses an *in vacuum* cryogenic specimen holder, and successful reconstruction was presented for a budding yeast cell in two dimensions (Huang *et al.*, 2009). The other uses a cryogenic gas-jet in a *non vacuum* environment, and two-dimensional reconstruction was demonstrated for *Deinococcus radiodurans* bacteria (Lima *et al.*, 2009). It is still technically difficult to perform the measurement and reconstruction of frozen-hydrated biological samples with high quality, and a marked improvement of the spatial resolution is yet to be demonstrated. Three-dimensional imaging of frozen-hydrated biological samples is also an issue to be tackled.

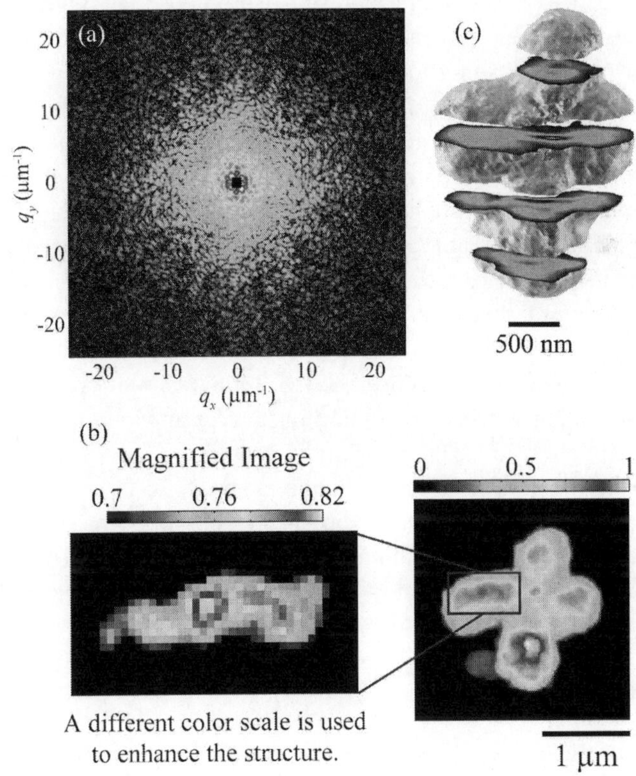

Figure 4.6 Observation of an unstained human chromosome in XDM. (a) The coherent diffraction pattern, (b) the two-dimensional reconstruction, and (c) the three-dimensional reconstruction are shown. In the magnified image of an arm region in (b), an axial wavy structure was observed first in an unstained chromosome. (Adapted from Nishino *et al.* Three-dimensional visualization of a human chromosome using coherent X-ray diffraction. *Physical Review Letters*, 102, art. no. 018101, © 2009, The American Physical Society.) See also Color Insert.

4.6 DISCUSSIONS AND OUTLOOK

The spatial resolution in biological imaging is often limited by the radiation damage. The limitation can be removed or lowered dramatically with XFELs by taking data before the sample is destroyed (Neutze *et al.*, 2000; Chapman, 2006b). This new possibility opens, because the pulse width of XFEL is an order of femtosecond or less, which is shorter than the time scale of the sample damaging processes, e.g., the Coulomb explosion of biomolecules. It is proposed to illuminate copies of a biomolecule with XFEL pulses, one by one in random orientations, to reconstruct the three-dimensional structure (Miao *et al.*, 2001; Huldt *et al.*, 2003; Fung *et al.*, 2009). This strategy is especially beneficial for the structural determination of membrane proteins, which play important roles in cell functions and drug design but are difficult to be crystallized. For larger biological objects, e.g., cells and organelles, it is not possible to prepare copies with an identical structure, because they each have different shapes. For such samples, stereo three-dimensional imaging is proposed to illuminate a sample simultaneously from a few different orientations by using an X-ray beam splitter (Sayre, 2008; Schmidt *et al.*, 2008).

XDM using synchrotron radiation sources and laboratory-based coherent light sources is important in promoting application works. Especially, for full three-dimensional imaging of cells or organelles, stable synchrotron radiation sources would be better suited. By carefully controlling the radiation dose for frozen-hydrated biological samples, XDM would become a powerful method in exploring cellular structures.

Acknowledgments

This work was supported by KAKENHI and the Promotion of XFEL Research of MEXT, Japan.

References

1. Abbey, B., Nugent, K. A., Williams, G. J., Clark, J. N., Peele, A. G., Pfeifer, M. A., De Jonge, M. and McNulty, I. (2008). Keyhole coherent diffractive imaging. *Nature Physics*, **4**, pp. 394–398.

2. Barakat, R. and Newsam, G. J. (1984). Necessary conditions for a unique solution to two-dimensional phase recovery. *Journal of Mathematical Physics*, **25**, pp. 3190–3193.

3. Chapman, H. N., Barty, A., Bogan, M. J., Boutet, S., Frank, M., Hau-Riege, S. P., Marchesini, S., Woods, B. W., Bajt, S., Benner, H., London, R. A., Plonjes, E., Kuhlmann, M., Treusch, R., Dusterer, S., Tschentscher, T., Schneider, J. R., Spiller, E., Moller, T., Bostedt, C., Hoener, M., Shapiro, D. A., Hodgson, K. O., Van der Spoel, D., Burmeister, F., Bergh, M., Caleman, C., Huldt, G., Seibert, M. M., Maia, F. R. N. C., Lee, R. W., Szoke, A., Timneanu, N. and Hajdu, J. (2006b). Femtosecond diffractive imaging with a soft-X-ray free-electron laser. *Nature Physics*, **2**, pp. 839–843.

4. Chapman, H. N., Barty, A., Marchesini, S., Noy, A., Hau-Riege, S. R., Cui, C., Howells, M. R., Rosen, R., He, H., Spence, J. C. H., Weierstall, U., Beetz, T., Jacobsen, C. and Shapiro, D. (2006a). High-resolution ab initio three-dimensional X-ray diffraction microscopy. *Journal of the Optical Society of America A — Optics Image Science and Vision*, **23**, pp. 1179–1200.

5. de la Cuesta, F. B., Wenger, M. P. E., Bean, R. J., Bozec, L., Horton, M. A. and Robinson, I. K. (2009). Coherent x-ray diffraction from collagenous soft tissues. *Proceedings of the National Academy of Sciences USA*, **106**, pp. 15297–15301.

6. Fienup, J. R. (1982). Phase retrieval algorithms: a comparison. *Applied Optics*, **21**, pp. 2758–2769.

7. Fung, R., Shneerson, V., Saldin, D. K. and Ourmazd, A. (2009). Structure from fleeting illumination of faint spinning objects in flight. *Nature Physics*, **5**, pp. 64–67.

8. Gerchberg, R. W. and Saxton, W. O. (1971). Phase determination for image and diffraction plane pictures in the electron microscope. *Optik*, **34**, pp. 275–284.

9. Henderson, R. (1995). The potential and limitations of neutrons, electrons and x-rays for atomic-resolution microscopy of unstained biological molecules. *Quarterly Reviews of Biophysics*, **28**, pp. 171–193.

10. Howells, M. R., Beetz, T., Chapman, H. N., Cui, C., Holton, J. M., Jacobsen, C. J., Kirz, J., Lima, E., Marchesini, S., Miao, H., Sayre, D., Shapiro, D. A., Spence, J. C. H. and Starodub, D. (2009). An assessment of the

resolution limitation due to radiation-damage in X-ray diffraction microscopy. *Journal of Electron Spectroscopy and Related Phenomena*, **170**, pp. 4–12.

11. Huang, X., Nelson, J., Lima, E., Marchesini, S., Miao, H., Neiman, A. M., Shapiro, D., Steinbrener, J., Stewart, A., Turner, J. J. and Jacobsen, C. (2009). Soft X-ray diffraction microscopy of a frozen hydrated yeast cell. *Physical Review Letters*, **103**, art. no. 198101.
12. Huldt, G., Szöke, A. and Hajdu, J. (2003). Diffraction imaging of single particles and biomolecules. *Journal of Structural Biology*, **144**, pp. 219–227.
13. Jiang, H., Ramunno-Johnson, D., Song, C. Amirbekian, B., Kohmura, Y., Nishiro, Y., Takahashi, Y., Ishikawa, T. and Miao, J. (2008). Nanoscale imaging of mineral crystals inside biological composite materials using x-ray diffraction microscopy. *Physical Review Letters*, **100**, art. no. 038103.
14. Lima, E., Wiegart, L., Pernot, P., Howells, M., Timmins, J., Zontone, F. and Madsen, A. (2009). Cryogenic X-ray diffraction microscopy for biological samples. *Physical Review Letters*, **103**, art. no. 198102.
15. Marchesini, S., Chapman, H. N., Hau-Riege, S. P., London, R. A., Szoke, A., He, H., Howells, M. R., Padmore, H., Rosen, R., Spence, J. C. H. and Weierstall, U. (2003b). Coherent x-ray diffractive imaging: applications and illustrations. *Optics Express*, **11**, pp. 2344–2353.
16. Marchesini S., He, H., Chapman, H. N., Hau-Riege, S. P., Noy, A., Howells, M. R., Weierstall, U. and Spence, J. C. H. (2003a). X-ray image reconstruction from a diffraction pattern alone. *Physical Review B*, **68**, art. no. 140101.
17. Miao, J. W., Charalambous, P., Kirz, J. and Sayre, D. (1999). Extending the methodology of X-ray crystallography to allow imaging of micrometre-sized non-crystalline specimens. *Nature*, **400**, pp. 342–344.
18. Miao, J. W., Hodgson, K. O., Ishikawa, T., Larabell, C. A., LeGros, M. A. and Nishino, Y. (2003). Imaging whole Escherichia coli bacteria by using single-particle x-ray diffraction. *Proceedings of the National Academy of Sciences USA*, **100**, pp. 110–112.
19. Miao, J. W., Hodgson, K. O. and Sayre, D. (2001). An approach to three-dimensional structures of biomolecules by using single-molecule diffraction images. *Proceedings of the National Academy of Sciences USA*, **98**, pp. 6641–6645.

20. Miao, J. W., Ishikawa, T., Johnson, B., Anderson, E. H., Lai, B. and Hodgson, K. O. (2002). High resolution 3D X-ray diffraction microscopy. *Physical Review Letters*, **89**, art. no. 088303.

21. Miao, J., Sayre, D. and Chapman, H. N. J. (1998). Phase retrieval from the magnitude of the Fourier transforms of nonperiodic objects. *Journal of the Optical Society of America A — Optics Image Science and Vision*, **15**, pp. 1662–1669.

22. Neutze, R., Wouts, R., van der Spoel, D., Weckert, E. and Hajdu, J. (2000). Potential for biomolecular imaging with femtosecond X-ray pulses. *Nature*, **406**, pp. 752–757.

23. Nishino, Y., Miao, J. W. and Ishikawa, T. (2003). Image reconstruction of nanostructured nonperiodic objects only from oversampled hard X-ray diffraction intensities. *Physical Review B*, **68**, art. no. 220101.

24. Nishino, Y., Takahashi, Y., Imamoto, N., Ishikawa T. and Maeshima, K. (2009). Three-dimensional visualization of a human chromosome using coherent X-ray diffraction. *Physical Review Letters*, **102**, art. no. 018101.

25. Pfeifer, M. A., Williams, G. J., Vartanyants, I. A., Harder, R. and Robinson, I. K. (2006). Three-dimensional mapping of a deformation field inside a nanocrystal. *Nature*, **442**, pp. 63–66.

26. Robinson, K., Vartanyants, I. A., Williams, G. J., Pfeifer, M. A. and Pitney, J. A. (2001). Reconstruction of the shapes of gold nanocrystals using coherent X-ray diffraction. *Physical Review Letters*, **87**, art. no. 195505.

27. Rodenburg, J. M., Hurst, A. C., Cullis, A. G., Dobson, B. R., Pfeiffer, F., Bunk, O., David, C., Jefimovs, K. and Johnson, I. (2007). Hard-X-ray lensless imaging of extended objects. *Physical Review Letters*, **98**, art. no. 034801.

28. Sayre, D. (1952). Some implications of a theorem due to Shannon. *Acta Crystallographica*, **5**, p. 843.

29. Sayre, D. (2008). Report on a project on three-dimensional imaging of the biological cell by single-particle x-ray diffraction. *Acta Crystallographica Section A*, **64**, pp. 33–35.

30. Schmidt, K. E., Spence, J. C. H., Weierstall, U., Kirian, R., Wang, X., Starodub, D., Chapman, H. N., Howells, M. R. and Doak, R. B. (2008). Tomographics femtosecond X-ray diffractive imaging. *Physical Review Letters*, **101**, art. no. 115507.

31. Shapiro, D., Thibault, P., Beetz, T., Elser, V., Howells, M., Jacobsen, C., Kirz, J., Lima, E., Miao, H., Neiman, A. M. and Sayre, D. (2005). Biological imaging by soft X-ray diffraction microscopy. *Proceedings of the National Academy of Sciences USA*, **102**, pp. 15343–15346.

32. Song, C., Jiang, H., Mancuso, A., Amirbekian, B., Peng, L., Sun, R., Shah, S. S., Zhou, Z. H., Ishikawa, T. and Miao, J. (2008). Quantitative imaging of single, unstained viruses with coherent X-Rays. *Physical Review Letters*, **101**, art. no. 158101.

33. Spence, J. C. H., Weierstall, U. and Howells, M. (2002). Phase recovery and lensless imaging by iterative methods in optical, X-ray and electron diffraction. *Philosophical Transactions of the Royal Society of London Series A — Mathematical Physical and Engineering Sciences*, **360**, pp. 875–895.

34. Thibault, P., Dierolf, M., Menzel, A., Bunk, O., David, C. and Pfeiffer, F. (2008). High-resolution scanning x-ray diffraction microscopy. *Science*, **321**, pp. 379–382.

35. Williams, G. J., Hanssen, E., Peele, A. G., Pfeifer, M. A., Clark, J., Abbey, B., Cadenazzi, G., de Jonge, M. D., Vogt, S., Tilley, L. and Nugent, K. A. (2008). High-resolution X-ray imaging of plasmodium falciparum-infected red blood cells. *Cytometry Part A*, **73**, pp. 949–957Williams, G. J., Pfeifer, M. A., Vartanyants, I. A. and Robinson, I. K. (2003). Three-dimensional imaging of microstructure in Au nanocrystals. *Physical Review Letters*, **90**, art. no. 175501.

36. Williams, G. J., Quiney, H. M., Dhal, B. B., Tran, C. Q., Nugent, K. A., Peele, A. G., Paterson, D. and de Jonge, M. D. (2006). Fresnel coherent diffractive imaging. *Physical Review Letters*, **97**, art. no. 025506.

Chapter 5

MODEL LYSOZYME CRYSTAL VERSUS AGGREGATE (UN)CONFINED FORMATION AS VIEWED BY RECENT THEORY AIDED BY COMPUTER EXPERIMENT

Jacek Siódmiak[a] and Adam Gadomski[b]
[a]*Department of Modeling of Physicochemical Processes*
[b]*Institute of Mathematics and Physics, University of Technology and Life Sciences in Bydgoszcz, Kaliskiego 7, 85-796 Bydgoszcz, Poland*
siedem@utp.edu.pl

The (dis)orderly aggregation of lysozyme is an intense field of research nowadays since such a model system contains all necessary paradigms to solve, preventing this way from the as yet applied trial-and-error method for obtaining the desired crystalline and other forms of lysozyme aggregates. The notions of nucleation–growth phase change, combined properly with the so-called viscoelastic phase separation, and opened toward its thermodynamic–kinetic, generally non-Markovian character, may improve considerably the overall description of the process against the impractical trial-and-error method. Both, seemingly simplified stochastic-in-nature analyses, carried out at a mesoscopic level, supported suitably

Synchrotron Radiation and Structural Proteomics
Edited by Eugenia Pechkova and Christian Riekel

www.panstanford.com

by submesoscopic-level computer simulations of this interface-controlled crystal formation process, may remedy and foster further actions toward more sophisticated and efficient solution to this long-standing, truly complex phase change and/or phase separation involving puzzle, both of them dealt with inseparably, what stands for a certain appreciable novelty of the present study. A final message that comes from the offered study so far states that a type of hybrid model combining sensitively some parts of analytical analyses with some well-designed computer simulation complementary co-analysis can assure a distinctive progress toward solving practically many of systems of interest.

5.1 INTRODUCTION

Protein molecules are made up of amino acids and possess amphiphilic nature when subjected to solvent, especially to water. Lysozyme is one protein molecule commonly met in nature and very prone to crystallize. The obtained crystals, depending on a broad spectrum of crystal formation conditions, have, in general, polymorphic character, expressed in terms of its unit cell, and its alterations: they are often termed non-Kossel crystals what makes them distinguishable from standard NaCl-type crystal forms.

Biologically motivated modeling becomes recently a necessary activity taken up by physicists toward obtaining first quantitative information about complex biosystems to be dealt with. Protein systems are those that are presently of central interest from the statistical mechanical and non-equilibrium thermodynamics points of view since they conform to basic research tasks underwent by both subdisciplines of statistical condensed matter physics (Basu and Chowdhury, 2007).

To be more specific, let us focus our attention on, in principle, isothermal binary mixtures of lysozyme and water molecules, having eventually some other minority ingredients involved, such as ion precipitants or impurities, thus a system as a whole very prone to (dis)ordered aggregation, and in particular, to non-Kossel crystal formation (Chernov, 1997).

When looking from the statistical mechanical point of view at such a binary lysozyme–water mixture, poured out in a three-dimensional (3D) laboratory vessel of big enough diameter, one is allowed mainly to look at interaction dynamics among the following: lysozyme molecules alone, lysozyme and water molecules, and finally, water molecules alone, postponing, however, other more specific and quite rarely occurring interactions, e.g., those between minority ingredients and the basic mixture's components, called thereafter the majority ingredients (components) of the mixture (Chernov, 2003).

In the system described above, one would reasonably expect to have both majority components in the amount of *ca.* 6.02×10^{23} per mole, thus a statistically big (Avogadro) number. As one may know, it is impossible to quantify readily all interactions in such immense system — therefore one is forced to resort to statistical mechanical description, with a suitably simplified version of the total system's energy (Hamiltonian) on which the Newton–Langevin-type complicated dynamics can be build up.

Then, the central limit theorem, as well as the equilibrium formalism of statistical Gibbs ensembles, would allow to get some ensemble (or, really rarely "equivalent" time) averages for describing the main system's, also dynamic, properties (Reif, 1967). To be more specific, however, let us note that there is quite a big difference in physical properties of both majority ingredients of the binary mixture addressed.

A lysozyme 14.7 kDa molecular weight biomolecule, in typical (dis)ordered aggregation conditions (temperature; pH), should be represented in its globular form by a positively charged sphere of radius of *ca.* 2 nm. Thus, its surface is somewhat like $0.25 \times 10^{-16}\,\text{nm}^2$.

From dimensional viewpoint, a water molecule is a dipole; its typical linear size is, however, 10 times smaller that that of lysozyme, being *ca.* 0.2 nm, i.e., as far as its hydrodynamic radius approximation is concerned. Thus, assuming crudely that this dipole is also a microsphere (and its projection onto the sphere's surface, a circle), and neglecting for simplicity the close-packing arguments on how to distribute most efficiently the two-dimensional (2D) microspheres (water molecules) over the sphere (lysozyme) surface (Shen *et al.*, 2005), one may say that we can have even one hundred water dipoles occluding one lysozyme molecule. The hydrophilic interactions between water molecules and the hydrophilic amino acids of the given protein assure spatial stabilization of the biomolecule and the conservation of its characteristic 3D (tertiary) structure.

It must be mentioned that the electrostatic interactions between charged biomolecules and dissolved additives, supporting the crystallization process, must be also taken into account. In the biopolimeric solutions, the most electrostatically active particles are polar amino acids and dissociated salts: NaCl (laboratory) and/or KCl (physiological environment) (Ducruix and Giegé, 1992). Electrostatic interactions are present between macromolecules and countercharged salt's particles and precipitants. This way, the net positive charge of lysozyme molecule is screened electrostatically by the surrounding crowd of counter-ions, leading eventually to a plausible model that involves the notion of electrostatic double layer, typically of Stern type. And, this model can be extended over the formation of aggregates of lysozyme molecules, irrespective whether they are formed as orderly or not (Gadomski and Siódmiak, 2002). It shall then be anticipated that the aggregates formed are typically full of water content, thus quite soft and semi-fragile (Booth *et al.*, 2002) to some extent; certain exceptions, for example, in the group of non-Kossel lysozyme crystals can also be mentioned (Pechkova *et al.*, 2005a).

The above statistical mechanical picture can be deteriorated markedly when the laboratory vessel is not big enough, so that the interaction of the binary solution cannot be neglected any longer, as could be a case of Petri dish of small enough height. It could also be an important case of introducing purposely some geometrical confinement into the system, such as the one coming from a nanotemplate, irrespective whether it is of Langmuir–Blodgett type, thus fluid (Pechkova and Nicolini, 2001) or of nanoporous solid nature (Dobrev *et al.*, 2005). It has been shown experimentally that the introduction of a geometrical confinement accelerates the (orderly) aggregation pace, in terms of both statistical mechanics and thermodynamics because of (i) reduction of the degrees of freedom involved in the system's evolution and (ii) decrease of the entropy production within the system. While point (i) looks obvious, though having, in addition, some aggregates' desorption involving hints at work (Siódmiak *et al.*, 2006), point (ii) requires, in general, a suitable non-equilibrium thermodynamic knowledge, anchored tightly to the second law of thermodynamics, and in addition, to the Gibbs (non-negative) entropy production principle.

Thus, when looking from the nonequilibrium thermodynamic point of view, one is directed to use extensively the notion of entropy, applied to the aqueous protein solution, thus to explore in depth the Gibbs principle to get the (asymptotic) growth rule, representative of such complex stochastic binary mixture, as well as to extract in a reasonable way the corresponding cessation-to-growth conditions, pointing to the final (dis)ordered structure formation (Gadomski, 2007).

In what follows, we would like to present in this review-type study, with some other related theoretical and experimental works in a background, our argumentation line, advocating for inventing and applying our non-equilibrium thermodynamic–kinetic analytical (stochastic) model, working at a mesoscopic level of description (next part), complemented by an in-parallel-designed computer sub-mesoscopic, coarse-grained model (following part), approaching readily some of those aspects of the former that are naturally hardly seen when being applied (Gadomski *et al.*, 2005). In general, our presentation and argumentation line concerns the formation of protein (lysozyme) (dis)orderly aggregates in terms of interface-controlled process. An auxiliary but relevant co-argumentation is thought to view the formation as a nucleation–growth phase transformation (Gadomski, 2000), see Fig. 5.4, seen from thermodynamic–kinetic point of view, intermingled, whenever appropriate, by phase separation, often confinement-induced, tendencies toward viscoelasticity of the emerging phase as a whole (Tanaka and Nishikawa, 2005). Concluding address and remarks will summarize the presented study.

5.2 LYSOZYME CRYSTAL VERSUS AGGREGATE FORMATION — THERMODYNAMIC–KINETIC APPROACH AT MESOSCALE

5.2.1 Nucleation

The important role of nucleation step in the overall lysozyme crystal formation has been emphasized in many, mostly experimental, studies; see, for example, Haas and Drenth, 2000. But it has not yet been addressed comprehensively due to, in general, extremely complicated, mainly kinetic, character of the process, *cf.* a discussion of internal degrees of freedom, and their influence on the (homogeneous) nucleation, in Reguera *et al.* (1998).

First, it has since long been realized that this process appears to be of thermodynamic–kinetic, thus truly complex character. For the case of homogeneous nucleation, to be clearly distinguishable from its heterogeneous counterpart, with a pivotal role of precipitants, co- and counter-ions, and impurities, geometrical confinements, etc., the thermodynamics is mainly subjected to creation of the nucleus' surface — such a creation comes from the equilibrium condition imposed on the free energy of the system (Huang, 1963), written commonly as $\delta\Phi(R) = 0$, where $\Phi(R)$ denotes the Gibbs' (free) energy of a spherical nucleus of radius R, and δ the respective variation. Thus, the thermodynamic factor as a whole can contribute substantially to the morphology of any ripe nucleus being created by the system, as well as can influence the presence of morphological Mullins–Sekerka (MS) type, and also other (non)linear interfacial instabilities, unavoidably present in the crystal formation, especially the one controlled by diffusion (Pimpinelli and Villain, 1998).

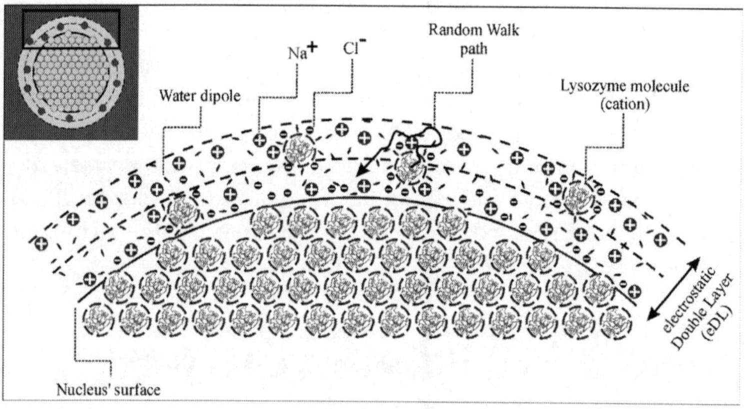

Figure 5.1 A zoomed piece of the spherical nucleus (upper left corner), with the electrical double layer surrounding it, the latter containing the most important electrostatic field emitting components (ions and dipoles); see also text for details. See also Color Insert.

Second, let us note that the complications of appropriate treatment of the nucleation step of lysozyme crystal formation may likely arise not only due to the heterogeneities (by the way, even small temperature variations can also make the nucleation heterogeneous) but also due to certain internal and interfacial cross-effects prevailing within the system in the course of time (Demei and Shizgal, 1993). The cross-effects mentioned are being specially effective right in the (bio)polymer nucleation stage, due to, in general, quite puzzling obedience, or not, of the system to the Laplace–Kelvin law (Dell'Isola *et al.*, 1995).

The cross-effects may point to the fact that both surface tension and nucleus' radius may depend on the subtleties of the local system's density profile, which, in turn, would prefer to depend on the interfacial concentration of the solution, thus the biopolymer-poor phase, in contrast to the biopolymer-rich phase that is simply the crystal phase. Such a natural quasi-separation of both phases, not necessarily under a deep temperature, but rather, here, a concentration "quench" (descent) at the interface, is able to induce certain additional viscoelastic effect, e.g., a (second) gradient of the accumulated stress field, or a gradient related to textural differences at the freshly formed surface and its internal (crystal interior) counterpart (Tanaka and Nishikawa, 2005). It may, in turn, influence the magnitude of the thermodynamic factor, termed the capillary (Gibbs–Thomson) length, which is involved in the Laplace–Kelvin law (Huang, 1963). Since the protein nucleation goes within the liquid–liquid immiscibility region one can take such a basic type of the capillary length as a first approximation, valid for correctly arranged lysozyme molecules, and then extend it to some additional curvature correction(s) (of Tolman type, see further) if they are a bit disorganized within the nucleus' volume.

According to our type of modeling of the nucleation, which emphasizes the creation of the electrical double layer, circumventing the nucleus of radius R_c, and causing the interfacial tension to change (Sear and Warren, 2002), the critical nucleus' radius reads (Gadomski and Siódmiak, 2002):

$$R_c = 2\Gamma_1 c_0 / (C - c_0), \tag{5.1}$$

cf. Fig. 5.1.

The above formula assumes that the nucleus can be considered as a drop of radius R_c (Huang, 1963), consisting of a certain very small number of lysozyme molecules, typically about four (Haas and Drenth, 2000) — it can, however, be treated according to the Laplace–Kelvin law when the capillary length Γ_1 fulfills the above stated obedience requirements. At this moment, let us note that a lysozyme crystal formation by hanging drop method, enhanced purposely by a fluid nanotemplate, assumes that the growing crystal tends to absorb mainly tetramers, just small but ripe nuclei (Pechkova *et al.*, 2005b; Siódmiak *et al.*, 2006), what has also been confirmed by another computer simulation (Ke *et al.*, 1998). As concerns the drop-shaped nucleus, it is reasonable to expect it, when the equilibrium surface concentration c_0, being typically small compared to the crystal density C, both quantities being proportional, according to the Raoult's law, to their corresponding osmotic pressures (Huang, 1963), are, for our biopolymer system, not largely distinguishable, so that $C \approx c_0$ may hold, though $C > c_0$ shall always be assumed. This, because of c_0 being typically small (Gadomski and Siódmiak, 2003a), leads clearly to very small supersaturation values of $c_0/(C - c_0)$, being effective during the diluted solution (random) nucleation events. Thus, to create a ripe enough viz thermodynamically stable nucleus one had to compensate this big dilution, or small supersaturation effect by the corresponding increase of the capillary length Γ_1, which, in turn, becomes feasible when the above mentioned cross-effects, due to the second-gradient theory, will be at work, also leading to the viscoelastic phase separation condition to be, at least partly, well approached, especially when a polynuclear system would emerge (Tanaka and Nishikawa, 2005) (Fig. 5.2).

Thus, the nucleus' existence, or system's tendency to avoid nucleus' breakage, will become surely decisive factors when the nucleus' surface tension σ_c, being proportional to Γ_1, will, for a given temperature and volume conditions, depend on the system's density profile which, in turn, depends closely upon the concentration of the nearby external phase surrounding the nucleus. This way the Laplace–Kelvin law will very likely be satisfied in a properly modified form (Dell'Isola *et al.*, 1995), involving the gradient trace, thus a necessary step toward growing the so-obtained nucleus will be made firmly. Then, a certain further exploration of the kinetic part of the nucleation process can be done by looking for an appropriate growth rule for the ripe, also mechanically (due to corresponding hydrophobic-type interactions between amino acid residues) stable (Chernov, 1997) nucleus' development. It implies that we have to turn our attention to the growth stage of the biopolymer crystal formation, the description of which, having done by means of analyzing a large number of experimental data, can be found elsewhere, especially in terms of the dynamics of layer growth in protein, mostly lysozyme, crystallization (Vekilov and Alexander, 2000). Thus, we are able to propose to consider the kinetics of the possibly time-dependent nucleation stage in terms of examining the ripe-nucleus growing effect, also with its viscoelastic peculiarities being addressed (Gadomski and Siódmiak, 2005).

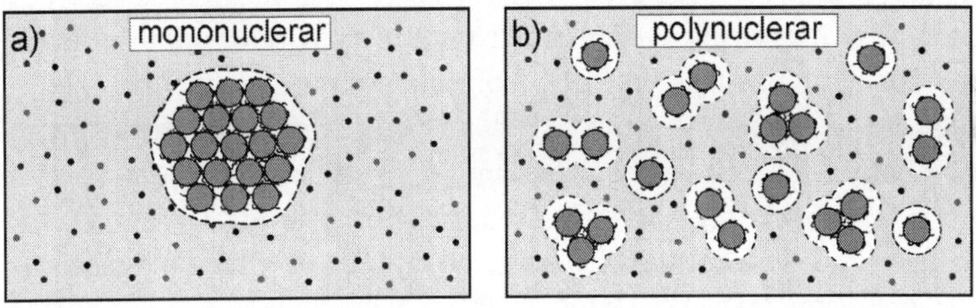

Figure 5.2 A mono- (left; denoted by a) and poly-nuclear (right; denoted by b) pathways of the single nucleus (a) and/or many nuclei (b) formation(s); the small colorful dots represent ions or dipoles; the dark bigger circles correspond to a picturesque representation of the lysozyme macroions, being sometimes "glued" together by some "entropic connectors" (thin lines), the latter being certain reactive pieces (amino acid residues) of the lysozyme molecule. The "white hallos" around each molecule, or molecular cluster, can be considered as some drawn prerequisites of the corresponding double layers. (Consult also a review by R. Mezzenga *et al.* (2005), especially some electron micrographs of liquid crystalline mesophases encountered in foods, namely Fig. 6 therein.)

It is worth mentioning that in Nanev and Tsekova (2000) a heterogeneous nucleation of lysozyme has been addressed experimentally by a method named the double-(thermal)-pulse technique, enabling to deal with nucleation and growth in a mutually independent way. This study confirmed that when nucleating lysozyme on a glass substrate, i.e., when subjected the nucleation to a 2D confinement, the resulting nuclei consist of three to four molecules, thus comparable to what has been estimated in Haas and

Drenth (2000), and utilized in Ke *et al.* (1998) and Siódmiak *et al.* (2006). Especially the latter study welcomes such an experimental result since in its computer model it has been assumed, similarly to the argumentation supporting another computational–physical study (Ke *et al.*, 1998), that the growing pace of the protein crystal can be increased when mainly the (ordered) tetramers desorb from a nanotemplate introduced to the system (Pechkova and Nicolini, 2001) — they are then absorbed by the growing crystal until a final structure is being formed. In addition, it can be stressed that the coming-from-thermodynamics driving force, namely the supersaturation, proposed to apply in Nanev and Tsekova (2000), and its influence, in the steady-state condition, on the nucleation rate, being determinable by the (small) number of molecules, constituting the nucleus, can be derived based on the so-called atomistic theory, initially developed to reveal the nucleation aspects in small molecule systems (Kashchiev, 2000). It is also worth pointing out that, as in our modeling, the necessity of two-field effect, or two-order parameter involvement, associated with the interface concentration as well as with the crystal's (aggregate) density, and the effects they both exert on the nucleation process, has also been considered by Oxtoby (2003).

Concluding this part, it should, however, be mentioned that maintaining, while based on the mechanistic–thermodynamic grounds described above, the nucleus maximally symmetrical or, idealistically speaking, sphere-shaped within the continuum-description approach at work, does not fully guarantee any emergence of "solid" non-Kossel crystals (Chernov, 1997), but some other more disorderly biomolecular aggregates are certainly allowed to occur (Siódmiak and Gadomski, 2006; Gadomski, 2007). It is simply due to the fact that the nucleation is a random (also, living) process, which means that the above offered argumentation line must always contain a stochastic character. This type of character can also be extended to the growth stage in which, however, the kinetic effects are allowed to prevail somehow over their thermodynamic counterpart, rendering inseparably the overall process thermodynamic–kinetic, also quite time-dependent viz non-Markovian in its basic character (Gadomski, 2007).

5.2.2 Growth

The growth step appears to be equally decisive in obtaining final output in a form of crystal or aggregate. While the nucleation step is more morphologically oriented, because it sets in the initial structure of the (dis)orderly aggregate, the growth step looks more kinetically oriented.

The most robust models proposed so far to reflect properly the kinetics of lysozyme (dis)orderly aggregation are, quite unexpectedly, very simple phenomenological models that unravel the kinetics in terms of phenomenological kinetic rules, invented to uncover most important experimental results (Pusey *et al.*, 1986; Chernov, 1997, 2003; Vekilov and Alexander, 2000; Chernov, 2001).

The phenomenological kinetic rules are always proposed in a form of the growth rule, stated in terms of the rate of change in time of the linear nucleus' characteristic, for example, the radius or a distance to the most representative growing surface of the crystal or aggregate, measured from its mass center (Honigmann, 1958), designated by R_{ch}. Such a (multiplicative) rule looks like (Vekilov and Alexander, 2000):

$$\frac{dR_{ch}}{dt} = \sigma_T v_{kin}(R_{ch}), \tag{5.2}$$

where σ_T denotes a main thermodynamic factor of the phase change, whereas $v_{kin}(R_{ch})$ is primarily responsible for the kinetic peculiarities of the phase change.

From all most known phenomenological models applied (Pusey *et al.*, 1986; Chernov, 1997, 2001) it is seen that σ_T is related to the main driving force of the phase change, namely the supersaturation (Nanev and Tsekova, 2000), while the structure of $v_{kin}(R_{ch})$ is put differently, depending on whether the model imposes a diffusional control on the system's behavior or it goes, based on experimental motivation, out of this type of control.

The diffusion-type control invokes more of less a MS mode morphological instability involving type of the modeling (Pimpinelli and Villain, 1998) whereas non-diffusional control is mainly the interface control — it is attributed to the screw dislocation driven growth of the terraces appearing on the crystal's surface — the so-called Burton–Cabrera–Frank (BCF mode) growing mode (Siódmiak and Gadomski, 2006); in case of non-crystalline aggregates such a picture is markedly deteriorated, leading to serious kinetic obstacles, such as bunch cascades at the surface, and the likes (Vekilov and Alexander, 2000).

While the thermodynamic factor σ_T will also be recovered by our interface-controlled type of modeling (Gadomski and Siódmiak, 2002), its kinetic counterpart, $v_{kin}(R_{ch})$, depending on whether the control is of diffusional or non-diffusional (nearby) external field nature, looks also differently (Pimpinelli and Villain, 1998).

Thus, in the former case it goes like $v_{kin}(R_{ch}) \cong 1/R_{ch}$, wherein the r.h.s. comes from counting explicitly the diffusional flux crossing the internal (interface) boundary. In the latter case, in turn, an explicit dependence on R_{ch} cannot be seen, presumably due to either non-standard, let us say, anomalous diffusional effects, such as those introduced by the nearby electrostatic field (Gadomski and Siódmiak, 2003a), or when the interfacial control prevails substantially, as in the case of BCF mode (Chernov, 1997).

When such a nonstandard confined-diffusion or purely interface-involving control happens to the system, which might also be due to aggregation acts, occurring at the nucleus' surface, one had to specify $v_{kin}(R_{ch}) \to \text{const}$ over the late-time course of the formation (Vekilov and Alexander, 2000).

From the above it follows that in both MS and BCF modes, one would expect $R_{ch} \cong t^{\varsigma}$ to hold asymptotically in the course of time t. A significant differentiation between these two modes manifests, however, when looking at the growth exponents ς.

In the MS mode one gets most typically (Pusey *et al.*, 1986) $\varsigma \approx 1/2$ while in the case of BCF mode one provides firmly (Vekilov and Alexander, 2000) $\varsigma \approx 1$.

It implies that the BCF mode is, when plotting onto log–log scale, roughly two times faster than the MS biomolecular-matter involving mode. Certainly, depending on details of the growing conditions, whether they are subjected to a confinement or not (Pechkova and Nicolini, 2001; Dobrev *et al.*, 2005), which is the ionic strength of the solution (Sear and Warren, 2002), or finally, which are the viscoelastic factors governing the growth (Chernov, 1997; Haas and Drenth, 2000; Gadomski and Siódmiak, 2005), then the growth exponent ς ranges typically between this two basic growing modes, thus $\varsigma \in [1/2; 1]$ mostly holds, though some exceptions can also be foreseen, mainly due to highly fluctuational growing conditions (Vekilov and Alexander, 2000; Gadomski and Siódmiak, 2002; Chernov, 2003).

A purely phenomenological extension of the kinetic (growth) rule can also be applied, such as the one of Pusey *et al.* (1986), namely:

$$\frac{dR_{ch}}{dt} = \sigma_T [v_{kin}(R_{ch})]^{\upsilon}, \tag{5.3}$$

wherein υ points to some nonlinear hydrodynamic modes, arising in the course of growth duration, i.e., if such modes would truly manifest one had to assume $\upsilon \neq 1$, thus having this way another fitting parameter at her/his disposal (Pusey *et al.*, 1986). Rising σ_T to some power in the above kinetic growth formula is rather not explicitly proposed; it could likely make sense for special-type aggregation processes such as 4 nm thick and R_{ch}-long fibril formation but within a certain range of temperature or pH values applied (Arnaudov and de Vries, 2005).

In what follows we wish to propose, in a concise form, a protein crystal growth model, motivated also by some experimental data on lysozyme (Gadomski *et al.*, 2005; Siódmiak and Gadomski, 2006; Siódmiak *et al.*, 2006; Gadomski, 2007), the foundations of which have deep non-equilibrium thermodynamic origin, and the kinetics follow certain landmarks of non-Markovianity, thus being, in general, also growth-history dependent (Santamaría-Holek *et al.*, 2007), while pointing indirectly this way to the aging effect from which many biomolecular aggregations suffer substantially (Gsponer and Vendruscolo, 2006).

In this model, which one may call readily a thermodynamic–kinetic approach to the crystal versus aggregate formation, both parts are taken inseparably — they appear in the model description together, and are also subjected to concrete specific conditions of the aggregation (Pechkova *et al.*, 2005b; Siódmiak *et al.*, 2006). The thermodynamic part is dedicated directly to the boundary conditions, and may be termed the boundary-condition proposal (B-CP). The kinetic part, in turn, is unambiguously dedicated to the macroions' (lysozyme molecules) velocity-correlation proposal, to be effective in the interfacial (active) zone (Vekilov and Alexander, 2000) — this is abbreviated thereafter as V-CP. A "synergistic" mode of both growing thermodynamic–kinetic supermodes emphasized, namely, when B-CP coincides most optimally with V-CP, will indicate which growth rule, and why, is followed by the system. Thus, by manipulating with B-CP and V-CP just "in parallel" and within the entropy production formalism proposed, we are able to derive certain either cooperation or confliction growth rules, or while put it differently, when an easy (facilitated) growing mode emerges, or when it is hampered by the encountered thermodynamic–kinetic circumstances. Such an offer we would like to see as an original as well as useful proposal toward complementing as yet applied, purely phenomenological kinetic laws (Pusey *et al.*, 1986; Chernov, 2003), serving to elucidate quite a large body of experimental data (Drenth, 2002), but without explaining in detail the thermodynamic–kinetic origin of the formations under study.

5.2.3 The B-CP Seen as an Extension of Gibbs–Thomson Boundary Condition

The boundary condition $c(R) = c_0[1+2\Gamma_1/R]$ due to Gibbs and Thomson (Huang, 1963) comes from a Taylor series-based linearization of $\exp[2\Gamma_1/R]$ with respect to the argument $2\Gamma_1/R$, with the capillary length $\mathbf{\Gamma}_1 = \gamma_R v_{m-c}/k_B T$, the argument due to small value of Γ_1 being typically distinctly less than one; notice that $k_B T$ represents the thermal energy, whereas γ_R represents the surface energy, while v_{m-c} denotes a volume subjected to one surface molecule, so that the capillary length brings additionally a molecular character; $c(R)c_0$ mean the concentrations at the curved, thus non-equilibrium, and flat (at equilibrium) interface, respectively. In fact, this standard equilibrium thermodynamic condition, prescribed customary at the curved interface, the one represented by twice the mean curvature $2/R$ (Gadomski and Siódmiak, 2002), comes from the Wulff's theory of equilibrium crystal's shapes, developed under a confinement of the total crystal's volume approaching a constant value, namely V_{crystal} = const., and under an equilibrium condition that $\sum_i \gamma_i A_i$ = min., i.e., when the products of crystal wall surface energies by the surface areas, when summed up, tend to a minimum.

An exact result but for n-wall symmetrical object, thus nearly for a sphere (droplet) of radius R, certainly in the limit of large n, is then as follows $\ln[c(R)/c_0 = 2\Gamma_1/R]$ (Honigmann, 1958).

This Gibbs–Thomson boundary condition is satisfied by a drop in an equilibrium with its supersaturated vapor (Huang, 1963). It is also applicable to small molecule (or, atoms/ions involving) precipitation and/or crystallization from an undercooled melt or supersaturated solution (Pimpinelli and Villain, 1998), also the one realized by the hanging drop method (Pechkova and Nicolini, 2001). It is, however, by no means applicable without appropriate extension(s) to lysozyme crystal versus aggregate formation. The reason for being inapplicable in such a standard form may be at least twofold. First, lysozyme (dis)orderly aggregates hardly attain a global minimum on the energy axis — they may better reside in a quasi-equilibrium state, if it happens to be the case, but quite often they are out of equilibrium, though in their mature stages, close to it. This is often due to a generic competition between folding and aggregation of proteins, and it may markedly, though slowly, change over time due to mostly environmental conditions (Gsponer and Vendruscolo, 2006). Second, according to the second-gradient theory presented while describing the nucleation stage (Dell'Isola *et al.*, 1995), a biomacromolecular system, such as that composed of lysozyme molecules of a few nanometers each, should manifest when passing between nano- and micrometer scales, a finite-size effect, pointing to a graininess of the crystal/aggregate surface, which invokes also a nontrivial contribution to the elasticity of the object's surface. This effect is detectable when a very natural extension of Gibbs–Thomson-type droplet's equilibrium thermodynamics is allowed to be applied, namely (Gadomski *et al.*, 2005):

$$c(R) = c_0\left(1 + \frac{2\Gamma_1}{R} + \frac{\Gamma_2{}^2}{\tilde{r}_{PMP}R}\right). \tag{5.4}$$

The additional r.h.s. term in the parenthesis, namely $\Gamma_2^2/\tilde{r}_{\text{PMP}}R$, is composed as follows. In the numerator the square of the Tolman length, Γ_2 (being roughly of the size comparable to Γ_1), appears— the Tolman length is responsible for accounting for certain biomolecule stiffness-to-elasticity effects (Gadomski *et al.*, 2005), thus to introduce readily the finite-size macromolecular effect, which explicitly manifests in the denominator by the molecule's size $\tilde{r}_{\mathrm{PMP}}$ of about 2–3 nm, thus, the signature of the graininess; presumably, it can implicitly account for a formation of hydrogen bonds between water and lysozyme molecule, thus, it can also be quantified by the hydrogen bond average strength (by the way, the hydrogen bonding is not a property of small molecule crystals or aggregates emerging from a solution — it is a distinguished property of protein formations grown within the aqueous solution limit). However, the proposed extension is still an extension of equilibrium type, though it makes a difference, mainly when applied at the nanometer-size scale (Gadomski *et al.*, 2005).

Another type of the proposed boundary condition, working effectively within quite a narrow realm of the parametric zone, appears to be a more toward out-of-equilibrium "disputable" extension of the form

$$c(R) = c_0\left(1 + \frac{2\Gamma_1}{R} + \sum_1^3 \alpha_i \frac{R^i - R_o^i}{R_o^i}\right). \tag{5.5}$$

First, it is thought of as a direct extension of the standard Gibbs–Thomson condition, pointing to quite a drastic "environmental" conditions (Gsponer and Vendruscolo, 2006) that may lead either to promote the natural course of the free-energy increase or they may contradict such a promotion by pronouncing some pinning effects, or by hampering the growth pace due to abnormal (confined) elastic behavior of some molecules, or certain groups of them, the latter belonging to the surface of the growing object (Gadomski and Siódmiak, 2005). This toward-non-equilibrium extension term reads

$$\sum_1^3 \alpha_i (R^i - R_O^i)/R_O^i$$

with the α_i-s, the coefficients pointing to some characteristic mechanism of elasticity, appropriate for three basic crystal surface epitaxial nucleation mechanisms (Pimpinelli and Villain, 1998), wherein i = 1, 2, 3 indicate the linear (by molecular rows), surface (typical epitaxy), and volume (thick film limit) nucleation mechanisms (Gadomski and Siódmiak, 2005; Siódmiak and Gadomski, 2006).

Our proposed model, due to incorporation of elastic surface-nucleation effects of auxetic/non-auxetic nature (Gadomski and Siódmiak, 2005), allowing for appearance of non-positive Poisson coefficients at the interface (dis)orderly aggregate versus surroundings, also includes a chance of examining certain "spurious" kinetic effects, such as formations of gels or other arrested kinetic states. It requires, however, an extension of the proposed modeling into nonlinear viscoelastic domain (Santamaría-Holek *et al.*, 2007, and references therein). Thus, in principle, such a procedure can be envisaged while entering the stochastic description of this modes coupling process, wherein the modes concern both viscoelasticity as well as late-time growing conditions of the process under consideration, as has been demonstrated in Santamaría-Holek *et al.* (2007).

Next, from the above formula it is seen that if one wished to arrive at a constant growth pace, thus when applying the kinetic criterion, one would be enforced to work within a narrow range of R-domain, which implies also rather more nanometer than micrometer size scale of the obtained crystals; otherwise, only the aggregates are plausible to obtain (Gadomski, 2007, and references therein). The kinetic criterion, delineated above, could roughly be equivalent to a very narrow crystallization window (Chernov, 1997, 2003) for the protein, and also, lysozyme non-Kossel crystals. Outside this window aggregates are exclusively expected to occur (Siódmiak and Gadomski, 2006). Before going into last extension proposed, let us remind that in the present extension the Tolman-type term, namely $\Gamma_2^2/\tilde{r}_{\mathrm{PMP}}R$, has been consequently omitted as being inapplicable when too strong viscoelastic effects of mutual interactions between crystal surface and nearby surroundings apply, i.e., if one is set out of the linear stability regime known for epitaxial growth as Grinfeld instability (Pimpinelli and Villain, 1998).

Last but not the least, a very natural extension of perturbed surface droplet's equilibrium thermodynamics into nonequilibrium occurs when

$$c(R) = C_0\left(1 + \frac{2\Gamma_1}{R} + \frac{\Gamma_2^{\,2}}{\tilde{r}_{PMP}R} - \beta_K \frac{dR}{dt}\right) \tag{5.6}$$

is presumed. This experimentally motivated condition, which has been first proposed by Goldenfeld in 1987 for accounting for the kinetics of round polycrystalline forms, named spherulites (Gadomski *et al.*, 2005, and references therein), contains a term that expresses directly a (local) kinetic departure from the equilibrium. This term, equipped with a minus sign, see above, looks like $-\beta_K(dR/dt)$, and quantifies for a local growth pace due to non-uniform evolution of the interface (β_K stands for a phenomenological coefficient). Notice, however, that $-\beta_K(dR/dt) \to 0$ when the typical non-Kossel crystal growth pace of roughly constant value applies. Therefore, when this term preserves over relatively long time scale, the non-Kossel crystal is not very close to its thermodynamic equilibrium, and the polycrystalline form eventually wins over its "purely" single-crystal counterpart. This departure from equilibrium can also be utilized when some defects of the temporarily formed crystal structure tend to dominate at the interface; then the coefficient β_K could be proposed as an indirect measure of the interfacial defects' concentration — then another model mentioned in Gadomski *et al.* (2005), named a UCJ (from the names of Ulman, Chalmers, and Jackson, known since 1964) solidifying-front crystal growth model ultimately applies.

When looking from a phenomenological (non-stochastic) point of view the deterministic growth rule that we propose to apply reads (Gadomski, 2007)

$$\frac{dR}{dt} = [\sigma(R)]^{-1} v(R), \tag{5.7}$$

where the near-surface solubility, being just the inverse of the supersaturation (Nanev and Tsekova, 2000), reads

$$\sigma(R) = \frac{C - c(R)}{c(R)}. \tag{5.8}$$

The verification of our deterministic growth rule has first been included in Gadomski and Siódmiak (2002) for experiments with lysozyme crystals growing up to a characteristic length of ~0.1 cm when it is allowed that the process is controlled mainly by incorporation of lysozyme molecules at the interface rather than by bulk diffusion. The obtained result is of the same quantitative weight as another estimation coming from applying a phenomenological growth rule, apparently within the BCF mode (Chernov, 1997). In addition, let us note that the deterministic rule, also according to Gadomski and Siódmiak (2002), can be rewritten with an inclusion of the critical nucleus' radius as

$$\frac{dR}{dt} = v(R)\frac{[c_o/(C-c_o)]R+R_c}{R-R_c}. \tag{5.9}$$

Let us briefly summarize here the main points of the derivation of the above deterministic growth rule (Gadomski and Siódmiak, 2002, 2003a), *cf.* Fig. 5.3. First, we take the mass of the growing object in two consecutive time steps, t and t_1, where $t_1 > t$, expressed via the corresponding densities (concentrations) of the object and of the surrounding field, marked in Fig. 5.3, $C(\vec{r})$ and $c(\vec{r})$, respectively. Second, we take the difference between the masses, Δm, and construct the ratio $\Delta m/\Delta t$, wherein $\Delta t = t_1 - t$. Then we go into the limit $\Delta t \to 0$. Further, we make use of the physical observation that $J = \Delta m/\Delta t$, where J is the global flux taken at the object's surface $\Sigma(t)$, containing the object's volume $V(t)$ as a whole. The above procedure is called thereafter the global mass-conservation law. Then, we make use of some simplifications, such as the spherical symmetry of the growing object, assume its density to be constant, $C(\vec{r})$ = const., utilize the (extended) Gibbs–Thomson-type thermodynamic internal boundary condition(s). Finally, we apply the formula for the mass-convective flux, appropriate for protein and/or colloid-type systems (Tanaka and Araki, 2007), which at a local level of mass-counting reads $j = v(\vec{r})c(\vec{r})$, with $v(\vec{r})$ being a $(\vec{r})$ (position)-dependent vector representation of $v(R)$, involved directly in the growth rule above. This procedure in total enables to write down formally the deterministic growth formula in the form presented above (Gadomski *et al.*, 1993).

Then, the asymptotic (late time) solution to the above, presumed that the lysozyme molecule velocity, $v(R)$, at the interface, is a constant parameter (Vekilov and Alexander, 2000; Gadomski and Siódmiak, 2002) reads

$$R(t) = \bar{R}t^{\nu_g}, \tag{5.10}$$

with a tendency of $v_g \to 1$ for crystalline formations (Gadomski, 2007); $\bar{R}$ stands for a kinetic set-point constant radius, roughly close to the equilibrium nucleation radius R_c described above. By the way, an extension to the above proposal can be seen in terms of the two thermodynamic lengths, Γ_1 and Γ_2. It looks like $R_c = 2\tilde{\Gamma}_1 c_0/(C - c_0)$, wherein $\tilde{\Gamma}_1 = \Gamma_1 + (\Gamma_1^2/2\, \tilde{r}_{PMP})$ (Gadomski *et al.*, 2005). This could then be considered as the second-gradient-type extension (Dell'Isola *et al.*, 1995) introduced for the purpose of properly approaching the nucleation stage, thus to appropriately account for the above mentioned kinetic set-point constant radius, $\bar{R}$.

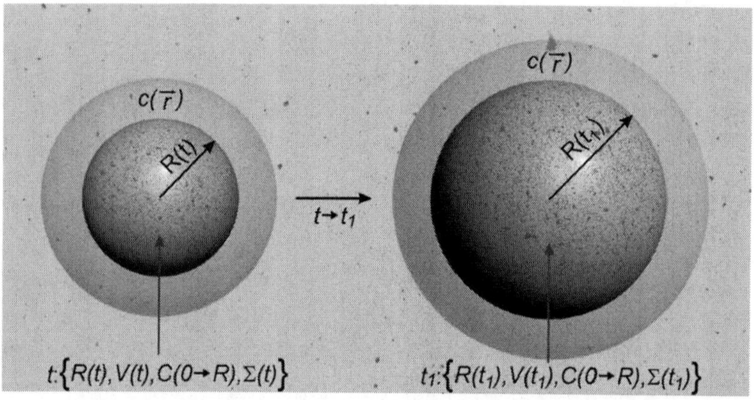

Figure 5.3 A growth of the (simplified) spherical nucleus in two consecutive time instants; for explanations of symbols used, see text, especially a sketch of the derivation of the deterministic growth rule ($0 \to R$ means: the crystal's radius range — interior of the crystal). See also Color Insert.

When looking from a stochastic point of view, in turn, the growth rule that we propose to apply can be provided by (Gadomski and Siódmiak, 2002; Gadomski, 2007)

$$\frac{dR}{dt} = [\sigma(R)]^{-1} V(t). \tag{5.11}$$

The only difference when compared to the deterministic rule relies on the replacement of $v(R)$ by $V(t)$ — the stochastic velocity of the incoming macroions (the statistical properties of $V(t)$ are determined by thermal fluctuations — see further), a much more realistic V-CP (Gadomski, 2007) to be met within the active zone of the crystal or aggregate growth (Gadomski and Łuczka, 2000; Vekilov and Alexander, 2000). In what follows we would like to sketch how does it look like, and which is the motivation that stays behind it.

5.2.4 The V-CP Seen as a Relevant Extension of the MS-Like Mode

While the MS mode expresses purely external field diffusional growth of the evolving object as a whole, its surface perturbation parameter decays exponentially with time (Pusey *et al.*, 1986; Gadomski *et al.*, 1993; Pimpinelli and Villain, 1998; Gadomski and Łuczka, 2000). In the case of our type of modeling, such a decay seems quite unrealistic; therefore, we would like to propose an interface-controlled growth, wherein the width of the interface is roughly of comparable size to the size of above mentioned perturbation (Fig. 5.1). The width, depending on the concrete from-solution crystallization conditions (Poon *et al.*, 2000), may change over time, also responding to a suitable concentration–temperature phase diagram zone (Haas and Drenth, 2000). Within the interface, however, a diffusional microscopic motion, just a random walk, in fact, of lysozyme macroions is observed (Vekilov and Alexander, 2000). Thus, the diffusion goes internally, so to say, being likely "intermittently" interrupted by the, let us say, almighty electrostatics (Poon *et al.*, 2000; Gadomski and Siódmiak, 2003a), settling up the rules of accretion and motion, or attachment and detachment (Siódmiak and Gadomski, 2006), if a molecule, or a molecular aggregate (Pechkova and Nicolini, 2001; Siódmiak *et al.*, 2006), is still not absorbed by the growing object. The time-dependent diffusion within the interface is given by the correlations, namely

$$D(t) = \int_0^t K(s)ds, \tag{5.12}$$

wherein the correlation function $K(s)$ is defined by some stochastic averages $<V(t) \geq 0$ and $K(t_1 - t_2) = <V(t_1)V(t_2)> \neq 0$ (a stationary Gaussian noise for the stochastic velocity field with nonzero temporal and algebraic correlation) (Gadomski, 2007), and where

$s := t_1 - t_2$ has to apply. This conditions look fairly complicated at first look, but point to the stationary ($<V(t) \geq 0$) Gaussian process. The truth is, however, more obvious: they simply establish the necessary and sufficient conditions for stating the diffusion coefficient $D(t)$ by means of the velocities of the lysozyme molecules walking at random in the interfacial zone. It is another, not Einstein-like definition of the diffusion coefficient, coming directly from the dissipation-fluctuation theorem — the formula, in turn, can be named the Green–Kubo relation (Gadomski, 2007 and refs. therein). It is worth to mention that only temporal correlations are discussed here, because the growth is considered within its (narrow) interface-controlled approximation, the spatial correlations of the velocity macroions' field are consequently postponed; they may be valid for another, more diffusion-type approximation, in turn. For the non-Kossel lysozyme crystal growth to prevail, an asymptotic correlational proposal has to look as follows

$$D(t) \propto t^{1-\nu}, \tag{5.13}$$

where $\nu \in (0,1)$ holds (Gadomski and Siódmiak, 2005). The limit of $\nu \to 0$ corresponds to some weak interaction, very small correlation strength parameter, describing thereby the average interactions within the binary system (Tanaka and Nishikawa, 2005), consisting of solute (lysozyme) and solvent viz water molecules (Gadomski and Siódmiak, 2003b). This stands also for the kinetic criterion, see above, of the non-Kossel crystal formation. Outside this limit a (dis)orderly aggregation occurs. The limit of $\nu \to 1$ corresponds to time-independent diffusion coefficient, which implies the standard diffusion condition (or, a normal Einstein-type random walk of a lysozyme molecule). It should be outlined that the stochastic growth rule $(dR/dt) \propto V(t)$ (where R is the sphere radius; for detailed description of the sphere-type approximation for description of the growth rule see Gadomski *et al.*, 1993) with the proportionality thermodynamic factor $[\sigma(R)]^{-1}$ is, in a mathematical sense, completely equivalent to the following Smoluchowski-type equation (Doi and Edwards, 1986)

$$\frac{\partial}{\partial t}P(R,t)+\frac{\partial}{\partial R}J(R,t)=0, \tag{5.14}$$

with the aggregating-matter flux

$$J(R,t)=-D(R,t)\frac{\partial}{\partial R}P(R,t)-\frac{D(R,t)}{k_BT}\frac{\partial\Phi}{\partial R}P(R,t) \tag{5.15}$$

and with the (Kramers' type) boundary conditions of the Smoluchowski equation

$$J(R=R_o,t)=J(R_\infty,t)=0 \tag{5.16}$$

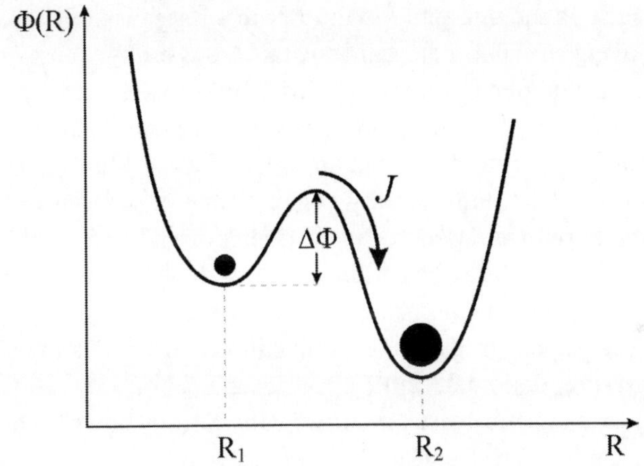

Figure 5.4 Surmounting (in a classical way) the so-called Kramers' barrier (Reguera *et al.*, 1998, and refs. therein) appears to be a necessary step in getting the crystal and/or molecular aggregate phase; for explanations of the symbols plotted, see text.

The overall diffusion function of the formation reads (Gadomski, 2007)

$$D(R,t)=D(t)/\sigma^2(R). \tag{5.17}$$

and represents by itself a competition effect between both proposals, namely the B-CP ("static") and its V-CP ("dynamic") counterpart (Gadomski and Siódmiak, 2005). This diffusion function, and the analyses staying behind it, enables someone to differentiate between orderly and disorderly aggregations. For example, an orderly aggregation would manifest when $D(R,t)$ diminished with time. This implies that the "static" condition $\sigma^2(R)$ wins over its "dynamic" counterpart $D(t)$. A disorder governs the system's behavior, in turn, when a converse reasoning applies, i.e., the velocity field is "wild" enough to overcome its "static" peaceful could-be antagonist, the latter being always in a MS-type description responsible for smoothening the surface of the growing object (Pimpinelli and Villain, 1998; Gadomski and Siódmiak, 2003a). When $\sigma^2(R)$ and $D(t)$ are cooperating, none of them being the winner of the aggregation versus phase separation game, a certain intermediate state clearly arises, such as the one reminiscent of the emergence of the hexatic phase during an order–disorder (crystal versus molten phase) phase change, presumably resembling that of Thouless–Kosterlitz type — in the case of lysozyme aggregation it is simply the viscoelastic liquid–liquid phase separation (Tanaka and Nishikawa, 2005; Wentzel and Gunton, 2007) or immiscibility region (Haas and Drenth, 2000). The suitably reformulated tochastic depiction of the crystallization phenomena, taken as a nucleation and-growth phase transformation change, could afford to discern the gelation processes, often occurring during lysozyme aggregation/crystallization process (Gadomski and Ausloos, 2006; Gadomski, 2007).

The last but not least quantity in the Smoluchowski equation above is the Gibbs free energy; it is provided by

$$\Phi(R) = -k_B T \ln[\sigma(R)] \tag{5.18}$$

and expresses truly non-equilibrium character until the late-time limit, fairly equivalent to a stationary state, applies. Notice that the equilibrium conditions can also be recovered while based on the function, namely when

$$\Phi'(R) = 0 \tag{5.19}$$

then a simple evaluation shows that

$$\ln[c(R)]\times\frac{1}{1-[1/(\sigma(R)+1)]}=0 \tag{5.20}$$

thus

$$\frac{1}{1-[1/(\sigma(R)+1)]}\xrightarrow[i>>t_o]{} const. \tag{5.21}$$

and

$$\ln[c(R)]'=0 \tag{5.22}$$

finally results (the prime denotes respective differentiation). This gives then exactly the equilibrium condition, the same that has been revealed above (see, the nucleation step) within the Wulff's approximation taken strictly at equilibrium; moreover, one had also $\ln[c(R)/c_0] = -2\Gamma_1/R^2 = 0$ which would imply that the explicit contributions of the second-order correction to the Gibbs–Thomson boundary condition should be forbidden — this contribution is simply to small to be effectively applied for the mature growing stage. Ultimately, let us report on recovering this way formally the equilibrium (certainly, very late-time, $t > t_0$) state, namely

$$\ln[c(R)]\xrightarrow[t>>t_o]{}\ln c_o = const. \tag{5.23}$$

and

$$c(R)\xrightarrow[t>>t_o]{} c_o. \tag{5.24}$$

It is worth introducing right at this stage the so-called Kramers' barrier, *cf.* Fig. 5.4, that is given by

$$\Delta\Phi(R):=\Phi(R(t_2))-\Phi(R(t_1)). \tag{5.25}$$

(To understand Fig. 5.4 formally, $R_1 \equiv R(t_1)$ and $R_2 \equiv R(t_2)$ have to be taken.) When $\Delta\Phi(R) < 0$, the aggregating system goes toward equilibrium; when $\Delta\Phi(R) > 0$, the system goes away from equilibrium, while at $\Delta\Phi\ (R) = 0$ it is supposed to be just in it (Hongmann, 1958; Gadomski and Siódmiak, 2005).

Let us point out that another thermodynamic–kinetic model of crystal growth of complex molecules from solution has also been proposed more than 20 years ago (Tiller, 1986). It takes also into account the presence of electrical double layers, this time mainly around the (protein) molecules, but around suitably small crystallites too, and shifts the crystal growth comprehension from far-field diffusion controlled to interface-controlled, i.e., somehow confined, thus exploring also the pathway that we have actually followed by our Smoluchowski-type of modeling, with the involvement of Kramers' barrier (Gadomski, 2007). The formulation by Tiller (1986) is, however, more — both phenomenological and dynamic — using mainly a typical Lennard–Jones-type attraction–repulsion potential's rationale for describing the dynamics in a colloid-type systems of interest. In our formulation, in turn, the emphasis is put more on kinetics, having, roughly speaking, the thermodynamic aspects of the two just compared models of the same (productive) entropy-involving character (Gadomski, 2007).

5.2.5 Cessation-to-Growth and Final-Structure Creation

Stationary solution to the above aggregate size-dependent Smoluchowski equation, i.e., when

$$\frac{\partial}{\partial t}P(R,t)=0, \tag{5.26}$$

which means practically that

$$J(R=R_\infty,t)=0, \tag{5.27}$$

where R_∞ is final, typically large enough size of the obtained aggregate or crystal, obtained when $c(\vec{r})$ becomes c_0, i.e., in the readily long-time limit) and implies that the corresponding probability of attaining a quasi-equilibrium form of the protein crystal/aggregate, P_∞, goes as (Doi and Edwards, 1986)

$$P_{\infty} \cong c_0/(C-c_0)=R_c/2\Gamma_1, \tag{5.28}$$

i.e., it is purely thermodynamically controlled, or equivalently, fully set in by the nucleation stage. Note that $0 < c_0 < C$ firmly applies (Gadomski and Siódmiak, 2003a). Let us simply remind the reader a "naked truth" coming out from the so-constructed model as a whole, see above. Namely, each experimenter would expect, when the crystal, or the aggregate, ceased to grow, what typically becomes effective after an appreciably long (stationarity invoking) time, that the concentration near the interface crystal-surrounding would take on a constant value. It is seen from the above probability-addressing formula that this is really the case. Moreover, the mentioned formula does not include any signature of even putative changes-in-time of the involved quantities.

Last but not least, the obtained formula, *cf.* the beginning of the chapter, in addition, explicitly bears the landmark of the nucleation radius — therefore one might conclude that the cessation-to-growth stage is somehow nucleation predetermined. The last sentence can also be viewed as a natural observation attributed to any phase change phenomenon in which a children phase grows at the expense of its parent counterpart. Notice that $0 < R_{\infty} \leq 2\Gamma_1$ must hold (Gadomski *et al.*, 2005). Bear in mind that at this crystal formation stage we refer to the stable, thermodynamically ripe nucleus of radius R_c — its stability is somehow guaranteed by the so-called second-gradient theory, first invented for microscopic bubbles (Dell'Isola *et al.*, 1995, and references therein). To sum up in part, it shall be ascertained that if $R_c \to 2\Gamma_1$ then the probability $P_{\infty} \to 1$, i.e., the more firmly one can attain the equilibrium crystal form (Honigmann, 1958; Chernov, 1997). Note that no kinetic subtleties are explicitly contained in the final form of P_{∞}. Thus, in other words, the crystal formation viewed by our type of modeling, and supported by helpful findings of the others (Haas and Drenth, 1999, 2000; Nanev and Tsekova, 2000; Poon *et al.*, 2000; Vekilov and Alexander, 2000; Sear and Warren, 2002; Chernov, 2003; Pechkova *et al.*, 2005a; Wentzel and Gunton, 2007), proclaims univocally that the nucleation stage appears to be the most important or decisive step, thus a careful preparation of the nucleus (Kashchiev, 2000) is really worth doing (Siódmiak and Gadomski, 2006; Siódmiak *et al.*, 2006). It is so indeed since it pre-determines the last stage of the formation (cessation to growth) when the (pre)final structure is going to emerge ultimately.

A more careful analysis of the nucleus' preparation, and then a realization based on coarse-grained molecular, carefully prepared building blocks, carried out mainly within the framework of BCF mode of growing a few types of lysozyme crystals, has since now on been shifted to the next part of our outlook on the lysozyme (dis)orderly aggregation seen preferentially from our theoretically oriented but also thoroughly experimentally motivated perspective (Chernov, 1997, 2001; Gadomski, 2007; Santamaría-Holek *et al.*, 2007).

5.3 LYSOZYME CRYSTAL VERSUS AGGREGATE FORMATION— COARSE-GRAINED APPROACH AT SUB-MESOSCALE BY MONTE CARLO SIMULATION

5.3.1 Coarse-Graining Procedure

Sub-mesoscopic simulations of the protein crystal growth are enormously demanding computationally. This situation is going on by virtue of very complex structure of single macromolecule, often consisted of hundreds of amino acids. Moreover, on account of complex composition of the growth milieu (water-based protein solution is a mixture of biomolecules, dissociated salts, pH stabilizers, and a number of precipitants conjuring up the growth process) and what follows very complex interactions including the following: (i) strong hydrophobic interactions (which is the driving force in protein crystallization process) between hydrophobic/hydrophilic groups of protein chain(s) and water molecules; (ii) electrostatic interactions between charged groups of protein chain(s) and dispersed in the solution salt's ions (Ducruix and Giegé, 1992).

For this reason, there is a need to jettison full atom or even amino acidal representation of the protein molecule on aid of advanced coarse-grained exposition of a single biomolecule.

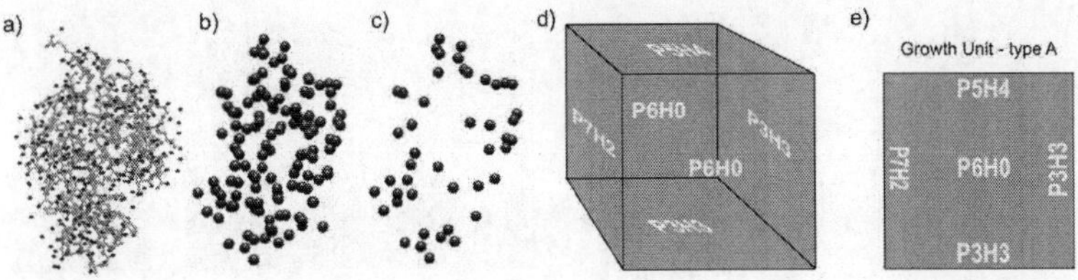

Figure 5.5 From the full atom representation of the lysozyme (PDB ID 193L) to the 2D growth unit: (a) Full atom representation of the lysozyme, (b) HP-amino acidic (hydrophobic–hydrophilic) representation — blue balls represent hydrophilic (P) whereas red balls represent hydrophobic (H) amino acids, (c) exterior part of the macromolecule, (d) 3D cubic representation. The net excess in a number of one of the monomer types determines the type of the wall (side) of the cubic box. (e) A 2D growth unit of A-type (reprinted with the permission from Siódmiak and Gadomski. Growing lysozyme crystals under variety of physicochemical conditions — a computer modelling. *Journal of Non-Crystalline Solids,* 354, pp. 4221–4226, © 2008, Elsevier). See also Color Insert.

In this picture, proteins are represented as rigid bodies (spheres or cubes) with specific active or inactive surface elements, where the degree of activation is determined by the local structure of the molecule under investigation (Lomakin *et al.*, 1999; Siódmiak and Gadomski, 2006). The coarse-graining procedure, which the task is emphasizing amphiphilic properties of the biomolecule's surface both in 3D and 2D depiction, for the lysozyme protein (Protein Data Bank [PDB] ID 193L) is presented in Fig. 5.5 (for more details, see following subsections and Siódmiak and Gadomski, 2006).

5.3.2 Mechanism of Growth of the Crystal's Surface

Most of the computer models of biomolecular crystal growth, in which the growth unit is prepared in the spirit of the coarse-graining procedure, reproduce/mimic the growth on one of the crystal's faces, e.g. (110) or (101) (Vekilov and Alexander, 2000). The growth mechanisms, almost all of them, more or less, base on the growth rules characteristic of the epitaxial growth where adatoms are deposited on a nucleus' surface.

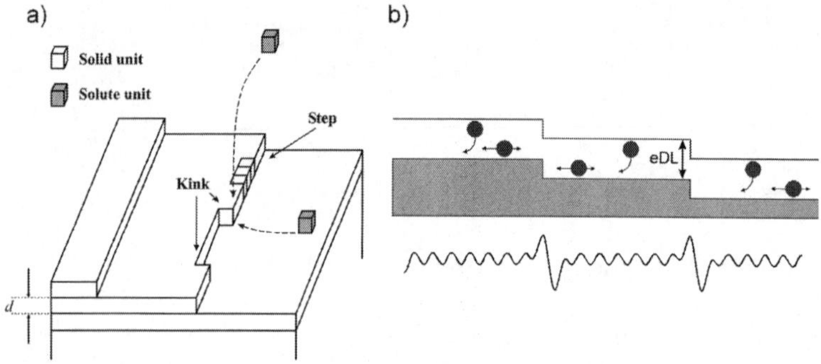

Figure 5.6 (a) The growth of a terrace. The kink positions are the most (from energetic point of view — occurrence of a Ehrlich–Schwoebel barrier) favorable places at the surface. Solute units aspire to be absorbed by the crystal. (b) Ehrlich–Schwoebel barrier experienced by the system at the edge of the terrace. eDL — electrostatic double layer (reprinted with permission from Liu *et al.* Prediction of crystal-growth morphology based on structural-analysis of the solid fluid interface. *Nature*, 374, 342–345, ©1995 Macmillan Publishers Ltd.).

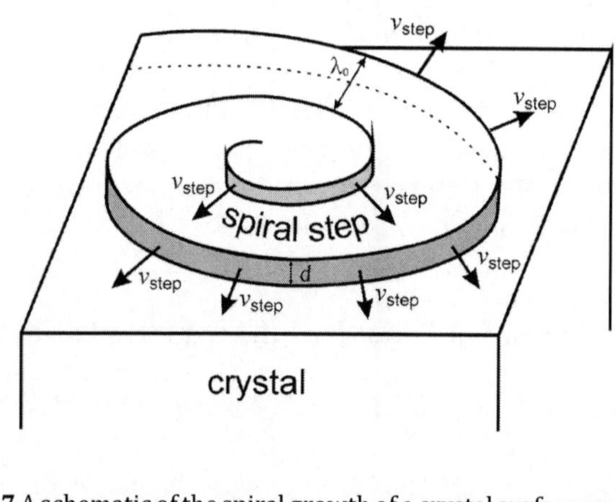

Figure 5.7 A schematic of the spiral growth of a crystal surface growing from solution — the BCF-type of growth (Burton *et al.*, 1951; Liu *et al.*, 1995). The spiral is conducted by the screw dislocation center and becomes roughly equivalent to concentric circuit steps of height d with a separation distance l0 (Żmija, 1987; Liu *et al.*, 1995) (reprinted with permission from Liu *et al.* Prediction of crystal-growth morphology based on structural-analysis of the solid fluid interface. *Nature*, 374, 342–345, ©1995 Macmillan Publishers Ltd.).

When the chemical reactants are controlled and the system parameters are set correctly, the depositing adatoms arrive at the surface with sufficient energy to move around on the surface and orient themselves to the nucleus/crystal arrangement of the already crystallized molecules (e.g., proteins). Thus, an epitaxial film deposited on a (110)-oriented surface of the nucleus will take on a (110) orientation.

It must be mentioned that adatoms which arrived at the surface can be absorbed by the crystal only in specific places, such as kinks (ledge) positions or point defects (Burton *et al.*, 1951), see Fig. 5.6a. One of the most frequent handicaps occurring on the crystal surface is a dislocation-type defect. In the case of dislocation-type defects, a long ledge is formed the height of which ranges from zero to one lattice constant's height.

Because the kink positions are the most (anticipating from energetic point of view occurrence of a Ehrlich–Schwoebel barrier [Schwoebel and Shipsey, 1966; Pimpinelli and Villain, 1998] see Fig. 5.6b) favorable places on the planer surface, adatoms (or macromolecules) aspire to be incorporated by the crystal in these places. This phenomenon occurs when the incoming material accumulate in the kink positions and so formed terrace grows (terrace's face propagates) to the direction perpendicular to its limbs. The addition of growth units along a dislocation growth step results in the formation of a hillock as shown in Fig. 5.7, and eventually the growth of the face as a whole, *cf.* Figs. 5.7 and 5.8.

The dislocation driven growth mechanism was well described by Burton, Cabrera, and Frank — BCF model (Burton *et al.*, 1951), as above consequently called thereafter the BCF mode of growing crystals. In this model the incorporation of adatoms into a surface site depend on many factors, including the adatom density, edge atom density, kink density, number of terraces, equilibrium vapor pressure, the impingement flux, the diffusivity of the impinging species, surface diffusity, and the binding energies of the adatoms, to mention but most important. The BCF mode assumes that the growth occurs on the surface with low concentration of dislocations and exchange of the material between the growing crystal and vapor phase is acceptable (adsorption and desorptions phenomena proceed simultaneously).

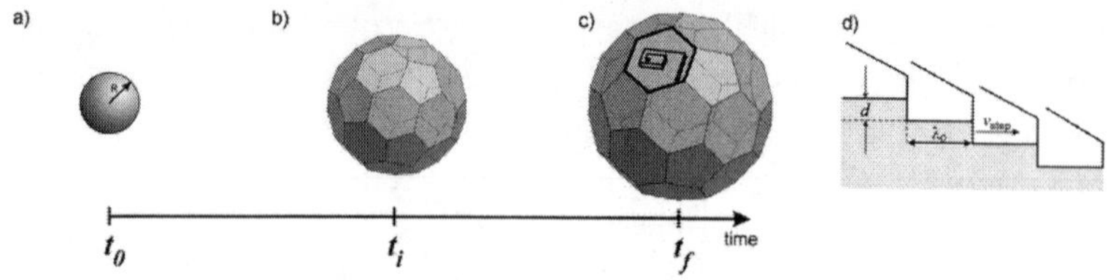

Figure 5.8 Formation of facets: (a) Spheroid-type crystal of radius *R*. (b) Flat crystalline surfaces formation. (c) Screw dislocation driven growth. (d) Section of the growing spiral.

5.3.3 From Spheroidal to Faceted Crystal Growth

In the deterministic as well as in the stochastic description one assumed that the crystal has spherical symmetry. Once a symmetrical spherical crystal grows (Libbrecht, 2005), see above, the expanding crystal develops facets because some crystalline surfaces accumulate material slower than others — this way an asymmetry is imposed on the system as whole. Condensing macromolecules are especially attracted to rounded surfaces (curvature effects) that are rough on atomic scales, because such areas present greater available molecular binding, Figs. 5.6a,b and 5.8a,b. The formation of facets — flat crystalline surfaces — is a nearly ubiquitous phenomenon in crystal growth. Faceting plays a major role in guiding the growth of protein crystals (Libbrecht, 2005). Molecularly flat regions — the facet surfaces — have fewer dangling chemical bonds and thus are less favorable attachment spots.

The microscopic growth process is characterized by the surface growth mechanisms of the protein crystal faces, including dislocation growth and 2D nucleation and growth. In dislocation driven growth (Pimpinelli and Villain, 1998), the growth occurs along screw dislocation defects on the crystal face. The addition of growth units along a dislocation growth step results in the formation of a hillock as shown in Fig. 5.8c, and eventually the growth of the face as a whole (Burton *et al.*, 1951). The growth rate of the faceted crystal is determined by the surface diffusivity of attracted macromolecules and the geometry of the steps, Fig. 5.8d. In this case the average growth rate of the spiral in the direction perpendicular to the surface could be identified with the growth rate of the spherical crystal (Siódmiak and Gadomski, 2006), *cf.* Figs. 5.7 and 5.8a,

$$V_{gr} \cong \frac{dR}{dt} = \frac{v_{step} \cdot d}{\lambda_0} \tag{5.29}$$

where v_{step} is a step propagation velocity parallel to the step, λ_0 is average distance between two steps, and d is a steps height; R stands (see above) for the radius of the spherical nucleus.

5.3.4 Computer Implementation of Spiral Growth

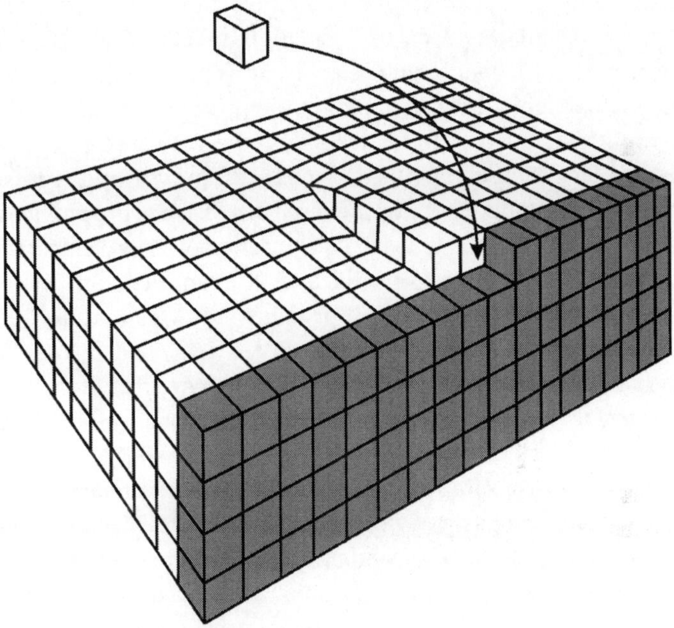

Figure 5.9 Lattice representation of the spiral formation. Gray layer represents the section of the spiral, *cf.* Fig. 5.8d.

Computer model of the lysozyme crystal growth is based on the implementation of the crystal's surface which can grow in a spiral way (strictly 3D structure) into a 2D lattice (Fig. 5.9). The Monte Carlo technique and the hydrophobic–polar (HP) approximation of the biopolymers (Larson *et al.*, 1985; Lau and Dill, 1989) are used to simulate the growth process. The most essential characteristic, properly defining this approximation, enables to use HP model with its onto-cube-walls projected (excess) HP properties, with a special emphasis placed on the outer "skin" region of the protein, which is the key feature of the proposed approximation to be applied in the present work for lysozyme biomolecules (Sióдmiak and Gadomski, 2006).

5.3.5 Growth Unit and Unit Cell Preparation

The simulation begins with an atomic-level analysis of the protein molecules under study, as taken from the RCSB PDB (Berman *et al.*, 2000, 2003), see Fig. 5.5a (Siódmiak and Gadomski, 2006). The HP (viz hydrophilic) representation of amino acid is applied to reduce a number of particles used in simulation, see Fig. 5.5b. Next, the exterior part of the macromolecule, Fig. 5c, is projected on the walls of the virtual cubic box, surrounding the protein, Fig. 5.5d. Finally, neglecting the most non-reactive sides, the 2D representation is proposed, Fig. 5.5e. Symbols on the sides of the growth unit represent degrees of hydrophobicity, i.e., the numbers of the hydrophobic and hydrophilic amino acids on each side. Interaction energies between each type of amino acids (HH, HP, and PP) can be taken from the well-known models of the lattice proteins (You-Quan *et al.*, 2005) and amount respectively to $E_{HH} = -2.3$, $E_{HP} = -1$, and $E_{pp} = 0$. The growth unit presented in Fig. 5.5e will be called A-type. B-type arises from the clockwise rotation of A-type unit by an angle of 90°. C-type arises from the clockwise rotation of B-type unit and D-type arises from the clockwise rotation of C-type unit, all of them again by 90° rotation. A unit cell, as a structure made up of the four 2D growth units, is a minimum energy configuration, see Fig. 5.10. The ABCD-type unit cell means that growth units are placed, in a spiral-like fashion, in the following positions: A — lower left, B — upper left, C — upper right, D — lower right. It can be seen that each growth unit is differently oriented in space. Moreover, unit cell has hydrophobic core and hydrophilic surface turned to the virtual solution.

5.3.6 The Growth of the Lattice Crystals

The growth procedure is similar to that specified by the well-known Frenkel–Kontorova (1938) and/or solid-on-solid (Abraham, 1986) models. In these models, virtual particles (adatoms or macromolecules) are placed above lattice points.

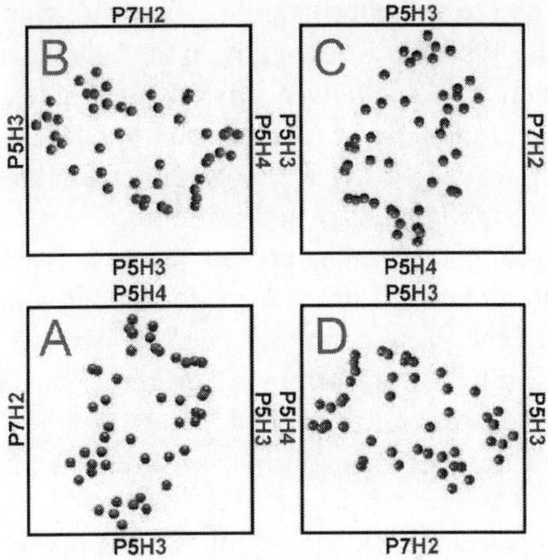

Figure 5.10 Non-Kossel-type (Chernov, 1997) ABCD unit cell (for growth unit type see text and Fig. 5.5c,e; growth units in ABCD unit cell: A — lower left, B — upper left, C — upper right, D — lower right). A hydrophobic core and also hydrophilic surface are visible (reprinted with the permission from Siódmiak and Gadomski. Growing lysozyme crystals under variety of physicochemical conditions — a computer modelling. *Journal of Non-Crystalline Solids,* 354, pp. 4221–4226, © 2008, Elsevier). See also Color Insert.

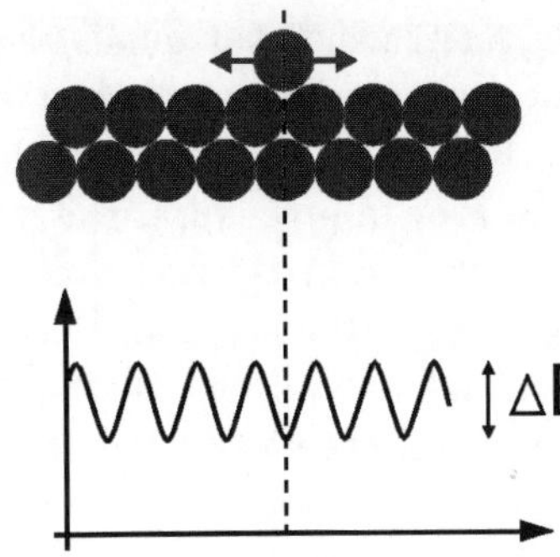

Figure 5.11 Energetic barrier between two neighboring points at the crystal's surface in the Frenkel–Kontorova (solid on solid) models. A commensurability of two neighboring crystal layers is revealed. See also Color Insert.

Particles hop to neighboring points. The direction of hoping is random and the hopping rate depends on the height of the energetic barrier between two neighboring positions (Fig. 5.11). To keep constant concentration of free particles on the surface instead of the particle which permanently enclose to the crystal new particle is deposited above random lattice point.

Another group of protein crystal models assumes that the probability of monomer attachment to the growing crystal is proportional to the protein volume fraction and the orientational factor representing the anisotropy of protein molecules (Kierzek *et al.*, 1999). The rate of detachment depended on the free energy of association of the given monomer in the lattice, as calculated from the buried surface area.

Also in Bratko and Blanch (2001), authors point at very important phenomenon accompanying the protein crystallization process viz refolding process which leads to increased aggregation. Their model is designed to examine the competition between intramolecular interactions leading to the native protein structure, and intermolecular association, resulting in the formation of aggregates of misfolded chains.

During the whole course of the presented simulation one observes exclusively the nucleus' surface and the growth units which are at the surface at each simulation step. In principle, molecules in the bulk are not taken into consideration what is consistent with the above presented mesoscopic modeling idea of the electrostatic double layer (see Fig. 5.6b) which surrounds the growing crystal and in which the most important processes seen in terms of surface formation take place (Fig. 5.1). This hint saves the processor time and also speeds up the growth procedure. The number of the free growth units which are present in a given moment at the surface is always considered to be proportional to the molecule concentration in a virtual solution ("bulk"), e.g., for bulk concentration of 6%, statistically 6% of the surface area is occupied by the growth units. In the beginning, a given number of the growth units, with their random orientations (A, B, C, or D), is placed by chance at the surface. There is no possibility that two movable growth units were put one over one because presented modeling is purposely confined to the one really nearest layer, only.

The movement direction of each growth unit placed at the surface is then randomly chosen: go to the left, go to the right, come off the surface, or stay at the same location. The movement probability, $p^{(\mathrm{m})}$, is consistent with the Metropolis Monte Carlo acceptance rule(s) (Metropolis *et al.*, 1953). This means that downhill transitions that lower the total energy are accepted with probability one and uphill transitions with probability proportional to the Boltzmann factor:

$$p^{(\mathrm{m})} = \begin{cases} 1 & \text{for} \quad \Delta E < 0 \\ e^{\frac{\Delta E}{k_B T}} & \text{for} \quad \Delta E > 0 \end{cases}. \tag{5.30}$$

There is a possibility that two movable growth units will come across each other. In this case, the movement is possible to occur after consideration of two components of the interaction energy: (i) interaction energy of the growth unit with the crystals surface and (ii) interaction energy of the growth unit with the neighboring movable growth unit(s). This way one can also observe some temporary aggregated forms of the growth units, e.g., 2D islands.

Every step is a combination of linear transition and rotation, i.e., the growth unit rolls over the crystal's (or, aggregate) surface. The growth unit becomes a part of the crystal when it is situated in a kink position and its orientation suits the other elements of the crystal unit cell (non-Kossel structural rule). Moreover, there is usually a substantial energy barrier. The reason for this barrier is that an exposed edge site has a higher potential than a corner (kink) site because a molecule sitting at an edge site has the smallest number of nearest neighbors. The resulting barrier is called the Ehrlich–Schwoebel barrier (Schwoebel and Shipsey, 1966; Pimpinelli and Villain, 1998), see Fig. 5.6b. Moreover, the kink site is the smallest potential site on the surface. In the case of energetically favorable move into a kink position and if the growth unit orientation does not suit the other elements of the crystal unit cell, the growth unit will not become a part of the crystal. It is because in the case of lysozyme and other molecular crystals molecules occupying different positions are identical, and thus, are characterized by the same chemical potential in solution, vapor, or melt. Correspondingly, the crystal should be characterized by one chemical potential averaged over the unit cell. The growth unit that would become a part of the crystal would still have a small chance to detach but only from the kink position. Detachment probability is proportional to the difference of interaction energies between the new and the old positions, namely $p^{-} \cong \exp(-\Delta E/k_B T)$. The new layer can start growing only when the bottom layer achieves a certain length equal to some λ_0. This is the minimum length for which the next twist of the spiral can arise, *cf.* Fig. 5.8 (Żmija, 1987).

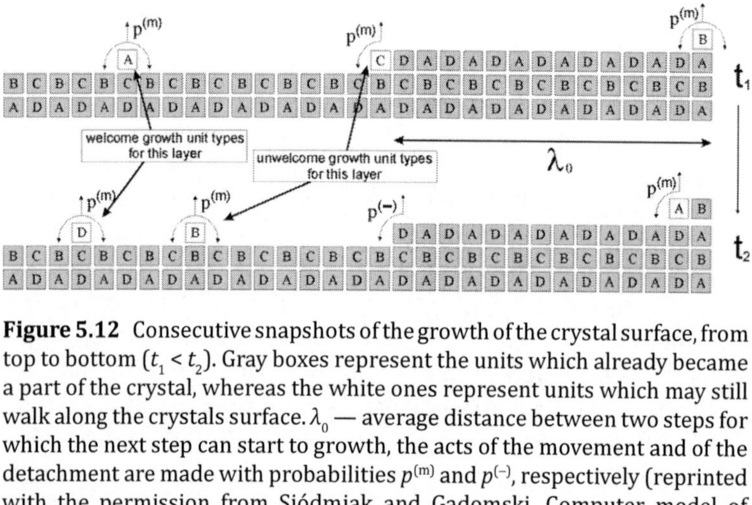

Figure 5.12 Consecutive snapshots of the growth of the crystal surface, from top to bottom ($t_1 < t_2$). Gray boxes represent the units which already became a part of the crystal, whereas the white ones represent units which may still walk along the crystals surface. λ_0 — average distance between two steps for which the next step can start to growth, the acts of the movement and of the detachment are made with probabilities $p^{(m)}$ and $p^{(-)}$, respectively (reprinted with the permission from Siódmiak and Gadomski. Computer model of biopolymer crystal growth and aggregation by addition of macromolecular units. A comparative study. *International Journal of Modern Physics C*, 17, pp. 1037–1053, © 2006, World Scientific Publishing Company).

Consecutive snapshots taken from applying the PDB-based computer model of the growing crystal surface, from top to bottom ($t_1 < t_2$), in the subsequent simulation steps are shown in Fig. 5.12 (Siódmiak and Gadomski, 2006). Gray boxes represent the units that already became a part of the crystal, whereas the white ones represent units being still able to walk along the crystal surface. Letters mean the growth unit type (orientation in the crystal's structure). Because the growth unit movement is site dependent, probabilities of the movements ($p^{(m)}$) and detachment ($p^{(-)}$) are also shown and are determined by the rules of the Metropolis Monte Carlo algorithm (Metropolis *et al.*, 1953). Moreover, the probability of growth unit motion, attachment, and detachment to/from the crystal surface are assumed to be proportional to the orientational factor representing the anisotropy of the molecule; for more details see Siódmiak and Gadomski (2006) and references therein.

5.3.7 Growth Rate and Morphological Phase Diagrams

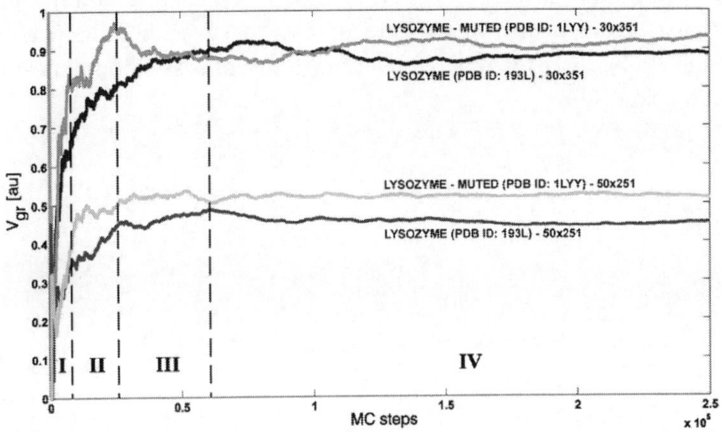

Figure 5.13 Certain most repeatable tendencies observed while examining the rate of the formation of mutant versus non-mutant lysozyme crystal, depending on the lattice size: $30(2R_M) \times 351(2R_M)$ and $50(2R_M) \times 251(2R_M)$ and temperature $T = 310$ K, *cf.* Table 5.1. Four regions are depicted on the plot: I — region of pre-nucleational effects, II — region of nucleus formation, III — region of non-stationary crystal growth, IV — region of stationary, nearly constant-tempo crystal growth. The as yet detected differences are mostly due to a finite lattice size effect. The curves have been obtained for the 6% protein concentration which is noticed to show optimal growing aggregation-oriented, thermodynamic–kinetic trends for both protein forms under study. The final values of the growth rates for the lattice size $30(2R_M) \times 351(2R_M)$ correspond to the appropriate values in Fig. 5.14 (reprinted with the permission from Siódmiak and Gadomski. Computer model of biopolymer crystal growth and aggregation by addition of macromolecular units. A comparative study. *International Journal of Modern Physics C*, 17, pp. 1037–1053, © 2006, World Scientific Publishing Company). See also Color Insert.

Figure 5.13 presents some repeatable tendencies observed while examining the tempo of the formation of mutant (PDB ID 193L) versus non-mutant (PDB ID 1LYY) lysozyme crystal (growth unit and unit cell preparation were made in the same way), depending on the lattice size: $30(2R_M) \times 351(2R_M)$ and $50(2R_M) \times 251(2R_M)$, where the first dimension is a lattice height, the second is a lattice width, and $2R_M$ is a lattice constant which is equivalent to the size of modeled growth unit.

The growth rate can be calculated using many experimental data viz: protein radius (physical as well as hydrophobic), protein concentration, surface and bulk diffusivity, temperature, and viscosity. Using these experimental data it is easy to convert MCsteps (*x*-axes in Fig. 5.13) to real (clock) time; for details see Siódmiak and Gadomski (2006).

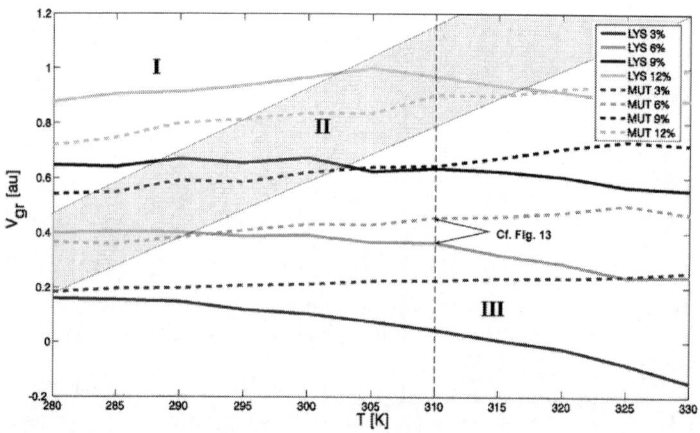

Figure 5.14 A comparative plot, revealing three regions of temperature behavior seen in terms of the tempo of the growing (poly)crystalline aggregates for the set of protein (non-mutant lysozyme) concentrations: 3, 6, 9, 12%. I — region below an optimal temperature for a given lysozyme concentration, II — crystallization window, III — region above an optimal temperature for a given lysozyme concentration. The data points for T = 310 K and concentration 6%, both for non-muted and muted lysozyme, correspond to the final values of the growth rates from Fig. 5.13 for lattice size $30(2R_M) \times 351(2R_M)$ (reprinted with the permission from Siódmiak and Gadomski, Computer model of biopolymer crystal growth and aggregation by addition of macromolecular units. A comparative study. *International Journal of Modern Physics C*, 17, pp. 1037–1053, © 2006, World Scientific Publishing Company). See also Color Insert.

The differences are attributed mostly to a finite lattice size effect. The curves have been obtained for the 6% protein concentration. Four regions are depicted on the plot: I — Region of pre-nucleational effects: At this stage the first growth units try to join the crystal surface. One can see that the growth rate fluctuations are very strong and the growth rate increases very rapidly with the increasing number of new terraces. II — Region of nucleus formation: At this stage the number of terraces tends to a constant value and depends on the lattice size. III — Region of non-stationary viz transient crystal growth: At the end of this stage the number of terraces is fixed and is approximately equal to h = (lattice width)/λ_0. IV — Region of stationary, close-constant-tempo crystal growth: a constant growth rate of lysozyme crystals is consistent with often experimentally observed behavior of lysozyme crystals grown from aqueous solution (Pusey *et al.*, 1986). Based on this result, one can conclude that the growth of lysozyme is controlled by the incorporation of lysozyme molecules to the surface of the crystal. In other words, the overall crystal formation has been mainly designed as an interface-controlled phenomenon. Investigations on several types of proteins (insulin, canavalin, and lysozyme) performed by other researchers (Monaco and Rosenberger, 1993; Forsythe and Pusey, 1994) have also suggested that the growth of such proteins is limited by interface-involved rather than by volume (in bulk) transport.

Figure 5.14 presents a comparative plot, roughly revealing three regions of temperature behavior seen in terms of the rate of growing of the (poly)crystalline aggregates for the set of protein (non-mutant lysozyme) concentrations: 3, 6, 9, 12%. For non-muted form of lysozyme (PDB ID 193L) several regions viz parametric windows (slots) have been found: I — Region below an optimal temperature for a given lysozyme concentration: In this range of temperature values the growth rate increases with increasing temperature. This behavior is associated with increasing mobility of the growth units but the temperature is still too small for dissolution effects to prevail. II — Crystallization slot, typically detectable by means of calculating the second virial coefficient from the state equation (Chernov, 1997): In this range of temperature values the growth rate conforms to an approximately constant value and the acts of attachment/detachment are balanced. A deflection point is only seen for the non-mutant form of the lysozyme; the mutant form, in turn, expresses a certain resistance against smooth crystallization and the deflection point is hardly visible for it (probably it could be seen above denaturation temperature). III — Region above an optimal temperature for a given lysozyme concentration: In this range of temperature the growth rate is decreasing because detachment starts dominating over the attachment. Outside the window for both variants one detects the protein aggregates that are going to be present for each of the concentration value chosen. The data points for $T = 310$ K and concentration 6%, both for non-muted and muted lysozyme, correspond to the final values of the growth rates from Fig. 5.13 for lattice size $30(2R_M) \times 351(2R_M)$ (Siódmiak and Gadomski, 2006).

A tendency of dissolving the crystal appears too — see the lowest curve (LYS 3%) with $V_{gr} < 0$ (due to enormous evaporation).

5.3.8 Crystal Growth Using Aggregates as the Growth Unit

A gob of theoretical as well as experimental works substantiate that the growth of (100) and (101) faces of tetragonal lysozyme crystal can be fully explained if not only monomeric growth units but also larger than monomer, e.g., tetramers, growth units are assumed (Durbin and Feher, 1991; Kierzek *et al.*, 1997, 2000; Nadarajah *et al.*, 1997; Forsythe *et al.*, 1999).

Tetrameric growth units can be obtained through some modification of the classical vapor diffusion method. A HEWL (hen egg white lysozyme) Langmuir–Blodgett thin film, prepared by a Langmuir–Schaeffer (LS) technique variation thereof (Ulman, 1991), was used as the template for the stimulation and rate increasing of lysozyme crystal growth (Pechkova and Nicolini, 2003). Monolayers of lysozyme were formed in a Langmuir Teflon trough by spreading 500 mL phosphate buffer (pH 6.5) solution with a HEWL concentration 4 mg/mL; 10 μM NaOH solution (pH 11) was used as a subphase. The subphase temperature was 22 °C. The formed film was compressed with a barrier speed of about 0.1 mm/s up to surface pressure of 18 mN/m and deposited by LS parallel shift technique onto the siliconazed cover glass slide. Obtained nanofilm was characterized by several experimental techniques such as circular dichroism, atomic force microscopy, and nanogravimetric methods (Pechkova and Nicolini, 2003) and utilized as a template for crystal growth in a common crystallization apparatus, placed in a contact with a protein solution drop. For other details, see Pechkova and Nicolini, 2001; Gadomski and Siódmiak, 2003a; and Pechkova *et al.*, 2005a,b.

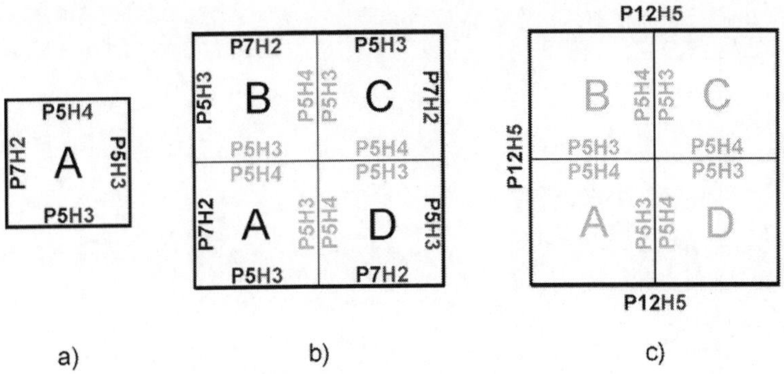

Figure 5.15 Different growth units: (a) monomeric growth unit of A-type, (b) energetically favorable (the lowest energy) configuration of four monomeric growth units, (c) generalized tetrameric growth unit. For details see Siódmiak *et al.* (2006). It can be seen that in the case of tetrameric growth unit (c) in comparison to monomeric growth unit (a) every side look the same what plays significant role in calculation of the movement probability (practically the movement is made without crossing of any energetic barrier because the geometrically smooth surface is also energetically smooth/flat) (Siódmiak *et al.* Computer model of a lysozyme crystal growth with/without nanotemplate — a comparison. *International Journal of Modern Physics C*, 17, pp. 1359–1366, © 2006, World Scientific Publishing Company). See also Color Insert.

Because in the computer modeling presented here a face of growth is not well defined (is it (100) or (101) face?) the incorporation of non-monomeric growth units was applied as a universal method (crystal's face independent) and is realized as follows. (The additional physical reason for the incorporation consists in some expected increase of the osmotic pressure of the solution.)

The tetrameric growth units look like the one drawn schematically in Fig. 5.15 (Siódmiak *et al.*, 2006). They are now established as both the units dispersed in the vicinity of crystal surface as well as the units finally incorporated by the crystal surface; see Siódmiak and Gadomski (2006) for details. Now, the growth unit is identical with unit cell. When compared to the monomers, they are allowed to perform their along-surface biased directionally Random Walk at practically zero energy cost, so that no energetic penalty is ascribed to such a rolling-over effect — this makes a basic difference between monomer- and tetramer-based crystal formation that one wished to perform.

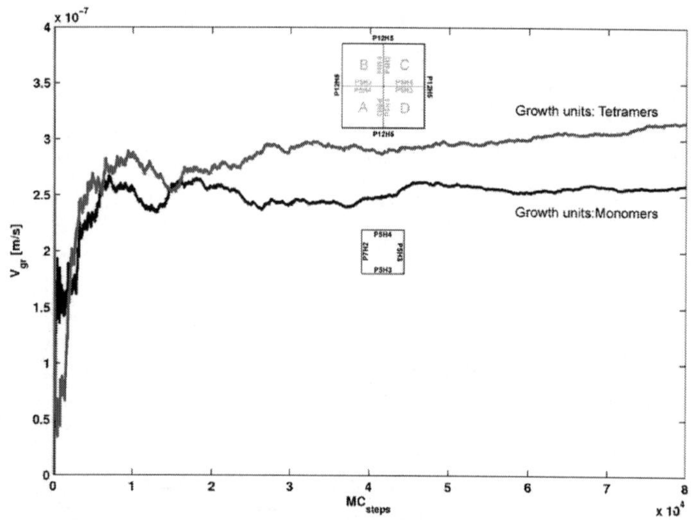

Figure 5.16 Simulated growth rate of lysozyme crystal obtained from the Langmuir–Blodgett–Schaefer-type template motivated technique (tetrameric growth units) — green (higher) line versus classical method (monomeric growth units) — blue (lower) line (Siódmiak *et al.* Computer model of a lysozyme crystal growth with/without nanotemplate — a comparison. *International Journal of Modern Physics C*, 17, pp. 1359–1366, © 2006, World Scientific Publishing Company). See also Color Insert.

The main result obtained from the comparative (monomeric vs. tetrameric growth unit) simulation is presented in Fig. 5.16. It can be seen that for a long time interval (here MCsteps) a difference in a value of the growth rate reaches 25%. This result can be mathematically proved after taking into consideration differences in the size, mass, and diffusity of monomeric as well as tetrameric growth unit; for details see Siódmiak *et al.* (2006).

From this comparison it follows that incorporating a tetrameric unit, Fig. 5.15, will result — without any special modification of the algorithm explored — in obtaining some acceleration mode of the under-confinement developing process (Pechkova and Nicolini, 2001; Pechkova *et al.*, 2005a,b). As shown in the experiment

(Pechkova *et al.*, 2005b), the acceleration can be quantified by a factor of 4/3 which follows in a very natural way from our type of modifying the recently introduced algorithm (Siódmiak and Gadomski, 2006). It should be clearly underlined that the obtained acceleration factor is only possible to occur when one assumed the non-monomeric growth units.

Let us also state clearly that the presented model does not account explicitly for a contribution of the structured water, detected experimentally in Pechkova and Nicolini (2003); a way of how to try to deal with it can perhaps be started from Gadomski and Łuczka (2000) by really incorporating the solute–solvent interaction conditions (Flory–Huggins parameter being involved) at a given temperature.

Some trials of incorporation into the model and some important — for protein crystal growth processes — physicochemical parameters are presented in following subsection.

5.3.9 Growing Lysozyme Crystals Under Variety of Physicochemical Conditions

The most important variables in the search of crystallization conditions are pH value, ionic strength, and temperature of the solution (Gilliland and Ladner, 1996; Pusey *et al.*, 2005). All parameters are strictly related to the virial coefficients, especially to the second virial coefficient (George *et al.*, 1997; Bonnete *et al.*, 1999; Chernov, 2003), B_{22}, which strongly influences the protein crystallization behavior (George and Wilson, 1994; George *et al.*, 1997). The temperature is directly used in presented model, i.e., it is used to calculate the probability of movement, $p^{(m)}$, and probability of the detachment of already crystallized growth unit, $p^{(-)}$. Because the pH and electrolyte concentration influence B_{22} which characterizes interaction energies between amino acids, interactions energies, E_{HH}, E_{HP}, and E_{PP}, must be changed to control and/or mimic pH or ionic strength of the solution (Rosenberger, 1996).

Because there is not unambiguous estimation of the interactions energies for model lattice proteins there are some possibilities of choosing various combinations of energy values. The chosen interaction energy values must be merely in agreement with a Miyazawa–Jernigan matrix (Miyazawa and Jernigan, 1985; Rosenberger, 1996;

Bahar and Jernigan, 1997). In this matrix effective inter-residue contact energies for protein in solution are estimated from the numbers of residue-residue contacts observed in crystal structures of the globular proteins. Splitting all 20 amino acids into two groups, H-hydrophobic and P-hydrophilic, it is possible to describe general rules of interaction energies within these two groups of residues. Effective inter-residue energies must fulfill the following inequalities (Miyazawa and Jernigan, 1985):

$$\begin{aligned} 2E_{\mathrm{HP}} &> E_{\mathrm{PP}} + E_{\mathrm{HH}} \\ E_{\mathrm{PP}} &\geq E_{\mathrm{HP}} > E_{\mathrm{HH}} \end{aligned} . \tag{5.31}$$

It is easy to check that inter-residue contact energies proposed by the well-known HP model, first introduced by Larson *et al.* (1985) for surfactant-containing systems, and then successfully applied by Lau and Dill (1989) to describe conformational behavior of lattice proteins, i.e., $E_{\mathrm{HH}} = -2.3$, $E_{\mathrm{HP}} = -1$, and $E_{\mathrm{PP}} = 0$, fulfill foregoing conditions. The use of the HP model implies automatically that the implicit solvent conditions have been assumed (Larson *et al.*, 1985; Lau and Dill, 1989).

Table 5.1 Number of the unit cell configurations (# UC types), unit cell's binding energies (UCe), and the growth rates for a various inter-residue energies (for the growth rate calculation, see Siódmiak and Gadomski, 2006).

E_{HH}	E_{HP}	E_{PP}	# UC types	UCe	V_{gr} [m/s]
−3.0	−1.0	0.0	1	−40.0	2.688×10^{-7}
−2.3	−1.0	0.0	1	−31.6	2.464×10^{-7}
−3.0	0.0	1.0	1	−20.0	2.697×10^{-7}
−3.0	1.0	1.0	16	−16.0	—
−3.0	0.0	1.5	1	−12.0	2.584×10^{-7}
−2.3	0.0	1.0	1	−11.6	1.946×10^{-7}
−3.0	1.0	1.5	1	−8.0	2.711×10^{-7}
−2.3	1.0	1.0	16	−7.6	—
−2.3	0.0	1.5	1	−3.6	2.712×10^{-7}
−2.3	1.0	1.5	1	0.4	2.191×10^{-7}

Some combinations of the interaction energies lead to the different values of the growth rate (for the growth rate calculation see Siódmiak and Gadomski, 2006) and to the various unit cell configurations, characterized by the same binding energy; see Table 5.1 and Fig. 5.17.

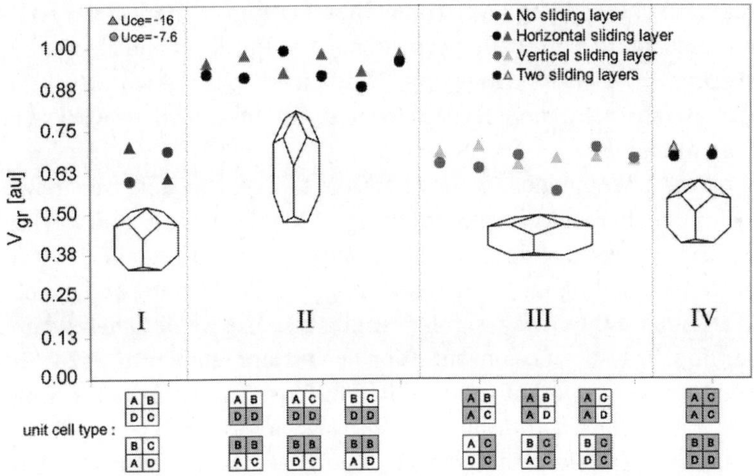

Figure 5.17 A (picturesque) diagram of the anisotropic growth. It can be seen that if the crystal is made up of unit cells which form the horizontal (or vertical) sliding layer (monolayer), group II (or group III), modeled crystal grows faster in a direction perpendicular to this layer and for a long-time period an elongated shape is observed. In the case when no sliding layer is formed, group I, or two horizontal or vertical sliding layers are formed, group IV, no elongated shape is admitted to the system. See also Color Insert.

Following can be seen that for unit cell binding energies: –7.6 and –16 two groups of velocities were obtained: "fast" and "normal" growth occurs. Analyzing these two groups characterized by different growth rates, it can be seen that in cases of $E_{HP} = E_{PP}$, 16 different types of the unit cell (different than ABCD), characterized by the same value of the binding energy, were obtained. Among them there are few for which crystals grow faster than for most types. Therefore, estimation of the mean value of the growth rate for a given set of "conditions" (when $E_{HP} = E_{PP}$) is not possible. In other cases, when $E_{HP} \neq E_{PP}$, one type of the unit cell was obtained, see Fig. 5.6, and the overall growth rate could be calculated.

A detailed (morphological) phase diagram showing the growth rate dependence on the unit cell configuration is shown in Fig. 5.17. In this diagram four groups of the unit cells can be distinguished. Two of them, I and IV, lead to rhomboid-like final shape of the crystal, while remaining two groups, II and III, lead to needle-like crystals. It can be seen that if the crystal is made up of unit cells which form the horizontal sliding layer (monolayer), group II, modeled crystal grows faster in a direction perpendicular to this layer and elongated shape is observed for a long time. The same results are seen in group III, where vertical monolayer is formed and the crystal grows faster horizontally.

This phenomenon occurs because the growth unit can move faster (smaller energetic cost of the movement) on the monolayer. In the case when no sliding layer is formed, group I, or two horizontal or vertical sliding layers are formed, group IV, no elongated shape is attained. By the way, similar results can also be obtained when varying the lysozyme concentration at constant temperature (24 °C) and fixed pH $\cong$ 4.6 but upon adding different amounts of the NaCl precipitant. Such a procedure leads to favoring either (101) or (110) planes grow predominantly in a tetragonal lysozyme crystal (Durbin and Feher, 1991).

This phenomenon can be explained also on the probabilistic grounds. The formation of the monolayer proceeds two times faster than the bilayer. In the case of monolayer, for example made up of growth units of type B, the probability that the B-type growth unit appears in the kink position, where the neighboring already crystallized growth unit is also of the B-type, is 1/4. In the case of the bilayer, for example made up of growth units of type A and C alternately, the probability that the A-type growth unit appears in the kink position, where the neighboring already crystallized growth unit is of the C-type, is 1/8. This probability is lower because of the kinked (edged) growth unit must be exactly of the C-type (not A-type) and the probability that in the case of A/C bilayer the kinked growth unit will be C-type is 1/2. Multiplying both probabilities one obtains exactly 1/8.

Extremely pH value and electrolyte concentration (ionic strength)-dependent shapes effect experimentally observed for the lysozyme crystals (Velev *et al.*, 1998) is shown in Fig. 5.18. For low values of the pH strictly 3D (rhomboid like) crystals are

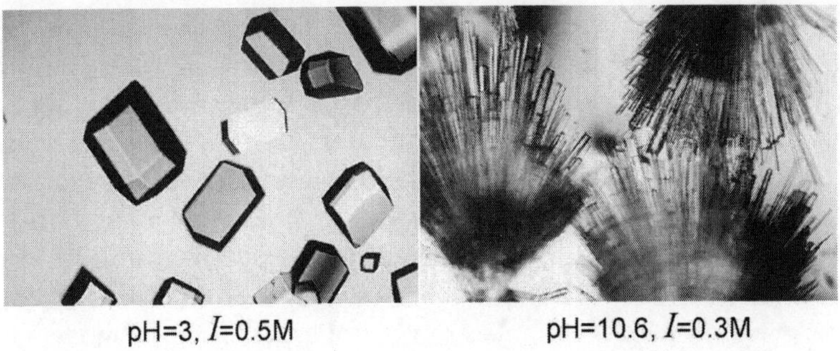

Figure 5.18 Extremely pH value and electrolyte concentration (ionic strength)-dependent shapes effect experimentally observed for the lysozyme crystals. For low values of the pH exactly 3D (rhomboid like) crystals are obtained, whereas for the high values of the pH significantly elongated (needle like) crystals are detected (reprinted with the permission from Velev *et al.* Protein interactions in solution characterized by light and neturon scattering. Comparison of lysozyme and chymotrypsinogen. *Biophysical Journal*, 75, 2682–2697, © 1998, Elsevier). See also Color Insert.

obtained, like in groups I and IV presented in Fig. 5.17, whereas for the high values of the pH significantly elongated (needle like) crystals are observed, like in groups II and III presented in Fig. 5.17.

5.4 CONCLUSIONS

Lysozyme crystal growth, and possibly other protein crystal growth types, such as the one of ribonuclease A, may be modeled by means of the sub-mesoscopic computer-aided approach proposed, in particular, when small aggregates, such as ordered tetramers coming from LS supports, are suitable for being incorporated in the crystal structure when the boundary conditions determine somehow a kinetic optimality and get ultimately the process accelerated markedly, thus playing a role of its intimate (chemical) catalysts

Non-equilibrium thermodynamic mechanism at a mesoscale appears to be amenable to model a wide class of growing processes, taking place in entropic environments, in which memory effects as well as boundary constraints are their basic growth-promoting landmarks toward hybrid (analytical plus computer aided) modeling. Therefore, a better comprehension of experimental material, and solid cooperation with experimenters, is needed for making substantial progress in this complex task, commonly named protein (lysozyme) crystal growth.

Sub-mesoscopic computer model in comparison to mesoscopic modeling gives us some opportunities of bringing out some details of complex structure of biomolecules. The knowledge of the structure of proteins is important since they often reveal anisotropic properties such as mechanical properties and surface charge distribution. A huge number of detailed information on particles, which reveal anisotropic properties, used in modeled system intrude very often on receiving some statistical information on mean values of thermodynamical parameter. Non-equilibrium thermodynamic gives us an opportunity to use an "averaged" particles in a model characterized by mean size, mean surface

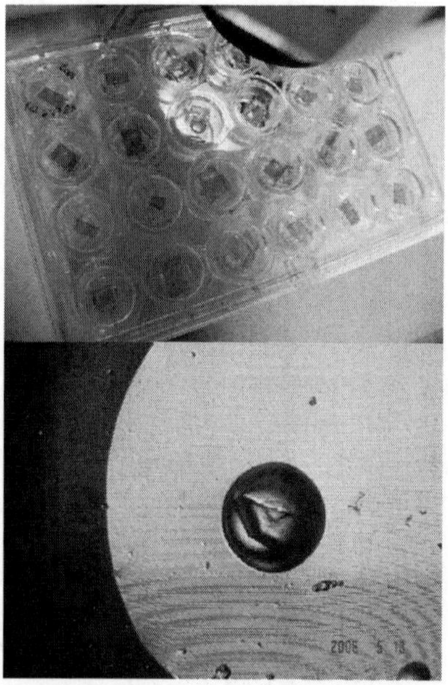

Figure 5.19 Lysozyme crystals grown out from orifices punched in kapton flat and porous membranes (Siwy *et al.*, 2003): top view, a setup consisting of 24 Limbro cells; bottom view, a magnified picture of a single crystal of lysozyme grown from the kapton membrane orifice (see, a paper "Could one effectively grow protein crystals: what comes out from theory and what from experiment?" at http://www.mischer-expo.de/achema/achema-2.htm, presented by A. Gadomski at Achema 2006, in Frankfurt on Main, and coauthored by C. B. Trame). See also Color Insert.

charge and its velocity what finally brings us closer to obtain an average size of the crystal/aggregate and the kinetics of the growth process in a long-time regime. The interchange of information between both presented types of modeling causes them to be in some (pronounced) sense complementary.

As a certain perspective, the hybrid model and some useful liaison with experimenters arise naturally for the future.

Acknowledgments

A.G. acknowledges a E.S.F. STOCHDYN traveling grant for visiting in April 2007 Nanoworld Institute of the University of Genova, where a part of the publishing material has been presented as an invited lecture during the XVIth E.L.B.A. Seminar, held at the Department of Biophysics, headed by Prof. Claudio Nicolini.

Let us cordially thank Dr. Ch. B. Trame (presently at Stanford University) for both fruitful discussions on kapton membrane-mediated lysozyme crystal formation as well as first experimental runs on the so-designed system, exemplified by Fig. 5.19 of this paper; see also http://www.mischer.com/achema/achema-2.htm (a lecture held at ACHEMA 2006 in Frankfurt on Main).

References

1. Abraham, D. B. (1986). *Phase Transitions and Critical Phenomena* 10 (eds. Domb, C. and Lebowitz, J. L.), p. 1. Academic, New York.
2. Arnaudov, L. N. and de Vries, R. (2005). Thermally induced fibrillar aggregation of hen egg white lysozyme. *Biophysical Journal*, **88**, pp. 515–526.
3. Bahar, I. and Jernigan, R. L. (1997). Inter-residue potentials in globular proteins and the dominance of highly specific hydrophilic interactions at close separation. *Journal of Molecular Biology*, **266**, pp. 195–214.
4. Basu, A. and Chowdhury, D. (2007). Modeling protein synthesis from a physicist's perspective: a toy model. *American Journal of Physics*, **75**, pp. 931–937.
5. Berman, H. M., Henrick, K. and Nakamura, H. (2003). Announcing the worldwide Protein Data Bank. *Nature Structural Biology*, **10**, p. 980.

6. Berman, H. M., Westbrook, J., Feng, Z., Gilliland, G., Bhat, T. N., Weissig, H., Shindyalov, I. N. and Bourne P. E. (2000). The Protein Data Bank. *Nucleic Acids Research*, **28**, pp. 235–242 (http://www.pdb.org).
7. Bonnete, F., Finet, S. and Tardieu, A. (1999). Second virial coefficient: variations with lysozyme crystallization conditions. *Journal of Crystal Growth*, **196**, pp. 403–414.
8. Booth, N. A., Chernov, A. A. and Vekilov, P. G. (2002). Step bunching in potassium dihydorgen phosphate crystal growth: phenomenology. *Journal of Material Research*, **17**, pp. 2059–2065.
9. Bratko, D. and Blanch, H. W. (2001). Competition between protein folding and aggregation: a three-dimensional lattice-model simulation. The Journal of Chemical Physics, **114**, pp. 561–569.
10. Burton, W. K., Cabrera, N. and Frank, F. C. (1951). The growth of crystals and the equilibrium structure of their surfaces. *Philosophical Transactions of the Royal Society London A*, **243,** pp. 299–358.
11. Chernov, A. A. (1997). Crystals built of biological macromolecules. *Physics Reports*, **288**, pp. 61–75.
12. Chernov, A. A. (2001). Crystal growth science between the centuries. *Journal of Materials Sciences — Materials in Electronics*, **12**, pp. 437–449.
13. Chernov, A. A. (2003). Protein crystals and their growth. *Journal of Structural Biology*, **142**, pp. 3–21.
14. Dell'Isola, F., Gouin, H. and Seppecher, P. (1995). Radius and surface tension of microscopic bubbles by 2nd gradient theory. *Comptes Rendus de l'Academie des Sciences Series II*, **320**, pp. 211–216.
15. Demeio, L. and Shizgal, B. (1993). Time dependent nucleation. II A semiclassical approach. *The Journal of Chemical Physics*, **98**, p. 5713.
16. Dobrev, D., Baur, D. and Neumann, R. (2005). Crystallization of lysozyme in pores of etched heavy-ion tracks. *Applied Physics A — Materials Science & Processing*, **80**, pp. 451–456.
17. Doi, M. and Edwards, S. F. (1986). *The Theory of Polymer Dynamics*. Clarendon Press, Oxford.
18. Drenth, J. (2002). *Principles of Protein X-ray Crystallography*. Springer, New York.
19. Ducruix, A. and Giegé, R. (1992). *Crystallization of Nucleic Acids and Proteins, A Practical Approach*. Oxford University Press, New York.

20. Durbin, S. D. and Feher, G. (1991). Simulation of lysozyme crystal growth by the Monte Carlo method. *Journal of Crystal Growth*, **110**, pp. 41–51.

21. Forsythe, E. L., Nadarajah, A. and Pusey, M. L. (1999). Growth of (101) faces of tetragonal lysozyme crystals: measured growth-rate trends. *Acta Crystallographica D — Biological Crystallography*, **55**, pp. 1005–1011.

22. Forsythe, E. and Pusey, M. L. (1994). The effects of temperature and NaCl concentration on tetragonal lysozyme face growth rates. *Journal of Crystal Growth*, **139**, pp. 89–94.

23. Frenkel, J. and Kontorova, T. (1938). The model of dislocation in solid body. *Zhurnal Eksperimental'noi i Teoreticheskoi Fiziki*, **8**, pp. 1340–1349.

24. Gadomski, A. (2000). Kinetic approach to the nucleation-and-growth phase transition in complex systems. *Nonlinear Phenomena in Complex Systems*, **3**, pp. 321–352.

25. Gadomski, A. (2007). Kinetic-thermodynamic effects accompanying model protein-like aggregation: the wave-like limit and beyond it. *Physica A — Statistical Mechanics and its Applications*, **373**, pp. 43–57.

26. Gadomski, A. and Ausloos, M. (2006). Part IV Condensed matter. In *The Logistic Map and the Route to Chaos — From the Beginnings to Modern Applications* (eds. Ausloos, M. and Dirickx, M.), p. 275. Springer, Berlin/Heidelberg.

27. Gadomski, A., Grzywna, Z. J. and Łuczka, J. (1993). The growing processes in diffusive and convective fields. *Chemical Engineering Science*, **48**, pp. 3713– 3721.

28. Gadomski, A. and Łuczka, J. (2000). On the kinetics of polymer crystallization: a possible mechanism. *Journal of Molecular Liquids*, **86**, pp. 237–247.

29. Gadomski, A. and Siódmiak, J. (2002). A kinetic model of protein crystal growth in mass convection regime. *Crystal Research and Technology*, **37**, pp. 281–291.

30. Gadomski, A. and Siódmiak, J. (2003a). A novel model of protein crystal growth: kinetic limits, length scales and the role of the double layer. *Croatica Chemica Acta*, **76**, pp. 129–136.

31. Gadomski, A. and Siódmiak, J. (2003b). *Proceedings of the 3rd International Symposium on Food Rheology and Structure.* Kerschensteiner Verlag GmbH, Zurich, p. 255.

32. Gadomski, A. and Siódmiak, J. (2005). On the elastic contribution to crystal growth in complex environments. *Physica Status Solidi B — Basic Solid State Physics*, **242**, pp. 538–549.

33. Gadomski, A., Siódmiak, J., Santamaría-Holek, I., Rubí, J. M. and Ausloos, M. (2005). Kinetics of growth process controlled by mass-convective fluctuations and finite-size curvature effects. *Acta Physica Polonica B*, **36**, pp. 1537–1559.

34. George, A., Chiang, Y., Guo, B., Arabshahi, A., Cai, Z. and Wilson, W. W. (1997). Second virial coefficient as predictor in protein crystal growth. *Methods in Enzymology*, **276**, pp. 100–110.

35. George, A. and Wilson, W. (1994). Predicting protein crystallization from a dilute-solution property. *Acta Crystallographica D — Biological Crystallography*, **50**, pp. 361–365.

36. Gilliland, G. L. and Ladner, J. E. (1996). Crystallization of biological macromolecules for X-ray diffraction studies. *Current Opinion in Structural Biology*, **6**, pp. 595–603.

37. Gsponer, J. and Vendruscolo, M. (2006). Theoretical approaches to protein aggregation. *Protein and Peptide Letters*, **13**, pp. 287–293.

38. Haas, C. and Drenth, J. (1999). Understanding protein crystallization on the basis of the phase diagram. *Journal of Crystal Growth*, **196**, pp. 388–394.

39. Haas, C. and Drenth, J. (2000). The interface between a protein crystal and an aqueous solution and its effects on nucleation and crystal growth. *Journal of Physical Chemistry B*, **104**, pp. 368–377.

40. Honigmann, R. (1958). *Equilibrium and Growth Forms of Crystals* — in German. Dr. Dietrich Steinkopff Verlag, Darmstadt.

41. Huang, K. (1963). *Statistical Mechanics*. John Wiley and Sons, Inc., New York.

42. Kashchiev, D. (2000). *Nucleation: Basic Theory with Applications*. Butterworth-Heinemann, Oxford.

43. Ke, S. C., DeLucas, L. J. and Harrison, J. G. (1998). Computer simulation of protein crystal growth using aggregates as the growth unit. *Journal of Physics D — Applied Physics*, **31**, pp. 1064–1070.

44. Kierzek, A. M., Pokarowski, P. and Zielenkiewiczm P. (1999). Lattice simulations of protein crystal formation. *Biophysical Chemistry*, **77**, pp. 123–137.

45. Kierzek, A. M., Pokarowski, P., Zielenkiewicz, P. (2000). Microscopic model of protein crystal growth. *Biophysical Chemistry*, **87**, pp. 43–61.

46. Kierzek, A. M., Wolf, W. M. and Zielenkiewicz, P. (1997). Simulations of nucleation and early growth stages of protein crystals. *Biophysical Journal*, **73**, pp. 571–580.

47. Larson, R. G., Scriven, L. E. and Davis, H. T. (1985). Monte Carlo simulation of model amphiphile-oil-water systems. *The Journal of Chemical Physics*, **83**, pp. 2411–2420.

48. Lau, K. F. and Dill, K. A. (1989). A lattice statistical mechanics model of the conformational and sequence spaces of proteins. *Macromolecules*, **22**, pp. 3986–3997.

49. Libbrecht, K. G. (2005). The physics of snow crystals. *Reports on Progress in Physics*, **68**, pp. 855–895.

50. Liu, X. Y., Boek, E. S., Briels, W. J. and Bennema, P. (1995). Prediction of crystal-growth morphology based on structural-analysis of the solid fluid interface. *Nature*, **374**, pp. 342–345.

51. Lomakin, A., Asherie, N. and Benedek, G. B. (1999). Aelotopic interactions of globular proteins. *Proceedings of the National Academy of Sciences USA*, **96**, pp. 9465–9468.

52. Metropolis, N., Rosembluth, A. W., Rosembluth, M. N., Teller, A. H. and Teller, E. (1953). Equation of state calculations by fast computing machines. *The Journal of Chemical Physics*, **21**, pp. 1087–1092.

53. Mezzenga, R., Schurtenberger, P., Burbidge, A. and Michel, M. (2005). Understanding foods as soft materials. *Nature Materials*, **4**, pp. 729–740.

54. Miyazawa, S. and Jernigan, R. L. (1985). Estimation of effective interresidue contact energies from protein crystal structures: quasi-chemical approximation. *Macromolecules*, **18**, pp. 534–552.

55. Monaco, L. A. and Rosenberger, F. (1993). Growth and etching kinetics of tetragonal lysozyme. *Journal of Crystal Growth*, **129**, pp. 465–484.

56. Nadarajah, A., Li, M. R. and Pusey, M. L. (1997). Growth mechanism of the (110) face of tetragonal lysozyme crystals. *Acta Crystallographica D — Biological Crystallography*, **53**, pp. 524–534.

57. Nanev, C. N. and Tsekova, D. (2000). Heterogeneous nucleation of hen-egg-white lysozyme — molecular approach. *Crystal Research and Technology*, **35**, 189–195.

58. Oxtoby, D. W. (2003). Crystal nucleation in simple and complex fluids. *Philosophical Transactions of the Royal Society of London Series A — Mathematical Physical and Engineering Sciences*, **361**, pp. 419–427.

59. Pechkova, E. and Nicolini, C. (2001). Accelerated protein crystal growth by protein thin film template. *Journal of Crystal Growth*, **2**, pp. 599–602.

60. Pechkova, E. and Nicolini, C. (2003). *Proteomics and Nanocrystallography*. Kluwer Academic Publishers, New York.

61. Pechkova, E., Fiordoro, S., Fontani, D. and Nicolini C. (2005b). Investigative crystal-growth mechanisms with and without LB template: protein transfer from LB to crystal. *Acta Crystallographica Section D — Biological Crystallography*, **61**, pp. 809–812.

62. Pechkova, E., Sivozhelezov, V., Tropiano, G., Fiordoro, S. and Nicolini C. (2005a). Comparison of lysozyme structures derived from thin-film-based and classical crystal. *Acta Crystallographica Section D — Biological Crystallography*, **61**, pp. 803–808.

63. Pimpinelli, D. K. and Villain, J. (1998). *Physics of Crystal Growth*. Cambridge University Press, New York.

64. Poon, W. C. K., Egelhaaf, S. U., Beales, P. A., Salonen, A. and Sawyer, L. (2000). Protein crystallization: scaling of charge and salt concentration in lysozyme solutions. *Journal of Physics Condensed Matter*, **12**, pp. L569–L574.

65. Pusey, M. L., Liu, Z. J., Tempel, W., Praissman, J., Lin, D., Wang, B. C., Gavira, J. A. and Ng, J. D. (2005). Life in the fast lane for protein crystallization and X-ray crystallography. *Progress in Biophysics & Molecular Biology*, **88**, pp. 359–386.

66. Pusey, M. L., Snyder, R. S. and Naumann, R. (1986). Protein crystal growth. Growth kinetics for tetragonal lysozyme crystals. *Journal of Biological Chemistry*, **261**, pp. 6524–6529.

67. Reguera, D., Rubí, J. M. and Perez-Madrid, A. (1998). Fokker–Planck equations for nucleation processes revisited. *Physica A — Statistical Mechanics and its Applications*, **259**, pp. 10–23.

68. Reif, F. (1967). *Statistical Physics*. McGraw-Hill, New York.

69. Rosenberger, F. (1996). Protein crystallization. *Journal of Crystal Growth*, **166**, pp. 40–54.

70. Santamaría-Holek, I., Rubí, J. M. and Gadomski, A. (2007). Thermokinetic approach of single particles and clusters involving

anomalous diffusion under viscoelastic response. *Journal of Physical Chemistry B*, **111**, pp. 2293–2298.

71. Schwoebel, R. L. and Shipsey, E. J. (1966). Step motion on crystal surfaces. *Journal of Applied Physics*, **37**, pp. 3682–3686.

72. Sear, R. P. and Warren, P. B. (2002). On the electrical double layer contribution to the interfacial tension of protein crystals. *Journal of Chemical Physics*, **117**, pp. 8074–8079.

73. Shen, M. Y., Davis, F. P. and Sali, A. (2005). The optimal size of a globular protein domain: a simple sphere-packing model. *Chemical Physics Letters*, **405**, pp. 224–228.

74. Siódmiak, J. and Gadomski, A. (2006). Computer model of biopolymer crystal growth and aggregation by addition of macromolecular units. A comparative study. *International Journal of Modern Physics C*, **17**, pp. 1037–1053.

75. Siódmiak, J. and Gadomski, A., (2008). Growing lysozyme crystals under variety of physicochemical conditions — a computer modelling. *Journal of Non-Crystalline Solids*, **354**, pp. 4221–4226.

76. Siódmiak, J., Gadomski, A., Pechkova, E. and Nicolini, C. (2006). Computer model of a lysozyme crystal growth with/without nanotemplate — a comparison. *International Journal of Modern Physics C*, **17**, pp. 1359–1366.

77. Siwy, Z., Apel, P., Baur, D., Dobrev, D. D., Korchev, Y. E., Neumann, R., Spohr, R., Trautmann, C. and Voss, K. (2003). Preparation of synthetic nanopores with transport properties analogous to biological channels. *Surface Science*, **532–535**, pp. 1061–1066.

78. Tanaka, H. and Araki, T. (2007). Spontaneous coarsening of a colloidal network driven by self-generated mechanical stress. *Europhysics Letters*, **79**, p. 58003.

79. Tanaka, H. and Nishikawa, Y. (2005). Viscoelastic phase separation of protein solution. *Physical Review Letters*, **9507**, p. 8103.

80. Tiller, W. A. (1986). Thermodynamic and kinetic considerations for crystal growth of complex molecules from solution. *Journal of Crystal Growth*, **76**, pp. 607–617.

81. Ulman, A. (1991). *An Introduction to Ultrathin Organic Films: From Langmuir-Blodgett to Self-Assembly*. Academic Press, Boston.

82. Vekilov, P. G. and Alexander, J. I. D. (2000). Dynamics of layer growth in protein crystallization. *Chemical Reviews*, **100**, pp. 2061–2089.

83. Velev, O. D., Kaler, E. W. and Lenhoff, A. M. (1998). Protein interactions in solution characterized by light and neturon scattering. Comparison of lysozyme and chymotrypsinogen. *Biophysical Journal*, **75**, pp. 2682–2697.

84. Wentzel, N. and Gunton, J. D. (2007). Liquid-liquid coexistence surface for lysozyme: role of salt type and salt concentration. *Journal of Physical Chemistry B*, **111**, pp. 1478–1481.

85. You-Quan, L., Yong-Yun, J., Jun-Wen, M. and Xiao-Wei, T. (2005). Medium effects on the selection of sequences folding into stable proteins in a simple model. *Physical Review E*, **72**, p. 021904.

86. Żmija, J. (1987). *Foundations of the Theory of Crystal Growth* — in Polish. PWN, Warszawa.

Chapter 6

OPTICAL TWEEZERS FOR TOUCHLESS SAMPLE MANIPULATION IN SYNCHROTRON RADIATION EXPERIMENTS

Silvia C. Santucci[a,b], Heinz Amenitsch[c], Dan Cojoc[a], and Christian Riekel[b]

[a]*CNR-INFM/ TASC, Area Science Park, SS14 Km 163.5, 34012 Basovizza (TS) Italy*

[b]*European Synchrotron Radiation Facility, B.P. 220, F-38043 Grenoble Cedex, France*

[c]*Austrian Academy of Sciences, Institute of Biophysics and Nanosystems Research, Schmiedlstr. 6, A-8042 Graz, Austria*

santucci@esrf.fr

We review progress in the development of optical tweezers as sample manipulators for third-generation synchrotron radiation sources. X-ray microbeam small- and wide-angle scattering experiments discussed in this chapter have been performed by using a custom optical tweezers setup. We also outline opportunities for future technical and scientific developments of optical tweezers at third-generation synchrotron radiation sources and X-ray free-electron lasers.

Synchrotron Radiation and Structural Proteomics
Edited by Eugenia Pechkova and Christian Riekel

www.panstanford.com

6.1 INTRODUCTION

The availability of intense microbeams and, more recently, nanobeams at third-generation synchrotron radiation (SR) sources allows probing local structures in hierarchically organized materials by small-angle and wide-angle X-ray scattering techniques (SAXS/WAXS) (Riekel, 2000; Paris, 2008; Riekel *et al.*, 2009). These techniques have been applied to a variety of synthetic polymers (Riekel and Davies, 2005) and biological materials (Paris, 2008; Riekel *et al.*, 2009) including protein microcrystals (Riekel *et al.*, 2005). The small beam dimensions have also allowed the reduction of the investigated object dimension. Indeed it has become possible studying membrane protein crystals of a few micrometer dimensions which are difficult to crystallize (Riekel *et al.*, 2005). Complementary imaging techniques such as X-ray microscopy and coherent X-ray diffraction imaging have contributed to an understanding of the structural properties of such systems (Huang *et al.*, 2009; Lima *et al.*, 2009).

It is well known in biological research that *single particle investigations* (cells and biomolecules) may provide a more accurate understanding of the relation between the macroscopic functionalities and the microscopic properties, with respect to the *averaged properties* measured over a large population (Carlo *et al.*, 2006; Knight, 2008). Such studies are best performed under *in vitro or in vivo* conditions to maintain macroscopic functionality. Although X-ray scattering techniques provide in principle excellent *in situ* capabilities for probing functional biological objects, radiation damage imposes usually experiments under cryoconditions (Lima *et al.*, 2009). A possible alternative is to distribute the radiation dose across multiple copies of an object as already done in protein microcrystallography (Riekel *et al.*, 2005). This calls for manipulation tools which allow both manipulating single objects under *in situ* conditions (e.g., a microfluidic cell) and provide sample replacement by handling a series of particles, one after another.

Sample manipulation is commonly performed through mechanical contact which may induce mechanical deformations.

Soft materials such as colloidal crystals, liposomes, or virus crystals will be particularly prone to such deformations. Indeed local phase transitions due to milli-Newton scale deformations have been observed by X-ray diffraction techniques in polymer fibers (Gourrier *et al.*, 2002). Moreover, the contact surface will also mask part of the sample with respect to the probing beam.

An alternative to mechanical manipulators are *optical tweezers* (OTs) which are based on the trapping capability of focused laser beams (Ashkin, 1970; Ashkin and Dziedzic, 1971; Ashkin *et al.*, 1986). Optical forces are three orders of magnitude lower than contact forces, being in the femto- to nano-Newton range (Wright *et al.*, 1994). OTs allow to manipulate, sort (MacDonald *et al.*, 2003), make interact (Dame *et al.*, 2006) separate (Dholakia *et al.*, 2007) biological and inorganic objects (Ashkin, 1980; Molloy and Padgett, 2002; Grier, 2003) of micron and submicron particles in fluidic environments, without mechanical contact and with complete spatial freedom and all-around visibility of the manipulated sample (Perch-Nielsen *et al.*, 2005). They also allow generating multiple traps, providing thus the possibility of distributing the radiation dose. An overview on the principles and numerous OT applications is given in the references (Lang and Block, 2003; Neuman and Block, 2004; Ashkin, 2006; Moffit *et al.*, 2008).

In the following, we will discuss aspects concerning the combination of OTs with X-ray microbeams at third-generation SR sources.

6.2 OPTICAL TWEEZING PRINCIPLES

The field of optical manipulations was pioneered by Ashkin and coworkers since the 1970s (Ashkin, 1970; Ashkin and Dziedzic, 1971; Ashkin *et al.*, 1986). In these groundbreaking experiments, the radiation pressure of a continuous-wave laser was used to accelerate, levitate, and trap (three-dimensional [3D] stable confinement) dielectric particles of micrometric size, in water and in air. Optical trapping was subsequently applied to atoms (Chu *et al.*, 1986), live bacteria, viruses (Ashkin *et al.*, 1987; Ashkin and Dziedzic, 1987), and laser atom cooling (Chu, 1998).

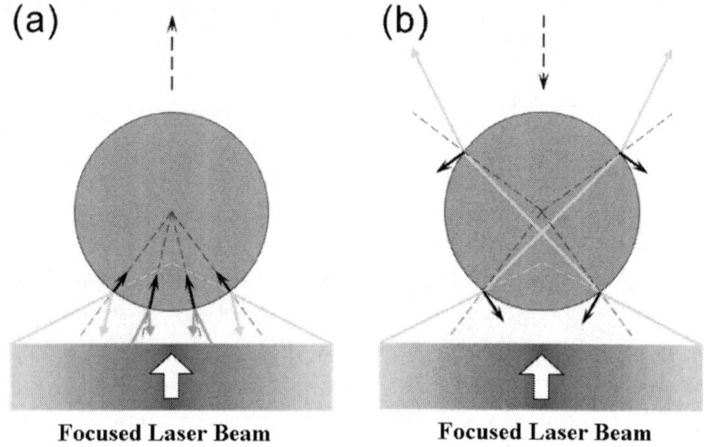

Figure 6.1 Schematic principle of optical trapping. A large particle of high refractive index is illuminated by a TEM_{00} Gaussian intensity profile laser beam focused by a high numerical aperture (NA) microscope objective. (a) The ideal case of a perfectly reflecting particle. Light rays (colored lines) are reflected at the surface. Dashed lines identify the normal directions. Change in light direction implies the transfer of linear momentum (black short arrows) to the particle. Due to the Gaussian profile, the particle is constrained along the laser axis and pushed in the propagation direction (long dashed arrow). (b) The ideal case of a perfectly transmitting particle. At every change of direction, linear momentum is transferred to the sphere (short black arrows). The high NA of the microscope objective makes the total momentum transfer oriented in the opposite verse of the light propagation (long dashed arrow). In practice, when 3D stable trapping is achieved both effects occur and balance one the other. See also Color Insert.

Optical trapping is due to the ability of light to exert pressure on matter (Maxwell, 1873). A schematic explanation in terms of geometrical optics can be given for particles with high refractive index (higher than the refractive index of the surrounding medium) and whose radius is much larger than the incident wavelength (the so-called Mie regime) (Lang and Block, 2003; Neuman and Block, 2004; Ashkin, 2006).[a] The trapping force can be schematized as resulting from the equilibrium between the *scattering force* and

[a] For a comprehensive treatment of optical trapping we recommend the recent book of A. Ashkin (Ashkin, 2006).

the *gradient force*. The scattering force is related to reflection and absorption and is proportional to the intensity of light. In the ideal case of a perfectly reflecting microsphere (Fig. 6.1a), light reflected at the surface transfers linear momentum to the particle (black arrows) and pushes it in the propagation direction of light. The gradient force is related to refraction, is proportional to the spatial intensity gradient of the incident light (beam profile), and acts in the direction of the gradient. In the case of a perfectly refracting microsphere (Fig. 6.1b), the momentum transferred by light to the particle (black arrows) acts oppositely to light propagation. In practice, when these opposite effects and the gravity are counterbalanced, optical trapping is realized.

The physical realization of trapping requires a TEM_{00} laser with Gaussian intensity profile providing a strong symmetric gradient that (i) constrains the particle along the axis of light propagation (restoring transversal constraint) and (ii) makes the gradient force to equilibrate the scattering push in the propagation direction of light (axial constraint).

To maximize the intensity gradient, microscope objectives with high numerical aperture (NA > 1) are used. For NA < 1, 3D confinement cannot be achieved as the particle will only be pushed along the beam direction. Stable trapping can, however, be realized by two counter-propagating lasers (Ashkin, 1970). Low refractive index particles, such as micron-sized bubbles, are pushed away from the laser beam (Ashkin, 1970). Such particles have nevertheless been trapped and handled by the doughnut profile of Laguerre–Gaussian beams (Gahagan and Swartzlander, 1996; Prentice *et al.*, 2004; Garbin *et al.*, 2005). While adequately shaped, low refractive index objects can be trapped by Gaussian beams, microfabricated ring-like objects have been trapped by using the radiation pressure exerted on their inner walls (Higurashi *et al.*, 1995). Even though trapped objects are usually dielectric and transparent such as glass beads (Ashkin, 1970) and oil droplets (Ashkin and Dziedzic, 1977) it has nevertheless been shown that highly absorbing samples, such as gold (Svoboda and Block, 1994a; Furukawa and Yamaguchi, 1998), silver (O'Neil and Padgett, 2000; Bosanac *et al.*, 2008), and copper (Gu and Morrish, 2002), can be trapped in the center of a doughnut profile laser (He *et al.*, 1995).

The optical trapping of a particle can be understood as confinement on the bottom of a potential well. The trapping force associated with this well is the force that has to be exerted to release the particle. A particle immersed in a medium of smaller refractive index n feels a trapping force (F) proportional to the incident laser power (W) (Ashkin, 1997):

$$F = Q\frac{nW}{c}, \tag{6.1}$$

where c is the speed of light in vacuum, and Q is the trapping efficiency factor that depends on the size, shape, material, and position of the particle with respect to the spatial profile of the beam. The optical forces are of the order of 1–100 pN (Svoboda and Block, 1994b). In order that optical forces are competitive with gravity, the mass of the sample has to be lower than 100 μg, which implies for silica beads a maximum diameter of ~100 μm. An asymmetric shape of the particle will be determining for its orientation in the trap. Cylindrical particles align their longer dimension along laser direction (Gauthier, 1997; Gauthier *et al.*, 1999). To orientate large asymmetrical particles in a controlled way, the use of multiple traps is necessary. Due to momentum transfer, a large birefringent sample will rotate around the focus of a polarized laser, unless it is stopped by supplementary traps. This has been demonstrated for objects such as lysozyme crystals (Singer *et al.*, 2004, 2006) and microfabricated devices (Higurashi *et al.*, 2001; Maruo and Inoue, 2007). Laser polarization rotation is actually a method to control the rotation of the sample (Friese *et al.*, 1998a,b,c; Padgett and Leach, 2008). The minimum displacement is commonly defined by the resolution power of the imaging system through the diffraction limit of 200–300 nm. By means of quantitative time-resolved imaging studies 10 nm of lateral resolution can be achieved (Cocker and Grier, 1996). The maximum displacement will depend on the working distance (WD) of the microscope objective, along the vertical axis, and on the field of view in the horizontal plane. A typical range is between 100 μm and 1 mm. Force ranges and length scales typical of optical manipulation fit, therefore, very well with SR experiments at the micrometer to nanometer scale, which provides a strong motivation for developing OTs at SR beamlines.

6.3 OTS AS SAMPLE MANIPULATORS

OTs belong to a larger class of devices that exert forces via light, magnetic (Gosse and Croquette, 2002), electric (Stuke *et al.*, 2005), acoustic (Wolf *et al.*, 2009), or other fields. In Table 6.1, different manipulators are compared, according to the force exerted on the samples, minimum controllable displacement and stiffness.

Table 6.1 Overview of manipulation techniques (adapted from Bustamante *et al.*, 2000). Cantilevers and microneedles exert physical contact while the other techniques exert an external field. Methods are compared according to maximum and minimum force exerted ($F_{max,}$ F_{min}), minimum controllable displacement (X), and stiffness.

Method	F_{min}–F_{max} [N]	X [m]	Stiffness [N/m]	Applications	Practical advantages
Cantilevers	10^{-11}–10^{-7}	10^{-10}	10^{-3}–10^{2}	Proteins Polysaccharides Bond strength	High spatial resolution; commercially available
Microneedles	10^{-12}–10^{-10}	10^{-9}	10^{-6}–1	Myosin motor force DNA/titin strength	Good operator control; soft spring constant
Flow field	10^{-13}–10^{-9}	10^{-8}	n.a.	DNA dynamics RNA polymerase	Rapid buffer exchange; simplicity of design
Magnetic field	10^{-14}–10^{-11}	10^{-8}	n.a.	DNA entropic elasticity Topoisomerase activity	Specificity to magnets; ability to induce torque
Photon field	10^{-13}–10^{-10}	10^{-9}	10^{-10}–10^{-3}	Protein motors Protein unfolding	Specific manipulation; high force resolution

Among all manipulation methods, OTs provide the lowest manipulation forces (pN range) that match well with forces intrinsic to biological samples. Moreover, as OTs with a micrometer-sized laser foci are used in an aqueous environment, a great variety of microfluidics applications become possible which would not be feasible by means of contact manipulators. An example for this is the selection of single objects among a collection in a fluidic reservoir while changing the flux of the surrounding solution (Ozkan *et al.*, 2003; Kovac and Voldman, 2007).

The combination of optical tweezing and microfluidics is also a unique tool in the study of single-cell dynamics (biochemical functionality as a function of stress) (Eriksson *et al.*, 2007). Since the trapping forces may be calibrated as a function of the displacement with high resolution, OTs have been used to characterize the correlations between forces and structural conformation in systems like molecular motors (Rice *et al.*, 2003) and DNA (Smith *et al.*, 1996; Huisstede et *al.*, 2007).

OTs are mainly limited by the accessible temperature range. The freezing of water and the dependence of the overall optical properties on temperature and pressure imply that the major part of OT applications is performed *in situ* at ambient conditions. Moreover, Brownian motion might be antagonistic to trapping, since thermal agitation could make the trapped particle escape from the optical well. In principle, a stiffer trap can be created by rising up the laser power (see Eq. 6.1), but an excessive laser power might cause radiation damage. Alternatively, an anti-Brownian electrokinetic method has been developed (Cohen and Moerner, 2006) to trap individual protein molecules in free solution. Care has to be taken in choosing the wavelength, but for most classes of systems there is a window of minimum radiation damage (Neuman and Block, 2004). Another way to avoid the laser damage is the use of functionalized beads; the sample of interest is covalently bonded to a functionalized bead which is manipulated by the laser. This approach has been adopted, for example, to stretch single molecules of DNA (Smith *et al.*, 1996).

6.4 OT SETUPS

The fundamental components of an OT setup are a laser, a microscope objective with NA ≥ 1, an imaging system, and steering/expanding optics components. A spatial light modulator (SLM) can be inserted between the laser and the microscope objective, to split the laser beam and generate multiple traps. The sample/s is/are held between two cover slips or in a capillary, in aqueous environment. A schematic diagram of an OT setup is shown in Fig. 6.2. To optimize the trapping of biological samples, an infrared laser with wavelength $\lambda \sim 1$ μm is advantageous in limiting the radiation damage (Vorobjev *et al.*, 1993; Neuman *et al.*, 1999; Gross, 2003). In non-biological applications a visible wavelength laser can also be used, provided it has a Gaussian intensity profile and it can deliver adequate power with enough stability. Typical setting of laser power lies between 100 mW to 1 W, depending on the desired trapping force (see Eq. 6.1) and on the desired number of the traps. The force generated is typically 1 pN per 10 mW of power that reaches the sample (Svoboda and Block, 1994b).

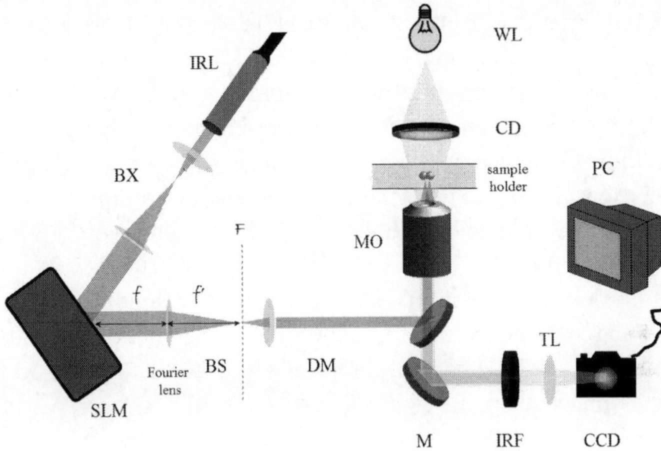

Figure 6.2 Scheme of an OT setup. In red: laser path; in yellow: optical imaging path. **Laser path**. The beam of a collimated infrared laser (IRL) is expanded by a telescopic couple of lenses (BX) and impinges onto a spatial light modulator (SLM). By means of diffractive optical elements, the SLM modulates the phase of the laser wavefront to split the beam and obtain the desired configuration in the Fourier plane (F). The modulated beam is transferred by the second lens of the telescope BS and the microscope objective (MO) in the focal plane of the MO (Dufresne and Grier, 1998). **Imaging path**. White light (WL) illuminates the sample through the condenser, passes through the dichroic mirror (DM), and is reflected by the mirror M onto a CCD through a tube lens (TL). To obtain clean images, a filter (IRF) blocks spurious IR reflections. See also Color Insert.

The microscope objective is the crucial component for the overall trapping efficiency, according to its transmittance and NA. Other determining parameters are the WD and the immersion medium. A short WD, even though compatible with a high NA, will impose constraints to the trapping depth. Immersion objectives provide the high NA required for trapping at a reasonably long WD (100 μm to 1 mm). Several methods have been developed for multiple trapping. The use of *diffractive optics elements* (DOEs) that generate multiple traps by modulating the phase of the laser wave front into multiple spots (diffraction orders) is very efficient and largely adopted nowadays (Cojoc *et al.*, 2004b; Spalding *et al.*, 2008). The DOE is calculated in real time using Fourier transform applied to a configuration of spots which represents the array of traps (Cojoc *et al.*, 2004c). For static applications, DOEs can be fabricated by lithographic techniques (Cojoc *et al.*, 2006). For dynamic multiple trapping, the use of SLM for computer generation of DOEs has nowadays become customary. In SLMs the phase modulation is performed by liquid crystal (nematic or ferroelectric) arrays of pixels which are remotely controlled via a PC. A sequence of DOEs sent to the SLM may be used to translate, rotate, push, and pull the trapped particle(s) (Cojoc *et al.*, 2004a), at a typical refresh frequency of 60 Hz.

Multiple trap arrays can also be generated by rapidly scanning a single beam between several traps (time sharing method) by means of acousto-optic deflectors (Guilford *et al.*, 2004). Compared to the SLM, the time sharing method is faster but limited to the generation of planar only arrays of traps.

6.5 OTS FOR X-RAY EXPERIMENTS

A typical OT setup is built by an inverted microscope modified in such a way that a laser beam can be introduced into the optical path before the objective. The microscope itself provides trapping, imaging, and sample manipulation. When projecting the integration of an OT setup into a synchrotron beamline as a sample manipulator, the geometrical environment will impose strict space constraints and possibly also weight limitations. Even more important are considerations about simultaneous transparency of the sample cell to the laser and to the X-rays. Optimization of scattering geometry, absorption, and background signal will call for a setup, built ad hoc for the experiments envisaged.

The first custom OT setup dedicated to X-ray scattering applications has been developed at the INFM-TASC laboratory in Trieste for SAXS/WAXS applications at the ESRF-ID13 beamline (Cojoc *et al.*, 2007). The experience acquired with this first generation of experiments has led to the development of a highly modular OT setup capable of adaptation to specific demands of different beamlines (Santucci *et al.*, 2010). In Fig. 6.3, a possible configuration for this new setup is shown. The setup is built as two modules, one to perform single trapping and rotations by means of a $\lambda/4$ plate, and a second module to perform multiple trapping by using an SLM.

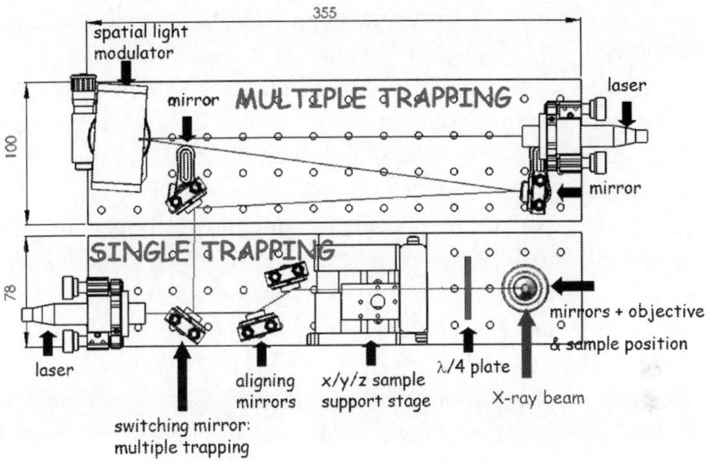

Figure 6.3 Top view of a project of the optical manipulator at the ESRF. Quotes are expressed in millimeters. The setup is organized in two separate modules, one for single trapping and rotation by means of a $\lambda/4$ plate of a single particle at a time, and an additive module for multiple trapping. The pink arrow corresponds to the incoming X-ray beam. The imaging path is not shown in this view. See also Color Insert.

The sample holder is a glass capillary of 100 μm squared section, and wall thickness of 50 μm. Round capillaries should be avoided since they introduce a spherical lens effect in the infrared beam. A water immersion microscope objective with NA = 1 and WD = 1 mm provides a good compromise between axial stability of the trap and the available working space. Due to its portability, the sample can be aligned offline with respect to the optical trap and then to be put on the beamline sample stage. The capillary is handled by means of motorized stages. The whole system is integrated into the ESRF beamline control software (SPEC, *Certified Scientific Software*) to synchronize optical manipulations and beamline operations.

For applications requiring extreme miniaturization, the trapping laser could be delivered by optical fibers (Liberale *et al.*, 2007).

6.6 SELECTED SAXS EXPERIMENTS INVOLVING OTS

6.1.1 Manipulation of Liposomes

Liposomes generally consist of spherical self-closed single or multi-bilayers of amphiphilic lipids (e.g., lecithin) (Fig. 6.4a). Microemulsions of liposomes are used for nano-encapsulation of drugs and vaccines (Caruthers *et al.*, 2007). This is just one example for the many technological applications of liposomes (Schramm, 2008).

Their encapsulation properties depend strongly on the characteristics of the bilayer structure (thickness and curvature), phase behavior, and on their propensity for pore formation, bilayer fusion, and formation of non-bilayer structures which is best studied for single particles.

At present, the trapping of clusters of tens of liposomes has been demonstrated for multilamellar palmitoyl-oleoyl-phosphatidylethanolamine (POPE) liposomes (Cojoc *et al.*, 2007).

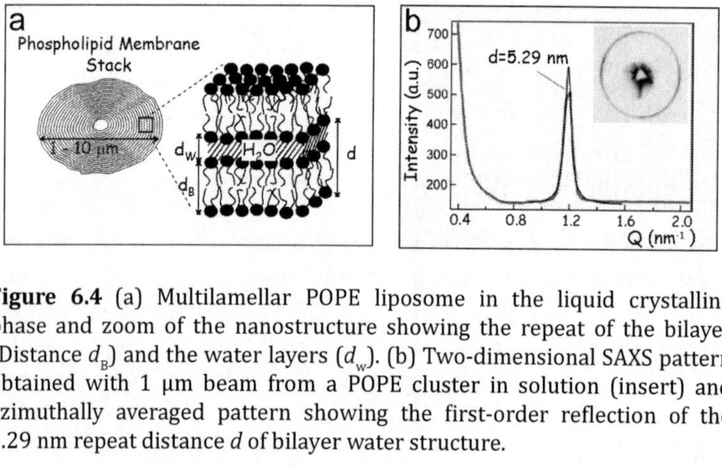

Figure 6.4 (a) Multilamellar POPE liposome in the liquid crystalline phase and zoom of the nanostructure showing the repeat of the bilayer (Distance d_B) and the water layers (d_w). (b) Two-dimensional SAXS pattern obtained with 1 μm beam from a POPE cluster in solution (insert) and azimuthally averaged pattern showing the first-order reflection of the 5.29 nm repeat distance *d* of bilayer water structure.

The micro-SAXS pattern from a trapped cluster shows a strong first-order reflection of the liquid crystalline phase (Fig. 6.4b). The trapped cluster could be localized in the capillary by raster micro-SAXS experiments with a 1 μm beam.

The composite pattern based on "pixels" corresponding to the area under the first-order reflection reveals clearly the location of the cluster (Fig. 6.5). Extrapolation of the peak to background ratio of 10 liposomes to the case of a single one seems to suggest that experiments on single liposome particles should be feasible which would allow studying interactions between two liposome particles. As a first step toward this goal it has been demonstrated that two POPE clusters can be trapped in two optical traps generated by a SLM (Fig. 6.1a) and studied by raster micro-SAXS (Cojoc *et al.*, 2007).

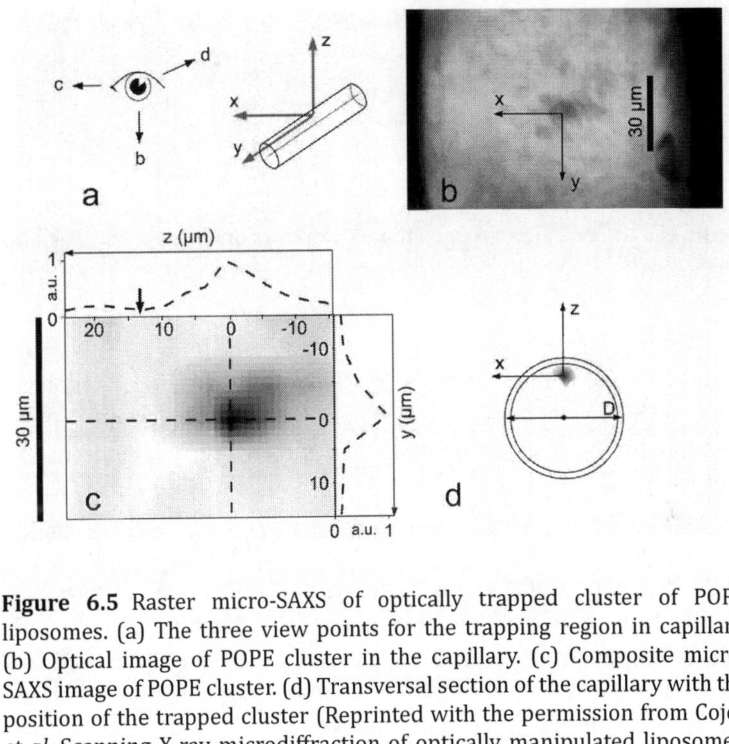

Figure 6.5 Raster micro-SAXS of optically trapped cluster of POPE liposomes. (a) The three view points for the trapping region in capillary. (b) Optical image of POPE cluster in the capillary. (c) Composite micro-SAXS image of POPE cluster. (d) Transversal section of the capillary with the position of the trapped cluster (Reprinted with the permission from Cojoc *et al.* Scanning X-ray microdiffraction of optically manipulated liposomes. *Applied Physics Letters*, 91, art. no. 234107, © 2007, American Institute of Physics).

6.1.2 Manipulation of Starch Granules

Potato starch granules form an onion-type shell structure which is composed of amorphous and semicrystalline layers with interspersed amorphous growth layers (Fig. 6.6).

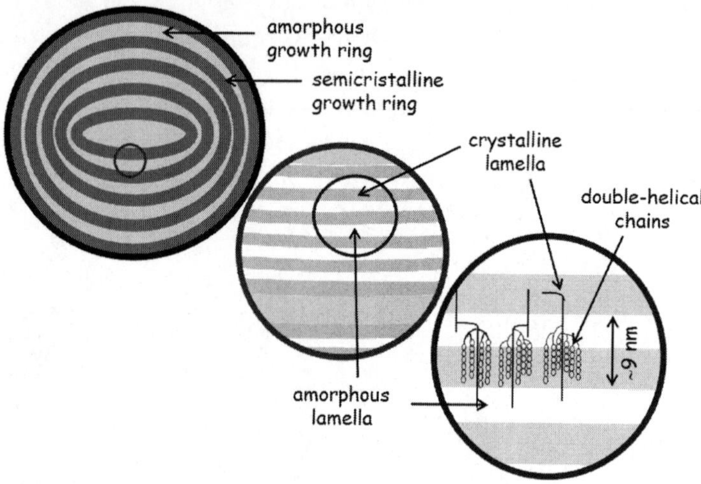

Figure 6.6 Hierarchical organization of starch granule (see text). See also Color Insert.

The layers consist mainly of branched amylopectin polysaccharide chains with a small fraction of linear amylose chains (Buléon *et al.*, 1998). The complex starch granule organization has been attributed to a model of branched clusters (Robin *et al.*, 1974; French, 1984) composed of double helical polysaccharide chains (Imberty and Perez, 1988; Popov *et al.*, 2009) (Fig. 6.6). The clusters form the crystalline lamella while the branching points and stems of the clusters form the amorphous lamella. The succession of amorphous and crystalline lamella with an average repeat of about 9 nm gives rise to a characteristic SAXS peak (Waigh *et al.*, 2000). The local starch organization has been studied *in situ* on single granules fixed on supports like grids (Buléon *et al.*, 1997) or glass fibers (Lemke *et al.*, 2004) by raster micro-SAXS/WAXS techniques. Starch granules of about 20 μm diameter have now been kept in free suspension in a liquid cell by OTs (Cojoc *et al.*, 2010) (Fig. 6.7a). Multiple optical traps generated by an SLM were required to stabilize particles of such a size. A raster SAXS image taken with a 4 × 4 μm^2 mesh is shown in Fig. 6.7b. Data collection times of 50 ms/pattern were required to limit radiation damage.

The OT setup offers the advantage that the surrounding liquid can be easily changed which could be used for studying the onset of enzymatic degradation of starch and thus to shed more light on microstructural changes accompanying the biochemistry of degradation (Buléon *et al.*, 1998).

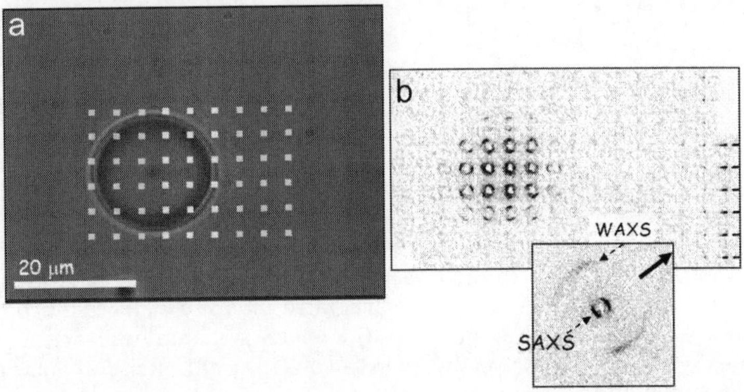

Figure 6.7 (a) Potato starch granule kept by multiple optical traps in a liquid cell. The position of the beam during the raster SAXS/WAXS experiment on a 4 μm × 4 μm mesh is indicated. (b) Composite image from raster scan and single SAXS/WAXS pattern. In the pattern, the solid arrow indicates the local fiber axis. The pixels of the composite scan contain only the SAXS peak. The horizontal streaks on the right border of the raster scan are due the capillary wall.

6.1.3 Aggregation of Mesoporous Materials

Optical traps due to the attractive forces can locally increase the concentration of particles, proteins, etc., and therefore act as nucleation center (Singer *et al.*, 2004) or lead to aggregation and eventually sedimentation. To demonstrate this effect a study on micrometer-sized mesoporous particles having an internal hexagonal mesostructure with a lattice constant of 3.57 nm was performed. This class of nanostructured materials are used for filtration, hosts for advanced materials like molecular quantum wires, adsorbents, and so forth (Soler-Illia, 2002). A typical diffraction pattern of an aggregate obtained with a 1 μm large X-ray beam is shown in Fig. 6.8a.

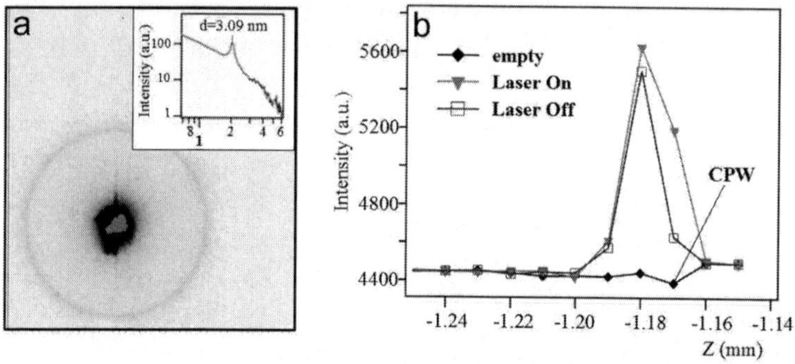

Figure 6.8 Scanning micro-SAXS of optically induced sedimentation of mesoporous aggregates. (a) Diffraction pattern and azimuthally integrated intensity of one aggregate showing the first-order of a hexagonal mesoporous structure. (b) The results of vertical micro-SAXS raster scan along the bottom of the capillary (integrated intensity of the diffraction pattern versus scanning position Z): before the trap was activated (empty), after the laser trap was switched on (laser on), after the trap was deactivated (laser off). CPW indicates roughly the position of the capillary wall. See also Color Insert.

A small quantity of silica particles was dispersed in water, which gives rise to a non-detectable signal with micro-SAXS. By means of the optical trap it was possible to aggregate several single particles, which sedimented at the bottom of the capillary and could be investigated by scanning micro-SAXS. The results of the scans are displayed in Fig. 8b, which demonstrates that the particles could be detected at the bottom of the capillary only after the activation of the optical trap and that they remain even fairly stable after the deactivation of the trap.

6.7 FUTURE TRENDS

The results presented above have given a first insight into the fascinating possibilities of using OT techniques for studying single supra-molecular assemblies with micron- and submicron-sized SR beams. We will in the following mention some possible trends in instrumentation, soft matter, and biology applications but OT techniques will find in our opinion, multiple other applications at third-generation SR sources and at the upcoming generation of ultra-high brilliance SR sources, such as X-ray free-electron laser (XFEL) and electron recoverage linac sources (Bilderback *et al.*, 2005).

6.7.1 OT Integration

OT techniques will become in the future only one amongst multiple sample manipulation, imaging, and characterization tools, integrated in the modular platform of a micro-SR beamline. This integration will rapidly advance as the size of SR beams and objects is reduced toward the nanoscale. One can also expect that laser beams will not only serve for particle manipulation but also for micro-Raman and microdissection applications. Indeed micro-Raman and micro-SAXS/WAXS have already been integrated to combine local molecular with local structural information (Davies *et al.*, 2006, 2009). Laser microdissection has been used in the laboratory for cutting down a high-performance polymer fiber (Davies *et al.*, 2007) and an insect sensor (Seidel *et al.*, 2008) to shapes which can be readily studied by transmission micro-SAXS/WAXS. In certain cases one might wish, however, to perform microdissection *in situ* on fragile systems, for example, to separate a more perfect domain from a protein microcrystal.

6.7.2 Particle shaping and assembly

Techniques for studying the shaping and assembly of particles are evolving to smaller volumes as SR beam sizes are reduced. This allows reducing the amount of sample needed (e.g., for "precious" proteins) but also the time scale of mixing processes. Indeed, conformational changes of proteins in solution (e.g., cytochrome C) can be studied by stopped-flow techniques with a time resolution limited to a few milliseconds (Panine *et al.*, 2006). The corresponding SAXS experiments with microfluidic mixing cells (Pollack *et al.*, 1999; Akiyama *et al.*, 2002) have reached sub-millisecond time resolution for turbulent mixing (Akiyama *et al.*, 2002). OT techniques could simplify the design of microfluidic systems by shaping the flow field, for example, to increase the turbulence. One could, however, also envisage to control the position of microdrops on surfaces by multiple traps and to initiate protein folding or aggregation by merging microdrops. A related technique without active positional control is based on the mixing of ballistic microdrops in flight (Graceffa *et al.*, 2009). The possibilities for studying particle assembly processes are multiple reaching from building mesoscopic matter, proteins, colloids, and soft crystals. Indeed the growth of optical lattices (Fournier *et al.*, 2005) and proteins crystals (Singer *et al.*, 2004) by OT techniques has already been demonstrated.

6.7.3 Particle Fractionation

Techniques for mass and size separation of macromolecules are usually based on the flow of a solution through a mesoscopic lattice made of silica beads or a gel. The integration of fractionation techniques into a beamline increases the sample throughput for protein solution scattering experiments. Indeed high-performance liquid chromatography has recently been integrated in a SAXS beamline for solution scattering studies of protein fractions (David and Perez, 2009). A further volume reduction requires microfluidic techniques. Fractionating techniques have been explored for colloidal transport in an OT array (Korda et al., 2002). A mesoscopic optical lattice could be adapted to the expected mass fractions by redefining the trap geometry by SLM techniques. A micro-SAXS raster scan at the exit of the post array, normal to the flow direction, should allow to record SAXS patterns for the different fractions.

6.7.4 Serial Crystallography

Protein microcrystallography has been developed for crystals which are difficult or impossible to grow to larger dimensions (Riekel *et al.*, 2005). Monochromatic data collection requires rotating protein crystals through a sequence of small angular steps (1–2°) to cover the full reciprocal space. Focused microbeams imply a particularly high local radiation dose. The radiation damage limit (Henderson, 1990) requires spreading the total data collection dose over a larger crystal or several microcrystals. For the smallest microcrystals (≤10 μm^3 volume) the radiation damage limit will be reached for only one X-ray "shot" per crystal. Covering the total reciprocal lattice will therefore require serial crystallography on multiple microcrystal copies. OT techniques would allow sorting microcrystals into a 3D array. By defining the beam position at one position of the array one could sequentially collect data from the microcrystals with an exchange of microcrystals in the beam by SLM holographic techniques. For monochromatic data collection a small rotation would be necessary for every microcrystal. Rotational motion could, however, be continuous if a fast detector captures the integrated patterns for small, random angular increments. Increasing the X-ray band-pass to a few percent would allow keeping the crystals fixed during data collection (Riekel *et al.*, 2005). Serial crystallography in combination with OT techniques is of particular interest for microcrystals which cannot be easily cryofrozen such as certain virus crystals.

6.7.5 Single Object Coherent Scattering

Multiple traps could also find applications in XFEL coherent X-ray diffraction imaging (Schroer *et al.*, 2008; Thibault *et al.*, 2008; Chapman, 2009) experiments. The principal difference to third-generation SR experiments will be the smaller size of objects, such as single cells or protein molecules, accessible only at XFEL sources. As objects will be destroyed by radiation damage after each XFEL pulse (Neutze *et al.*, 2000), serial data collection of objects at different orientations will be necessary to reconstruct the 3D structural information. Current XFEL strategies aim at using particle beams by acoustically modulated jets (Chapman, 2009). Multitrapping OT techniques could become an attractive alternative as its sample sorting and positioning features are an advantage as compared to the random sample selection and positioning features involved in particle streams.

Acknowledgments

The authors are grateful for support in the development of an advanced OT system through the ESRF. S. C. Santucci acknowledges the funding of her position through the ESRF and the FP6 SAXIER project. We acknowledge the input of E. Di Fabrizio (Università della Magna Graecia, BIONEM Lab, Catanzaro, Italy) in starting this project. E. Ferrari (TASC), V. Garbin (TASC), B. Marmiroli (IBN), B. Sartori (IBN), M. Rappolt (IBN), M. Burghammer

(ESRF), E. Di Cola (ESRF) and Sebastian Schoeder (ESRF) contributed very efficiently to this project. We acknowledge technical support by C. Morello (Elettra), F. Salvador (TASC), and L. Lardiere (ESRF). We are grateful to F. Sette (ESRF) and G. Ruocco (INFM-SOFT) for continuous support.

References

1. Akiyama, S., Takahashi, S., Kimura, T., Ishimori, K., Morishima, I., Nishikawa, Y. and Fujisawa, T. (2002). Conformational landscape of cytochrome c folding studied by microsecond-resolved small-angle x-ray scattering. *Proceedings of the National Academy of Sciences* USA, **99**, 1329–1334.
2. Ashkin, A. (1970). Acceleration and trapping of particles by radiation pressure. *Physical Review Letters*, **24**, 156–159.
3. Ashkin, A. (1980). Applications of laser-radiation pressure. *Science*, **210**, pp. 1081–1099.
4. Ashkin, A. (1997). Forces of a single-beam gradient laser trap on a dielectric sphere in the ray optics regime. *Methods in Cell Biology*, **55**, pp. 1–27.
5. Ashkin, A. (2006). *Optical Trapping and Manipulation of Neutral Particles Using Lasers*. World Scientific Publishing Co. Pte. Ltd., Singapore, pp. 1–940.
6. Ashkin, A. and Dziedzic, J. M. (1971). Optical levitation by radiation pressure. *Applied Physics Letters*, **19**, pp. 283–285.
7. Ashkin, A. and Dziedzic, J. M. (1977). Observation of resonances in the radiation pressure on dielectric spheres. *Physical Review Letters*, **38**, pp. 1351–1354.
8. Ashkin, A. and Dziedzic, J. M. (1987). Optical trapping and manipulation of viruses and bacteria. *Science*, **235**, pp. 1517–1520.
9. Ashkin, A., Dziedzic, J. M., Bjorkholm, J. E. and Chu, S. (1986). Observation of a single-beam gradient force optical trap for dielectric particles. *Optics Letters*, **11**, pp. 288–290.
10. Ashkin, A., Dziedzic, J. M. and Yamane, T. (1987). Optical trapping and manipulation of single cells using infrared laser beams. *Nature*, **330**, pp. 769–771.

11. Bilderback, D. H., Elleaume, P. and Weckert. E. (2005). Review of third and next 2 generation synchrotron light sources. *Journal of Physics* B, **38**, pp. S773–S797.

12. Bosanac, L., Aabo, T., Bendix, P. M. and Oddershede, L. B. (2008). Efficient optical trapping and visualization of silver nanoparticles. *Nano Letters*, **8**, pp. 1486–1491.

13. Buléon, A., Colonna, P., Planchot, V. and Ball, S. (1998). Starch granules: structure and biosynthesis. *International Journal of Biological Macromolecules*, **23**, pp. 85–112.

14. Buléon, A., Pontoire, B., Riekel, C., Chanzy, H., Helbert, W. and Vuong, R. (1997). Crystalline ultrastructure of starch granules revealed by synchrotron radiation microdiffraction mapping. *Macromolecules*, **30**, pp. 3952–3954.

15. Bustamante, C., Macosko, J. C. and Wuite, G. J. L. (2000). Grabbing the cat by the tail: manipulating molecules one by one. *Nature*, **1**, pp. 130–136.

16. Carlo, D. D. and Lee., L. P. (2006). Single-cell enzyme concentrations, kinetics, and inhibition analysis using high-density hydrodynamic cell isolation arrays. *Analytical Chemistry*, **78**, pp. 4925–4930.

17. Caruthers, S. D., Wickline, S. A. and Lanza, G. M., (2007). Nanotechnological applications in medicine. *Current Opinion in Biotechnology*, **18**, pp. 26–30.

18. Chapman, H. N. (2009). X-ray imaging beyond the limits. *Nature Materials*, **8**, pp. 299–301.

19. Chu, S. (1998). Nobel Lecture: the manipulation of neutral particles. *Review of Modern Physics*, **70**, pp. 685–706.

20. Chu, S., Bjorkholm, J. E., Ashkin, A. and Cable, A. (1986). Experimental observation of optically trapped atoms. *Physical Review Letters*, **57**, pp. 314–317.

21. Cocker, J. C. and Grier, D. G. (1996). Methods of digital video microscopy for colloidal studies. *Journal of Colloid and Interface Science*, **179**, pp. 298–310.

22. Cohen, A. E. and Moerner, W. E. (2006). Suppressing Brownian motion of individual biomolecules in solution. *Proceedings of the National Academy of Sciences USA*, **103**, pp. 4362–4365.

23. Cojoc, D., Amenitsch, H., Ferrari E., Santucci S. C., Sartori B., Rappolt, M., Marmiroli, B., Burghammer, M. and Riekel, C. (2010). Local X-ray

structural analysis of optically manipulated biological samples. *Applied Physics Letters*, accepted for publication.

24. Cojoc, D., Cabrini, S., Ferrari, E., Malureanu, R., Danailov, M. B. and Di Fabrizio, E. (2004a). Dynamic multiple optical trapping by means of diffractive optical elements. *Microelectronic Engineering*, **73–74**, pp. 927–932.

25. Cojoc, D., Emiliani, V., Ferrari, E., Garbin, V. and Di Fabrizio, E. (2004b). Dynamic multiple beads manipulation on x-y-z directions. In *Optical Trapping and Optical Micromanipulation* (eds. Dholakia, K. and Spalding, G. C.), Vol. 5514, pp. 126–136. SPIE, Denver, USA.

26. Cojoc, D., Ferrari, E., Cabrini, S., Malureanu, R., Danailov, M. B., Carpentiero, A., Prasciolu, M., Kumar, R., Businaro, L. and Di Fabrizio, E. (2004c). Design and implementation of optical tweezers arrays using diffractive optical elements. In *Sixth International Conference on Correlation Optics* (ed. Angelsky, O. V.), Proceedings of SPIE, Vol. 5477, pp. 281–292, SPIE Press, Bellingham WA.

27. Cojoc, D., Ferrari, E., Garbin, V., Cabrini, S., Carpentiero, A., Prasciolu, M., Businaro, L., Burchard, K. and Di Fabrizio, E. (2006). Wave front engineering by means of diffractive optical elements for applications in microscopy. In *Seventh International Conference on Correlation Optics* (ed. Angelsky, O. V.), Vol. 6254, art. no. 625416.

28. Cojoc, D., Ferrari, E., Garbin, V., Di Fabrizio, E., Amenitsch, H, Rappolt, M., Sartori, B., Laggner, P., Burghammer, M. and Riekel, C. (2007). Scanning X-ray micro-diffraction of optically manipulated liposomes. *Applied Physics Letters*, **91**, art. no. 234107.

29. Dame, R. T., Noom, M. C. and Wuite, G. J. L. (2006). Bacterial chromatin organization by H-NS protein unravelled using dual DNA manipulation. *Nature*, **444**, pp. 387–390.

30. David, G. and Perez, X. (2009). Combined sampler robot and high-performance liquid chromatography: a fully automated system for biological small-angle X-ray scattering experiments at the Synchrotron SOLEIL SWING beamline. *Journal of Applied Crystallography*, **42**, pp. 892–900.

31. Davies, R., Burghammer, M. and Riekel, C. (2006). Simultaneous mRaman and mXRD: probing the deformation of a single high-performance fiber. *Macromolecules*, **39**, pp. 4834–4840.

32. Davies, R., Burghammer, M. and Riekel, C. (2007). On-axis X-ray microscopy of the crystallographic microstructure within a single high-performance fibre. *Macromolecules*, **40**, pp. 5038–5046.

33. Davies, R., Burghammer, M. and Riekel, C. (2009). A combined microRaman and microdiffraction setup at the European Synchrotron Radiation Facility ID13 beamline. *Journal of Synchrotron Radiation*, **16**, pp. 22–29.

34. Dholakia, K., Lee, W. M., Paterson, L., MacDonald, M. P., McDonald, R., Andreev, I., Mthunzi, P., Brown, C. T. A., Marchington, R. F. and Riches, A. C. (2007). Optical separation of cells on potential energy landscapes: enhancement with dielectric tagging. *IEEE Journal of Selected Topics in Quantum Electronics*, **13**, pp. 1646–1654.

35. Dufresne, E. R. and Grier, D. G. (1998). Optical tweezer arrays and optical substrates created with diffractive optics. *Review of Scientific Instruments*, **69**, pp. 1974–1977.

36. Eriksson, E., Scrimgeour, J., Graneli, A., Ramser, K., Wellander, R., Enger, J., Hanstrop, D. and Goksor, M. (2007). Optical manipulation and microfluidics for studies of single cell dynamics. *Journal of Optics A: Pure and Applied Optics*, **9**, pp. S113–S121.

37. Fournier, J. M., Rohner, J., Jacquot, P., Johann, P., Mias, S. and Salathé, R. P. (2005). Assembling mesoscopic particles by various optical schemes. In *Optical Trapping and Optical Micromanipulation II* (eds. Dholakia, K. and Spalding, G. C.), Vol. 5930, art. no. 5930Y. SPIE, Bellingham, WA.

38. French, D. (1984). Organization of starch granules. In: *Starch, Chemistry and Technology* (eds. Whistler, R.L., Bemiller, J. N. and Paschall, E. F.), Academic Press, Orlando (Florida, USA), pp. 183–247.

39. Friese, M. E. J., Nieminen, T. A., Heckenberg, N. R. and Rubinsztein-Dunlop, H. (1998a). Optical torque controlled by elliptical polarization. *Optics Letters*, **23**, pp. 1–3.

40. Friese, M. E. J., Nieminen, T. A., Heckenberg, N. R. and Rubinsztein-Dunlop, H. (1998b). Optical alignment and spinning of laser-trapped microscopic particles. *Nature*, **394**, pp. 348–350.

41. Friese, M. E. J., Nieminen, T. A., Heckenberg, N. R. and Rubinsztein-Dunlop, H. (1998c). Optical alignment and spinning of laser-trapped microscopic particles. *Nature*, **395**, pp. 621.

42. Furukawa, H. and Yamaguchi, I. (1998). Optical trapping of metallic particles by a fixed Gaussian beam. *Optics Letters*, **23**, pp. 216–218.

43. Gahagan, K. T. and Swartzlander, G. A. (1996). Optical vortex trapping of particles. *Optics Letters*, **21**, pp. 827–829.

44. Garbin, V., Cojoc, D., Ferrari, E., Proietti, R. Z., Cabrini, S. and Di Fabrizio, E. (2005). Optical micro-manipulation using Laguerre-Gaussian beams. *Japanese Journal of Applied Physics*, **44**, pp. 5773–5776.

45. Gauthier, R. C. (1997). Theoretical investigation of the optical trapping force and torque on cylindrical micro-objects. *Journal of The Optical Society of America B — Optical Physics*, **14**, pp. 3323–3333.

46. Gauthier, R. C., Ashman, M. and Grover, C. P. (1999). Experimental confirmation of the optical-trapping properties of cylindrical objects. *Applied Optics*, **38**, pp. 4861–4869.

47. Gosse, C. and Croquette, V. (2002). Magnetic tweezers: micromanipulation and force measurement at the molecular level. *Biophysical Journal*, **82**, pp. 3314–3329.

48. Gourrier, A., García-Gutiérrez, M. C. and Riekel, C. (2002). Combined microindentation and synchrotron radiation microdiffraction applied to polymers. *Macromolecules*, **35**, pp. 8072–8077.

49. Graceffa, R., Burghammer, M., Davies, R., Ponchut, C. and Riekel, C. (2009). Studying macromolecular solutions without wall effects by stroboscopic small-angle X-ray scattering. *Applied Physics Letters*, **94**, art. no. 062902.

50. Grier, D. (2003). A revolution in optical manipulation. *Nature*, **424**, pp. 810–817.

51. Gross, S. P. (2003). Application of optical traps in vivo. In *Biophotonics. Part B. Methods in Enzymology* (eds. Marriott, G. and Parker, I.), Vol. 361, pp. 162–174. Academic Press, FL, USA.

52. Gu, M. and Morrish, D. (2002). Three-dimensional trapping of Mie metallic particles by the use of obstructed laser beams. *Journal of Applied Physics*, **91**, pp. 1606–1612.

53. Guilford, W. H., Tournas, J. A., Dascalu, D. and Watson, D. S. (2004). Creating multiple time-shared laser traps with simultaneous displacement detection using digital signal processing hardware. *Analytical Biochemistry*, **326**, pp. 153–166.

54. He, H., Friese, M. E. J., Heckenberg, N. R. and Rubinsztein-Dunlop, H. (1995). Direct observation of transfer of angular momentum to absorptive particles from a laser beam with a phase singularity. *Physical Review Letters*, **75**, pp. 826–829.

55. Henderson, R. (1990). Cryo-protection of protein crystals against radiation damage in electron and X-ray diffraction. *Proceedings*

of the Royal Society of London Series B — Biological Sciences, **241**, pp. 6–8.

56. Higurashi E., Ohguchi, O. and Ukita, H. (1995). Optical trapping of low-refractive-index microfabricated objects using radiation pressure exerted on their inner walls. *Optics Letters*, **20**, pp. 1931–1933.

57. Higurashi, E., Sawada, R. and Ito, T. (2001). Optically driven angular alignment of microcomponents made of in-plane birefringent polyimide film based on optical angular momentum transfer. *Journal of Micromechanics and Microengineering*, **11**, pp. 140–145.

58. Huang, X., Nelson, J., Kirz, J., Lima, E., Marchesini, S., Miao, H., Neiman, A. M., Shapiro, D., Steinbrener, J., Stewart, A., Turner, J. J. and Jacobsen, C. (2009). Soft X-Ray diffraction microscopy of a frozen hydrated yeast cell. *Physical Review Letters*, **103**, art. no. 198101.

59. Huisstede, J. H. G., Subramaniam, V. and Bennink, M. L. (2007). Combining optical tweezers and scanning probe microscopy to study DNA–protein interactions. *Microscopy Research and Technique*, **70**, pp. 26–33.

60. Imberty, A. and Pérez, S. (1988). A revisit to the three-dimensional structure of B-type starch. *Biopolymers*, **27**, pp. 1205–1221.

61. Knight, A. E. (2008). *Single Molecule Biology*. Elsevier, Oxford.

62. Korda, P. T., Taylor, M. B. and Grier, D. G. (2002). Kinetically locked-in colloidal transport in an array of optical tweezers. *Physical Review Letters*, **89**, art. no. 128301.

63. Kovac, J. R. and Voldman, J. (2007). Intuitive, image-based cell sorting using optofluidic cell sorting. *Analytical Chemistry*, **79**, pp. 9321–9330.

64. Lang, M. J. and Block, S. M. (2003). Laser-based optical tweezers. *American Journal of Physics*, **71**, pp. 201–215.

65. Lemke, H., Burghammer, M., Flot, D., Roessle, M. and Riekel, C. (2004). Structural processes during starch granule hydration by synchrotron radiation microdiffraction. *Biomacromolecules*, **5**, pp. 1316–1324.

66. Liberale, C., Minzioni, P., Bragheri, F., De Angelis, F. and Di Fabrizio, F. (2007). Miniaturized all-fibre probe for three-dimensional optical trapping and manipulation. *Nature Photonics*, **1**, pp. 723–727.

67. Lima, E., Wiegart, L., Pernot, P., Howells, M., Timmins, J., Zontone, F. and Madsen, A. (2009). Cryogenic X-Ray diffraction for biological samples. *Physical Review Letters*, **103**, art. no. 198102.

68. MacDonald, M. P., Spalding, G. C. and Dholakia, K. (2003). Microfluidic sorting in an optical lattice. *Nature*, **426**, pp. 421–424.

69. Maruo, S. and Inoue, H. (2007). Optically driven viscous micropump using a rotating microdisk. *Applied Physics Letters*, **91**, art. no. 084101.

70. Maxwell, J. C. (1873). *Treatise on Electricity and Magnetism*. Clarendon Press, Oxford, UK.

71. Moffitt J. R., Chemia, Y. R., Smith S. B. and Bustamante, C. (2008). Recent advances in optical tweezers. *Annual Review of Biochemistry*, **77**, pp. 205–228.

72. Molloy, J. E. and Padgett, M. J. (2002). Lights, action: optical tweezers. *Contemporary Physics*, **43**, pp. 241–258.

73. Neuman, K. C. and Block, S. M. (2004). Optical trapping. *Review of Scientific Instruments*, **75**, pp. 2787–2809.

74. Neuman, K. C., Chadd, E. H., Liou, G. F., Bergman, K. and Block, S. M. (1999). Characterization of photodamage to *Escherichia coli* in optical traps. *Biophysical Journal*, **77**, pp. 2856–2863.

75. Neutze, R., Spoel, R. V. D., Weckert, D. and Hadju, J. (2000). Potential for biomolecular imaging with femtosecond X-ray pulses. *Nature*, **406**, pp. 753–757.

76. O'Neil, A. T. and Padgett, M. J. (2000). Three-dimensional optical confinement of micron-sized metal particles and the decoupling of the spin and orbital angular momentum within an optical spanner. *Optics Communications*, **185**, pp. 139–143.

77. Ozkan, M., Wang, M., Ozkan, C., Flynn, R., Birkbeck, A. and Esener, S. (2003). Optical manipulation of objects and biological cells in microfluidic devices. *Biomedical Microdevices*, **5**, pp. 61–67.

78. Padgett, M. and Leach, J. (2008). Rotation of particles in optical tweezers. In *Structured Light and Its Applications* (ed. Andrews D. L.), pp. 237–248, Academic Press, FL, USA.

79. Panine, P., Finet, S., Weiss, T. M. and Narayan, T. (2006). Probing fast kinetics in complex fluids by combined rapid mixing and small-angle X-ray scattering. *Advances in Colloid and Interface Science*, **127**, pp. 9–18.

80. Paris, O. (2008). From diffraction to imaging: new avenues in studying hierarchical biological tissues with x-ray microbeams. *Biointerphases*, **3**, pp. FB16–FB26.

81. Perch-Nielsen, I. R., Rodrigo, P. J. and Gluckstad, J. (2005). Real-time interactive 3D manipulation of particles viewed in two orthogonal observation planes. *Optics Express*, **13**, pp. 2852–2857.

82. Pollack, L., Tate, M. W., Darnton, N. C., Knight, J. B., Gruner, S. M., Eaton , W. A. and Austin, R. H. (1999). Compactness of the denaturated state of a fast-folding protein measured by submillisecond small-angle x-ray scattering. *Proceedings of the National Academy of Sciences USA*, **96**, pp. 10115–10117.

83. Popov, D., Buléon, A., Burghammer, M., Chanzy, H., Montesanti, N., Putaux, J. L., Potocki-Veronese G. and Riekel, C. (2009). Crystal structure of A-amylose: a revisit from synchrotron microdiffraction analysis of single crystals. *Macromolecules*, **42**, pp. 1167–1174.

84. Prentice, P. A., MacDonald, M. P., Frank, T. G., Cuschieri, A., Spalding, G. C., Sibbett, W., Campbell, P. A. and Dholakia, K. (2004). Manipulation and filtration of low index particles with holographic Laguerre-Gaussian optical trap arrays. *Optics Express*, **12**, pp. 593–600.

85. Rice, S. E., Purcell, T. J. and Spudich, J. A. (2003). Building and using optical traps to study properties of molecular motors. In *Biophotonics. Part B* (eds. Marriott, G. and Parker, I.), Vol. 361, pp. 112–133. Academic Press, FL, USA.

86. Riekel, C. (2000). New avenues in X-ray microbeam experiments. *Reports on Progress in Physics*, **63**, pp. 233–262.

87. Riekel, C., Burghammer, M., Davies, R., Gebhardt, R. and Popov, D. (2009). Fundaments of soft condensed matter scattering and diffraction with microfocus techniques. In *Applications of Synchrotron Light to Non-Crystalline Diffraction in Materials and Life Sciences*. (Series Lecture Notes in Physics) (eds. García-Gutiérrez, M. C., Nogales, A., Gómez, M. and Ezquerra, T. A.), Vol. 776. Springer-Verlag, Heidelberg.

88. Riekel, C., Burghammer, M. and Schertler, G. (2005). Protein crystallography microdiffraction. *Current Opinion in Structural Biology*, **15**, pp. 556–562.

89. Riekel, C. and Davies, R. (2005). Applications of synchrotron radiation micro-focus techniques to the study of polymer and

biopolymer fibers. *Current Opinion in Colloid & Interface Science*, **9**, pp. 396–403.

90. Robin, J. P., Mercier, C., Charbonniere, R. and Guilbot, A. (1974). Lintnerized Starches. Gel filtration and enzymatic studies of insoluble residues from prolonged acid treatment of potato starch. *Cereal Chemistry*, **51**, pp. 389–405.

91. Santucci, S. C., Cojoc, D. Amenitsch, H., Marmiroli, B., Sartori, B., Burghammer, M., Schoeder, S. Di Cola, E. and Riekel, C. (2010). An optical tweezers setup for synchrotron radiation experiments on fragile, soft and biological matter. *In preparation*.

92. Schramm, L. L. (2008). *Dictionary of Nanotechnology, Colloid and Interface Science*. Wiley-VCH, Weinheim, pp. 1–298.

93. Schroer, C. G., Boye, P., Feldkamp, J. M., Patommel, J., Schropp, A., Schwab, A., Stephan, S., Burghammer, M., Schoder, S. and Riekel, C. (2008). Coherent X-ray diffraction imaging with nanofocused illumination. *Physics Review Letters*, **101**, art. no. 090801.

94. Seidel, R., Gourrier, A., Burghammer, M., Riekel, C., Jeronimidis, G. and Paris, O. (2008). Mapping fibre orientation in complex-shaped biological systems with micrometre resolution by scanning X-ray microdiffraction. *Micron*, **39**, pp. 198–205.

95. Singer, W., Nieminen, T. A., Gibson, U. J., Heckenberg, N. R. and Rubinsztein-Dunlop, H. (2006). Orientation of optically trapped nonspherical birefringent particles. *Physical Review E*, **73**, art. no. 021911.

96. Singer, W., Rubinsztein-Dunlop, H. and Gibson, U. (2004). Manipulation and growth of birefringent protein crystals in optical tweezers. *Optics Express*, **12**, pp. 6440–6445.

97. Smith, S. B., Cui, Y. J. and Bustamante, C. (1996). Overstretching B-DNA: the elastic response of individual double-stranded and single-stranded DNA molecules. *Science*, **271**, pp. 795–799.

98. Soler-Illia, G. J. A. A., Sanchez, C., Lebeau, B. and Patarin, J. (2002). Chemical strategies to design textured materials: from microporous and mesoporous oxides to nanonetworks and hierarchical structures. *Chemical Reviews*, **102**, pp. 4093–4138.

99. Spalding, G. C., Courtial, J. and Di Leonardo, R. (2008). Holographic optical tapping. In *Structured Light and Its Applications* (ed. Andrews, D. L.), pp. 139–168. Academic Press, FL, USA.

100. Stuke, M., Mueller, K., Mueller, T., Hagedom, R., Jaeger, M. and Fuhr, G. (2005). Laser-direct-write creation of three-dimensional OREST

microcages for contact-free trapping, handling and transfer of small polarizable neutral objects in solution. *Applied Physics A — Materials Science & Processing*, **81**, pp. 915–922.

101. Svoboda, K. and Block, S. M. (1994a). Optical trapping of metallic Rayleigh particles. *Optics Letters*, **19**, pp. 930–932.

102. Svoboda, K. and Block, S. M. (1994b). Biological applications of optical forces. *Annual Review of Biophysics and Biomolecular Structure*, **23**, pp. 247–285.

103. Thibault, P., Dierolf, M., Menzel, A., Bunk, O., David, C. and Pfeiffer, F. (2008). High-resolution scanning X-ray diffraction microscopy. *Science*, **321**, pp. 379–382.

104. Vorobjev, I. A., Liang, H., Wright, W. H. and Berns, M. W. (1993). Optical trapping for chromosome manipulation: a wavelength dependence of induced chromosome bridges. *Biophysical Journal*, **64**, pp. 533–538.

105. Waigh, T. A., Kato, K. L., Donald, A. M., Clarke, C. J. and Riekel, C. (2000). Side-chain liquid-crystalline model for starch. *Starch/Stärke*, **52**, pp. 450–460.

106. Wolf, S. E., Leiterer, J., Kappl, M., Emmerling, F. and Tremel, F. (2009). Early homogenous amorphous precursor stages of calcium carbonate and subsequent crystal growth in levitated droplets. *Journal of the American Chemical Society*, 130, pp. 12342–12347.

107. Wright, W. H., Sonek, G. J. and Berns, M. W. (1994). Parametric study of the forces on microspheres held by optical tweezers. *Applied Optics*, **33**, pp. 1735–1748.

Chapter 7

MOLECULAR MODELING TO FACILITATE PROTEIN CRYSTALLIZATION

Victor Sivozhelezov[a,b], Eugenia Pechkova[a], and Claudio Nicolini[b]

[a]*Nanoworld Institute, Fondazione EL.B.A., University of Genoa Medical School, Italy*

[b]*Institute of Cell Biophysics, Russian Academy of Sciences, Pushchino, Moscow Region 142290, Russia*

manuscript@ibf.unige.it

7.1 INTRODUCTION

Protein crystallography remains the principal method for solving atomic structures of proteins and their complexes, especially for large proteins whose structures are unattainable using nuclear magnetic resonance (NMR). Protein crystallization remains a bottleneck of the method, for either resistance of many proteins to crystallize or difficulty in obtaining well-diffracting crystals.

In spite of considerable progress reached recently for improvement of crystallization by altering the phase state of proteins, such as Langmuir–Blodgett film assisted crystallization for soluble proteins (Pechkova and Nicolini, 2004) or lipidic cubic phase crystallization method for membrane proteins (Landau and Rosenbusch, 1996), the process of crystallization remains largely empirical, as evident from development of more and more expansive crystallization screens, which allow to quickly test large

Synchrotron Radiation and Structural Proteomics
Edited by Eugenia Pechkova and Christian Riekel

www.panstanford.com

numbers of solution conditions, which have previously been found to be useful for crystallization. From structural genomics initiatives, it follows that the success rate for producing X-ray quality crystals has been estimated to be within 20%–50% for those proteins that can be expressed in soluble form (Hui and Edwards, 2003; Fogg *et al.*, 2006).

It is becoming understood that, while proteins obey the general theories of crystallization (Chernov *et al.*, 2004), there are details in protein crystallization beyond the scope of those theories (Dumetz *et al.*, 2008). Such details are related to complex shapes and anisotropy of physicochemical properties of protein structures, in which small structural changes may cause large changes in phase behavior.

Particularly, it was long known that hemoglobin, with a single mutation, shows a different crystallization behavior (Vekilov *et al.*, 2002). It was also recently found that a single mutation of gamma-crystallin inverts the temperature dependence of solution/crystal phase transition, namely crystallization occurs upon elevation rather than decrease in temperature (McManus *et al.*, 2007); see also comment in Thurston *et al.* (2007). In such cases, drastic differences in crystallization behavior are observed, but three-dimensional (3D) structures of the mutant proteins remain practically the same as the native ones.

Since molecular details of proteins are addressed by atomic-scale molecular modeling techniques, it can be expected to provide insights into protein interactions crucial in improving protein crystallization. Applications of those techniques are discussed in this mini review/discussion paper. First, properties of proteins relevant to their crystallization behavior are summarized (Section 7.2). Then, phase behavior of protein solutions is discussed (Section 7.3). In both sections, theoretical difficulties are underlined making molecular modeling essential for optimizing protein crystallization.

Importantly, molecular modeling is not frequently used for optimization of crystallization of proteins, because modeling should typically start from an already existing crystal structure. Such examples are discussed in the Section 7.4. In the following two sections, examples will be discussed in detail for optimizing crystallization of poorly crystallizable proteins. In such cases suggestions for crystallization improvement must use, instead of crystal structures, the theoretical models, which are often quite crude.

We will show that molecular modeling can be used even in such very difficult cases, and we exemplify this with the poorly soluble, membrane-attached cytochrome P450scc, and the transmembrane protein cephalopod rhodopsin.

7.2 SPECIFIC PROPERTIES OF PROTEINS RELEVANT TO CRYSTALLIZATION

To understand why proteins are specific in their crystallization behavior, it is instructive to recall the physicochemical properties of proteins differing from those synthetic polymers, small molecules, and colloid particles, which might underlie the distinction of proteins in their crystallization behavior.

In terms of statistical physics of polymers (Grosberg and Khokhlov, 1994), proteins are characterized by a large number of interactions among side chains located far away along the main polymer chain (protein backbone). This leads to emergence of secondary and tertiary (3D) structures characterized by distinct energy minima in conformation space, which are as a rule absent in synthetic polymers. Thus, the distinct 3D structures of proteins resemble small organic molecules.

On the other hand, the polymeric nature of proteins is reflected in the protein flexibility reflected in the fact that even small, atomic-scale, perturbations, such as ones that may be caused by changes of salt concentration or temperature, may change position of energy minima in the protein's conformational space. Such perturbations will not necessarily destroy major interactions responsible for the 3D structure eventually observed by crystallography or NMR, but may drastically affect interactions between protein molecules, and hence crystal nucleation.

Another distinction of proteins from small molecules is their size, making the range of interactions in not very dilute solution comparable to the size of the protein, which is typical for colloidal particles. Protein crystallization therefore is a two-stage process, namely the system passes through a denser liquid phase during transition from dilute solution to crystal (Vekilov, 2005).

However, the already mentioned sensitivity of properties of protein surfaces to even minor perturbations, as well as the non-uniform nature of the proteins geometries and surfaces, distinguishes proteins from colloidal particles as well.

For water-soluble proteins, their 3D structure is determined mainly by existence of a hydrophobic nucleus, whose destruction leads to denaturation of the protein. Thus, even powders of intact proteins are not always possible to obtain by simple evaporation, and techniques such as lyophilzation must be used. For crystallization of proteins, restrictions are even more stringent, considering also that nucleation rate depends not only on supersaturation, but also on the rate of supersaturation increase (Garcia-Ruiz, 2003).

Hence, protein crystals, contrary to crystals of small molecules e.g., salt, always contain a relatively large amount of water. Protein crystallization therefore should be heavily dependent on interactions between proteins in a mostly aqueous medium. This is largely relevant also with respect to membrane proteins, since even transmembrane proteins possess extensive cytoplasmic domains. The water–protein interactions can be either hydrophobic or polar/electrostatic, considering also that the major contributions to hydrogen bonds are electrostatic.

While the hydrophobic forces are generally recognized to have entropic nature at room temperature (Chandler, 2005), the same is true for electrostatic interactions in aqueous medium, which is often missed in molecular modeling studies of proteins. Entropy is contributing to electrostatic interactions in water simply because the liquid water's dielectric constant is inversely proportional to absolute temperature (Cantor and Schimmel, 1980).

It is not surprising that thermodynamics of protein crystallization, unlike the enthalpy-driven crystallization of salts, was found to be entropy-driven (Vekilov *et al.*, 2002). Then, the strategies for optimizing crystallization should include optimization of entropic effects, for example, at the expense of water expulsion from the crystal into bulk solvent.

All of the above-discussed protein properties underlie rational approaches to protein crystallization, of which molecular modeling is the most helpful. However, it must be noted that a hypothesis of the (presumably evolutionary) "negative design" of proteins for crystallization has recently been put forward saying that the surface properties of proteins have been selected to prevent native-state aggregation and crystallization *in vivo* (Doye *et al.*, 2007).

If this hypothesis is true, then random or quasi-random mutagenesis of surface residues should tend to improve protein crystallization, regardless of the above-discussed intrinsic protein properties. Indeed, in the only example available so far, DNA shuffling allowed to obtain several mutants forming high-quality crystals while the crystallizable protein remained functional (Keenan *et al.*, 2005). Note that the examples (D'Arcy *et al.*, 1999) quoted by Doye *et al.* (2007) in support of the hypothesis are irrelevant since they belong to rational rather than random mutagenesis for protein crystallization improvement.

7.3 PHASE BEHAVIOR OF PROTEINS UNDER CRYSTALLIZATION CONDITIONS

Since crystallization is a phase transition, it is phenomenologically represented by phase diagrams, which are graphs of equilibrium conditions between the thermodynamically distinct phases. Each point in the phase space corresponds to values of several parameters determining the free energy of the system, such as pressure and temperature. In phase diagrams of solutions, at least one axis is usually reserved for the parameter determining composition of the solution. In case of an aqueous solution of a protein, this parameter is concentration of protein, typically expressed as its volume fraction. For protein crystallization, the other axis normally corresponds to either concentration of the precipitant, e.g., salt, or temperature. In terms of thermodynamics, phase transitions correspond to crossing the solubility lines, at which free energies of the two phases are equal.

However, this is necessary but not sufficient for crystallization to occur, since first-order phase transitions, such as protein crystallization, proceed from the solution phase via nucleation followed by growth of the newly formed crystalline phase, and so supersaturation of protein solution is required to form nuclei of supercritical size, the driving force of nucleation being fluctuations.

At high levels of supersaturation, liquid–liquid phase separation may occur into protein-rich and protein-depleted phases, the protein-rich phase being metastable with respect to crystalline phase (Vekilov, 2005). Besides, formation of amorphous aggregates of protein molecules is likely to occur when supersaturation is increased, eventually leading to formation of the amorphous precipitate, leaving a narrow "crystallization slot" in the phase space.

This phenomenon was found to correlate with the second virial coefficient of the osmotic pressure of proteins, which describes pairwise intermolecular interaction in solution (George *et al.*, 1997), namely for large positive values of the virial coefficient, crystallization will not occur on observable time scales, whereas for large negative values, amorphous precipitation will occur.

Similarity was noted in phase behavior of crystallization process to any association process involving proteins, namely temperature dependences of nucleation rates were found to be have the same functional form ("scaling law"), and determined by relative distance of the temperature from the spinodal temperature (Pullara *et al.*, 2007).

While the fluctuations driving protein crystallization were assumed in the classical phase transition theories to be occurring along the single order parameter, namely density (Garcia-Ruiz, 2003), the situation is different for protein crystallization (Vekilov, 2005), where only specific orientation of protein molecules can eventually produce nuclei of the ordered crystalline phase. Such specific orientations can be identified via atomic-scale molecular modeling.

7.4 KNOWN CRYSTAL CONTACTS IN PROTEINS AND THEIR OPTIMIZATION

Improving the already existing crystal contacts is based on the observation that in most of cases where surface mutagenesis has led to a structure determination, the sites of mutation were located at crystal contacts. In Charron *et al.* (2002) known crystal surfaces of aspartyl-tRNA synthetase-1 from *Thermus thermophilus* were modified by mutating amino acids involved in packing contacts in the orthorhombic lattice. Addition of potential contacts leads to well-shaped crystals of same space group and cell parameters than wild-type crystals. Instead, when lattice contacts were disturbed, crystallization did not occur or yielded poor quality crystals.

Another direction of where molecular modeling usage is artificial introduction of metal ions into the protein. This was successfully applied to apo acyl carrier protein from *Escherichia coli* (Qiu and Janson, 2004), which already has a high content of carboxyl groups suitable for metal chelation. Generally, molecular modeling can be used for engineering surface Asp and Glu mutants in metal-chelating positions, and then crystallizing the protein in presence of metal ions such as Zn^{2+}, with the same metal ions used also for anomalous signals to solve the phase problem.

Special attention is paid by mutating the residues that rarely appear in crystal contacts such as lysines. Thus, the surface-entropy reduction approach (Derewenda and Vekilov, 2006), in which single or preferably multiple adjacent lysines and/or other flexible residues are mutated to alanine to reduce the entropic cost of crystallization. In another approach lysines are replaced with residues more frequently found in crystal contacts, such as arginine or glutamine (Dasgupta *et al.*, 1997). Obviously these approaches as noted in Anstrom *et al.* (2005) are conflicting, because alanine has no positive or negative preference for crystal contacts (Dasgupta *et al.*, 1997), while arginine and glutamine have side-chain entropies comparable to those of lysine.

A very different view on the role of surface residues in protein crystallization may follow from the observation that especially favorable packing arrangements may come from residues not involved in crystal contact, but instead forming dense networks of salt bridges, as is the case for beta- and gamma-crystallins (Chirgadze, 1992).

It is obvious that only detailed molecular modeling complete with free energy calculations may resolve the issue, since as follows from the section 7.2, thermodynamic and structural properties of proteins are strongly interrelated, with contributions of each amino acid not necessarily additive.

7.5 POORLY SOLUBLE PROTEIN CYTOCHROME P450SCC

Cytochromes P450scc are contained in animal adrenal cortex (Lambeth *et al.*, 1986), which has exceptional biomedical importance because it is the key enzyme of steroid metabolism, with the superfamily of cytochromes P450, sharing the same structural fold, being quite diverse, both in function and amino acid sequence (Nicolini *et al.*, 2001). Among members of those superfamily, CytP450scc is very difficult to crystallize in such a manner as to obtain well-diffracting crystals.

Some qualitative comparisons are available of the crystallization behavior of P450s (Li and Poulos, 2004), indicating that while the crystallization techniques used so far for P450 crystal growth are not much different from those utilized for other proteins, special protein engineering strategies have been developed to generate soluble, homogeneous membrane-bound P450 samples amendable for crystallization.

Newly determined P450 structures also provide convincing evidence that P450 enzymes are highly dynamic and flexible. Common structural elements found in all P450s have been identified that undergo large conformational changes to allow substrate access and product release. In addition, flexible regions may enable the active site to adapt to the binding of substrates of different size, shape, and polarity.

To optimize P450scc crystallization for obtaining the necessary entire diffraction data set, it is necessary to take into account that the plethora of the crystallized cytochromes P450 and the corresponding number of reported 3D atomic structures suggest the use of surface mutagenesis, specifically recommended for the case when the obtained crystals do not have the sufficient quality, but crystallizable homologs are available (Dale *et al.*, 2003).

The general recommendation is to mutate residues predicted by homology models (if available) to be solvent-exposed to those favorable for crystal contact formation. Statistically, those desirable residues are arginine and glutamine (Dasgupta *et al.*, 1997; Baud and Karlin, 1999).

In this work, we attempt to propose the mutations that are expected to facilitate crystallization of CytP450scc according to the above indications, but with two essential amendments. First, we use two very different homology models to map the assumedly surface-exposed residues. One is the classical CytP450scc model based on CytP450cam (Vijayakumar and Salerno, 1992), and the other is our model based on the crystallographic structure of CytP4502B4 (Sivozhelezov and Nicolini, 2005).

The two models differ in two major points: first, the CytP450cam model reflects the well-known "closed" conformation of CytP450, from which it is not essentially known how the substrate reaches its active site, and second, is in the degree of sequence identity with the target, which is about 12% for the CytP450cam model and 21% for the CytP4502B4 model.

The second amendment follows from P450scc being an integral membrane protein. It is very likely to possess solvent-exposed hydrophobic areas in such a number and overall size that it would be prone to aggregate in an indiscriminate manner instead of forming regular structures such as crystals. The relation of the hydrophobic effect to random nature of the formed aggregates can be considered theoretically proven. Indeed, the greater the contiguity of the hydrophobic patches on the monomer surface, the less ordered the aggregates become, despite the opportunities for rearrangement offered by a reversible pathway (Patro and Przybycien, 1996).

Here, we combine the above concept with the notion that it is the electrostatic forces that primarily steer the molecules to each other in aqueous solution, even though the eventual energetics of protein–protein contact may be determined by the hydrophobic effect. A detailed discussion on the comparative role of hydrophobic effect and polar interactions in protein–protein contact may be found in Kumar and Nussinov (2002).

In the same vein, we calculated the solvent-exposed hydrophobic areas on the protein and the electrostatic potential of the protein, and mapped the electrostatic potential onto the hydrophobic surfaces and identified the amino acid residues forming the contiguous hydrophobic patches on the surface of CytP450scc, moreover located in areas of neutral electrostatic potential, to be considered them as candidates for crystallization-optimizing point mutations.

Thus, the approach used herein is based on displaying and comparing hydrophobic regions that are not screened by electrostatic potentials, and should therefore be negatively affecting crystallization of proteins by favoring irregular aggregation of proteins. To assess its relevance to the protein crystallizability, calculations were initially performed for one of the proteins most readily crystallizable in a broad range of conditions, namely the lysozyme.

Figure 7.1 shows that contiguous hydrophobic patches cover only a small fraction of the protein surface. One of them is quite large (left part of Fig. 7.1, left and right part of Fig. 7.1, right) and can contribute to the random nature of lysozyme aggregation prior to lysozyme crystal nucleation.

This random aggregation indeed happened (Skouri *et al.*, 1991), which confirms the approach adopted herein. Overall, a prevailing fraction of lysozyme surface is either free of hydrophobic regions, or bears an electric potential at the hydrophobic surfaces. Moreover, large fractions of lysozyme surfaces are free of exposed hydrophobic amino acids, so most lysozyme surface can contribute to forming crystal contacts. Thus, the approach adopted in this paper suggests that lysozyme should be a readily crystallizable protein, which is in fact observed.

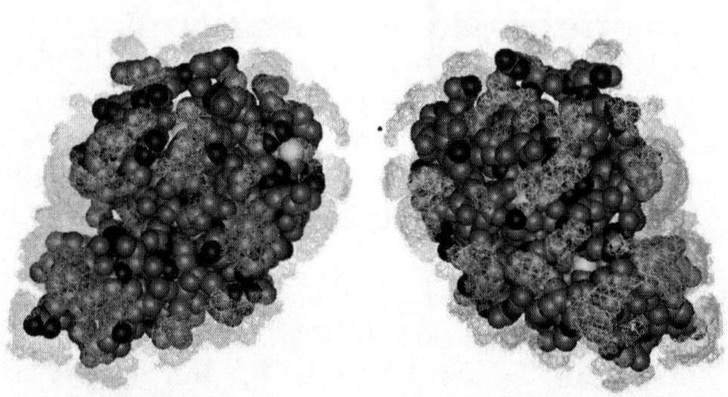

Figure 7.1 Hydrophobic patches surrounding the lysozyme molecule shown in CPK representation. The patches are colored according to electrostatic potential: dark gray, positive; light gray, negative; almost white, neutral. The two images differ by 180° rotation around the vertical axis. See also Color Insert.

However, lysozyme cannot serve as a reference crystallizable protein to be compared with cytochrome P450scc since it is completely soluble, it lacks any homology with the cytochrome, and its molecular weight is three times smaller.

Therefore we analyzed the protein, which was the template of our homology model, cytochrome P4502B4 (Fig. 7.2). It shows the highest homology with cytochrome P450scc while it is successfully crystallized and membrane-attached.

As for lysozyme, most of the protein surface is not covered by white patches, although some of them are quite large.

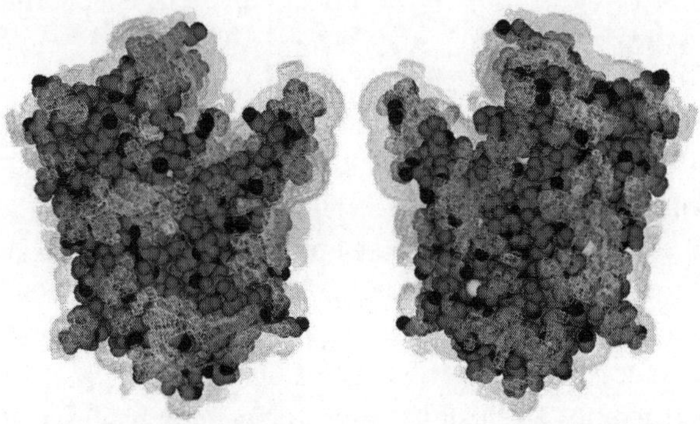

Figure 7.2 Hydrophobic parches around cytochrome P4502B4. See also Color Insert.

Consider next the structure of cytochrome P450scc predicted by homology with cytochrome P450cam (Vijayakumar and Salerno, 1992) shown in Fig. 7.3. We see only small fractions of contiguous pieces of hydrophilic surface outside hydrophobic patches.

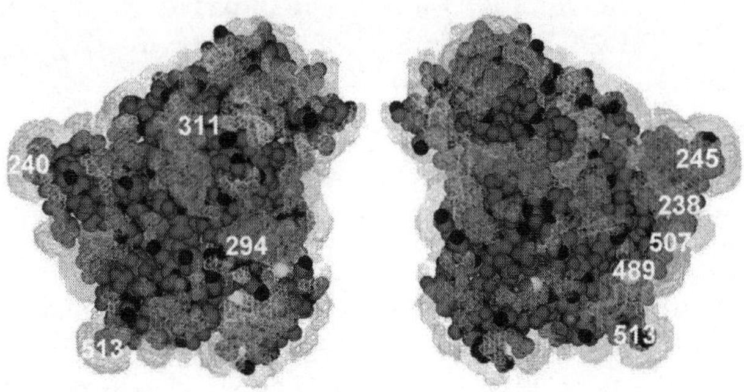

Figure 7.3 Hydrophobic patches around the molecule of cytochrome P450scc, according to the model based on cytochrome P450cam. See also Color Insert.

Many hydrophobic patches are occupied by positive or negative electrostatic potential, mostly caused by amino acids of lysine and glutamate, which are not prone to form crystalline contacts (Dasgupta *et al.*, 1997; Baud and Karlin, 1999).

The most prominent hydrophobic area with neutral electrostatic potential is limited by the amino acids 238–245 in the sequence of P450scc (left and right edges of Fig. 7.3).

The structure of cytochrome P450scc predicted by homology with cytochrome P4502B4 (Fig. 7.4) shows even more striking difference from the pattern exhibited by cytochrome P4502B4. Indeed, almost the entire surface of the protein is covered by the white patches.

According to the logics of Patro and Przybycien (1996), no ordering should be observed in the aggregates formed by the P450scc molecules, and thus very little chance for P450scc molecules in their native form to yield well-diffracting crystals. Our approach therefore suggests three options for altering the structure of P450scc to assist crystallization. The first and foremost is site directed mutagenesis. This approach has already a considerable record of success reviewed in Dale *et al.* (2003). These successful examples include, for instance, exhaustive mutagenesis of all 29 possibly exposed hydrophobic residues of catalytic domain of HIV integrase, resulting in just one mutant giving well-diffracting crystals (Dyda *et al.*, 1994).

More rational approach (termed "crystal engineering") involved theoretical predictions and resulted in crystallizability introduced or improved by inducing smaller number of mutations. For example nine mutations resulted in improved crystallizability on the case of 24 kDa fragment of DNA gyrase (D'Arcy *et al.*, 1999) while only one mutation was required for adding the propensity to crystallize in the case of human leptin (Zhang *et al.*, 1997). This is the approach actually applied herein. The size and occurrence of electrostatically unscreened hydrophobic patches (Fig. 7.3) for the cytochrome P4502B4-based cytochrome P450scc model is larger than for P4502B4 (Fig. 7.2). Some regions of the P450scc surface are completely covered by patches.

We identified the hydrophobic groups contributing to the patches. Residues appearing among such hydrophobic groups in both our model (open conformation, based on cytochrome

P4502B4) and the literature model (closed conformation, based on cytochrome P450cam) are thus predicted to be hampering crystallization by two very different models, and therefore must be considered for mutations facilitating crystallization.

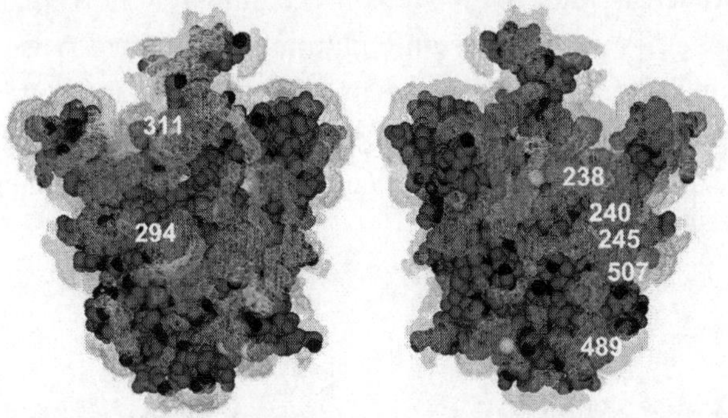

Figure 7.4 Hydrophobic patches around the molecule of cytochrome P450scc, according to the model based on cytochrome P4502B4. See also Color Insert.

They are marked in Figs. 7.3 and 7.4, thus indicating their position with respect to structure of P450scc, according to either of the models. Thus, eight amino acid residues are proposed to be mutated: Tyr238, Met240, Val245, Phe294, Leu311, Ile489, Phe507, and Phe513.

It must be noted that all of the identified residues are well away of the heme and cannot therefore participate in the presumably hydrophilic interaction site of the P450scc with its electron transfer partner, adrenodoxin (Hlavica *et al.*, 2003, and references therein). Therefore, this site, which is a likely source of crystal contact, will not be damaged by the proposed mutations. For the same reason, no damage is expected by those mutations to the electron-accepting functionality.

As for substrate access to the active site, the situation is more complex. Two of the residues proposed for mutation, Tyr238 and Met240, may be on the way of the substrate in the channel present in the CytP4502B4-based model, but their influence on the passage of the substrate through the channel needs to be addressed by separate molecular dynamic studies as done in Wade *et al.* (2004) and references therein.

Following Dasgupta *et al.* (1997) and Baud and Karlin (1999), each of the proposed residues should be replaced either by arginine or glutamine, the latter two being most prone to crystal contact formation.

However, at this point we can recommend the replacement of the specified residue by only glutamines, for two reasons: (1) glutamine is sterically more similar to any of the mutated residues than arginine; and (2) arginine bears a hydrophobic alkane chain that can diminish the effect of removing a part of the uncharged hydrophobic surface.

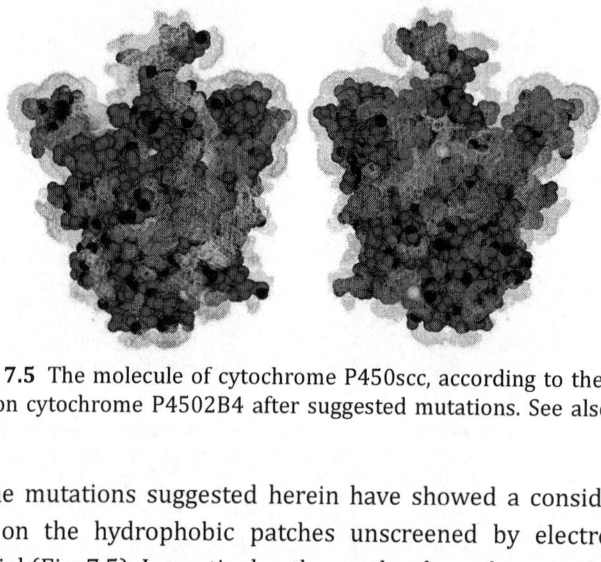

Figure 7.5 The molecule of cytochrome P450scc, according to the model based on cytochrome P4502B4 after suggested mutations. See also Color Insert.

The mutations suggested herein have showed a considerable effect on the hydrophobic patches unscreened by electrostatic potential (Fig. 7.5). In particular, the patches have decreased in size but, even more importantly, there are fewer contiguous patches in case of the mutant protein. Moreover, those patches become comparable to those for P4502B4 in size. Therefore, the proposed mutation should improve crystallization, even though the comparisons herein are drawn in a qualitative manner.

The second option for optimizing crystallization of the P450scc following from results herein is to diminish the hydrophobic effect by utilizing the dimeric nature of the template protein in its crystal. Indeed, if P450scc becomes organized in a dimer similarly to P4502B4, then the hydrophobic residues hampering crystallization may form a stable interface within a P450scc dimer that would, as for P4502B4, form an asymmetric crystal unit.

This requires modeling of P450scc in dimeric form using again P4502B4 as the template, and then identifying residues whose mutation will stabilize the P450scc dimer. In this manner, dimer interaction can be reinforced *ab initio* in the nanofilm template, increasing its order, stability, and the crystallization potential.

7.6 MEMBRANE PROTEIN CEPHALOPOD RHODOPSIN

7.6.1 Properties of Cephalopod Rhodopsin Relevant to Its Crystallization

Crystallization of cephalopod rhodopsins can become a basis of general paradigm of membrane protein crystallization because cephalopod rhodopsins are exceptionally well ordered in their membranes *in vivo*, but so far have neither 3D structures nor suitable crystals. In two dimensions (2D), a cephalopod rhodopsin forms 2D (but not 3D) crystals similar to those of bovine rhodopsin (bovR), allowing to obtain only approximate structures from crude (8 Å resolution) X-ray density map (Davies *et al.*, 2001). Meanwhile, their homologs include bacteriorhodopsin (bR) which has been nicknamed "lysozyme of membrane proteins" because of its exceptional ability to crystallize in a wide variety of conditions, so it is used to test every new crystallization technique (Caffrey, 2003). Its even closer homolog, bovR, has also been crystallized, both in 2D and 3D, and atomic resolution structures were obtained (Palczewski *et al.*, 2000; Teller *et al.*, 2001; Okada *et al.*, 2002), even though it does not form regular arrays *in vivo*.

This high ordering *in vivo* compared with poor ordering *in vitro* implies large but so far unexplored, crystallization potential in cephR, and the molecular manipulation techniques to be used in the process of crystallization are likely to eventually provide a general paradigm for membrane protein crystallization.

Considering that techniques of molecular manipulation using Langmuir–Blodgett technology are already described for cephR and other photosensitive proteins (Maxia *et al.*, 1995; Pepe and Nicolini, 1996), a 3D model of cephR using bovR as a template is built and compared with the structure of bR (Sivozhelezov and Nicolini, 2006). Therein, this comparison is discussed in terms of crystallization of cephR.

CephR, bovR, and bR all belong to the structural superfamily of rhodopsin-like proteins sharing the overall seven transmembrane helix topology of bR except for some details in distances and relative orientations of the helices (Faulon *et al.*, 2003). This allows to use bR as the primary template for all structure/function studies with eventual applications, basing on the huge body of such structure/function data, reviewed in Hirai and Subramaniam (2003).

Structurally, cephR differs from bovR in presence of a ~90 amino acid long, additional C-terminal cytoplasmic domain rich in proline residues. Very likely, that domain is responsible for the regularity of packing in the membrane, and presents a major advantage with respect to crystallization because cephR membranes or films can be expected to have intrinsically higher degree of ordering compared to those of bovR, or even bR. The other structural difference is that the isomerized retinal chromophore remains bound to the metarhodopsin protein after activation by light. The latter fact should again be advantageous for cephR over bR for crystallization because bovR is actually destroyed each time it responds to the light quantum and has to be reconstituted from opsin and retinal. The photoreversibility of cephR is effected via the special protein retinochrome, highly homologous to rhodopsin (Pepe, 2001).

Although both in cephalopod and in bovR the light energy is captured by 11-cis to all-trans photoisomerization of a covalently bound prosthetic group (chromophore) retinal, and both rhodopsins belong to G-protein-coupled proteins, mechanisms of transducing the initial optical signal to the eventual electrical signal differ between the two proteins. Indeed, in both cases, the electrical signal results from ionic channel opening, but the opening in the case of bovR is performed by cyclic GMP synthesized from GMP by phosphodiesterase, which in turn is triggered by G-protein transducin capturing the conformational change in rhodopsin, while in case of cephR, the ionic channel is opened by inositol 1,4,5-triphosphate (IP3) synthetized by phospholipase C, which is triggered in turn by the Gq protein capturing the conformational change in rhodopsin (Pepe, 1999, 2001). In terms of crystallization trials, this suggests the use of Gq protein fragments but not transducin-like fragments to enforce 3D structure of cephR.

7.6.2 Tools for Modeling the 3D Structure of Membrane Proteins

While classical homology modeling is still successfully used (Miedlich *et al.*, 2004), two other approaches were implemented, one based on intraprotein hydrogen bond optimization (Pogozheva *et al.*, 1997) and the other based on first principles of transmembrane protein assembly (Trabanino *et al.*, 2004).

As for the hydrogen bond optimization approach, it seems to be working well if experimentally derived constraints are used, but the experimental data underlying these constraints are themselves under discussion for cephR. Besides, the presence of retinal within the transmembrane part of the molecule could hamper correct hydrogen bond assignment, which is crucial for accuracy of the method. The first principles approach is more attractive considering also that it has already been tested on bovR. Since it is mostly based on assigning the transmembrane helices, we started by predicting positions of the helices, and then manually adjusting the resulting alignment of cephR versus bR using sequence identity and similarity.

7.6.3 Estimate of cephR Model Quality by Comparison with Its Template bovR

The model reproduces the positions of the residues surrounding the chromophore (retinal) correctly, particularly the Lys306 residue providing the Schiff base connection of the protein to the retinal. The model shows an even higher alpha helical content than its template (203 versus 190 alpha helical residues), which is exceptional in comparative modeling. This could be a basis for higher organization of cephR into 2D arrays compared to the bovR, but the experimental data (Davies *et al.*, 2001) show that it happens only *in vivo* but not *in vitro*. Definitely some factor other than the rhodopsin structure itself is responsible for the high ordering of cephR *in vitro*, and, as further analysis will suggest, this could be interaction with another transmembrane protein present in membranes together with cephR.

The sequence identity level in the eventual model was 24.4%, which is above the average 20% quoted for comparisons of vertebrate versus invertebrate visual pigments and therefore supports the reliability of our model. Since all visual pigments described to date have distinctive features (Nathans, 1992), quality of the model can be assessed by the model's ability to reproduce those features. The first is a lysine in the middle of the seventh putative transmembrane segment, corresponding to Lys296 in bovR, which is the site of covalent binding of the chromophore via a retinylidene Schiff base. In our case, it is Lys306. The next feature is a pair of cysteines corresponding to Cys110 and Cys187 in bovR, which are presumed to form a disulfide bond connecting the first and second extracellular loops. These are present in our model as Cys109 and Cys187. Another quality indicator is presence of the sequence (Glu/Asp)-Arg-Tyr, or a close match to this sequence, at the beginning of the second cytosolic loop, in out case it is Asp133, Arg134, andTyr135. Finally, one or more serine or threonine residues should be located in the cytosolic carboxyl terminus, which in bovR are the sites of light-dependent phosphorylation by rhodopsin kinase. In our model, those are threonines 329, 330, and 336. Identity of the counterion is another important issue with respect to both the structure/function relation ships and biotechnology applications. Indeed, in both cephR and bR, the conformational rearrangements required for function (proton pumping for bR, signal transduction for cephR) demand precedent storage of energy (Bryl, 2003, and references therein), which is in both cases achieved by separating the charge of the retinal/lysine Schiff base from its counterion, acting also as a starting point for proton pumping in bR. Our model predicts that the position equivalent to the bovR counterion (Glu113) is occupied by Tyr112. This tyrosine always remains neutral (Nakagawa *et al.*, 1999) so its role as the counterion is ruled out. However, our model predicts that the residue Glu181 is within 3 Å of the Schiff base nitrogen and can therefore serve as the counterion. This is in perfect agreement with the recent mutational analyses proving that it is the Glu181 that is the counterion in invertebrate rhodopsins (Terakita *et al.*, 2004) while the vertebrate rhodopsin counterion (Glu113 in bovR) was acquired later in the course of evolution.

7.6.4 Comparison of the cephR Model with bR

Using the known Vector Alignment Search Tool (Madej *et al.*, 1995), we have located and listed (Table 7.1) the proteins with structures available in RCSB Protein Data Bank (PDB; Berman *et al.*, 2000, 2003; http://www.pdb.org) that are similar to the model obtained for CephR. The structures of bovR were excluded because the CephR model was based on that protein, so the similarity was expected by definition. NMR structures were also excluded because they do not fully reproduce the alpha helices compared to the crystal structures, while they are extremely important in transmembrane proteins. All of the found structures are those of bR, with numbers of overlapped residues of 175–189, all within the transmembrane domains of CephR and bR, and 7%–10% amino acid identity. The only other protein that we found to be structurally similar to our model of CephR was the transmembrane cytochrome oxidase (not shown). This similarity is possibly related to similarity of function between bR and the oxidase in that both of them involve transfer of charged species, whether electron or proton, across the membrane.

Table 7.1 Part I. Entries of PDB similar in 3D structures to the model of octopus rhodopsin. Bovine rhodopsin entries have been omitted being a trivial case. Columns are PDB id code with chain number, number of amino acids residues matching between the octR model and the given protein, VAST score describing overall similarity of secondary structures, RMS deviation of the matching residues in octR and the given protein, and per cent amino acid identity.

PDB id	Aligned length	VAST score	RMSD	% Id	Description
1PXR A	204	8.1	4.2	7.4	Structure of Pro50ala mutant of bacteriorhodopsin
1PY6 B	203	8.0	4.3	6.9	Bacteriorhodopsin crystallized from bicells
1S53 B	202	8.0	4.2	6.9	Thr46ser bacteriorhodopsin
1Q5J B	201	8.0	4.2	7.0	Crystal structure of bacteriorhodopsin mutant P91a crystallized from bicelles
1S52 A	201	8.3	4.2	7.5	Thr24val bacteriorhodopsin
1TN5 B	201	8.0	4.2	7.0	Structure of bacterorhodopsin mutant K41p
1S53 A	201	8.0	4.2	7.0	Thr46ser bacteriorhodopsin
1TN5 A	201	8.0	4.2	7.0	Structure of bacterorhodopsin mutant K41p
1PXR B	200	8.1	4.2	7.0	Structure of Pro50ala mutant of bacteriorhodopsin
1KME B	199	8.1	4.2	8.0	Crystal structure of bacteriorhodopsin crystallized from bicelles
1KME A	193	8.1	4.3	8.3	Crystal structure of bacteriorhodopsin crystallized from bicelles
1BRX	192	8.1	4.2	8.9	Bacteriorhodopsinlipid complex
1VJM A	188	9.4	4.3	9.6	Deformation of helix c in the low-temperature L intermediate of bacteriorhodopsin
1BRR B	188	9.7	4.2	9.0	X-ray structure of the bacteriorhodopsin trimerlipid complex
1R84 A	187	9.4	4.2	9.1	Nmr structure of the 13-Cis-15-Syn retinal in dark-adapted bacteriorhodopsin
1QM8 A	187	8.0	4.3	11.2	Structure of bacteriorhodopsin at 100 K
1JGJ A	186	9.1	4.0	7.0	Crystal structure of sensory rhodopsin Ii at 2.4 Å: insights into color tuning and transducer interaction
1QHJ A	186	9.4	4.1	8.6	X-ray structure of bacteriorhodopsin grown in lipidic cubic phases

The notable point of the results in Table 7.1 is that apart from the classical structure of ground-state native bR, many mutant structures were found among proteins structurally similar to our model of CephR.

Besides, structures are occurring of the excited intermediate state of bR, the so-called L-state. All of those similarities may be explained by the fact that the alpha helices in our model of cephR are generally more twisted kinked than in bR, while both the L-state and the mutants of bR have twists in the helices.

Table 7.1 Part II. See previous page.

PDB id	Aligned length	VAST score	RMSD	% Id	Description
1P8U A	186	8.1	4.3	10.2	Bacteriorhodopsin N' intermediate at 1.62 Å resolution
1TN0 B	185	9.5	4.0	8.1	Structure of bacterorhodopsin mutant A51p
1TN0 A	185	9.5	3.9	8.1	Structure of bacterorhodopsin mutant A51p
1E0P B	185	9.6	4.2	8.1	L intermediate of bacteriorhodopsin
1E0P A	185	9.4	4.1	8.6	L intermediate of bacteriorhodopsin
1PXS A	185	8.1	4.3	7.6	Structure of Met56ala mutant of bacteriorhodopsin
1C8R A	185	8.1	4.4	9.2	Bacteriorhodopsin D96m bR state At 2.0 Å resolution
1AP9	185	8.0	4.3	10.8	X-ray structure of bacteriorhodopsin from microcrystals grown in lipidic cubic phases
2BRD	184	7.5	4.1	9.2	Crystal structure of bacteriorhodopsin in purple membrane
1BRR C	183	9.7	4.0	8.7	X-ray structure of the bacteriorhodopsin trimerlipid complex
1BRR A	182	9.6	3.9	8.2	X-ray structure of the bacteriorhodopsin trimerlipid complex
1QKP A	178	9.4	4.1	9.0	High-resolution X-ray structure of an early intermediate in the bacteriorhodopsin photocycle
1M0K A	175	8.0	4.0	9.7	Bacteriorhodopsin K intermediate at 1.43 Å resolution
1BRD	144	9.1	3.4	6.2	Bacteriorhodopsin

The finding that the L-state of bR may be more similar to our model than the ground-state of bR required additional testing. To that end, we performed an attempt to superimpose our model onto coordinates derived from crystals of both ground-state and L-state bR obtained under identical conditions and reported in the same paper (Royant *et al.*, 2000), PDB id code 1E0P. Only the N-terminal residues of bR (up to Ala 14, part of transmembrane Helix 1) of the ground-state bR matches our model (residues Val34–Gly43 of CephR), while only the L-state bR matches our model with respect to the remaining residues (from

Leu (Helix 1) to Helix Leu221 (Helix 7) of bR versus Val44 (Helix 1) to Ser317 (Helix 7) of CephR). Therefore, even though the L-state and the ground-state of bR differ very little, our model is reproducibly more similar to the L-state bR than to the ground-state bR. The match between our CephR model and the bR structure, however, has an exception of Helix 4 in cephR and the given protein, and per cent amino acid identity.

The difference is in the distance between Helix 4 and the rest of the transmembrane helix domain, which is several Angstrom larger in cephR compared to bR. Most likely this is responsible for larger conformational mobility of cephR, required by its function as a G-protein-coupled sensor versus the more rigid structure of bR, required by its proton pump (energy conversion) function. The mismatch of the position of Helix 4 provides the source of most differences among the cephR and bR structures.

However, the fact that we distinguished two forms of bR with respect to their similarity with cephR shows that it may be functionally important. Indeed, Helix 3 is twisted in our model of cephR. In contrast, this helix is twisted only in the L-state of bR but straight in ground-state, as well as K-state of bR. This feature suggests that Helix 3 of cephR may be intrinsically less prone to conformational changes upon excitation than the corresponding Helix C of bR while most of the conformational mobility of cephR is mediated by Helix 4.

Considering that the ground-state isomers of retinal are 11-cis for cephR and all-trans in bR, we attempted to accommodate an all-trans retinal in the cephalopod opsin by means of molecular modeling. We inserted all-trans retinal into the native cephR retinal cavity. This allowed checking if cephR could bind all-trans retinal in the same way it binds its native ground-state 11-cis retinal.

7.6.5 Binding of All-Trans Retinal to Cephalopod Opsin and Its Relevance for cephR Crystallization

Cephalopod opsin should readily recombinate with all-trans retinal (Sivozhelezov and Nicolini, 2006). The superimposed structures reported therein allow to see that cephR opsin can accommodate retinal both in its native-cis confirmation and the all-trans conformation. This is supported by the experimental data (Koutalos *et al.*, 1989) showing that, while both bovR and cephR are easily regenerated with their native 11-cis-retinal, cephR can be additionally regenerated with all-trans retinal and moreover the 13-cis retinal which is the native ground-state chromophore for bR.

It is further supported by the recent finding that rhodopsin from lancelet (amphioxus), a primitive chordate, is, like cephR and unlike bovR, able to bind all-trans retinal, and mutational analyses revealed that Trp265 is responsible for this property (Tsukamoto *et al.*, 2005). We found that the pairwise homologies between rhodopsins from lancelet and cephalopod, as well as bovine and cephRs, are similar (about 30% identity) but the 275th amino acid position in our cephR model equivalent to the 265th position of lancelet rhodopsin is also occupied by tryptophan (data not shown).

In contrast, there are no tryptophans at the corresponding position of bovR or positions close in its amino acid sequence. On one hand, this adds to the validity of our cephR model since it predicts not only the experimentally known effect, i.e., the ability of cephR to bind all-trans retinal, but also its cause, i.e., the tryptophan at 275th position. On the other hand, none of the mutations proposed in this paper are at 275th position so they are unlikely to affect this property. Besides, the mutation of the neighboring Tyr278 to Trp proposed herein is unlikely to affect the conformation of Trp275 because both Tyr and Trp residues have aromatic side chains.

However, the conformations of the ionone ring in the model of all-trans reconstituted cephR and the 11-cis dark state cephR are not easily interconvertible within the protein because of contacts with protein side chains. This is supported by the fact that intraprotein conversion of 11-cis to all-trans conformation in any rhodopsin requires the entire range of the photoprocess, which has the time scale of milliseconds. Upon activation by light, followed by transition through batho and meso to lumi-rhodopsin, the retinal molecule dissociates from cephR to form the free retinal and the cephalopod opsin, so cephR regenerates again only after the retinal has been conversed into the 11-cis configuration. However, our data together with the data of Koutalos *et al.* (1989) as well as evidence for presence of alternative binding sites for retinal on opsin (Schadel *et al.*, 2003) suggest that, unless the all-trans retinal is quickly transformed into 1-cis retinal upon release into the lipidic matrix or bound to some transporter protein delivering it to and from the isomerization site, it will be promptly sequestered by the same cephalopod opsin, thus leading to the inactive form of cephR and terminating the visual cycle. In reality, the visual cycle is not terminated, which implies that the all-trans retinal release from cephR is followed by its uptake by a protein other than rhodopsin.

One such protein could be the retinochrome, the retinal isomerization enzyme (Pepe and Cugnoli, 1992). However, the body of the experimental data available for invertebrate rhodopsins suggest that the retinochrome is located away from the rhodopsin, but linked to it via a shuttle protein RALBP (Pepe *et al.*, 1990). We considered the distance from the site of retinal release from cephR at which retinal uptake by RALBP must take place. And we found that RALBP protein must be located very near (less than 20 Å away) from the rhodopsin (Sivozhelezov and Nicolini, 2006). Importantly, in other invertebrates like honey bee, such a retinal-binding protein has also been detected and moreover is able to perform the function of retinochrome (Pepe *et al.*, 1990).

Thus, another protein in addition to cephR (possibly RALBP) should be located in the immediate vicinity to cephR *in vivo*. In addition with stoichiometry of cephR with respect to this retinal-binding protein, which should allow fast uptake of retinal released from cephR *in vivo*, cephR alone is not sufficient with respect to forming 2D or 3D structures; therefore, adding RALBP to otherwise non-crystallizable cephR preparations must be considered.

7.7 CONCLUSIONS

The last two sections of this review show that, even when crystal structure of the protein is not available, molecular modeling can identify ways to optimize crystallization of such "problem" protein. For the poorly soluble protein such as cytochrome P450scc, the message is to replace, by site directed mutagenesis, the especially disadvantageous hydrophobic residues by polar ones. For a membrane protein, in addition to the known approaches, a solution could be to add another protein to the crystallization medium, namely the protein possibly responsible for formation of the ordered structures *in vivo*.

For soluble proteins discussed in Sections 7.2–7.4, the key factor precluding reliable predictions of crystallization behavior is presence of large amounts of water in the crystal. On the one hand, water is essential in maintaining the 3D structure of protein. On the other hand, numbers of water molecules around proteins in crystals can be essentially reduced while preserving the crystallizability of the protein, as follows from two general observations, that thermostable proteins tend to crystallize better than their mesophilic homologs, and that crystals of thermostable proteins contain less water molecules than their those of their mesophilic homologs (Pechkova *et al.*, 2007). Therefore, the recipe for improved protein crystallization is to try keeping water content of the immediate vicinity of the protein as low as allowed by requirement for integrity of the given protein.

Acknowledgments

This project was supported by grants FIRB RBPR05JH2P from MIUR to Claudio Nicolini of University of Genova, to Fondazione Elba by MIUR for "Funzionamento," and by a FIRB International Grant on Proteomics and Cell Cycle (RBIN04RXHS) from MIUR to CIRSDNNOB-Nanoworld Institute of the University of Genova.

References

1. Anstrom, D. M., Colip, L., Moshofsky, B., Hatcher, E. and Remington, S. J. (2005). Systematic replacement of lysine with glutamine and alanine in *Escherichia coli* malate synthase G: effect on crystallization. *Acta Crystallographica F — Structural Biology and Crystallization Communications*, **61**, pp. 1069–1074.

2. Baud, F. and Karlin, S. (1999). Measures of residue density in protein structures. *Proceedings of the National Academy of Sciences USA*, **96**, pp. 12494–12499.

3. Berman, H. M., Henrick, K. and Nakamura, H. (2003). Announcing the worldwide Protein Data Bank. *Nature Structural Biology*, **10**, p. 980.
4. Berman, H. M., Westbrook, J., Feng, Z., Gilliland, G., Bhat, T. N., Weissig, H., Shindyalov, I. N. and Bourne, P. E. (2000). The Protein Data Bank. *Nucleic Acids Research*, **28**, 235–242.
5. Bryl, K. (2003). Visual and archaeal rhodopsins: similarities, differences and controversy. *Cellular and Molecular Biology Letters*, **8**, pp. 285–296.
6. Caffrey, M. (2003). Membrane protein crystallization. *Journal of Structural Biology*, **142**, pp. 108–132.
7. Cantor, C. and Schimmel, P. (1980). *Biophysical Chemistry*. W. H. Freeman, New York.
8. Chandler, D. (2005). Interfaces and the driving force of hydrophobic assembly. *Nature*, **437**, pp. 640–647.
9. Charron, C., Kern, D. and Giege, R. (2002). Crystal contacts engineering of aspartyl-tRNA synthetase from *Thermus thermophilus*: effects on crystallizability. *Acta Crystallographica D — Biological Crystallography*, **58**, pp. 1729–1733.
10. Chernov, A., Segre, P. and Holmes, A. (2004). Crystallization physics in biomacromolecular solutions. In Crystal Growth: F*rom Fundamentals to Technology*, pp. 95–114. Elsevier, Amsterdam, Netherlands.
11. Chirgadze, Y. N. (1992). Plane charge clusters structural invariants of eye lens gamma-crystallins: functional role and evolutionary conservatism. *Molecular Biology Moscow*, **26**, pp. 940.
12. Dale, G., Oefner, C. and D'Arcy, A. (2003). The protein as a variable in protein crystallization. *Journal of Structural Biology*, **142**, pp. 88–97.
13. D'Arcy, A., Stihle, M., Kostrewa, D. and Dale, G. (1999). Crystal engineering: a case study using the 24 kDa fragment of the DNA gyrase B subunit from *Escherichia coli. Acta Crystallographica D — Biological Crystallography*, **55**, pp. 1623–1625.
14. Dasgupta, S., Iyer, G. H., Bryant, S. H., Lawrence, C. E. and Bell, J. A. (1997). Extent and nature of contacts between protein molecules in crystal lattices and between subunits of protein oligomers. *Proteins*, **28**, pp. 494–514.
15. Davies, A., Gowen, B. E., Krebs, A. M., Schertler, G. F. and Saibil, H. R. (2001). Three-dimensional structure of an invertebrate rhodopsin

and basis for ordered alignment in the photoreceptor membrane. *Journal of Molecular Biology*, **314**, pp. 455–463.

16. Derewenda, Z. S. and Vekilov, P. G. (2006). Entropy and surface engineering in protein crystallization. *Acta Crystallographica D — Biological Crystallography*, **62**, pp. 116–124.

17. Doye, J. P., Louis, A. A., Lin, I. C., Allen, L. R., Noya, E. G., Wilber, A. W., Kok, H. C. and Lyus, R. (2007). Controlling crystallization and its absence: proteins, colloids and patchy models. *Physical Chemistry Chemical Physics,* **9**, pp. 2197–2205.

18. Dumetz, A., Chockla, A., Kaler, E. and Lenhoff, A. (2008). Protein phase behavior in aqueous solutions: crystallization, liquid-liquid phase separation, gels, and *aggregates. Biophysical Journal*, **94**, pp. 570–583.

19. Dyda, F., Hickman, A. B., Jenkins, T. M., Engelman, A., Craigie, R. and Davies, D. R. (1994). Crystal structure of the catalytic domain of HIV-1 integrase: similarity to other polynucleotidyl transferases. *Science*, **266**, pp. 1981–1996.

20. Faulon, J. L., Sale, K. and Young, M. (2003). Exploring the conformational space of membrane protein folds matching distance constraints. *Protein Science*, **12**, pp. 1750–1761.

21. Fogg, M. J., Alzari, P., Bahar, M., Bertini, I., Betton, J. M., Burmeister, W. P., Cambillau, C., Canard, B., Corrondo, M. A., Coll, M., Daenke, S., Dym, O., Egloff, M. P., Enguita, F. J., Geerlof, A., Haouz, A., Jones, T. A., Ma, Q., Manicka, S. N., Migliardi, M., Nordlund, P., Owens, R. J., Peleg, Y., Schneider, G., Schnell, R., Stuart, D. I., Tarbouriech, N., Unge, T., Wilkinson, A. J., Wilmanns, M., Wilson, K. S., Zimhony, O. and Grimes, J. M. (2006). Application of the use of high-throughput technologies to the determination of protein structures of bacterial and viral pathogens. *Acta Crystallographica D — Biological Crystallography*, **62**, pp. 1196–1207.

22. Garcia-Ruiz, M. J. (2003). Nucleation of protein crystals. *Journal of Structural Biology*, **142**, pp. 22–31.

23. George, A., Chiang, A., Guo, B., Arabshahi, A., Cai, Z. and Wilson, W. (1997). Second virial coefficient as predictor in protein crystal growth. *Methods in Enzymology*, **276**, pp. 100–110.

24. Grosberg, A. Yu. and Khokhlov, A. R. (1994). *Statistical Physics of Macromolecules*. American Institute of Physics Press, New York.

25. Hirai, T. and Subramaniam, S. (2003). Structural insights into the mechanism of proton pumping by bacteriorhodopsin. *FEBS Letters*, **545**, pp. 2–8.

26. Hlavica, P., Schulze, J. and Lewis, D. F. (2003). Functional interaction of cytochrome P450 with its redox partners: a critical assessment and update of the topology of predicted contact regions. *Journal of Inorganic Biochemistry*, **96**, pp. 279–297.

27. Hui, R. and Edwards, A. (2003). High-throughput protein crystallization. *Journal of Structural Biology*, **142**, pp. 154–161.

28. Keenan, R., Siehl, D., Gorton, R. and Castle L. (2005). DNA shuffling as a tool for protein crystallization. *Proceedings of the National Academy of Sciences USA*, **102**, pp. 8887–8892.

29. Koutalos, Y., Ebrey, T. G., Tsuda, M., Odashima, K., Lien, T., Park, M. H., Shimizu, N., Derguini, F., Nakanishi, K. and Gilson, H. R. (1989). Regeneration of bovine and cephalopod opsins in situ with natural and artificial retinals. *Biochemistry*, **28**, pp. 2732–2739.

30. Kumar, S. and Nussinov, R. (2002). Close-range electrostatic interactions in proteins. *ChemBioChem*, **3**, pp. 604–617.

31. Lambeth, J. D. (1986). Cytochrome P450scc a review of the specificity and properties of the cholesterol binding site. *Endocrine Research*, **12**, pp. 371–392.

32. Landau, E. M. and Rosenbusch, J. P. (1996). Lipidic cubic phases: a novel concept for the crystallization of membrane proteins. *Proceeding of the National Academy of Sciences USA*, **93**, pp. 14532–14535

33. Li, H. and Poulos, T. L. (2004). Crystallization of cytochromes P450 and substrate-enzyme interactions. *Current Topics in Medicinal Chemistry*, **4**, pp. 1789–1802.

34. Madej, T., Gibrat, J. F. and Bryant, S. H. (1995). Threading a database of protein cores. *Proteins*, **23**, pp. 356–369.

35. Maxia, L., Radicchi, G., Pepe, I. M. and Nicolini, C. (1995). Characterization of Langmuir-Blodgett films of rhodopsin. Thermal stability studies. *Biophysical Journal*, **69**, pp. 1440–1446.

36. McManus, J. J., Lomakin, A., Ogun, O., Pande, A., Basan, M., Pande, J. and Benedek, G. B. (2007). Altered phase diagram due to a single point mutation in human {gamma} D-crystallin. *Proceedings of the National Academy of Sciences USA*, **104**, pp. 16856–16861.

37. Miedlich, S. U., Gama, L., Seuwen, K., Wolf, R. M. and Breitwieser, G. E. (2004). Homology modeling of the transmembrane domain of the human calcium sensing receptor and localization of an allosteric binding site. *Journal of Biological Chemistry*, **279**, pp. 7254–7263.

38. Nakagawa, M., Iwasa, T., Kikkawa, S., Tsuda, M. and Ebrey, T. G. (1999). How vertebrate and invertebrate visual pigments differ in their mechanism of photoactivation. *Proceedings of the National Academy of Sciences USA*, **96**, pp. 6189–6192.

39. Nathans, J. (1992). Rhodopsin: structure, function, and genetics. *Biochemistry*, **31**, pp. 4923–4931.

40. Nicolini, C., Erokhin, V., Ghisellini, P., Paternolli, C., Ram, M. K and Sivozhelezov, V. (2001). P450scc engineering and nanostructuring for cholesterol sensing. *Langmuir*, **17**, pp. 3719–3726.

41. Okada, T., Fujiyoshi, Y., Silow, M., Navarro, J., Landau, E. M. and Shichida, Y. (2002). Functional role of internal water molecules in rhodopsin revealed by X-ray crystallography. *Proceedings of the National Academy of Sciences USA*, **99**, pp. 5982–5987.

42. Palczewski, K., Kumasaka, T., Hori, T., Behnke, C. A., Motoshima, H., Fox, B. A., Le Trong, I., Teller, D. C., Okada, T., Stenkamp, R. E., Yamamoto, M. and Miyano, M. (2000). Crystal structure of rhodopsin: a G protein-coupled receptor. *Science*, **289**, pp. 739–745.

43. Patro, S. Y. and Przybycien, T. M. (1996). Simulations of reversible protein aggregate and crystal structure. *Biophysical Journal*, **70**, pp. 2888–2902.

44. Pechkova, E. and Nicolini, C. (2004). Protein nanocrystallography: a new approach to structural proteomics. *Trends in Biotechnology*, **22**, pp. 117–122.

45. Pechkova, E., Sivozhelezov, V. and Nicolini, C. (2007). Protein thermal stability: the role of protein structure and aqueous environment. *Archives of Biochemistry and Biophysics*, **466**, pp. 40–48.

46. Pepe, I. M. (1999). Rhodopsin and phototransduction. *Journal of Photochemistry and Photobiology B*, **48**, pp. 1–10.

47. Pepe, I. M. (2001). Recent advances in our understanding of rhodopsin and phototransduction. *Progress in Retinal and Eye Research*, **20**, pp. 733–759.

48. Pepe, I. M. and Cugnoli, C. (1992). Retinal photoisomerase: role in invertebrate visual cells. *Journal of Photochemistry and Photobiology B*, **13**, pp. 5–17.

49. Pepe, I. M., Cugnoli, C. and Schwemer, J. (1990). Rhodopsin reconstitution in bleached rod outer segment membranes in the presence of a retinal-binding protein from the honeybee. *FEBS Letters*, **268**, pp. 177–179.

50. Pepe, I. M. and Nicolini, C. (1996). Langmuir-Blodgett films of photosensitive proteins. *Journal of Photochemistry and Photobiology B*, **33**, pp. 191–200.

51. Pogozheva, I. D., Lomize, A. L. and Mosberg, H. I. (1997). The transmembrane 7-alpha-bundle of rhodopsin: distance geometry calculations with hydrogen bonding constraints. *Biophysical Journal*, **72**, pp. 1963–1985.

52. Pullara, F., Emanuele, A., Palma-Vittorelli, M. and Palma, M. (2007). Protein aggregation/crystallization and minor structural changes: universal versus specific aspects. *Biophysical Journal*, **93**, pp. 3271–3278.

53. Qiu, X. and Janson, C. (2004). Structure of apo acyl carrier protein and a proposal to engineer protein crystallization through metal ions. *Acta Crystallographica D — Biological Crystallography*, **60**, pp. 1545–1554.

54. Royant, A., Edman, K., Ursby, T., Pebay-Peyroula, E., Landau, E. M. and Neutze, R. (2000). Helix deformation is coupled to vectorial proton transport in the photocycle of bacteriorhodopsin. *Nature*, **406**, pp. 645–648.

55. Schadel, S. A., Heck, M., Maretzki, D., Filipek, S., Teller, D. C., Palczewski, K. and Hofmann, K. P. (2003). Ligand channeling within a G-protein-coupled receptor. The entry and exit of retinals in native opsin. *Journal of Biological Chemistry*, **278**, pp. 24896–24903.

56. Sivozhelezov, V. and Nicolini, C. (2005). Homology modeling of cytochrome P450scc and the mutations for optimal amperometric sensor. *Journal of Theoretical Biology*, **234**, pp. 479–485.

57. Sivozhelezov, V. and Nicolini, C. (2006). Theoretical framework for octopus rhodopsin crystallization. *Journal of Theoretical Biology*, **240**, pp. 260–269.

58. Skouri, M., Delsanti, M., Munch, J. P., Lorber, B. and Giege, R. (1991). Dynamic light scattering studies of the aggregation of lysozyme under crystallization conditions. *FEBS Letters*, **295**, pp. 84–88.

59. Teller, D. C., Okada, T., Behnke, C. A., Palczewski, K. and Stenkamp, R. E. (2001). Advances in determination of a high-resolution three-dimensional structure of rhodopsin, a model of G-protein-coupled receptors (GPCRs). *Biochemistry*, **40**, pp. 7761–7772.

60. Terakita, A., Koyanagi, M., Tsukamoto, H., Yamashita, T., Miyata, T. and Shichida, Y. (2004). Counterion displacement in the molecular

evolution of the rhodopsin family. *Nature Structure & Molecular Biology*, **11**, pp. 284–289.

61. Thurston, G. M. (2007). Protein anisotropy turns solubility on its head. *Proceedings of the National Academy of Sciences USA*. **104**, pp. 18877–18878.

62. Trabanino, R. J., Hall, S. E., Vaidehi, N., Floriano, W. B., Kam, V. W. and Goddard, W. A. (2004). First principles predictions of the structure and function of g-protein-coupled receptors: validation for bovine rhodopsin. *Biophysical Journal*, **86**, pp. 1904–1921.

63. Tsukamoto, H., Terakita, A. and Shichida, Y. (2005). A rhodopsin exhibiting binding ability to agonist all-trans-retinal. *Proceedings of the National Academy of Sciences USA*, **102**, pp. 6303–6309.

64. Vekilov, P. (2005). Two-step mechanism for the nucleation of crystals from solution. *Journal of Crystal Growth*, **275**, pp. 65–76.

65. Vekilov, P., Feeling-Taylor, A., Yau, S. and Petsev, D. (2002). Solvent entropy contribution to the free energy of protein crystallization. *Acta Crystallographica D*, **58**, pp. 1611–1616.

66. Vijayakumar, S. and Salerno, J. C. (1992). Molecular modeling of the 3-D structure of Cytochrome P-450scc. *Biochimica et Biophysica Acta*, **1160**, pp. 281–286.

67. Wade, R. C., Winn, P. J., Schlichting, I. and Sudarko, J. (2004). A survey of active site access channels in cytochromes P450. *Journal of Inorganic Biochemistry*, **98**, pp. 1175–1182.

68. Zhang, F., Basinski, M. B., Beals, J. M., Briggs, S. L., Churgay, L. M., Clawson, D. K., DiMarchi, R. D., Furman, T. C., Hale, J. E., Hsiung, H. M., Schoner, B. E., Smith, D. P., Zhang, X. Y., Wery, J. P. and Schevitz, R. W. (1997). Crystal structure of the obese protein leptin-E100. *Nature*, **387**, pp. 206–209.

Chapter 8

GOLD NANOPARTICLE THIN FILMS ON GLASS: INFLUENCE OF FILM THICKNESS AND ANNEALING TIME

Stephan V. Roth[a], Harald Walter[b], Rainer Gehrke[a], Markus Schenk[a,c], and Peter Müller-Buschbaum[d]

[a]*HASYLAB at Deutsches-Elektronen Synchrotron, Notkestr. 85, D-22607 Hamburg, Germany*

[b]*CSEM SA, Mattenstrasse 22 CH 4016 Basel, Switzerland*

[c]*Max-Planck-Institut für Quantenoptik, Hans-Kopfermann-Str. 1, 85748 Garching*

[d]*Technische Universität München, Physik Department LS E13, James-Franck-Str.1, D-85747 Garching, Germany*

stephan.roth@desy.de

We investigate *ex situ* the kinetics of a thin, thermally evaporated gold layer on glass below and at the percolation thickness. Using grazing incidence ultra-small-angle X-ray scattering (GIUSAXS), we follow the domain formation during annealing well below the bulk melting point of gold. We observe a strong dewetting of the gold layer for thicknesses near the critical percolation thickness. We compare our findings with optical spectroscopy, which corroborates the GIUSAXS data.

Synchrotron Radiation and Structural Proteomics
Edited by Eugenia Pechkova and Christian Riekel

www.panstanford.com

8.1 INTRODUCTION

Controlling the morphology of metal nanoparticle layers is of high technological and scientific interest. The broad field of applications of such layers includes catalysts (Valden *et al.*, 1998; Renaud *et al.*, 2003), solar cells (Westphalen *et al.*, 2000), biorecognition (Elghanian *et al.*, 1997), magnetic data storage (Lin *et al.*, 2005), and optoelectronic devices (Bauer *et al.*, 2003). The nanoparticle layer morphology is tuned to the specific application. This means either a fully established island geometry or a percolated or smooth system (Bauer *et al.*, 2003; Roth *et al.*, 2003, 2006b, 2007; Walter *et al.*, 2006). Thus, one can create full-scale homogenous contacts (Kaune *et al.*, 2009), color effects for anti-forging devices (Bauer *et al.*, 2003), or increase the efficiency of organic solar cells (Catchpole and Polman, 2008). The latter is done by incorporating such noble metal nanoparticles into the photoactive organic film (Metwalli *et al.*, 2008).

Especially the two latter applications make use of the plasmon resonance in noble metal nanoparticles. This resonance stems from collective oscillations of the free electron gas in external electromagnetic fields (Sun and Xia, 2003) in the confined geometry of the nanoparticle layer. The oscillation leads to pronounced absorption bands in the visible light (Hulteen *et al.*, 1997). Being a resonant phenomenon, it depends strongly on the nanoparticle layer structure and morphology. This influences the width and position of the plasmon resonance in metal layers. The size of the metal particles ranges from 1 nm to 100 nm.

Essential for contact application are the questions of percolation and thermal stability of the metal layer. This means that an electrically conducting path exists in the metal layer. Moreover, for annealing well below T_m ripening and diffusion must be taken into account (Revenant *et al.*, 2004).

8.2 ROUTES TO NANOSTRUCTURING

To produce a nanostructured metal layer, a multitude of preparation methods exists. Favored methods are creation from liquid colloidal solution or vacuum deposition. In solution casting, the solvent evaporates and the colloidal nanoparticles are deposited on the surface (Bigioni *et al.*, 2006; Roth *et al.*, 2007, 2009; Siffalovic *et al.*, 2008). The colloidal particles can be, e.g., spheres or triangles (Roth *et al.*, 2009). Large-scale arrangements over several millimeters can be achieved (Siffalovic *et al.*, 2008; Roth *et al.*, 2010). To create regular arrangements, the self-assembly processes are exploited (Xia *et al.*, 2000) during evaporation. The ordering takes place at the triple phase boundary contact line (Roth *et al.*, 2007) and is strongly governed by thermodynamic fields and capillary effects (Haw *et al.*, 2002; Müller-Buschbaum *et al.*, 2006b).

Vacuum deposition is another route employed in roll-to-roll applications (Bauer *et al.*, 2003; Walter *et al.*, 2006). Vacuum deposition includes thermal evaporation, sputter deposition, or pulsed laser deposition (Roth *et al.*, 2003, 2006b; Biswas *et al.*, 2006; Röder *et al.*, 2008) to create large-scale nanostructured composite films for the above-mentioned applications. The methods strongly differ in the thermal energy of the metal atoms being deposited. In the case of sputter deposition, this energy is around 1 eV, while for thermal evaporation the kinetic energy is around 25 meV. The installed structures are different, as is seen in the case of polymer–metal films (Roth *et al.*, 2003, 2006b), e.g., during thermal evaporation larger structures are installed.

Further steps of nanostructuring include thermal annealing (Lin *et al.*, 2005; Kashem *et al.*, 2009), leading to hierarchically organized structures on multiple length scales. The structures installed strongly depend on the initial layer thickness. Processes like dewetting, Vollmer–Weber growth, and ripening may occur (Lazzari *et al.*, 1999; Revenant *et al.*, 2004).

8.3 SAMPLE PREPARATION

In the approach described here, we investigated a basic two-layer system, namely gold (Au) on an amorphous substrate (glass) (Parrill *et al.*, 1993; Naudon *et al.*, 2000). We used thermal evaporation in a vacuum chamber (tectra GmbH). The glass slides had a dimension of 25 mm × 30 mm × 1 mm and were coated successively using a rotation stage in the vacuum chamber. The vacuum pressure was 4×10^{-5} mbar. The gold target was heated using a current of $I = 3.5$ A and a voltage of $U = 78$ V, applied for a certain period t_{evap}. The target-to-sample distance was about 350 mm. Different gold layer thicknesses d were achieved using different evaporation times t_{evap}. The following gold layer thicknesses d were prepared: $d = 3$, 5, and 8 nm (d_1, d_2, d_3). The critical thickness for percolation is around $d_c = 7 \pm 1$ nm for noble metals such as Ag and Au (Charton *et al.*, 2001, 2004; Walter *et al.*, 2002; Roth *et al.*, 2003, 2004, 2006b). Thus, d covers the range well below to just above the percolation threshold. During thermal evaporation, the gold atoms self-assemble and form islands (Levine *et al.*, 1989; Parrill *et al.*, 1993; Naudon *et al.*, 2000). At the critical thickness, percolation is observed and the optical and electrical properties change significantly.

To investigate morphological and structural changes as a function of film thickness, we further applied low-temperature annealing (Hulteen *et al.*, 1997) at $T = 300$ °C, which is well below the Au bulk melting temperature $T = 1064$ °C. To do so, a sample of each thickness d was prepared for seven different annealing times on different glass slides. The chosen annealing times were as follows: $t_{anneal} = 10$ min, 20 min, 1 h, 3 h, 7 h, and 24 h.

8.4 EXPERIMENTAL METHODS

To correlate the structure of the thin gold layers with its optical properties, we used grazing incidence ultra-small-angle X-ray scattering (GIUSAXS) and optical spectroscopy.

8.4.1 Grazing Incidence Ultra-Small-Angle X-Ray Scattering (GIUSAXS)

While local surface information is easily obtained by atomic force microscopy, a statistically more significant characterization of the large-scale samples becomes accessible by using scattering techniques. Although gold is a strongly scattering material, a thin film of gold (in the range of some nanometers) might give rise to a weak signal in a transmission geometry. By the use of a reflection geometry the surface sensitivity is enhanced (Müller-Buschbaum,

2003; Müller-Buschbaum *et al.*, 2009). We investigated the samples using GIUSAXS using the ultra-small-angle option uniquely offered at the BW4 beamline at HASYLAB/DESY (Müller-Buschbaum *et al.*, 1997; Roth *et al.*, 2006a). This combination of reflection geometry and USAXS setup is especially suited to investigate large-scale structural and morphological changes at length scales up to an upper limit significantly larger than 1 μm (Müller-Buschbaum *et al.*, 2006a). The GIUSAXS — carried out at the beamline BW4 of HASYLAB — used a wavelength of $\lambda = 0.138$ nm with a sample-to-detector distance $D_{\mathrm{SD}} = 12.87$ m and an angle of incidence $\alpha_{\mathrm{i}} = 0.631°$. The beam size was 400×400 μm^2, defined by two high-quality entrance cross-slits ensuring a high signal-to-background ratio. Most part of the scattering pathway was under high vacuum conditions to reduce background caused by air scattering. Samples were placed in a special designed vacuum GISAXS chamber and measured *ex situ* for different annealing times and film thicknesses, respectively.

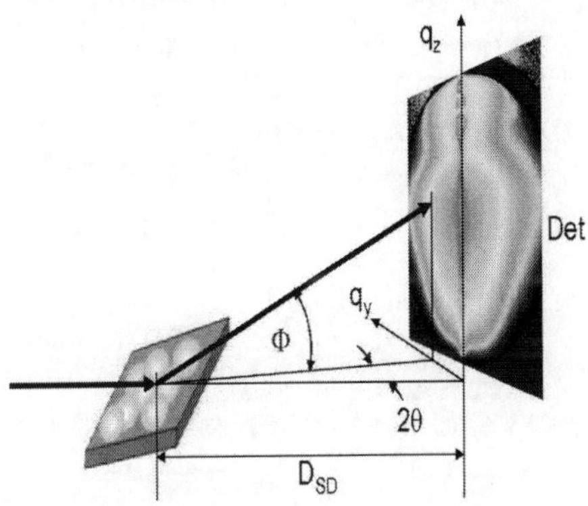

Figure 8.1 Geometry of the GIUSAXS experiment. $\Phi = \alpha_{\mathrm{i}} + \alpha_{\mathrm{f}}$ denotes the sum of incident and exit angle, 2θ is the scattering angle, and q_y, q_z denote the reciprocal space coordinate system. $D_{\mathrm{SD}} = 12.87$ m is the sample-to-detector distance. Det denotes a two-dimensional detector. See also Color Insert.

The geometry of the GIUSAXS experiment is shown in Fig. 8.1. The sample is mounted on a two-axis goniometer. q_y and q_z denote a reciprocal space coordinate system with q_y being parallel to and q_z being perpendicular to the sample surface. q_y and q_z are related to the incidence angle α_{i}, the exit angle α_{f}, and to the out-of-plane angle θ via (Sinha *et al.*, 1988; Salditt *et al.*, 1995):

$$q_v = \frac{2\pi}{\lambda}\sin(2\theta)\cos(\alpha_f),\ q_z = \frac{2\pi}{\lambda}\sin(\Phi),$$

with $\Phi = \alpha_i + \alpha_f$.

The intensity distribution $I\ (q_v, q_z)$ was recorded with a two-dimensional (2D) detector. We used a 2D multi-wire proportional chamber gas detector with an effective pixel size of $400 \times 400\ \mu m^2$, because of its high sensitivity and high dynamic range.

8.4.2 Optical Spectroscopy

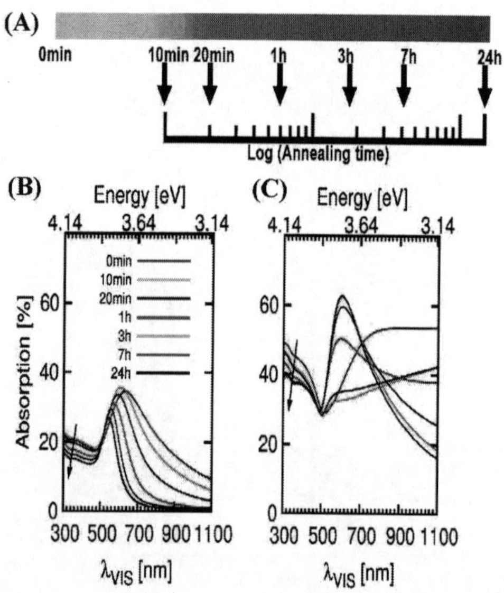

Figure 8.2 (a) Annealing times used (logarithmic scale) and pictorial representation of the color change during annealing. The color bar shows the transition from blue to red in the visible wavelength regime (red-shift) due to the break up of the coalesced gold layer during annealing for a deposited mass thickness of d_3 = 8 nm. (b) Optical absorption spectra for gold mass thickness d_1 = 3 nm: The spectra indicate a granular island structure. Upon annealing a shift of the maximum is observed, indicating a transition to a monodisperse size distribution. The arrow indicates the increasing annealing time. (c) Optical absorption spectra for gold mass thickness d_2 = 8 nm: A clear transition from a coalescent, smooth gold layer to a granular film is observed. Included is an energy axis to indicate that energetic aspects are correlated with the structures. See also Color Insert.

Suited methods to characterize the optical properties are ellipsometry (Körstgens *et al.*, 2010) and optical spectroscopy. In the case presented here, we used optical spectroscopy to investigate the extinction curves in the wavelength range $\lambda_{vis} = 300$–1100 nm. For this purpose, we used a Perkin Elmer spectrometer for these optical spectroscopic measurements in transmission geometry.

This allows for correlating *ex situ* the structural changes during low-temperature annealing with the characteristic plasmon resonance frequencies of the nanoparticle layers. Already from optical inspection a clear color change of the nanoparticle layers can be observed, depending on the gold layer thickness and annealing time. As an example, we present optical micrographs for a nominal gold layer thickness of $d = 8$ nm in Fig. 8.2a.

8.5 DATA ANALYSIS

To analyze the structural changes in the nanoparticle morphology during low-temperature annealing, we first present the GIUSAXS analysis. On the basis of this quantitative analysis, we derive a model to describe nanoparticle growth during low-temperature annealing (see described in the next section). The 2D GIUSAXS signal $I(q_y, q_z)$ can be characterized by two perpendicular intensity line cuts, the so-called detector and out-of-plane cut. In the detector cut, the intensity of the signal along the line ($\alpha_i = $ const, α_f variable) is followed. Here correlations perpendicular to the sample surface can be investigated, e.g., layer thickness, nanoparticle height, and roughness correlations (resonant diffuse scattering) (Müller-Buschbaum, 2003). One remarkable feature is the Yoneda peak, which occurs at the critical angle of the material under consideration at $\alpha_i, \alpha_f = \alpha_c$.

The out-of-plane cut gives the intensity as a function of the out-of-plane scattering vector $q_y = (2\pi/\lambda)\sin(2\theta)\cos(\alpha_f)$, with 2θ being the out-of-plane scattering angle and α_f the exit angle with respect to the sample surface. Here, correlations parallel to the sample surface, such as distance, domain sizes, and nanoparticle radii, can be detected (Müller-Buschbaum, 2003; Roth *et al.*, 2003, 2006b, 2007).

Figure 8.3a shows the detector cuts of the d = 8 nm gold nanoparticle layer for different annealing times. For comparison, data from the bare substrate have been included as well to illustrate the influence of the added gold layer. With increasing annealing time of t_{anneal} = 0 min, 10 min, 20 min, 1 h, 3 h, 7 h, and 24 h, the scattering pattern is changing. The Yoneda peak in the data taken at 0 min at $\Phi = 1.05°$ can be attributed to the gold nanoparticle layer. For t_{anneal} < 1 h, only one broad Yoneda peak is visible. The shoulder for $\Phi < 0.9°$ originates from the Yoneda peak of glass at $\Phi_{c,1}$. The Yoneda peak of gold slightly changes its position and intensity, compared to the broad shoulder. This already indicates a roughening of the gold layer. For $t_{anneal} \geq 1$ h, however, the detector cuts change dramatically. At t_{anneal} = 1 h, the Yoneda peak at $\Phi = 1°$ has strongly decreased in intensity, the intensity is shifted to $\Phi < 0.9°$. For t_{anneal} > 1 h, only one broad Yoneda peak with its position near $\Phi_{c,1} = 0.83°$ is visible. Its shape does no longer change with increasing annealing time. When comparing with the detector cut of the bare substrate, one can clearly observe that the detector cut for t_{anneal} > 1 h is dominated by the contribution from the glass substrate. As seen in Fig. 8.3b, the Yoneda peak of the glass substrate is located at $\Phi_{c,1} = 0.83°$.This behavior indicates a strong roughening for $t_{anneal} \leq$ 1 h and negligible further change in the structure for t_{anneal} >1 h.

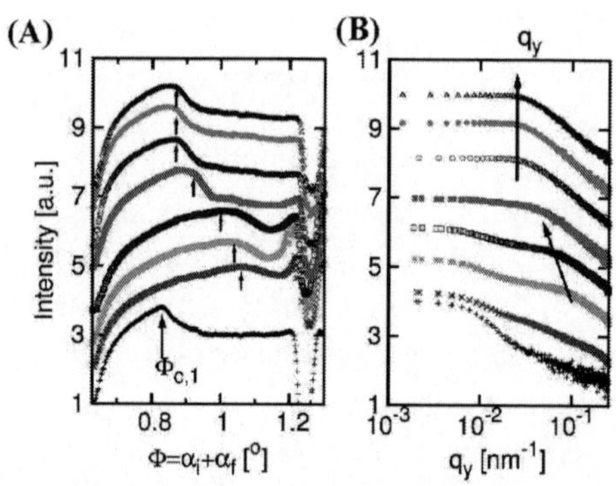

Figure 8.3 GIUSAXS data for d_3 = 8 nm; from bottom to top: glass, 0 min, 10 min, 20 min, 1 h, 3 h, 7 h, and 24 h annealing time. The curves are shifted from the bottom to the top along the intensity axis for better visibility. (a) Detector cuts for q_y = 0. $\Phi_{c,1}$ is the position of the Yoneda peak of glass. The small arrows indicate the evolution of the Yoneda peak of gold. (b) Out-of-plane cuts for at α_f = 0.43°. The arrows indicate the evolution of the side maxima, from which ζ is deduced. See also Color Insert.

Figure 8.3b shows the corresponding out-of-plane scans at $\alpha_f = 0.43°$ near the critical angle of gold at the chosen wavelength ($\alpha_f = 0.5°$). Again data of the bare substrate are included to demonstrate the absence of most prominent in-plane lengths in the bare substrate. Already without annealing ($t_{anneal} = 0$ min) a weak shoulder can be seen in the data, corresponding to a small and less well-defined lateral structure. For the quantitative analysis we use the simple relation $\xi(t_{anneal}) = 2\pi/q_y^*(t)$, where q_y^* denotes the position of the side maximum, to calculate the most prominent in-plane length *scale* ξ as a function of annealing time. For $t_{anneal} = 0$ min, it results in $\xi(0 \text{ min}) = 36$ nm.

With increasing annealing time the side maximum shifts to smaller q_y-values, indicating an increasing most prominent in-plane length $\xi(t_{anneal}) = 2\pi/q_y^*$. These results corroborate the coarsening of the nanoparticle layer deduced from the analysis of the detector cuts (see above). This is induced by coalescence of the nanoparticles and leads to an increase in the most prominent in-plane length. In detail, Fig. 8.4 shows the most prominent in-plane length $\xi(t_{anneal})$ as a function of annealing time for $d_1 = 3$ nm, $d_2 = 5$ nm, and $d_3 = 8$ nm.

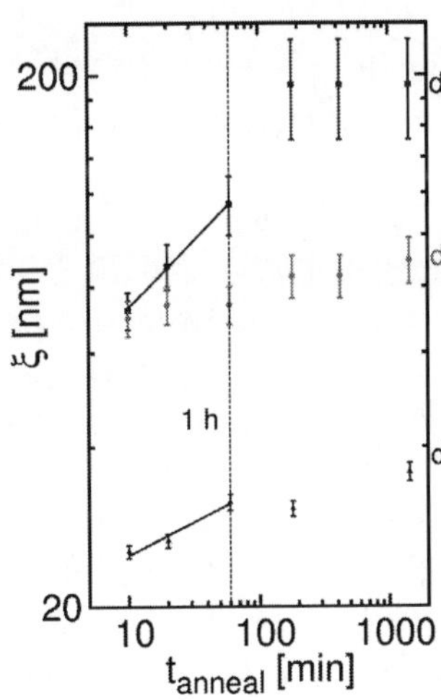

Figure 8.4 Evolution of the most prominent in-plane length ξ as a function of annealing time for different thicknesses $d_1 = 3$ nm, $d_2 = 5$ nm, $d_3 = 8$ nm. The solid lines correspond to power laws $\xi \propto t_{anneal}^{0.25\pm0.01}$ for d_3 and $\xi \propto t_{anneal}^{0.12\pm0.02}$ for d_1. See also Color Insert.

Depending on the initially deposited layer thickness, the temporal behavior of ξ differs significantly. For $d_3 = 8$ nm, two different time regimes can be identified:

1) 0 min < $t_{anneal} \leq 1$ h: $\xi(t_{anneal})$ follows a power law, $t^{0.25\pm0.01}$. This is *strongly sub-diffusional* and indicates that the growth is not diffusion-limited, as the exponent of the power law is well below 0.5. The power law is a signature of coalescence of nanoparticles during annealing due to a high mobility of the initially deposited nanoparticles (Lazzari *et al.*, 2007). Following Lazzari *et al.* (2007), we can interpret ξ as a domain size.
2) $t_{anneal} > 1$ h: $\xi(t_{anneal})$ stays constant. This means, coalescence and growth of the nanoparticles are suppressed. The structure has reached an equilibrium, stable state. This observation is in agreement with the detector cuts in Fig. 8.3a. A coalescence behavior is expected for deposited thickness $d \geq 6$ nm (Lazzari *et al.*, 2007).

For lower deposited mass thickness of $d_1 = 3$ nm, again, we find a coalescence behavior for $t_{anneal} \leq 1$ h with $\xi \sim t^{0.12\pm0.02}$, indicating homogenous nucleation (Lazzari *et al.*, 2007) and ξ = constant for $t_{anneal} > 1$ h. Surprisingly, for $d_2 = 5$ nm, $\xi(t_{anneal})$ is nearly constant, but shows a step at $t_{anneal} = 1$ h, which corresponds to the cross over time for $d_1 = 3$ nm and $d_3 = 8$ nm.

To correlate the structure derived from GIUSAXS with the optical properties of the binary systems, we performed additional optical spectroscopic measurements. These measurements allow for correlating the temperature dependence with a shift in the plasmon resonance frequency (Bauer *et al.*, 2003). In Fig. 8.2 we show the absorption spectra of the Au/glass system for two mass thicknesses $d_1 = 3$ nm (Fig. 8.2b) and $d_3 = 8$ nm (Fig. 8.2c) measured at perpendicular illumination direction. For $d_1 = 3$ nm, the optical spectra show a single resonance only. The peak shifts to lower values ($\lambda_{vis} > 680$ nm to $\lambda_{vis} > 540$ nm) and becomes sharper during annealing. These observations are fully consistent with the spectral evolution induced by diameter changes for spherical metal nanoparticles (Sun and Xia, 2003).

However, for d_3 = 8 nm we observe a distinct change in the optical spectra as a function of annealing time. For t_{anneal} < 1 h, no distinct plasmon band is observed. Instead, high absorption in the spectra for λ_{vis} > 600 nm is seen (Bauer *et al.*, 2003). This indicates a coalesced thin film. For t_{anneal} > 1 h and increasing annealing time, the intensity distribution in the spectra changes. A strong peak around λ_{vis} > 540 nm (the typical gold plasmon resonance (Bauer *et al.*, 2003)) occurs. This resonance proves the existence of distinct, well-separated nanoparticles ("domains"), as seen in Fig. 8.5. This transition from a coalescent thin film to isolated nanoparticles is also seen in Fig. 8.2 with the transition from bluish-grey to red color of the samples.

8.6 DISCUSSION

We summarize our findings in the kinetics model in Fig. 8.5. At room temperature before starting annealing (at t_{anneal} = 0 min), the initial deposition of gold leads to a self-assembled nanoparticle structure, as sketched in Fig. 8.5 by the usual bimodal distribution (Roth *et al.*, 2003).

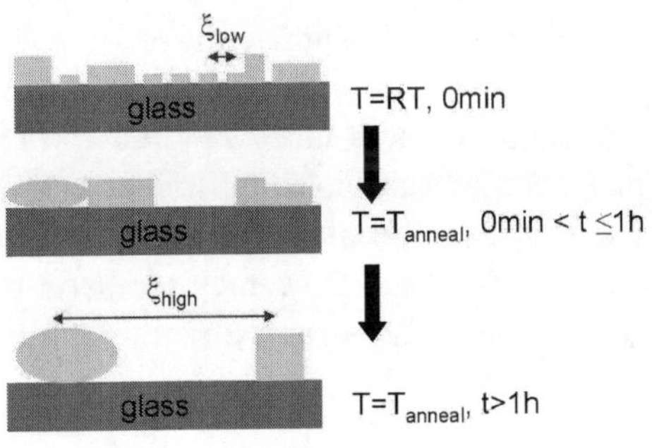

Figure 8.5 Model used for the GIUSAXS data evaluation of the structural evolution for d = 8 nm. ξ_{low} denotes the value for ξ at 0 min (before starting annealing), ξ_{high} the value for t > 1 h annealing. See also Color Insert.

For annealing times t_{anneal} ≤ 1 h (at T = 300°C), we observe coalescence. This result is supported both by GIUSAXS and optical interferometry analysis. For large annealing times t_{anneal} > 1 h (at T = 300°C) and for all three investigated mass thicknesses

(d = 3, 5, and 8 nm) a constant most prominent in-plane length ξ is observed. This leads us to the conclusion that the separation between the by now large-sized nanoparticles does not allow further coalescence or growth.

Using GIUSAXS and optical spectroscopy, we are able to deduce the structure–function relationship of the annealed structure in terms of optical properties and optically non-visible nanoscopic length scales. GIUSAXS clearly allows for determining lateral length scales using the simple relation $\xi(t_{\text{anneal}}) = 2\pi/q_{y}^{*}$. From Figs. 8.3b and 8.5, it becomes clear that for annealing times > 1 h, the nanoparticles' distance is no longer affected by annealing. Some minor changes are visible in the optical spectra; see Fig. 8.2b,c. By combining these results, we can expect that only nanoparticle shape changes, which would affect the plasmon resonances, take place in this time regime: In Fig. 8.2b,c the optical spectra converge for $t_{\text{anneal}} > 1$ h.

Implications for the applications are manifold. Temperature stability clearly is an issue in sensor applications or anti-counterfeiting, to name just a few. In sensing applications like surface-enhanced Raman scattering (SERS) (Jiang *et al.*, 2009; Mistark *et al.*, 2009) the signal detection and interpretation depends strongly on the shape and distance of the metal nanoparticles used as tracers (Garrell, 1989). When using our route to nanostructuring, the advantage is that after applying sufficiently long annealing times a stable structure is reached. Therefore, this method ensures a proper operation of application devices up to T = 300°C! In combination with modern high-temperature polymers[a] this method could open additional routes in anticounterfeiting, where one could enlarge the security range of the features: Even high-temperature treatment would not destroy the security features and thus leaves the product uniquely identifiable.

8.7 CONCLUSION

To summarize, we have presented a combined *ex situ* investigation of the kinetics of gold nanoparticles on glass using optical spectroscopy and GIUSAXS. We investigated the behavior on large lateral length scales (domains) of the thin gold layers as a function of annealing

[a] http://www.freepatentsonline.com/EP0315851.html

time at T = 300°C and the deposited mass thickness. We clearly observe coalescence of the deposited gold nanoparticles. All three investigated thicknesses below and in the vicinity of the critical percolation thickness show a similar behavior, with a non-diffusion-limited growth for annealing times $t_{anneal} \leq 1$ h and a constant equilibrium regime for $t_{anneal} > 1$ h. Furthermore, we correlated the layer structure with its optical spectra, which is of importance for applications, such as SERS and anti-counterfeiting.

Acknowledgments

Portions of this research were carried out at the light source DORIS III at HASYLAB/DESY. DESY is a member of the Helmholtz Association (HGF). The authors would like to thank R. Döhrmann for his help at the beamline BW4.

References

1. Bauer, G., Hassmann, J., Walter, H., Haglmüller, J., Mayer, C. and Schalkhammer, T. (2003). Resonant nanocluster technology — from optical coding and high quality security features to biochips. *Nanotechnology*, **14**, pp. 1289–1311.
2. Bigioni, T. P., Lin, X.-M., Nguyen, T. T., Corwin, E. I., Witten, T. A. and Jaeger, H. M. (2006). Kinetically driven self assembly of highly ordered nanoparticle monolayers. *Nature Materials*, **5**, pp. 265–270.
3. Biswas, B., Karulkar, P. C., Eilers, H., Norton, M. G., Skorski, D., Davitt, C., Greve, H., Schürmann, U., Zaporojtchenko, V. and Faupel, F. (2006). Vapor phase deposition of nanostructured polymer-metal composites for advanced technology applications. *Vacuum Technology and Coating*, **7**, pp. 54–59.
4. Catchpole, K. R. and Polman, A. (2008). Plasmonic solar cells. *Optics Express*, **16**, pp. 21793–21800.
5. Charton, C. and Fahland, M. (2001). Optical and electrical properties of sputtered Ag films on PET webs. *Surface and Coatings Technology*, **142**, pp. 175–180.
6. Charton, C. and Fahland, M. (2004). Electrical properties of Ag films on polyethylene terephthalate deposited by magnetron sputtering. *Thin Solid Films*, **449**, pp. 100–104.

7. Elghanian, R., Storhoff, J. L., Mucic, R. C., Letsinger, R. L. and Mirkin, C. A., (1997). Selective colorimetric detection of polynucleotides based on the distance-dependent optical properties of gold nanoparticles. *Science*, **277**, pp. 1078–1081.

8. Garrell, R. L. (1989). Surface-enhanced Raman spectroscopy. *Analytical Chemistry*, **61**, pp. 401A–411A.

9. Haw, M. D., Gillie, M. and Poon, W. C. K. (2002). Effects of phase behavior on the drying of colloidal suspensions. *Langmuir*, **18**, pp. 1626–1633.

10. Hulteen, J. C., Patrissi, C. J., Miltner, D. L., Crosthwait, E. R., Oberhauser, E. B. and Martin, C. R. (1997). Changes in the shape and optical properties of gold nanoparticles contained within alumina membranes due to low-temperature annealing. *Journal of Physical Chemistry B*, **101**, pp. 7727–7731.

11. Jiang, X., Zhang, L., Wang, T. and Wan, Q. (2009). High surface-enhanced Raman scattering activity from Au-decorated individual and branched tin oxide nanowires. *Journal of Applied Physics*, **106**, art. no. 104316.

12. Kashem, M. M. A., Perlich, J., Diethert, A., Wang, W. N., Memesa, M., Gutmann, J. S., Majkova, E., Capek, I., Roth, S. V., Petry, W. and Müller-Buschbaum, P. (2009). Array of magnetic nanoparticles via particle co-operated self-assembly in block copolymer thin film. *Macromolecules*, **42**, pp. 6202–6208.

13. Kaune, G., Ruderer, M. A., Metwalli, E., Wang, W., Couet, S., Schlage, K., Röhlsberger, R., Roth, S. V. and Müller-Buschbaum, P. (2009). In situ GISAXS study of gold film growth on conducting polymer films. *ACS Applied Materials & Interfaces*, **1**, pp. 353–362.

14. Körstgens, V., Wiedersich, J., Meier, R., Perlich, J., Roth, S. V., Gehrke, R. and Müller-Buschbaum, P. (2010). Combining imaging ellipsometry and grazing incidence small angle X-ray scattering for in situ characterization of polymer nanostructures. *Analytical and Bioanalytical Chemistry*, **396**, pp. 139–149.

15. Lazzari, R., Jupille, J. and Borensztein, Y. (1999). In situ study of a thin metal film by optical means. *Applied Surface Science*, **142**, pp. 451–454.

16. Lazzari, R., Renaud, G., Jupille, J. and Leroy, F. (2007). Self-similarity during growth of the Au/TiO2(110) model catalyst as seen by the scattering of x-rays at grazing-angle incidence. *Physical Reviews B*, **76**, art. no. 125412.

17. Levine, J. R., Cohen, J. B., Chung, Y. W. and Georgopoulos, P. (1989). Grazing-incidence small-angle X-ray-scattering — new tool for studying thin-film growth. *Journal Applied Crystallography*, **22**, pp. 528–532.

18. Lin, Y., Boker, A., He, J. B., Sill, K., Xiang, H. Q., Abetz, C., Li, X. F., Wang, J., Emrick, T., Long, S., Wang, Q., Balazs, A. and Russell, T. P. (2005). Self-directed self-assembly of nanoparticle/copolymer mixtures. *Nature*, **434**, pp. 55–59.

19. Metwalli, E., Couet, S., Schlage, K., Röhlsberger, R., Körstgens, V., Ruderer, M., Wang, W., Kaune, G., Roth, S. V. and Müller-Buschbaum, P. (2008). In situ GISAXS investigation of gold sputtering onto a polymer template. *Langmuir*, **24**, pp. 4265–4272.

20. Mistark, P. A., Park, S., Yalcin, S. E., Lee, D. H., Yavuzcetin, O., Tuominen, M. T., Russell, R. P. and Achermann, M. (2009). Block-copolymer-based plasmonic nanostructures. *ACS Nano*, **3**, pp. 3987–3992.

21. Müller-Buschbaum, P. (2003). Grazing incidence small-angle X-ray scattering: an advanced scattering technique for the investigation of nanostructured polymer films. *Analytical Bioanalytical Chemistry*, **376**, pp. 3–10.

22. Müller-Buschbaum, P. (2009). A basic introduction to grazing incidence small angle X-ray scattering. In *Applications of Synchrotron Light to Scattering and Diffraction in Materials and Life Sciences* (eds. Ezquerra, T. A., Garcia-Gutierrez, M., Nogales, A. and Gomez, M.), Vol. 776, pp. 61–90. Springer, Berlin,

23. Müller-Buschbaum, P., Bauer, E., Maurer, E., Schlögl, K., Roth, S. V. and Gehrke, R. (2006a). Route to create large-area ordered polymeric nanochannel arrays. *Applied Physics Letters*, **88**, art. no. 083114.

24. Müller-Buschbaum, P., Bauer, E., Pfister, S., Roth, S. V., Burghammer, M., Riekel, C., David, C. and Thiele U. (2006b). Creation of multi-scale stripe-like patterns in thin polymer blend films. *Europhysics Letters*, **73**, pp. 35–41.

25. Müller-Buschbaum, P., Vanhoorne, P., Scheumann, V. and Stamm, M. (1997). Observation of nano-dewetting structures. *Europhysics Letters*, **40**, pp. 655–660.

26. Naudon, A., Babonneau, D., Thiaudiere, D., Lequien, S. (2000). Grazing-incidence small-angle X-ray scattering applied to the characterization of aggregates in surface regions. *Physica B*, **283**, pp. 69–74.

27. Parrill, J. R. L., Georgopoulos, P., Chung, Y. W. and Cohen, J. B. (1993). GISAXS — glancing incidence small-angle X-Ray-scattering. *Journal de Physique IV*, **3**, pp. 411–417.

28. Renaud, G., Lazzari, R., Revenant, C., Barbier, A., Noblet, M., Ulrich, O., Leroy, F., Jupille, J., Borensztein, Y., Henry, C. R., Deville, J.-P., Scheurer, F., Mane-Mane, J. and Fruchart, O. (2003). Real-time monitoring of growing nanoparticles. *Science*, **300**, pp. 1416–1419.

29. Revenant, C., Leroy, F., Lazzari, R., Renaud, G., and Henry, C. R. (2004). Quantitative analysis of grazing incidence small-angle x-ray scattering: Pd/MgO(001) growth. *Physical Review B*, **69**, art. no. 035411.

30. Röder, J., Faupel, J. and Krebs, H.-U. (2008). Growth of polymer-metal nanocomposites by pulsed laser deposition. *Applied Physics A-Materials Science & Processing*, **93**, pp. 863–867.

31. Roth, S. V., Autenrieth, T., Grübel, G., Riekel, C., Burghammer, M., Hengstler, R., Schulz, L. and Müller-Buschbaum, P. (2007). In situ observation of nanoparticle ordering at the air-water-substrate boundary in colloidal solutions using X-ray nanobeams. *Applied Physics Letters*, **91**, art. no. 091915.

32. Roth, S. V., Burghammer, M., Riekel, C., Müller-Buschbaum, P., Diethert, A. and Panagiotou, P. (2003). Self-assembled gradient nanoparticle-polymer multilayers investigated by an advanced characterization method: microbeam grazing incidence X-ray scattering. *Applied Physics Letters*, **82**, pp. 1935–1937.

33. Roth, S. V., Döhrmann, R., Dommach, M., Kuhlmann, M., Kröger, I., Gehrke, R., Walter, H., Schroer, C., Lengeler, B. and Müller-Buschbaum, P. (2006a). Small-angle options of the upgraded ultrasmall-angle X-ray scattering beamline BW4 at HASYLAB. *Review of Scientific Instruments*, **77**, art. no. 085106.

34. Roth, S. V., Kuhlmann, M., Walter, H., Snigirev, A., Snigireva, I., Lengeler, B., Schroer, C., Burghammer, M., Riekel, C. and Müller-Buschbaum, P. (2009). Colloidal silver nanoparticle gradient layer prepared by drying between two walls of different wettability. *Journal of Physics-Condensed Matter*, **21**, art. no. 264012.

35. Roth, S. V., Müller-Buschbaum, P., Burghammer, M., Walter, H., Panagiotou, P., Diethert, A. and Riekel, C. (2004). Microbeam grazing incidence small angle X-ray scattering — a new method to investigate heterogeneous thin films and multilayers. *Spectrochimica Acta Part B-Atomic Spectroscopy*, **59**, pp. 1765–1773.

36. Roth, S. V., Rothkirch, A., Autenrieth, T., Gehrke, R., Wroblewski, T., Burghammer, M. C., Riekel, C., Schulz, L., Hengstler, R. and Müller-Buschbaum, P. (2010). Spatially resolved investigation of solution cast nanoparticle films by X-ray scattering and multidimensional data set classification. *Langmuir*, **26**, pp. 1496–1500.

37. Roth, S. V., Walter, H., Burghammer, M., Riekel, C., Lengeler, B., Schroer, C., Kuhlmann, M., Walther, T., Sehrbrock, A., Domnick, R. and Müller-Buschbaum, P. (2006b). Combinatorial investigation of the isolated nanoparticle to coalescent layer transition in a gradient sputtered gold nanoparticle layer on top of polystyrene. *Applied Physics Letters*, **88**, art. no. 021910.

38. Salditt, T., Metzger, T. H., Brandt C., Klemradt, U. and Peisl, J. (1995). Determination of the static scaling exponent of self-affine interfaces by nonspecular x-ray-scattering. *Physical Review B*, **51**, pp. 5617–5627.

39. Siffalovic, P., Majkova, E., Chitu, L., Jergel, M., Luby, S., Capek, I., Satka, A., Timmann, A. and Roth, S. V. (2008). Real-time tracking of superparamagnetic nanoparticle self-assembly. *Small*, **4**, pp. 2222–2228.

40. Sinha, S. K., Sirota, E. B., Garoff, S. and Stanley, H. B. (1988). X-ray and neutron scattering from rough surfaces. *Physical Review B*, **38**, pp. 2297–2311.

41. Sun, Y. and Xia, Y. (2003). Gold and silver nanoparticles: a class of chromophores with colors tunable in the range from 400 to 750 nm. *Analyst*, **128**, pp. 686–691.

42. Valden, M., Lai, X. and Goodman, D. W. (1998). Onset of catalytic activity of gold clusters on titania with the appearance of nonmetallic properties. *Science*, **281**, 1647–1650.

43. Walter, H., Bauer, G., Domnick, R. and Hassmann, J. (2002). In 45th Annual Technical Conference Proceedings ISSN, 0737-5921, p. 443.

44. Walter, H., Bauer, G., Domnick, R., Jakopic, G. and Leitner, A. (2006). Role of granular structure in metal layers on the optical properties of absorbing mirrors. *Optical engineering*, **45**, art. no. 103801.

45. Westphalen, M., Kreibig, U., Rostalski, J., Lüth, H. and Meissner, D. (2000). Metal cluster enhanced organic solar cells. *Solar Energy Materials and Solar Cells*, **61**, pp. 970–105.

46. Xia, Y., Gates, B., Yin, Y. and Lu, Y. (2000). Monodispersed colloidal spheres: old materials with new applications. *Advanced Materials*, **12**, pp. 693–712.

SECTION III
STRUCTURAL PROTEOMICS

Chapter 9

ATOMIC STRUCTURE AND RADIATION RESISTANCE OF LANGMUIR–BLODGETT PROTEIN CRYSTALS

Eugenia Pechkova[a], Sean McSweeney[b], and Claudio Nicolini[a]

[a]*Nanoworld Institute, Fondazione EL.B.A., University of Genoa Medical School, Italy*

[b]*ESRF, Grenoble, France*

manuscript@ibf.unige.it

The structural changes associated to increasing exposure of intense X-ray synchrotron radiation for a wide range of model proteins are summarized in this review for two distinct methods of crystallization. The two types of crystals are respectively grown by the Langmuir–Blodgett (LB) nanotemplate method and by the classical hanging drop method to quantify their distinct radiation resistance. Changes in parameters like B factor and reflection intensity versus absorbed dose, along with changes in electron density maps, were monitored for both types of crystals. The six model proteins were studied and compared using four different beamlines, namely, the ID-13, ID14-2, ID23-1, and ID29 of the ESRF in Grenoble, keeping in mind the fundamentals knowledge of radiation damage being acquired in the last decade. Consistently crystals grown by LB nanotemplate-based method proved to be significantly more radiation stable when compared to crystals grown by the classical hanging drop method.

Synchrotron Radiation and Structural Proteomics
Edited by Eugenia Pechkova and Christian Riekel

www.panstanford.com

9.1 INTRODUCTION

The increase of interest to protein crystals was related to their importance for determining the macromolecule structures by X-ray crystallography. Other methods (like NMR, etc.) have serious limitations in terms of protein size and can be used with great difficulties for structure determination of the complexes, which are particularly important for the understanding of protein functions. With the recent rapid progress in macromolecular microcrystallography using synchrotron radiation (Cusack *et al.*, 1998; Riekel, 2000), it is clear that X-ray crystallography will remain the most important structure determination method for the foreseeable future. Most macromolecular structures are nowadays determined using synchrotron radiation, profiting from the brilliance and tune ability, both in wavelength and size, of the X-ray beam. The development of structural proteomics is limited mostly by the problem of (i) initial crystal production and (ii) by the quality of diffraction data collected — in particular it is important to limit the influence of X-ray radiation damage. The biggest challenge in protein crystallography is that only a small fraction of proteins have been crystallized until now (Protein Structure Initiative, NIGMS-NIH, Bethesda, USA). Despite recent advances in the field of protein crystallization, crystal growth remains the slowest step and critical in determination of protein structure. The crystallization problem may be partially solved by the use of high throughput nanodrop robotic crystallization systems, which significantly increases the number of the crystallization trials and decreases the amount of protein required, thus allowing full automation of large-scale crystallization experiments. However, still in many cases, protein crystallization remains the bottleneck to protein structure determination, and many scientifically and industrially important proteins have not been crystallized to date. Almost all approaches to protein crystallization are based on the classical crystallization methods (Rosenberger, 1996) (e.g., vapor diffusion, batch), varying crystallization conditions. However, these methods often give rare and non-reproducible results, first of all to the big and/or partly non-soluble proteins (i.e., membrane proteins) and require a long empirical search for the optimal crystallization conditions, since for every protein the specific crystallization conditions have to be determined. For these reasons protein crystallization is often called art instead of science (Pechkova and Nicolini, 2002). This demands the exploration of novel crystallization techniques and a microscopic understanding of all steps involved in crystallization processes.

The second problem — radiation damage — still remains a critical issue, especially with the development of intense third-generation synchrotron sources, like ESRF. Though exposure times for collection of complete datasets at these sources are very short, the X-ray beam from undulator beamlines induces substantial radiation damage even to cryocooled crystals after a few minutes of irradiation. Upon primary photon absorption which can cause the breakage of covalent bonds (primary effect), an avalanche of radicals will be formed and propagated throughout the crystal (secondary effect), leaving a fingerprint on the macromolecule (Burmeister, 2000; Ravelli and McSweeney, 2000; Weik *et al.*, 2000). Primary effects of radiation damage depend linearly on absorbed radiation dose and cannot be avoided by cryocooling (Garman, 2003; Ravelli and Garman, 2006). The secondary effect of free radical migration is reduced at cryotemperatures (Garman, 1999). The advent of cryoprotection has lead to the rapid growth in the number of structures solved using synchrotron radiation. There are a number of strategies which have been attempted to minimize radiation damage problem, including cryocooling techniques (Kuzay *et al.*, 2001; Nicholson *et al.*, 2001), free radical scavengers (Murray and Garman, 2002), and beam defocusing, which reduces the dose but needs larger crystal volume for maintaining the same signal (Walsh *et al.*, 1999). In this context microcristallography can offer the solution of the problem for the strong dependence of the photoelectronic path from the crystal size and for the subsequent correlated reduction of the radiation damage (Nave and Hill, 2005; Cowan and Nave, 2008).

However, the main problem remains unsolved — damage still occurs and is becoming especially problematic for data collection from protein microcrystals at micro-focused synchrotron radiation beams (Riekel *et al.*, 2000). Thus, the need to produce well-ordered, diffracting, and radiation stable crystals becomes increasingly important and new techniques able to solve this problem are sorely needed.

Langmuir–Blodgett (LB) nanotemplate crystallization method was first introduced by Pechkova and Nicolini in 2001. This method, based on LB nanotechnology (Troitsky *et al.*, 1996; Nicolini, 1997; Nicolini and Pechkova, 2006a), namely the LB protein monolayer deposited on the glass slide in the common hanging drop method, causes the acceleration of the nucleation and the crystal growth rate. A LB nanotemplate was successfully used for the crystallization of protein, which was not crystallizable by the classical vapor diffusion method (Pechkova and Nicolini, 2004, and the chapter by Nicolini *et al.* in this volume). This technique can easily be automatized and the protein nanotemplate proven to be utilizable with existing robotic system (manuscript in preparation). Moreover, LB-based crystallization process has previously proved to produce better crystals in term of resolution and data quality both for model and target protein (Pechkova *et al.*, 2003; Pechkova and Nicolini, 2004). Moreover, radiation stability in terms of reflection intensity of LB lysozyme crystal was confirmed. (Pechkova *et al.*, 2004).

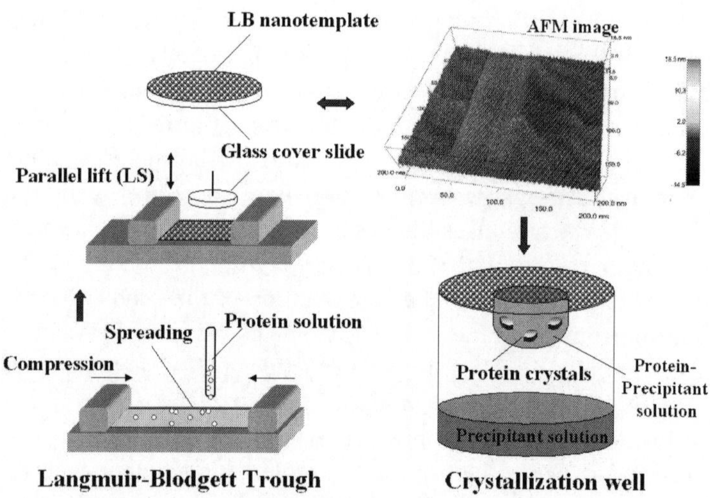

Figure 9.1 LB nanotemplate crystallization method. See also Color Insert.

In summary, the modification of the classical hanging drop vapor diffusion method by utilization of LB nanotemplate (Fig. 9.1) has proven recently capable

(a) to induce crystallization of proteins so far impossible such as ribosomal proteins aIF2 α and β (Pechkova *et al.*, 2008), cytochrome P450scc (Pechkova and Nicolini, 2004), and human protein kinase CK2a (Pechkova *et al.*, 2003);
(b) to accelerate of crystal growth to larger size or to optimize their properties (Pechkova and Nicolini, 2001; Pechkova *et al.*, 2005a,c); and
(c) to obtain crystals with new properties such as radiation resistance (Pechkova *et al.*, 2004, 2009b) and unique domain organization (Nicolini *et al.*, 2010).

However, the additional information are needed to solve still numerous doubts and open questions for the most complete characterization of LB-based protein crystal nucleation and growth, to allow its subsequent routine utilization in yet structurally unsolved protein systems capable to overcome the present limits and the bottleneck of protein crystallography.

9.2 PROTEIN CRYSTALLIZATION BY LB NANOTEMPLATE METHODS

We chose different model proteins — low molecular weight protein, easily crystallizable by classical vapor diffusion method. Lysozyme from chicken egg white, proteinase K from *Tritirachium album*, thaumatin from *Thaumatococcus daniellii*, thermolysin from *Bacillus thermoproteoliticus*, and ribonuclease A and insulin from bovine pancreas were purchased by Sigma and used without further purification.

Crystals of these proteins were grown in parallel using two different methods, namely LB nanotemplate method and classical hanging drop method. The LB technique in its variation, a Langmuir–Schaeffer (LS) method used in different fields of science and technology (Nicolini, 1997; Nicolini and Pechkova, 2006), was utilized.

In this method, a protein nanofilm can be engineered onto solid substrates using LB trough. The first step is to bring the protein molecules to air–water interface of LB trough teflon bath, filled with distilled water, purified with MilliQ system to 18.2 MΩ cm or appropriate buffer solution (a subphase) (Pechkova *et al.*, 2003;

Pechkova and Nicolini, 2006). The concentration of protein solution for spreading is 10–100 mg/ml, depending on the nature of the protein and number of films to be prepared. After filtration with Millex HV (Syringe-Driven Filter Unit, Millipore) this solution has to be spread over the water surface by a Hamilton syringe (100 μl) without use of any dispersant. Protein molecules will form a monolayer on the air–water interface of the LB through bath.

The second step is compression of the protein monolayer immediately after spreading with two teflon barriers to obtain highly packed, ordered protein monolayer on the air–water interface. The compression should continue slowly until the monolayer reaches the desired surface pressure (Troitsky *et al.*, 1996; Nicolini, 1997). Compression speed could be chosen in the interval of 10–100 cm/min, depending on the trough size and the nature of the protein. The monolayers of proteins studied were compressed with the speed of 30–70 cm/min. The speed of compression also very much depends on the protein nature (e.g., solubility: the more soluble protein is, the faster the trough barriers should be moved) and can have influence on its surface pressure–area (π–A) isotherm form. Surface pressure–area isotherms measurements with displacement transducer of Wilhelmy plate (Troitsky *et al.*, 1996) are routinely performed to characterize the protein monolayer at the air–water interface compressed in a wide range of mN per meter. At the low pressures, the layer so not compressed enough and is rather disordered, while at the high pressures breakage of monolayer can occur. The following parameters can be varied in a wide range with the goal to reach the optimal isotherm: protein concentration, compression velocity, and subphase composition.

The third step is the transfer of protein monolayer from the subphase (water or buffer) surface onto a solid support. The transfer was performed at the surface pressure, which corresponds to the highly packed system. The choice of surface pressure of deposition is very important — this surface pressure should correspond to the highly packed ordered monolayer. However, to verify this issue, the further characterization of LB protein film is necessary after deposition. For the model proteins studied, surface pressure was chosen from 15 to 25 mN/m, depending on the particular protein.

This transfer is performed by touching the support in vertical (LB) or in parallel to the subphase surface according to LS technique (horizontal lift) (Nicolini, 1997). In this process, the support choice is also important. In the case of nanotemplates preparation for protein crystallization, siliconized circle glass cover slide (Hampton Research, USA) were used as a support. The slides were washed in distilled water, drayed in gaseous nitrogen flux, and used as substrate for the protein thin film deposition by LS method. The transferred protein monolayer has to be dried in the gaseous nitrogen flux. After this, second layer deposition can be performed onto the first one and also dried in the same way. We usually used one to two monolayers for the protein nanotemplate crystallization experiment, but an infinite number of monolayer scan be deposited and different film composition including altering protein–polymers layer can be engineered and used in different areas of science and technology (Nicolini, 1997; Nicolini and Pechkova, 2006). In the case of fast-degradable proteins or oxigenofobic proteins, protective-plate method can be utilized for protein preservation during the deposition (Troitsky *et al.*, 1996). Atomic force microscopy (AFM), circular dichroism (Nicolini *et al.*, 1993), nanogravimetry (Facci *et al.*, 1993), and microGISAX measurements were performed to characterize protein deposition and order in this thin film templates. It was found that these film posses the following characteristics:

- excellent overall film packing and order, which can be estimated by AFM measurements (Pechkova *et al.*, 2007b);
- uniformity and reproducibility of deposition. This characterization can be performed by nanogravimetry method, which is used to estimate the surface density of every protein monolayer (Pechkova and Nicolini, 2004);
- thermostability up to 150 °C with protein structure and function preservation (Facci *et al.*, 1996; Pechkova *et al.*, 2007a);
- storage stability up to several years (Nicolini, 1997); and
- long range order, in particular after the heat treatment (10 min at 150 °C) (Pechkova *et al.*, 2009a).

As it is shown in Fig. 9.1, highly packed and ordered protein monolayer transferred from the air–water interface onto the siliconized glass cover slide can be used as a template for modification of classical vapor diffusion hanging drop method (Pechkova and Nicolini, 2002, 2004). During the crystallization procedure, the following parameters were varied: precipitant nature and concentration, protein monolayer surface pressure during template preparation, number of the protein monolayers in protein nanobiofilm template. All experiments were carried out at the controlled temperature 20 °C.

Table 9.1 Different protein crystallization conditions used for both LB and classical hanging drop vapor diffusion method. The cryoprotectant used during the X-ray diffraction data collection is indicated for every protein.

Protein	Protein solution	Reservoir solution	Drop	Cryoprotectant
Proteinase K	20 mg/ml in 25 mM HEPES buffer pH 7	400 mM Na/ K-tartrate in 25 mM HEPES buffer pH 7	1:1	4M TMAO
Ribonuclease A	10 mg/ml in 50 mM Na-acetate pH 5.5	1.75 M $(NH_4)_2SO_4$ 2M NaCl in 100mM Na-acetate pH 5.5	1:1	Mother liquid
Thaumatin	15 mg/ml in 100 mM ADA buffer pH 6.5	1 M Na/K in 100 mM ADA buffer tartrate pH 6.5	1:1	30% glycerol, 0.7M Na/K in 100 mM ADA buffer tartrate pH 6.5
Thermolysin	100 mg/ml in 50 Mm MES pH 6.0	35% saturated $(NH_4)_2SO_4$	1:1 with 1M NaCl in 50mM MES pH 6.0	Dry paraton-N (0.5h in vacuum centrifuge)
Insulin (Zn-free)	18 mg/ml in 50 mM Na_2 HPO_4 pH 10.4 1 mM EDTA	400 mM Na_2HPO_4 pH 10.4 10 mM EDTA	1:1	30% glycerol in 400 mM Na_2HPO_4 pH 10.4 10 mM EDTA
Lysozyme	40 mg/ml in 50 mM NaAc buffer pH 4.5	0.9 M NaCl	1:1	20% glycerol in 0.9 M NaCl

The same crystallization conditions were used for both the LB and classical methods, to study the influence of the LB nanotemplate on the crystallization process. Protein solution (3 μl) was mixed with equal volume of the reservoir solution and equilibrated against 1 ml of reservoir solution at the room temperature. The crystallization conditions used for both LB nanotemplate and classical hanging drop vapor diffusion method for different model proteins are summarized in Table 9.1. Under the same crystallization condition the protein crystals appears earlier and grow to bigger dimensions when the LB template was used, in comparison with classical hanging drop method.

This phenomenon was for the first time observed for chicken egg white lysozyme (Pechkova and Nicolini, 2001). The crystal growth in the LB nanotemplate method is indeed accelerated already at the first step of crystallization, which was confirmed by the microGISAX measurements on ID13 ESRF beamline (Pechkova *et al.*, 2005b) both for model protein (chicken egg white lysozyme; Pechkova and Nicolini, 2006) and target protein (bovine cytochrome P450scc, Nicolini and Pechkova, 2006b). This acceleration could be explained by the presence of specific aggregates of proteins, which can detach from the LB film and becomes the protein nucleation points in the drop solution.

This fact can also explain the nanotemplate crystallization of proteins, which cannot be crystallized by classical methods (human protein kinase CK2α catalytic subunit, bovine cytochrome P450scc, archaea initiation factors IFα and IFβ). The well-ordered, organized protein nanotemplate can supply the ordered aggregates to the protein solution, which easily became the new nucleation points, helping the system to overcome the free energy barrier of crystallization. The *in situ* micro and nano GISAX method was applied for the detailed study of this process in the real time (Nicolini *et al.*, 2010).

9.3 X-RAY DATA COLLECTION AND ANALYSIS

After the several hours or few days, both classical and LB crystals of all model protein achieve their maximum size, usually big enough to be analyzed by synchrotron radiation beamline. The crystal of approximately equal size could be chosen for comparative analysis of classical and LB method (Table 9.2).

A classical and LB crystal was cryocooled to 100 K using its mother liquor containing cryoprotectants, indicated in Table 9.1. All X-ray data were collected to a high resolution at the ID13, ID14-2, ID23-1, ID29 beamlines, ESRF (Grenoble, France), using a CCD detector.

Table 9.2 LB and classical protein crystals analyzed at the different ESRF beamlines.

Beamline at ESRF	Protein crystal and method of preparation	Number of crystal being studied	Crystal size (μm)	Beam size (μm)	Total radiation dose (E+07Gray)
ID13	Lysozyme classical	2	400 × 200 × 100	20 × 20	188
ID13	Lysozyme LB	2	750 × 375 × 190	20 × 20	186
ID29	Thaumatin classical	1	100 × 200 × 150	100 × 100	94.3
ID29	Thaumatin LB	3	100 × 300 × 150	100 × 100	94.3
ID23-1	Proteinase K classical	1	200 × 100 × 150	80 × 60	4.41
ID23-1	Proteinase K LB	1	200 × 100 × 150	80 × 60	4.41
ID14-2	Proteinase K classical	1	200 × 300 × 100	160 × 200	0.06
ID14-2	Proteinase K LB	3	160 × 200 × 50	160 × 200	0.08
			300 × 200 × 200	100 × 100	0.04
			350 × 300 × 300	100 × 100	0.12
ID14-2	Ribonuclease classical	1	140 × 140 × 50	140 × 140	0.08
ID14-2	Ribonuclease LB	1	140 × 140 × 50	140 × 140	0.1
ID14-2	Thermolysin classical	1	300 × 80 × 80	160 × 130	0.08
ID14-2	Thermolysin LB	1	1200× 120 × 120	160 × 130	0.07

The dimensions of classical protein crystal, analyzed as the reference, were similar to those grown LB nanotemplate method (Table 9.2). In case of the parallel beam (e.g., ID14), the beam size was chosen to cover almost all crystal (Figs. 9.2 and 9.3). In the case of the non-parallel beam (e.g., ID23), the beam size was smaller than a crystal (Table 9.2). In both cases, the beam size has the same dimension both for classical and LB crystals to cover the same fraction of the LB and classical crystal volume (Fig. 9.2). The choice of beam size depends not only on the protein crystal dimensions, but also on its shape. In Fig. 9.3, crystals of very different shapes (ribonuclease A and thermolysin crystals), mounted in the nylon loops, are together shown with the beam position suitable for these experiments.

The choice of beamsize is rather important issue in the radiation damage experiment, since a change in the beam size alters the background scattering. The crystal diffraction intensity and the size and profile of the diffraction spots are affected when the beam is smaller than the crystal. Therefore, it is necessary to consider the effect of the beam size on the data statistics (Bourenkov and Popov, 2005).

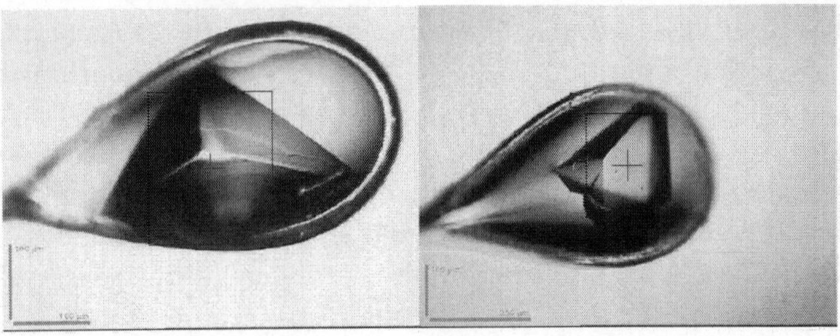

Figure 9.2 Proteinase K classical (left) and LB (right) crystal in the nylon loop in the cryogenic stream. The blue square contour indicates the ID 14-2 beam position. See also Color Insert.

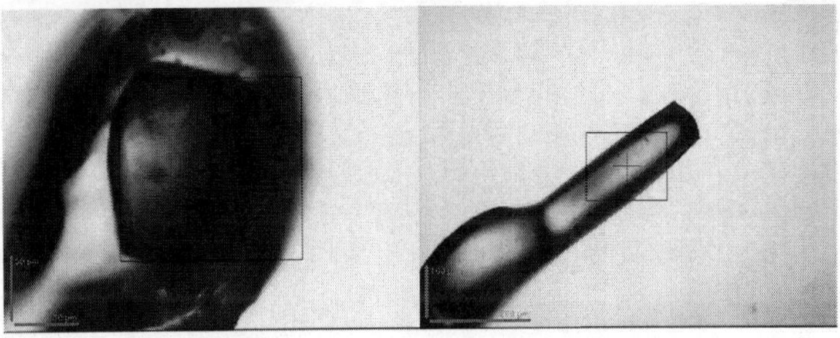

Figure 9.3 LB crystal of ribonuclease A (left) and thermolysin (right) mounted in the nylon loop in the cryogenic stream. The blue square contour indicates the ID14-2 beam position. See also Color Insert.

Both for LB and classical crystal, X-ray diffraction datasets were collected at high resolution. The first dataset was collected on a freshly frozen crystal. Every data set composed of about 130–180 subsequent rotation images. In between these data sets the rotating crystal was exposed to non-attenuated beam. Several LB crystals were analyzed: in some cases data collection was performed before and after burning, and in other cases the LB crystal, after initial data collection, was subjected to three or more subsequent steps of burning with data collection session after each burn (e.g., subsequent four data set were collected at the

ID23-1 for the proteinase K LB and classical crystals after 200, 300, 500, and 700 s of X-ray exposure). Diffraction studies of 20 different protein crystals, prepared either by LB or classical method, are summarized in Table 9.2. All data sets obtained were processed with MOSFLM and the CCP4 suite (Collaborative Computation project, Number 4, 1994). The protein model taken from the RCSB Protein Data Bank (Berman *et al.*, 2000, 2003; PDB codes are 1PTK for proteinase K, 1RQW for thaumatin, 1KEI for thermolysin, 2BLP for ribonuclease A, and 2YVB for lysozyme) was refined against the classical and LB crystals data sets using REFMAC5 (Murshudov *et al.*, 1999) to obtain model phases for statistical comparisons. Electron density maps were inspected at same contour level using COOT, and images were rendered using Raster3d for comparing electron density maps for two types of crystals.

The experiment was done by following experimental scheme: firstly, the classical crystal diffraction data sets at high resolution was collected. Next the data sets were collected after one or more consequent X-ray burning steps. The same was done for the crystals, grown by LB nanotemplate method. The model structure obtained from the first step of data collection was further used to refine the coordinates and isotropic B factors of all the atoms in protein for further data collection steps after each step of high X-ray dose irradiation for both classical and LB crystals.

9.4 RADIATION DOSE CALCULATION

The accumulated dose received by a sample depends on the fluence (photons mm^{-2}) received by the relevant portion of it during the X-ray exposure. The fluence will depend on the flux density (photons s^{-1} mm^{-2}) and the exposure time. The flux density is likewise determined from the beam size and flux (photons s^{-1}). The accurate documentation of the flux and beam sizes for various protein crystallography beamlines is therefore very important. A reliable and convenient means of measuring the flux using pin diodes is recently given by Owen *et al.* (2009). Although the pin diodes interrupt the beam, they can be used to calibrate the ionization chambers often used during the X-ray exposures (Fig. 9.4). It is also necessary to remember that the incident X-ray photon flux decays with the beam current, which could be changed with the time.

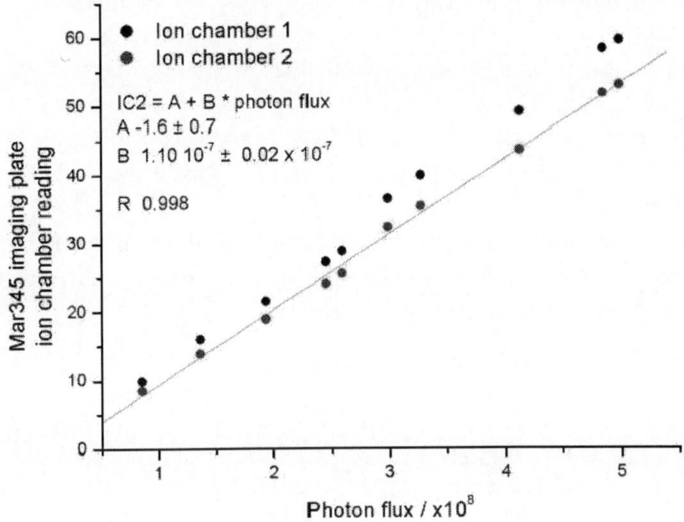

Figure 9.4 Example of ion camber calibration utilized by Garman group. This calibration was produced using a silicon photodiode (Hamamatsu, model S3204-, diode thickness of 500 μm), valid only for the generator used in T5, LMB, Oxford (not anymore in use) (data reproduced with permission of Elspeth Garman from http://lmb.bioch.ox.ac.uk/www/garman/lab_tools.html#guide9). See also Color Insert.

The beam size itself is also not simple to define unless a top hat profile, combined with accurately measured apertures, is available. A Gaussian beam (for example) will deposit energy non-uniformly into the crystal, causing for instance differential cell expansion and varying degrees of specific structural damage through the sample. This phenomenon exacerbates the problems caused by radiation damage since different parts of the crystal are being affected by varying amounts (Garman and Nave, 2009). Finally, an estimate of the dose deposited by the incident beam in the crystal is required.

For every step of X-ray burning, as well as for the data collection session, the radiation dose (Gray) for the exposed crystal volume was calculated using the program RADDOSE (Murray *et al.*, 2004). The latest development of this program is recently described by Paithankar *et al.* (2009).

An estimate of the total dose was computed assuming the photon flux density (photons s^{-1} μm^{-2}) value for the ESRF beamlines as it is shown in Table 9.3. For a data collection step dose can be given as:

$$D(\text{Gray}) = \mu E t I_0,$$

where μ = mass absorption coefficient (cm^2/g), E = energy of the X-ray (keV), t = time of exposure (s), and I_0 = flux of the X-ray beam (photons/s/nm^2).

The RADDOSE program takes into consideration total linear attenuation coefficient of the crystal, which is estimated from the sum of atomic cross sections of all atoms.

Table 9.3 ESRF beamlines characteristics.

Beam line	Energy range	Typical beam size, μm²	Res. limit	Typical exp time, s	Maximum phi speed rotation, s/degree	No. of passes	Typical transmission for MAD/SAD	Photon flux density at 200 mA photons/s μm²
ID13	0.934 Å (13.274 keV)	5 × 5 20 × 20	0.95	1	0.3	3	Usually 100%	1E+10
ID14-2 fixed energy	0.933 Å (13.289 keV)	100 × 100	1.0	1	0.3	3	Usually 100%	1E+06
ID23-1 MAD energy	0.62 to 2.5 Å (20 to 5.0 keV)	40 × 30	0.64	0.1	0.1	1	1–5%	8E+08
ID29 MAD	0.62 to 2.4 Å (20 to5.2 keV)	60 × 60	0.64	0.1	0.1	1	1–5%	6E+09

The use of mass attenuation coefficient instead of mass absorption coefficient leads to a little overestimation of the dose.

In principle, the effective exposure can vary with changes in the synchrotron ring current, but we assume that these dose estimates should be seen as an upper limit, not taking into account any decay of the ring current and beam intensity during data collection in the time mode.

Both LB and classical crystals were subjected to four subsequent steps of X-ray burn with data collection sessions after each burn.

The radiation dose absorbed by crystals during data collections and the burning sessions (the total dose) on different ESFR beamlines are presented in Table 9.2. It is worth to notice that radiation dose for typical data sets (in average 130 frames, which corresponds to the 13 s of exposure) was much less than the doses after every X-ray burning (e.g., 90, 200, or 300 s of exposure) and total dose which is several times higher than those for typical dataset, approaching double of the Henderson limit (Henderson, 1990) and 4/3 of new Owen *et al.* (2006) dose limit.

9.4.1 RADIATION DAMAGE QUANTIFICATION

Accurately quantifying radiation damage is essential to understanding the mechanisms by which it occurs and to evaluating possible approaches to reducing its effects. Metrics for radiation sensitivity usually used for radiation damage studies include diffraction resolution, mosaicity, unit cell size, integrated reflection intensities, and Wilson B factors. Pairwise R factor (Rd) (Diederichs, 2006) between identical and symmetry-related reflections occurring on different images, as a function of frame number, can be also used to visualize radiation damage quantitatively (Tripathi *et al.*, in preparation).

After protein structure refinement, different parameters can be plotted against radiation dose for comparison of LB nanotemplate-grown crystal and classical crystal radiation damage stability. These parameters include overall structure resolution, intensity to sigma *ratio*, unit cell volume, and number of unique reflections.

For both the LB and the classical crystals, highly redundant complete datasets were collected at high resolution before and after every X-ray "burn." During this data collection, the differences in the diffraction limit was observed: indeed, the diffraction limit of the LB crystal after every step of X-ray "burn" is significantly better, than those of the classical crystal.

Data statistics showed the changes observed in R_{merge} and the intensities (I/sigma) values of classical and LB crystals before and after every burn. The increase of such parameters as unit-cell volume and mosaicity can indicate that the classical crystal had suffered from radiation damage during the X-ray "burn." Instead, the LB crystal demonstrates similar values before and after "burn."

The gradual changes of the different parameters of LB and classical crystal from the first to the last step of burning were plotted against dose. For example, the resolution of the subsequently collected data set is decreasing faster in case of the classical crystal, compared to the LB crystal, although both crystals initially diffract to the same resolution. The same phenomenon is observed with the number of unique reflections (Pechkova *et al.*, 2009b). This dependence is consistent with that previously reported for lysozyme (Pechkova and Nicolini, 2004).

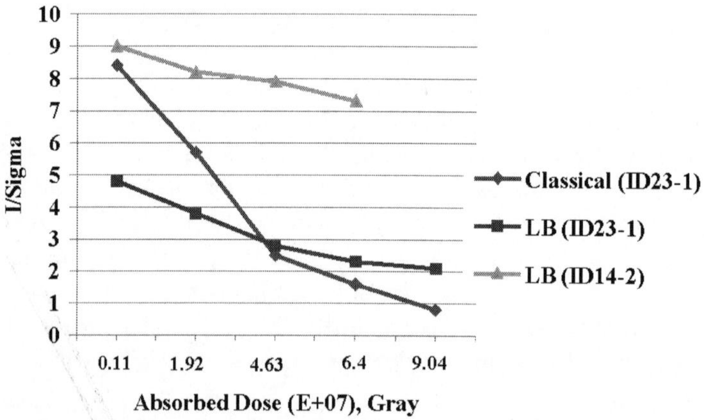

Figure 9.5 Intensity per sigma versus adsorbed dose for proteinase K LB and classical crystals at ID23-1 ESRF beamline. See also Color Insert.

Interestingly, even in the case of initially less intense diffraction signal of LB crystal, the signal did not significantly decrease as a function of absorbed dose. The classical crystal displays different behavior, even in case of initially of higher signal-to-noise, the ratio decreases dramatically with the absorbed dose. Thus, we observe that the signal-to-noise ratio in the data collected from LB crystal decays very slowly in comparison of classical crystal (Fig. 9.5).

Through several burning steps and data collection sessions the change in R_{merge} was negligible, pointing to data quality of the crystal, grown by LB nanotemplate method. It can be observed that for randomly chosen reflections points with initially equal intensity measured from the classical crystal the intensity is subject to significant decay, while LB crystal intensity of diffraction decrease more slowly.

Moreover, with increased adsorbed dose, the unit-cell volume of classical crystal is clearly increasing from one dataset to another, while for the LB crystal unit-cell volume remains practically the same, even if this parameter is not a conclusive marker for radiation damage (Ravelli *et al.*, 2003).

After all classical and LB structures refinements, we compared the electron density maps for some of the radiation damage sensitive amino acid residues (Burmeister, 2000; Ravelli and McSweeney, 2000; Weik *et al.*, 2000). Comparisons of electron density maps for the carboxylic acid group of aspartic acid residue of classical versus LB crystal structures contoured at 2 sigma are presented in Fig. 9.6. It appears that in case of LB structure, the decrease in the electron density map is slight while the classical crystal structure suffers more from the radiation damage.

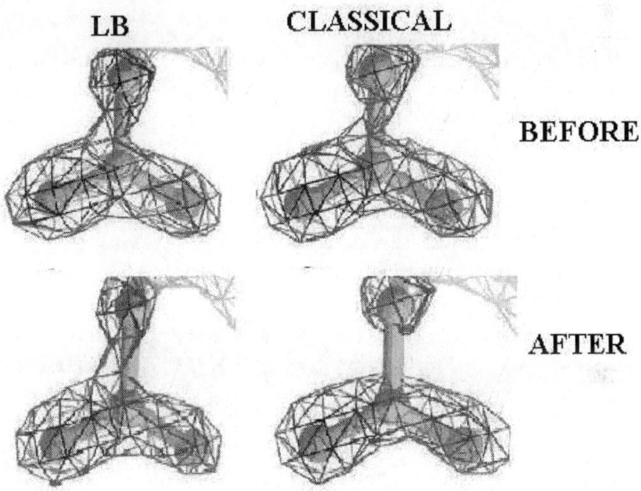

Figure 9.6 Comparison of electron density for carboxylic group of aspartic acid of LB and classical thaumatine crystal before and after high radiation dose exposure at ID 29 ESRF beamline. See also Color Insert.

A change in electron density for the classical crystals is already visible after the fist step of X-ray burn of the crystal. The same effect is also observed for electron density of the disulphide bond of cysteine residues (Fig. 9.7) and carboxylic acid group of glutamic acid residues (Fig. 9.8) as well as for the methyl group of methionine residues (Fig. 9.9) of all studied proteins.

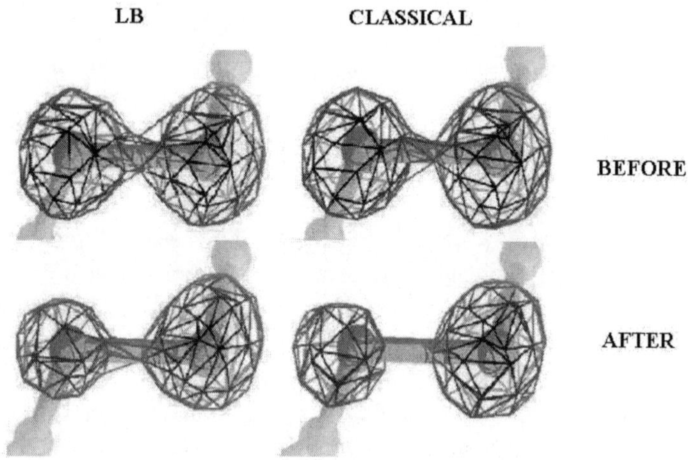

Figure 9.7 Comparison of electron density for the disulphide bond of 71 cysteine residue of LB and classical thaumatine crystal before and after high radiation dose exposure at ID 29 ESRF beamline. See also Color Insert.

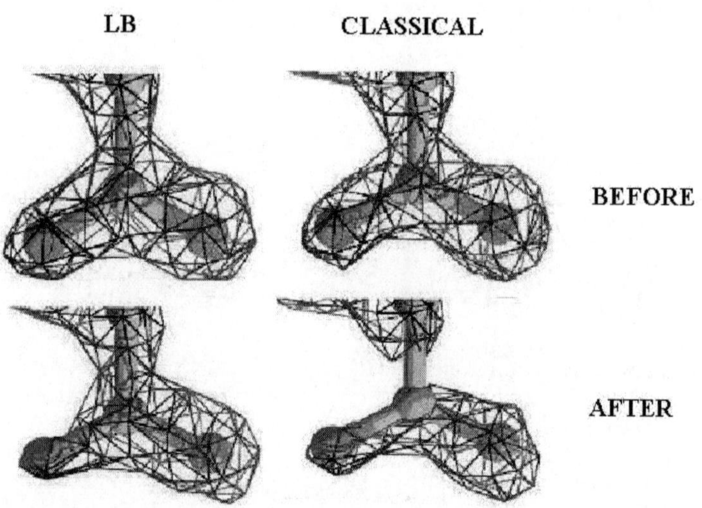

Figure 9.8 Comparison of electron density for carboxylic group of glutamic acid residue at 42 position in amino acid chain of LB and classical thaumatine crystal before and after high radiation dose exposure at ID 29 ESRF beamline. See also Color Insert.

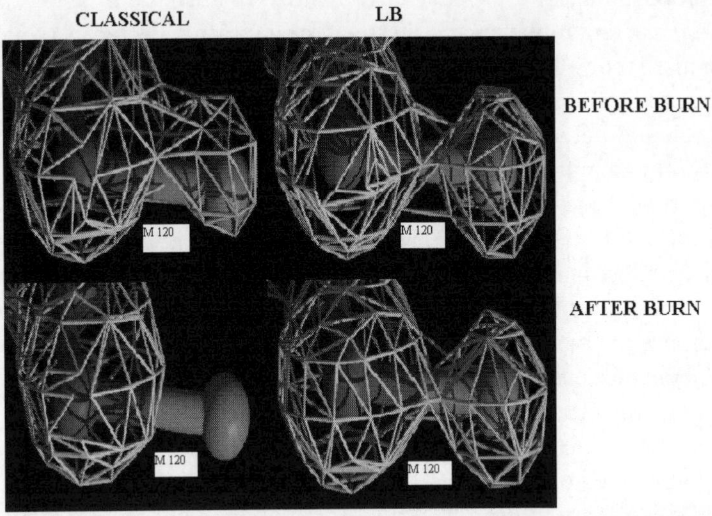

Figure 9.9 Comparison of electron density for the methyl group of methionine residue at 120 position in amino acid chain of LB and classical thermolysine crystal before and after high radiation dose exposure at ID 14-2 ESRF beamline. See also Color Insert.

9.4.2 B Factor Calculation

The B factor is a measure of the effective diameter of an atom's electron density. Various disorders (static and thermal) can effectively "spread out" the electron density of a given atom, causing an increase in its B factor. In other words, B factor indicates the positional spread of each atom in protein. The B factor is related to the *rms* error in an atom's position and, for this reason, B factors are related to resolution. Large movements of other atoms would be suppressed at 100 K because the amorphous solvent present in protein crystals at cryotemperatures is a glass (i.e., has the structure of a liquid but with rigidly bound atoms). However, local flexibility will be present even at 100 K, so small movements could occur. Such movements are a pre-requisite for identifying damage by techniques such as X-ray diffraction.

In the case of comparison of successive measurements of the same set of reflections with our first reference set, it is reasonable to choose relative isotropic B factors as a metric sets. There are two main advantages to using relative instead of absolute B factors as metrics of radiation damage. Firstly, if the data are collected in a small angular wedge to minimize errors from non-uniform damage, there may not be enough reflections to give the intensity statistics needed to reliably determine the absolute B factors from a Wilson plot. Secondly, sometimes the aim is to determine B factors for crystals after they cease to diffract beyond the useful resolution limit needed for Wilson statistics (<3 Å). This is possible with relative B factors, since reflections are compared using SCALEIT with the identical reflections in subsequent sets (Kmetko *et al.*, 2006). For example, to compare the data sets collected from different proteins, the relative B factor on dose can be plotted against radiation dose, showing a simple linear dependence. Usually, plots of relative B factors versus dose are in fact considerably smoother than those of absolute B factors (Kmetko *et al.*, 2006).

In the case of comparison of radiation damage of LB and classical protein crystals, relative B factors for each residue were calculated before and after intense X-ray burning. The alternative way to present this result is to sum atomic B factors for each residue to obtain a total B factor for every residue and then average it over number of each type of residue present in the protein. Relative increase in B factor after high X-ray dose is:

$$B_r = (B_a - B_b)/B_b,$$

where B_r = relative increase of B factor for residues of protein, B_a = average B factor for residues of protein after high X-ray dose, and B_b = average B factor for residue of each protein before high X-ray dose.

Average increase of B factors were calculated averaged for each residue for both type of crystals (LB and classical). Figure 9.10 shows the increase in average B factor of residues in classical and LB-based ribonuclease A crystals. The average B factor increase for each residue is more in case of classical crystal compared to LB crystal, indicating more damage to classical crystal. Similar changes were observed for proteinase K, thermolysin, and thaumatine crystals, proving the reproducibility of observation. In all cases, the effect is quite more evident using the relative B factors (Fig. 9.11).

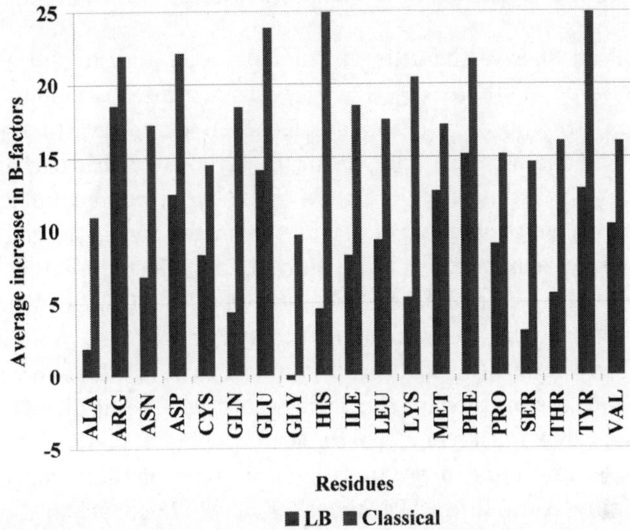

Figure 9.10 Increase in average B factors calculated for side chains of each residue for ribonuclease A. The effect is quite more evident using the relative B factor. See also Color Insert.

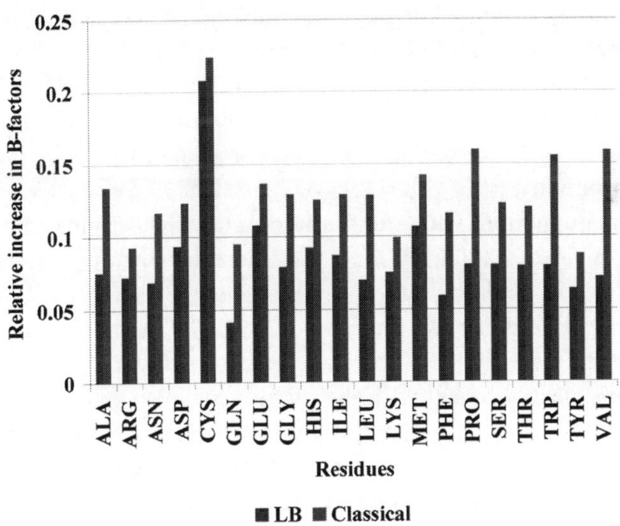

Figure 9.11 Relative B factors associated with side chains of the protein backbone for LB and classical crystal at same dose of 2.43 MGy, for proteinase K protein (ID14). For higher dose the difference is quite more striking (Pechkova *et al.*, 2009). See also Color Insert.

9.5 CONCLUSION AND FUTURE PERSPECTIVE

Overall analysis of the different model protein crystals grown by two different methods suggests that crystal grown by the LB-based method are more radiation stable when compared to crystals grown by the classical hanging drop method. Typical parameters such as diffraction limit, signal-to-noise ratio, number of unique reflections, unit-cell volume, as well as B factor during the course of burning experiment seem to be more mildly affected in LB crystals than classical crystals (see Pechkova *et al.*, 2009). A structural evaluation of the electron density maps shows less pronounced decarboxylation of aspartic and glutamic acid residues in case of LB-based crystals than in case of classical crystals. The investigation of a possible reason for the radiation stability of the LB-based crystal is presently in progress even if some possible suggestion raised in recent times (Pechkova *et al.*, 2005c; Pechkova *et al.*, 2007c) can be offered. Previously, dehydration has been observed to lead to better ordered protein crystals, significantly improving the diffraction limit (Heras *et al.*, 2003; Kuo *et al.*, 2003; Bowler *et al.*, 2006, and references therein). Further, we have shown (Pechkova *et al.*, 2007c), by data mining, the systematically lower water content in crystals of thermophilic protein compared to their mesophilic counterparts. For the case of thioredoxin we also experimentally showed that the low water content in the thermophilic homolog and in the LB film of mesophilic homolog corresponds to their higher thermal stability.

In combination with our earlier data on lysozyme crystals (Pechkova *et al.*, 2004, 2005c), this indicates that the decrease in surface or inner bound water may be the factor connecting experimentally observed changes in protein stability made in solution, thin LB film, and LB nanotemplate-grown crystals. This explanation is consistent with the differences in cell dimensions of LB and classical crystal, shown in previous report (Pechkova *et al.*, 2009b). Particularly, it was shown that the unit-cell volume in case of LB crystal is initially less than in the case of classical one, and this difference remains also through all X-ray burning steps, even if the unit-cell dimensions are increased due to the radiation damage. As shown before for lysozyme (Pechkova *et al.*, 2004) the significant radiation resistance of protein crystals could then be derived from the protein thermostable LB patches acting as nucleation centers. A systematic study undertaken at the ID13, ID14,ID23, ID29 confirms indeed that water content in LB-based protein crystal decreases with respect to "classical" along with increase in alpha helix content, thereby suggesting that decrease in bound water could be one the sources of radiation resistance in crystals.

9.6 RCSB PROTEIN DATA BANK ACCESSION CODES

3DE3, 3DE4, 3DE5, 3DE6, 3DE7, 3D9Q, 3DDZ, 3DE0, 3DE1, 3DE2,3DW3, 3DWE; 3DVQ, 3DVR, 3DVS, 3DW1

Proteinase K crystal structures by classical vapor diffusion and LB nanotemplate method after before and after several steps of high X-ray dose exposure on ESRF ID23-1 and ID14-2.

3DNZ, 3DZP, 3DZR, 3E0A

Thaumatin by classical vapor diffusion and LB nanotemplate method after before and after high X-ray dose exposure on ESRF ID29.

3DNZ, 3DO2, 3DO1, 3DO0

Thermolysin by classical vapor diffusion and LB nanotemplate method after before and after high X-ray dose exposure on ESRF ID14-2.

3I6F, 3I6J, 3I6H, 3I67

Ribonuclease A by classical vapor diffusion and LB nanotemplate method before and after high X-ray dose exposure on ESRF ID14-2.

Acknowledgments

This work was supported by an International FIRB-MIUR grant on Functional Proteomics and Cell Cycle to the CIRSDNNOB-Nanoworld Institute of University of Genoa, by a PNR-MIUR grant on Biocatalysis to the University of Genoa and to the Fondazione El.B.A., and by a yearly MIUR Grant for Funzionamento to Fondazione El.B.A. This work was also supported by the grant FIRB RBPR05JH2P from MIUR to Claudio Nicolini of University of Genova.

References

1. Berman, H. M., Henrick, K. and Nakamura, H. (2003). Announcing the worldwide Protein Data Bank. *Nature Structural Biology*, **10**, pp. 980.

2. Berman, H. M., Westbrook, J., Feng, Z., Gilliland, G., Bhat, T. N., Weissig, H., Shindyalov, I. N and Bourne, P. E. (2000). The Protein Data Bank. *Nucleic Acids Research*, **28**, pp. 235–242.

3. Bourenkov, G. P. and Popov, A. N. (2005). A quantitative approach to data-collection strategies. *Acta Crystallographica D — Biological Crystallography*, **62**, pp. 58–64.

4. Bowler, M. W., Montgomery, M. G., Leslie, A. G. W. and Walke, J. E. (2006). Reproducible improvements in order and diffraction limit of crystals of bovine mitochondrial F1-ATPase by controlled dehydration. *Acta Crystallographica D — Biological Crystallography*, **62**, pp. 991–995.

5. Burmeister, W. P. (2000). Structural changes in a cryo-cooled protein crystal owing to radiation damage. *Acta Crystallographica D — Biological Crystallography*, **56**, pp. 328–341.

6. Collaborative Computational Project, Number 4. (1994). The CCP4 suite: programs for protein crystallography. *Acta Crystalographica D — Biological Crystallography*, **50**, pp. 760–763.

7. Cowan, J. A. and Nave C. (2008). The optimum conditions to collect X-ray data from very small samples. *Journal of Synchrotron Radiation*, **15**, pp. 458–462.

8. Cusack, S., Belrhali, H., Bram, A., Burghammer, M., Perrakis, A. and Riekel, C. (1998). Small is beautiful: protein micro-crystallography. *Nature Structural Biology*, **5**, pp. 634–637.

9. Diederichs, K. (2006). Some aspects of quantitative analysis and correction of radiation damage. *Acta Crystallographica D — Biological Crystallography*, **62**, pp. 96–101.

10. Facci, P., Erokhin, V. and Nicolini, C. (1993). Nanogravimetric gauge for surface density measurements and deposition analysis of LB films. *Thin Solid Films*, **230**, pp. 86–89.

11. Facci, P., Erokhin, V. and Nicolini, C. (1994). Scanning tunneling microscopy of a monolayer of reaction centers. *Thin Solid Films*, **243**, pp. 403–406.

12. Garman, E. (1999). Cool data: quantity and quality. *Acta Crystallographica D — Biological Crystallography*, **55**, pp. 1641–1653.

13. Garman, E. (2003). "Cool" crystals: macromolecular cryocrystallography and radiation damage. *Current Opinion in Structural Biology*, **13**, pp. 545–551.

14. Garman, E. and Nave, C. (2009). Radiation damage in protein crystals examined under various conditions by different methods. *Journal of Synchrotron Radiation*, **16**, pp. 129–132.

15. Henderson, R. (1990). Cryoprotection of protein crystals against radiation-damage in electron and X-ray diffraction. *Proceedings of the Royal Society of London Series B — Biological Sciences*, **241**, pp. 6–8.

16. Heras, B., Edeling, M. A., Byriel, K. A., Jones, A., Raina, S. and Martin, J. L. (2003). Dehydration converts DsbG crystal diffraction from low to high resolution. *Structure*, **11**, pp. 139–145.

17. Kmetko, J., Husseini, N. S., Naides, M., Kalinin, Y. and Thorne, R. E. (2006). Quantifying X-ray radiation damage in protein crystals at cryogenic temperatures. *Acta Crystallographica D — Biological Crystallography*, **62**, pp. 1030–1038.

18. Kuo, A., Bowler, M. W., Zimmer, J., Antcliff, J. F. and Doyle, D. A. (2003). Increasing the diffraction limit and internal order of a membrane protein crystal by dehydration. *Journal of Structural Biology*, **141**, pp. 97–102.

19. Kuzay, T. M., Kazmierczak, M. and Hsieh, B. J. (2001). X-ray beam/biomaterial thermal interactions in third-generation synchrotron sources. *Acta Crystallographica D — Biological Crystallography*, **57**, pp. 69–81.

20. Murray, J. and Garman, E. (2002). Investigation of possible free-radical scavengers and metrics for radiation damage in protein cryocrystallography. *Journal of Synchrotron Radiation*, **9**, pp. 347–354.

21. Murray, J. W., Garman, E. F. and Ravelli R. B. G. (2004). X-ray absorbtion by macromolecular crystals: the effects of wavelenght and crystal composition on absorbed dose. *Journal of Applied Crystallography*, **37**, pp. 513–522.

22. Murshudov, G. N., Vagin, A. A., Lebedev, A., Wilson, K. S. and Dodson, E. J. (1999). Efficient anisotropic refinement of macromolecular

structures using FFT. *Acta Crystallographica D — Biological Crystallography*, **55**, pp. 247–255.

23. Nave, C. and Hill, M. A. (2005). Will reduced radiation damage occur with very small crystals? *Journal of Synchrotron Radiation*, **12**, pp. 299–303.

24. Nicholson, J., Nave, C., Fayz, K. B. and Garman, E. (2001). Modelling heating effects in cryocooled protein crystals. *Nuclear Instruments & Methods in Physics Research. Section A, Accelerators, Spectrometers, Detectors and Associated Equipment*, **467–468**, pp. 1380–1383.

25. Nicolini, C. (1997). Protein monolayer engineering: principles and application to biocatalysis. *Trends in Biotechnology*, **15**, pp. 395–401.

26. Nicolini, C., Erokhin, V., Antolini, F., Catasti, P. and Facci, P. (1993). Thermal stability of protein secondary structure in Langmuir-Blodgett films. *Biochimica et Biophysica Acta*, **1158**, pp. 273–278.

27. Nicolini, C. and Pechkova, E. (2006a). Nanostructured biofilms and biocrystals. *Journal of Nanoscience and Nanotechnology*, **6**, pp. 2209–2236.

28. Nicolini, C. and Pechkova E. (2006b) Structure and growth of ultrasmall protein microcrystals by synchrotron radiation: I. μGISAXS and μdiffraction of P450scc. *Journal of Cellular Biochemistry*, **97**, pp. 544–552.

29. Nicolini, C., Riekel, C. and Pechkova, E. (2010). Growth and organization of Langmuir-Blodgett protein crystals via *in situ* GISAXS, laser-microdissection, nanodiffraction, Raman s pectroscopy and atomic force microscopy. In *Synchroton Radiation and Structural Proteomics* (eds. E. Pechkova and C. Riekel), Vol. 3, Ch. 15, pp. 383–407. Pan Stanford Series on Nanobiotechnology, London, New York, Singapore.

30. Owen, R. L., Holton, J. M., Schulze-Briese, C. and Garman, E. F. (2009). Determination of X-ray flux using silicon pin diodes. *Journal of Synchrotron Radiation*, **16**, pp. 143–151.

31. Owen, R. L., Rudino-Pinera, E., Garman, E. F. (2006). Experiment determination of radiation dose for cryocooled crystals. *Proceedings of the National Academy of Sciences USA*, **103**, pp. 4912–4917.

32. Paithankar, K. S., Owen, R. L. and Garman, E. F. (2009). Absorbed dose calculations for macromolecular crystals: improvements to RADDOSE. *Journal of Synchrotron Radiation*, **16**, pp. 152–162.

33. Pechkova, E., Fiordoro, S., Fontani, D. and Nicolini, C. (2005a). Investigating crystal growth mechanisms with and without LB template: protein transfer from LB to crystal. *Acta Crystallographica D — Biological Crystallography*, **61**, pp. 809–812.

34. Pechkova, E., Innocenzi, P., Malfatti, L., Kidchob, T., Gaspa, L., Nicolini, C. (2007a). Thermal stability of lysozyme Langmuir–Schaefer films by FTIR spectroscopy. *Langmuir*, **23**, pp. 1147–1151.

35. Pechkova, E. and Nicolini, C. (2001). Accelerated protein crystal growth onto the protein thin film. *Journal of Crystal Growth*, **231**, pp. 599–602.

36. Pechkova, E. and Nicolini C. (2002). From art to science in protein crystallization by means of thin film technology. *Nanotechnology*, **13**, pp. 460–464.

37. Pechkova, E. and Nicolini, C. (2004). Protein nanocrystallography: a new approach to structural proteomics. *Trends in Biotechnology*, **22**, pp. 117–122.

38. Pechkova, E. and Nicolini, C. (2006). Structure and growth of ultrasmall protein microcrystals by synchrotron radiation: II. μGISAX and microscopy of lysozyme. *Journal of Cellular Biochemistry*, **97**, pp. 553–560.

39. Pechkova, E., Roth, S., Burghammer, M., Fontani, D., Riekel, C. and Nicolini, C. (2005b). μGisaxs and protein nanotemplate crystallization: methods and instrumentation. *Journal of Synchrotron Radiation*, **12**, pp. 713–716.

40. Pechkova, E., Sartore, M., Giacomelli, L. and Nicolini, C. (2007b). Atomic force microscopy of protein films and crystals. *Review of Scientific Instruments*, **78**, pp. 093704-1–093704-7.

41. Pechkova, E., Sivozhelezov, V. and Nicolini, C. (2007c). Protein thermal stability: the role of protein structure and aqueous environment. *Archives of Biochemistry and Biophysics*, **466**, pp. 40–48.

42. Pechkova, E., Sivozhelezov, V., Tropiano, G., Fiordoro, S. and Nicolini, C. (2005c). Comparison of lysozyme structures derived from thin film-based and classical crystals. *Acta Crystallographica D — Biological Crystallography*, **61**, pp. 803–808.

43. Pechkova, E., Tripathi, S. and Nicolini, C. (2009a). microGISAXS of LB protein films: effect of temperature on long range order. *Journal of Synchrotron Radiation*, **16**, pp. 330–335.

44. Pechkova, E., Tripathi, S., Ravelli, R., McSweeney, S. and Nicolini, C. (2009b). Radiation stability of proteinase K crystals grown by LB nanotemplate method. *Journal of Structural Biology*, **168**, pp. 409–418.

45. Pechkova, E., Tropiano, G., Riekel, C. and Nicolini C. (2004). Radiation stability of protein crystals grown by nanostructured templates: synchrotron microfocus analysis. *Spectrochimica Acta B — Atomic Spectroscopy*, **59**, pp. 1687–1693.

46. Pechkova, E., Vasile, F., Spera, R. and Nicolini, C. (2008). Crystallization of alpha and beta subunits of IF2 translation initiation factor from archabacteria *Sulfolobus solfataricus*. *Journal of Crystal Growth*, **310**, pp. 3767–3770.

47. Pechkova, E., Zanotti, G. and Nicolini, C. (2003). Three-dimensional atomic structure of a catalytic subunit mutant of human protein kinase CK2. *Acta Crystallographica D — Biological Crystallography*, **59**, pp. 2133–2139.

48. Ravelli, R. B. G. and Garman, E. (2006). Radiation damage in macromolecular cryocrystallography. *Current Opinion in Structural Biology*, **16**, pp. 624–629.

49. Ravelli, R., Leiros, H., Pan, B., Caffrey, M. and McSweeney, S. (2003). Specific radiation damage can be used to solve macromolecular crystal structures. *Structure*, **11**, pp. 217–224.

50. Ravelli, R. B. G. and McSweeney, S. M. (2000). The "Fingerprint" that X-rays can leave on structures. *Structure*, **8**, pp. 315–328.

51. Riekel, C. (2000). New avenues in X-ray microbeam experiments. *Reports on Progress in Physics*, **63**, pp. 233–262.

52. Riekel, C., Burghammer, M. and Muller, M. (2000). Microbeam SAX experiments and their combination with microdiffraction. *Journal of Applied Crystallography*, **33**, pp. 421–423.

53. Rosenberger, F. (1996). Protein crystallization. *Journal of Crystal Growth*, **166**, pp. 40–54.

54. Troitsky, V. I., Sartore, M., Berzina, T. S., Nardelli, D. and Nicolini, C. (1996). Instrument for depositing Langmuir-Blodgett films composed of alternating monolayer using a protective layer of water. *Review of Scientific Instruments*, **67**, pp. 4216–4223.

55. Walsh, M. A., Dementieva, I., Evans, G., Sanishvili, R. and Joachimiak, A. (1999). Taking MAD to the extreme: ultrafast protein structure

determination. *Acta Crystallographica D — Biological Crystallography*, **55**, pp. 1168–1173.

56. Weik, M., Ravelli, R. B. G., Kryger, G., McSweeney, S., Raves, M. L., Harel, M., Gros, P., Silman, I., Kroon, J. and Sussman, J. L. (2000). Specific chemical and structural damage to proteins produced by synchrotron radiation. *Proceedings of the National Academy of Sciences USA*, **97**, pp. 623–628.

Chapter 10

TOWARD THE UNDERSTANDING OF MOLECULAR ASPECTS OF *HELICOBACTER PYLORI cag*-PAI

Alessandro Angelini, Laura Cendron, Anke Seydel, Nicola Barison, Roberto Battistutta, and Giuseppe Zanotti

[a]*Department of Chemistry, University of Padua, Via Marzolo 1, 35131 Padua, Italy*

[b]*Venetian Institute of Molecular Medicine (VIMM), Via Orus 2, 35129 Padua, Italy*

giuseppe.zanotti@unipd.it

Helicobacter pylori, the pathogen that infects the stomach of about half the world's population, is associated with a spectrum of gastric diseases, ranging from gastritis to peptic ulcers and gastric cancer. Pathogenic strains of *H. pylori* use a type IV secretion system (T4SS) to deliver the toxin CagA into the human host cells. The T4SS, along with the toxin itself, is coded into a genomic insert, termed the *cag* pathogenicity island (*cag*-PAI). The latter contains about 30 open reading frames, most of which the exact function is not well characterized or totally unknown. In this paper, we summarize the methodology used to systematically investigate the function and structure of the proteins coded by the *cag*-PAI genes.

Synchrotron Radiation and Structural Proteomics
Edited by Eugenia Pechkova and Christian Riekel

www.panstanford.com

10.1 INTRODUCTION

Every year, *Helicobacter pylori* chronically infects the gastric mucosa of millions of people worldwide: it has been estimated that over 50% of the world population carries this infection (Covacci *et al.*, 1999), making it one of the most common. *H. pylori* has been associated with the development of several diseases, such as chronic gastritis, gastric and duodenal ulcer, gastric adenocarcinoma, and mucosa-associated lymphoma (Parsonnet *et al.*, 1991; Parsonnet, 1994; Moss and Sood, 2003). Approximately 10% of individuals who are infected during childhood will develop peptic ulcer disease; more dramatically, about 1% of the infected individuals develop gastric neoplasia: this concerns over 600,000 individuals worldwide every year (Pisani *et al.*, 1999).

Many of the defined *H. pylori* virulence factors, including flagella, the acid neutralizing enzyme urease, the neutrophil-activating protein HP-NAP, and the vacuolating cytotoxin VacA, are found in all strains (Monack *et al.*, 2004). Strains isolated from patients with severe diseases possess the so-called *cag* pathogenicity island (*cag*-PAI) and are defined as type I strains (Censini *et al.*, 1996). The *H. pylori* pathogenicity island was originally named *cag* (cytotoxin-associated gene) since it was thought to be associated with the expression of VacA. It was later shown that both factors, VacA and *cag*-PAI, are independent of each other, even though *cag*-negative strains often do not express VacA (Tummuru *et al.*, 1994). At the genetic level, the *cag*-PAI is a 40 kb chromosomal region, which was horizontally transferred from an unknown source, possibly about one million years ago, into the *H. pylori* chromosome. It is flanked by a 31 bp direct repeat, which contains the recombination site and corresponds to the last nucleotides of the glutamate racemase (*glr*) gene (Censini *et al.*, 1996; Akopyants *et al.*, 1998) (Fig. 10.1).

Different numbers of insertion sequences are associated with the *cag*-PAI, depending on the strain. Often the *cag*-PAI is split into two sub-regions, *cag*I and *cag*II, interrupted by two IS605 insertion elements. In terms of virulence, strains with many insertions often resemble more type II strains (less virulent, not carrying the *cag*-PAI) than other type I strains (Jiang *et al.*, 1996). Following the discovery of *H. pylori*, it was noted that this pathogen has an extraordinary

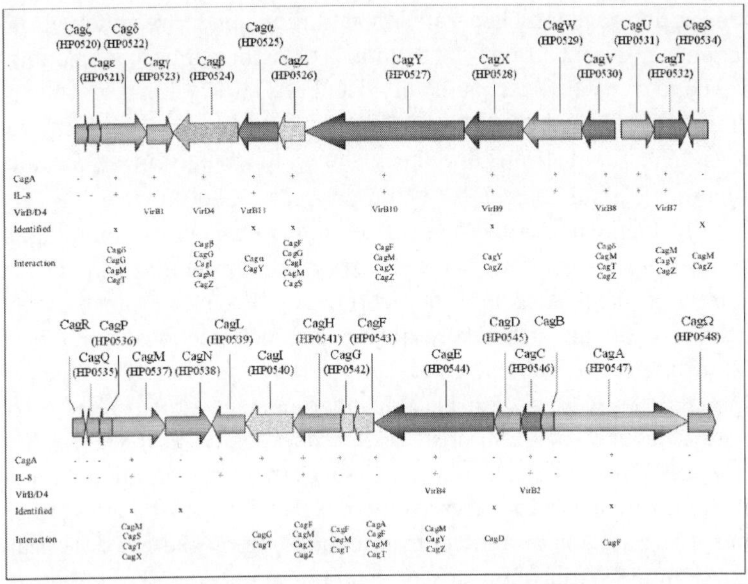

Figure 10.1 Graphical representation of the *cag*-PAI from *H. pylori* strain CCUG 17874 showing relative gene size and orientation. Most of the *cag* genes are probably involved in the assembly of the type IV secretion system that translocates the protein CagA into the gastric epithelial cells. Seven genes (marked in orange) show similarity to components of the VirB/D4 type IV secretion system of the plant pathogen *A. tumefaciens*. Proteins encoded by the island are involved in two major processes, the induction of interleukin-8 (IL-8) production by gastric epithelial cells and the translocation of CagA from the bacterium into host cells. All genes depicted by orange and green arrows are essential for CagA translocation, whereas lighter shades of red and green indicate genes that are not involved in this process but are however important for the IL-8 secretion. The purple arrows indicate genes that are required neither for translocation of CagA nor for IL-8 secretion. Based on mutagenesis study (Fischer *et al.*, 2001) the gene products essential for CagA delivery (Cag^{PY}) and IL-8 induction are indicated with "+," whereas the non-essential genes are indicated with "–." The proteins which have been detected *in vivo* by two-dimensional gel electrophoresis and mass spectrometry (MS) analysis (Busler *et al.*, 2006; Smith *et al.*, 2007) are indicated with "x." Interactions among *cag*-PAI encoded proteins have been detected using a yeast two-hybrid system (Busler *et al.*, 2006) and electron microscopy (Rohde *et al.*, 2003; Tanaka *et al.*, 2003). See also Color Insert.

genetic heterogeneity generated by mutation and/or recombination between different strains due to natural transformation, conjugation, or phage transfer (Suerbaum and Josenhans, 2007). It has also been shown that individuals can be colonized by multiple strains (Taylor *et al.*, 1995), and strains have been shown to change during chronic infection (Kersulyte *et al.*, 1999; Kuipers *et al.*, 2000).

The highly plastic *cag*-PAI region represents a perfect paradigm in the *H. pylori* genome. Not only strains differ significantly in the number of *cag*-PAI genes they carry, but the island can be lost, either completely or partially, during chronic infections of humans (Bjorkholm *et al.*, 2001; Kraft *et al.*, 2006) or experimentally infected animals (Sozzi *et al.*, 2001). Many patients have been shown to simultaneously carry *cag*-positive and *cag*-negative bacteria (Figura *et al.*, 1998). It has therefore been proposed that a dynamic population of *cag*-positive and *cag*-negative strains can exist in one individual, and that can expand and contract depending on the physiological and immune status of the host, and on the requirements of different niches within the stomach (Covacci and Rappuoli, 1998). Since the first description of the *cag*-PAI (CCUG 17874) in 1996 (Censini *et al.*, 1996), the full *cag*-PAI has been sequenced in three other complete genome sequences (*H. pylori* 26695, J99, and HPAG1) (Tomb *et al.*, 1997; Alm *et al.*, 1999; Oh *et al.*, 2006). Depending on the clinical strain, there are between 27 and 30 open reading frames predicted within the *cag* region. One of these open reading frames encodes the immunodominant antigen CagA, which is localized at the 3′ end of the island (Covacci *et al.*, 1993; Tummuru *et al.*, 1994). CagA was identified as the first protein of the *cag*-PAI and was shown to be a major virulence factor (Parsonnet *et al.*, 1997). Moreover, CagA is so far the only *H. pylori* macromolecule known to be translocated into the gastric epithelial cells by a type IV secretion system (T4SS) codified by the *cag*-PAI (Odenbreit *et al.*, 2000). Inside the host cell, CagA toxin induces multiple changes which cause the development of ulcers and cancer (Bourzac and Guillemin, 2005). Biopsies from patients with severe gastric diseases including chronic active gastritis, peptic ulcer disease, mucosa-associated lymphoid tissue lymphoma, and gastric cancer contain the *cagA* gene in more than 90% of all the cases, establishing a direct correlation of the presence of the *cagA* gene with diseases (Covacci *et al.*, 1997; Blaser, 1998; Covacci and Rappuoli, 1998).

Infection of cultured gastric epithelial cells with *cag*-PAI-positive strains results in interleukin-8 (IL-8) induction and a drastic cellular elongation. The discovery of mutations that can block CagA delivery but not IL-8 induction has led to the hypothesis that a second secreted effector molecule could be responsible for IL-8 induction. Recent experiments have implicated peptidoglycan, acting through the cytoplasmic receptor Nod1, as the secreted factor that induces IL-8 (Viala *et al.*, 2004), while others have hypothesized that in some strain backgrounds CagA itself is responsible for some IL-8 induction (Brandt *et al.*, 2005).

As mentioned above, the *cag*-PAI of *H. pylori* encodes proteins with homologies to structural and functional components of T4SS of other Gram-negative bacteria (Censini *et al.*, 1996; Akopyants *et al.*, 1998; Fischer *et al.*, 2001). T4SS are multi-component membrane-spanning transport systems ancestrally related to the conjugation system and can be divided according to their function into three categories: DNA transfer, DNA uptake and release, and effectors translocation. The T4SS of *H. pylori*, translocating CagA, belongs to the last group. This group is shared by several prominent plant and animal pathogens, such as *Agrobacterium tumefaciens*, *Brucella suis*, *Bartonella henselae*, and *Legionella pneumophila* (Yeo and Waksman, 2004; Christie *et al.*, 2005). All the pathogens mentioned here deliver effector molecules, such as protein-coupled DNA or proteins, to their target eukaryotic cell through what is thought to be a needle-like surface organelle.

From the genes encoded by the *cag*-PAI, 18 are required for translocation of CagA into gastric epithelial cells (Fischer *et al.*, 2001; Bourzac and Guillemin, 2005), but only four have sequence homologies to genes that encode components of the VirB/D4 T4SS of *A. tumefaciens*, the best-studied model of the T4SS, and that has become the standard reference of comparison for effectors translocation T4SS in Gram-negative bacteria. Other five Cag proteins have some features in common with *A. tumefaciens* T4SS components, and have been proposed to be functional homologues of these latter. The remaining genes are unique to the *H. pylori cag*-PAI, and very little is known about their expression and function. Previous studies have determined the expression *in vivo* of eight *cag*-PAI proteins (CagA, CagD, CagM, CagS, CagX, CagZ, Cagα, and Cagδ) (Jungblut *et al.*, 2000;

Backert *et al.*, 2005; Busler *et al.*, 2006; Smith *et al.*, 2007). Three of them (CagY, CagX, and CagT) were also detected by electron microscopy and reported to be associated with the T4SS structure (Rohde *et al.*, 2003; Tanaka *et al.*, 2003).

A first step in understanding the functions of individual members of the *cag*-PAI has been knock-out/mutagenesis studies, but many questions about the functions of individual *cag*-PAI genes remain unanswered. The *cag*-PAI genes, whose deletion does not alter CagA delivery and IL-8 secretion in tissue cultured cells, may encode delivered effector proteins with unknown activities in infection. Other pathogens that employ a T4SS in their infection process, such as *A. tumefaciens*, *L. pneumophila*, and *B. henselae*, deliver multiple effector proteins into host cells (Schulein *et al.*, 2005; Vergunst *et al.*, 2005). For instance, eight secreted effectors of *L. pneumophila* have been identified, but only one of them was found to be required for growth of this pathogen within infected cells, suggesting that the effectors have more subtle or redundant functions in infection. In *B. henselae* seven contiguous genes located within a T4SS-encoding PAI were predicted to produce proteins with a common C-terminal domain that mediated their secretion into host cells, which was validated with the Cre/*lox* system. Interestingly, the combined set of effectors is required for host cell cytoskeleton changes and one of these effectors is phosphorylated within the host cells, reminiscent of *H. pylori* CagA (Schulein *et al.*, 2005). Moreover, mutations that resulted in intermediate and variable levels of CagA delivery and IL-8 induction are promising candidates for proteins important in T4SS assembly and function. These genes may encode accessory proteins required for substrate translocation (such as chaperones), may encode additional substrates themselves, or simply encode proteins with redundant functions in T4SS. For example, CagL has been very recently discovered to have a crucial role in mediating CagA translocation, by acting as a molecular link between the type IV secretion apparatus of *H. pylori* and host cell integrins. Indeed, CagL was demonstrated to occur on the bacteria surface together with the secretion apparatus, to bind to $\alpha_5\beta_1$ integrin receptors on gastric epithelial cells, thus anchoring the bacterial secretion system to the host cells surface and finally making the CagA injection possible (Kwok *et al.*, 2007).

To date, crystal structures of some T4SS components have been published, including homologues of hexameric NTPases VirB11 (Cagα from *H. pylori*) (Yeo *et al.*, 2000; Savvides *et al.*, 2003) and VirD4 (TrwB from *Escherichia coli* R388) (Gomis-ruth and Coll, 2001), the VirB5 protein (TraC from pKM101 plasmid) (Yeo *et al.*, 2003), VirB8 (form *B. suis* and *A. tumefaciens*) (Terradot *et al.*, 2005; Bailey *et al.*, 2007), and VirB10 (ComB10 from *H. pylori*) (Terradot *et al.*, 2005). The NMR structure of a complex between a domain of the VirB9 homologue TraO bound to a VirB7-like domain of TraN from pKM101 plasmid have been also recently described (Bayliss *et al.*, 2007). The model of the hexameric structure for *A. tumefaciens* VirB4 has been recently proposed (Middleton *et al.*, 2005). Until now, apart from the ATPase Cagα VirB11 homologue, high-resolution structure data of Cag proteins are available only for the proteins CagS, CagZ (Cendron *et al.*, 2004, 2007), and CagD (in preparation).

The goal of our project is to determine most of the Cag protein structures by X-ray crystallography, to elucidate the role of *cag*-PAI in the *H. pylori* pathogenesis, to furnish valuable information for vaccine production, and to provide insights into the mechanism of bacterium pathogenesis. In this review we will describe the strategies adopted to produce recombinant proteins from 16 open reading frames encoded by the *cag*-PAI and to determine their structures by X-ray crystallography. Considering the substantial lack of biochemical info about Cag proteins, the targets selection was based on predicted properties, in particular concerning secondary structures, folding, and hydrophobicity. The major obstacle turned out to be poor expression and low solubility. From the nine soluble proteins obtained, four were crystallized and three structures have been determined so far.

10.2 INTERACTIONS AMONG *CAG*-PAI PROTEINS

Based on what is known about the T4SS in other bacterial species (Cascales and Christie, 2003; Yeo and Waksman, 2004), it may be presumed that numerous protein–protein interactions occur among *H. pylori* Cag proteins to form a type IV secretion apparatus. The protein–protein interaction map of *H. pylori* was the first reported for a prokaryotic organism (Rain *et al.*, 2001). Surprisingly, no interactions between Cag proteins were detected, while several others were found to connect them to members of other cell pathways or proteins with unknown functions. Examples are the interaction between Cagα (a VirB11 homologue) and HP1451, CagY (a VirB10 homologue) and HP0149, CagM and HP0261, Cagδ and HP1414, CagI and HP1489, and CagA and HP0496. The role of the HP proteins is largely unknown, and they are not considered to be component of the *H. pylori* T4SS. The interactions shown for CagY, Cagδ, and CagI could not be confirmed on isolated proteins *in vitro*, due to the toxicity or insolubility of one or both partners (Terradot *et al.*, 2004), whereas the interactions of CagM and also the one of CagA were confirmed by pull-down assays (Terradot *et al.*, 2004). Very recently, the interaction between Cagα and HP1451 was structurally characterized. The biological significance of those interactions is at the moment unclear, but these results suggest that the assembly and function of the Cag-type secretion system of *H. pylori* might depend or require other previously unidentified proteins. This makes it more difficult to understand the complex functioning of the *H. pylori*'s T4SS.

Moreover, a recent study has identified some interactions among Cag proteins using a yeast two-hybrid system and biochemical analyses (Busler *et al.*, 2006). Several of the interactions between *H. pylori* Cag proteins correspond to homologous interaction between components of the *A. tumefaciens* VirB/D4 T4SS. For example, homotypic interactions were detected for both *H. pylori* Cagβ and Cagα, similar to what has been observed for the *A. tumefaciens* homologues (VirD4 and VirB11, respectively) (Yeo and Waksman, 2004). The C-terminal portion of *H. pylori* CagY has been shown to interact with CagX, similarly to the *A. tumefaciens* homologues (VirB10 and VirB9, respectively) (Das and Xie, 2000).

Generally speaking, the most common methods to identify protein–protein interactions result difficult to be applied to membrane complexes, especially when the expression of the proteins of interest is tightly regulated and/or the abundance of the components is not so high to allow a simple immunoprecipitation. In *H. pylori*, these limits are stigmatized by the peculiar nature of its peptidoglycan and outer membranes, so that any protocols for isolating and/or treating the cellular fractions need to be adapted and finely adjusted.

10.3 PROJECT DESIGN, METHODS OF CLONING, PROTEIN EXPRESSION, SOLUBILITY, PURIFICATION, AND CRYSTALLIZATION TRIALS

10.3.1 Project Design

Significant amount of protein, usually in the order of milligrams, are required for structural studies and, generally, the protein must be active, soluble, stable, highly pure, and concentrated. Two major bottlenecks exist when addressing this issue: the first is poor solubility or aggregation and the other is protein crystallization. Here we will briefly describe the strategy adopted for cloning, protein expression, purification, and crystallization tests.

Combined bioinformatics and literature information were used to guide initial attempts to identify the optimal boundaries (starting and ending residues of the constructs) of a protein or a protein domain. The target proteins were ranked according to several predicted properties. Each sequence was analyzed to estimate the following features: amino acids content, molecular weight, isoelectric point, absorption coefficients at 276–280 nm, presence of cysteines, and stability index. The bioinformatics analysis allowed to predict the presence of transmembrane domains or putative signal sequence for secretion. Finally, programs for secondary and tertiary structure prediction of proteins were used, as well as bioinformatics tools, to predict the presence of folded and unfolded regions in the protein of interest. The majority of the software used are available from the ExPASy server (ExPASy ProtParam tools; www.expasy.org).

10.3.2 Cloning for Protein Expression

Genes of interest were cloned with an affinity tag. For each protein, various expression constructs may be needed to test for optimal expression, solubility, and crystallization. The selected genes were amplified from the *H. pylori* pathogenic type I strain CCUG 17874 (Genebank numbers: AF282853 and AF282852). In most cases, full-length *cag* genes were cloned. In the cases of genes predicted with secretion signal sequences, only the portion of the gene encoding the mature protein was cloned. This was done to minimize the possibility that the hydrophobic signal sequence might increase aggregation. In the cases where *E. coli* did not express big or multi-domain proteins, a smaller fragment of the target protein was selected to improve the expression levels. Also, the solubility of a poorly soluble (or insoluble) protein, in some cases, was improved by selecting a soluble domain for expression. The expression vectors used were predominantly pET vectors (Novagen), although we used other constructs, carrying different fusion tags, such as pGEX vectors (GE Healthcare) and pMal vectors (New England Biolabs) (hexahistidine, glutathione S-transferase (GST), and maltose-binding protein [MBP], respectively). These plasmids have a high copy number, because they contain the ColE1 origin of replication (originating from the pBR322 vector). In the pET vectors the gene of interest is under the control of an IPTG inducible T7 RNA polymerase promoter, recognized only by this specific polymerase and not by the *E. coli* polymerases, ensuring thus a tight control. A different strategy, using the TOPO-cloning kit (Invitrogen), was also successfully applied. Prior to proceeding with protein expression the coding regions of all constructs obtained were confirmed by DNA sequencing.

10.3.3 Expression of Proteins

E. coli remains the preferred host for heterologous expression of recombinant proteins for structural studies. Moreover, the same *E. coli* is a Gram-negative bacterium, like *H. pylori,* and can thus be considered the best choice. The ligated sequenced products were directly transformed into *E. coli* BL21(DE3) cells (Invitrogen), which were used for preliminary small-scale expression tests. In case of low or no expression, other *E. coli* strains like BL21 (DE3) STAR (Invitrogen) and the CodonPlus (DE) RP/RIL (Stratagene) or BL21 (DE3) Rosetta (Novagen) were tested. The *E. coli* C41 and C43 strains were instead used to produce membrane proteins and the *E. coli* B834 strain cells (Avidis), that are methionine-auxotroph engineered bacteria cells, have been used for Selenomethionine incorporation and subsequent crystallographic studies.

10.3.4 Solubility Studies

Protein solubility represents a major bottleneck for production of proteins for X-ray crystallography. A suitable lysis buffer should contain as much additives as possible to keep each protein folded and soluble. In case of poorly soluble proteins, solubility was optimized by using several screening buffers, varying the pH value, salt concentration or adding detergents, reducing agents, and cofactors. If inclusion bodies were present, different strategies were implemented. These included varying the conditions to decrease the rate of protein translation or enhance protein folding or denaturation and refolding. Improving the solubility of recombinant proteins in *E. coli* commonly involves changing some of the expression conditions. Factors such as temperature, changes of the *E. coli* expression strain, different parameters, or induction condition can be manipulated. In particular, to decrease the amount of protein found in inclusion bodies, the induction was carried out at a temperature of 20–25 °C. When high expression levels of insoluble protein were obtained and no significant improvement was achieved by the before mentioned strategies, refolding procedures were attempted. The protein was solubilized from inclusion bodies using denaturing buffers supplemented with 8 M urea or 6 M guanidine-HCl, the latter with more success.

10.3.5 Protein Purification

Cells were lysed by French press or sonication. To increase the lysis efficiency and reduce the solution viscosity, lysozyme and DNase were added to the lysis buffer. All lysis buffers also contained a protease inhibitor cocktail. Protein solubility was checked by SDS-PAGE and Western blot. The tag was used to isolate the recombinant fusion protein by affinity chromatography. As the majority of our constructs were hexahistidine tagged, the first chromatographic step of purification was standard immobilized metal affinity chromatography. Other affinity tags such as MBP and the GST implied the use of amylose-resin (New England Biolabs) and glutathione-sepharose (GE Healthcare) as affinity matrices, respectively. In spite of its usefulness, the affinity tag may sometimes interfere with folding, oligomerization, protein localization (especially in the case of membrane proteins), and finally protein crystallization. As many of these tags are large proteins, it is often necessary to remove them. For these reasons, all the vectors used contain affinity tags with a protease site between the tag and the construct of interest. The best results were obtained with TEV protease, from the tobacco etch virus, which shows high specificity, is relatively easy to produce in large amounts and in most cases leaves the lowest number of residues at the N-terminus, after cleavage.

However, the use of MBP or GST tags requires a further purification step to separate them from the protein of interest. Indeed, after proteolytic digestion of the fusion proteins isolated by affinity chromatography, a size exclusion or an ion-exchange purification step is the technique of choice, depending on the *pI* and the molecular weight of the components. When using MBP or GST tags, the removal of fusion partners quite often caused a significant loss of solubility and precipitation. This behavior was less frequently observed in constructs containing hexahistidine, since its capacity to keep tagged proteins in solution is *per se* low. If necessary, other purification steps, such as hydrophobic interaction or ion-exchange chromatography, were applied to obtain the desired homogeneity.

To check the purity and the oligomerization state of the isolated protein, the sample is concentrated by ultrafiltration and loaded onto a gel filtration column, selected on the base of the monomer size. This generally constitutes the last purification step, because it allows to keep the sample in well-defined buffer, which represents the best compromise between a protein solubilizing buffer and the right amount of salts and organic substances to interfere less with the crystallization reagents.

If the tagged protein was recovered from inclusion bodies, in denaturing reducing conditions, refolding procedures were also tried. The most successful technique turned out to be the so-called on-column method, which implies the loading of the denatured hexahistidine tagged protein on the Nickel affinity matrix and its renaturing while it is still bound to the resin, before the usual elution gradient with growing imidazole concentrations. For example, this method allowed to refold the SeMet derivative of hexahistidine-CagS, which resulted totally unsoluble after Selenomethionines introduction. Generally, we proceeded with characterization if the protein was >90% pure. All the proteins were then concentrated using 5 to 30 mg ml^{-1} as a starting concentration for crystallization tests. The first crystallization results are then used as a guide to suggest the optimal protein concentration and the most promising precipitants for further trials.

10.3.6 Protein Characterization

All purified proteins were characterized by UV/VIS spectroscopy, circular dichroism, analytical gel filtration (high-performance liquid chromatography [HPLC]), dynamic light scattering (DLS), and mass spectroscopy (MS), before crystallization trials were attempted. The MS data were used to determine the precise protein molecular weight and the presence of degraded protein. Analytical gel filtration (HPLC) and DLS were instead used to evaluate the homogeneity and eventual oligomeric form of the samples. Other characterization studies that were employed include UV/VIS spectroscopy, fluorescence spectroscopy, Western blot analysis, and native gel electrophoresis, and, if an enzymatic activity was predicted, functional assays were performed.

10.3.7 Protein Crystallization

Optimal crystallization conditions are unpredictable. Nowadays, the only way to search through a large number of variables that may affect crystal growth (pH, precipitant, salt, protein concentration, additives, buffer, temperature, detergents, organics, etc.) is to conduct a sparse matrix search of promising crystallization conditions. Both hanging drop and sitting drop vapor-diffusion methods were used. The crystallization experiments were performed manually or using an Oryx8 crystallization robot (Douglas Instruments). In a typical experiment, 0.2 μl screening solution was added to 0.2 μl protein solution on 96-well crystal plates; the reservoir wells contained 100 μl screening solution. The use of nanodroplet high-throughput crystallization allows for numerous conditions to be sampled while keeping the required protein quantities to a minimum. Conditions that yielded crystals were optimized both with the same robot, increasing the drop volume, and with manually prepared conditions, using 24-well Linbro plates (Hampton Research), either in hanging or sitting arrangement. Drops combining different ratios of protein and precipitant volumes (1–4 μl of each) were allowed to equilibrate against 400–500 μl reservoir solution.

10.3.8 Data Collection and Structure Determination

Diffraction data were measured at the synchrotron radiation sources of ESRF (Grenoble, France) or ELETTRA (Trieste, Italy). Since all the proteins studied in this work do not present structural homologues, MAD (multiple anomalous dispersion) or SAD (single anomalous dispersion) were used to determine the three-dimensional structures.

10.4 STRUCTURAL STUDIES

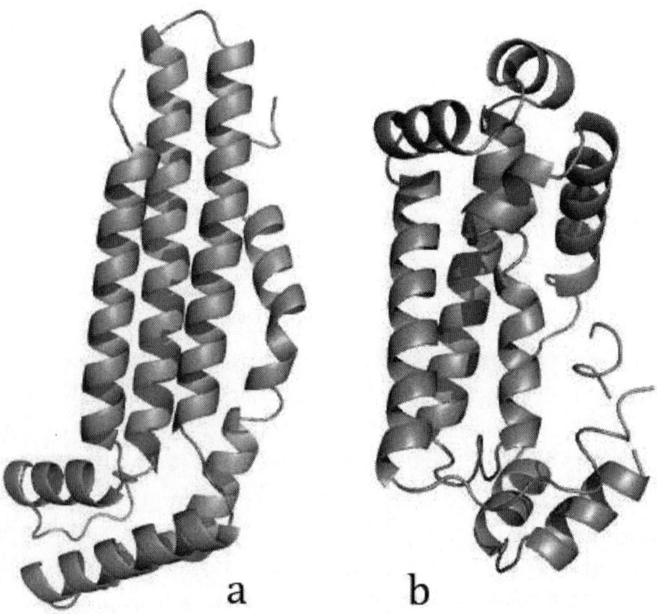

Figure 10.2 Cartoon view of the crystal structure of CagZ (a, left) and CagS (b, right). See also Color Insert.

As can be seen from Table 10.1, 16 proteins in the 27–30 that constitute the *cag*-PAI have been cloned and expressed. Three cases were successful and the three-dimensional structure determined:

this is the case for CagZ, a protein consisting of a single compact L-shaped domain composed of seven α-helices (Cendron *et al.*, 2004) (Fig. 10.2a), CagS, a compact single domain protein with an all-α structure (Cendron *et al.*, 2007) (Fig. 10.2b), and CagD, whose structure have been only recently solved. Crystals have been grown for two domains of protein CagV and the structure determination process is under way. X-ray diffraction data have been collected for both domains, at a resolution around 3.0 Å, but data phasing has not yet been successful.

Table 10.1 Proteins of *Helicobacter pylori cag*-PAI under study. Protein name, SwissProt, TrEMBL, or PIR (Protein Information Resource) accession number. Protein function, information on putative function, and localization of the protein. *A. tumefaciens* homologue, name of the homologue in *A. tumefaciens*, if existing. Status, current status of the project: Cn (cloned); Ex (expressed); Sp (soluble protein); Pp (purified protein); Cy (crystals); Sv (structure solved).

Protein name (*H. pylori* 26696 gene)	Protein function	*A. tumefaciens* homologue	Status	Comment
CagC/25 (hp0546)	Putative pilin	VirB2?	Cn	Toxic/no expression
CagD/24 (hp0545)	Cytoplasm — inner membrane Unknown function protein		Sv	Three-dimensional structure
CagF/22 (hp0543)	Hypothetical chaperone CagA interaction		Pp	Partial degradation/ crystallization trials
CagG/21 (hp0542)	Adhesin		Ex	Toxic/little expression and degradation
CagI/19 (hp0540)	Hypothetical membrane Protein, unknown function		Ex	Inclusion bodies with discrete degradation
CagL/18 (hp0539)	Adhesin			

(*Contd.*)

Table 10.1 (*Contd.*)

Protein name (*H. pylori* 26696 gene)	Protein function	*A. tumefaciens* homologue	Status	Comment
CagM/16 (hp0537)	Structural component		Pp	Inhomogeneous aggregates
CagN/17 (hp0538)	Outer membrane Bound Unknown function		Pp	Crystallization trials
CagS/ (hp0534)	Cytoplasm Unknown function		Sv	2G3V
CagT/ (hp0532)	Structural component Focal circular arrangement at the membrane anchoring of the pilus	VirB7	Ex	Insoluble or unstable constructs
CagU/11 (hp0531)			Cn	Toxic/little expression/ insoluble
CagV/10 (hp0530)	Structural component Membrane bound	VirB8	Cy	Structure determination in progress
CagX/8 (hp0528)	Structural component	VirB9	Ex	Insoluble
CagY/7 (hp0527)	Structural component Pilus surface	VirB10	Cn	Toxic/little expression
CagZ/6 (hp0526)	Cytoplasm		Sv	1S2X
Cagγ/4 (hp0523)	Putative transglycosylase	VirB1	Ex	Insoluble

References

1. Akopyants, N. S., Clifton, S. W., Kersulyte, D., Crabtree, J. E., Youree, B. E., Reece, C. A., Bukanov, N. O., Drazek, E. S., Roe, B. A. and Berg, D. E. (1998). Analyses of the cag pathogenicity island of *Helicobacter pylori*. *Molecular Microbiology*, **28**, pp. 37–53.

2. Alm, R. A., Ling, L. S. L., Moir, D. T., King, B. L., Brown, E. D., Doig, P. C., Smith, D. R., Noonan, B., Guild, B. C., deJonge, B. L., Carmel, G., Tummino, P. J., Caruso, A., Uria-Nickelsen, M., Mills, D. M., Ives, C., Gibson, R., Merberg, D., Mills, S. D., Jiang, Q., Taylor, D. E., Vovis, G. F. and Trost, T. J. (1997). Genomic-sequence comparison of two unrelated isolates of the human gastricpathogen *Helicobacter pylori*. *Nature*, **397**, pp. 176–180.

3. Backert, S., Kwok, T., Schmid, M., Selbach, M., Moese, S., Peek, R. M., Konig W, Meyer, T. F. and Jungblut, P. R. (2005). Subproteomes of soluble and structure-bound *Helicobacter pylori* proteins analyzed by two-dimensional gel electrophoresis and mass spectrometry. *Proteomics*, **5**, pp. 1331–1345.

4. Bailey, S., Ward, D., Middleton, R., Grossmann, J. G. and Zambryski, P. C. (2006). Agrobacterium tumefaciens VirB8 structure reveals potential protein-protein interaction sites. *Proceedings of the National Academy of Sciences U.S.A.*, **103**, pp. 2582–2587.

5. Bayliss, R., Harris, R., Coutte, L., Monier, A., Fronzes, R., Christie, P. J., Driscoll, P. C. and Waksman, G. (2007). NMR structure of a complex between the VirB9/VirB7 interaction domains of the pKM101 type IV secretion system. *Proceedings of the National Academy of Sciences U.S.A.*, **104**, pp. 1673–1678.

6. Bjorkholm, B. M., Oh, J. D., Falk, P. G., Engstrand, L. G. and Gordon, J. I. (2001). Genomics and proteomics converge on *Helicobacter pylori*. *Current Opinion in Microbiology*, **4**, pp. 237–245.

7. Blaser, M. J. (1998). Science, medicine, and the future — *Helicobacter pylori* and gastric diseases. *Britsh Medical Journal*, **316**, pp. 1507–1510.

8. Bourzac, K. M. and Guillemin, K. (2005). *Helicobacter pylori*-host cell interactions mediated by type IV secretion. *Cellular Microbiology*, **7**, pp. 911–919.

9. Brandt, S., Kwok, T., Hartig, R., Konig, W. and Backert, S. (2005). NF-kappa B activation and potentiation of proinflammatory responses by the *Helicobacter pylori* CagA protein. *Proceedings of the National Academy of Sciences U.S.A.*, **102**, pp. 9300–9305.
10. Busler, V. J., Torres, V. J., McClain, M. S., Tirado, O., Friedman, D. B. and Cover, T. L. (2006). Protein-protein interactions among *Helicobacter pylori* Cag proteins. *Journal of Bacteriology*, **188**, pp. 4787–4800.
11. Cascales, E. and Christie, P. J. (2003). The versatile bacterial type IV secretion systems. *Nature Reviews Microbiology*, **1**, pp. 137–149.
12. Cendron, L., Seydel, A., Angelini, A., Battistutta, R. and Zanotti G. (2004). Crystal structure of CagZ, a protein from the *Helicobacter pylori* pathogenicity island that encodes for a type IV secretion system. *Journal of Molecular Biology*, **340**, pp. 881–889.
13. Cendron, L., Tasca, E., Seraglio, T., Seydel, A., Angelini, A., Battistutta, R., Montecucco, C. and Zanotti G. (2007). The crystal structure of CagS from the *Helicobacter pylori* pathogenicity island. *Proteins-Structure Function and Bioinformatics*, **69**, pp. 440–443.
14. Censini, S., Lange, C., Xiang, Z., Crabtree, J. E., Ghiara, P., Borodovsky, M., Rappuoli, R. and Covacci, A. (1996). Cag, a pathogenicity island of *Helicobacter pylori*, encodes type I-specific and disease associated virulence factors. *Proceedings of the National Academy of Sciences U. S. A.*, **93**, pp. 14648–14653.
15. Christie, P. J., Atmakuri, K., Krishnamoorthy, V., Jakubowski, S. and Cascales, E. (2005). Biogenesis, architecture, and function of bacterial type IV secretion systems. *Annual Review of Microbiology*, **59**, pp. 451–485.
16. Covacci, A., Censini, S., Bugnoli, M., Petracca, R., Burroni, D., Macchia, G., Massone, A., Papini, E., Xiang, Z. Y., Figura, N. and Rappuoli, R. (1993). Molecular characterization of the 128-kda immunodominant antigen of *Helicobacter-pylori*-associated with cytotoxicity and duodenal-ulcer. *Proceedings of the National Academy of Sciences U.S.A.*, **90**, pp. 5791–5795.
17. Covacci, A., Falkow, S., Berg, D. E. and Rappuoli, R. (1997). Did the inheritance of a pathogenicity island modify the virulence of *Helicobacter pylori*? *Trends in Microbiology*, **5**, pp. 205–208.
18. Covacci, A. and Rappuoli, R. (1998). *Helicobacter pylori*: molecular evolution of a bacterial quasi-species. *Current Opinion in Microbiology*, **1**, pp. 96–102.

19. Covacci, A., Telford, J. L., Del Giudice, G., Parsonnet, J. and Rappuoli, R. (1999). *Helicobacter pylori* virulence and genetic geography. *Science*, **284**, pp. 1328–1333.

20. Das, A. and Xie, Y. H. (2000). The Agrobacterium T-DNA transport pore proteins VirB8, VirB9, and VirB10 interact with one another. *Journal of Bacteriology*, **182**, pp. 758–763.

21. Figura, N., Vindigni, C., Covacci, A., Presenti, L., Burroni, D., Vernillo, R., Banducci, T., Roviello, F., Marrelli, D., Riscontri, M., Kristodhullu, S., Gennai, C. and Vaira, D. (1998). CagA positive and negative *Helicobacter pylori* strains are simultaneously present in the stomach of most patients with non-ulcer dyspepsia: relevance to histological damage. *Gut*, **42**, pp. 772–778.

22. Fischer, W., Puls, J., Buhrdorf, R., Gebert, B., Odenbreit, S. and Haas R. (2001). Systematic mutagenesis of the *Helicobacter pylori* cag pathogenicity island: essential genes for CagA translocation in host cells and induction of interleukin-8. *Molecular Microbiology*, **42**, pp. 1337–1348.

23. Gomis-Ruth, F. X. and Coll, M. (2001). Structure of TrwB, a gatekeeper in bacterial conjugation. *International Journal of Biochemistry & Cell Biology*, **33**, pp. 839–843.

24. Jiang, Q., Hiratsuka, K. and Taylor, D. E. (1996). Variability of gene order in different *Helicobacter pylori* strains contributes to genome diversity. *Molecular Microbiology*, **20**, pp. 833–842.

25. Jungblut, P. R., Bumann, D., Haas, G., Zimny-Arndt, U., Holland, P., Lamer, S., Siejak, F., Aebischer, A. and Meyer, T. F. (2000). Comparative proteome analysis of *Helicobacter pylori*. *Molecular Microbiology*, **36**, pp. 710–725.

26. Kersulyte, D., Chalkauskas, H. and Berg, D. E. (1999). Emergence of recombinant strains of *Helicobacter pylori* during human infection. *Molecular Microbiology*, **31**, pp. 31–43.

27. Kraft, C., Stack, A., Josenhans, C., Niehus, E., Dietrich, G., Correa, P., Fox, J. G., Falush, D. and Suerbaum, S. (2006). Genomic changes during chronic *Helicobacter pylori* infection. *Journal of Bacteriology*, **188**, pp. 249–254.

28. Kuipers, E. J., Israel, D. A., Kusters, J. G., Gerrits, M. M., van der Ende, A., van der Hulst, R. W. M., Wirth, H. P., Hook-Nikanne, J., Thompson, S. A. and Blaser, M. J. (2000). Quasispecies development of *Helicobacter pylori* observed in paired isolates obtained years apart from the same host. *Journal of Infectious Diseases*, **181**, pp. 273–282.

29. Kwok, T., Zabler, D., Urman, S., Rohde, M., Hartig, R., Wessler, S., Misselwitz, R., Berger, J., Sewald, N., Konig, W. and Backert, S. (2007). Helicobacter exploits integrin for type IV secretion and kinase activation. *Nature*, **449**, pp. 862–864.

30. Middleton, R., Sjolander, K., Krishnamurthy, N., Foley, J. and Zambryski, P. (2005). Predicted hexameric structure of the Agrobacterium VirB4 C terminus suggests VirB4 acts as a docking site during type IV secretion. *Proceedings of the National Academy of Sciences U.S.A.*, **102**, pp. 1685–1690.

31. Monack, D. M., Mueller, A. and Falkow, S. (2004). Persistent bacterial infections: the interface of the pathogen and the host immune system. *Nature Reviews Microbiology*, **2**, pp. 747–765.

32. Moss, S. F. and Sood, S. (2003). *Helicobacter pylori*. *Current Opinion in Infectious Diseases*, **16**, pp. 445–451.

33. Odenbreit, S., Puls, J., Sedlmaier, B., Gerland, E., Fischer, W. and Haas, R. (2000). Translocation of *Helicobacter pylori* CagA into gastric epithelial cells by type IV secretion. *Science*, **287**, pp. 1497–1500.

34. Oh, J. D., Kling-Backhed, H., Giannakis, M., Xu, J., Fulton, R. S., Fulton, L. A, Cordum, H. S., Wang, C. Y., Elliott, G., Edwards, J., Mardis, E. R., Engstrand, L. G. and Gordon, J. I. (2006). The complete genome sequence of a chronic atrophic gastritis *Helicobacter pylori* strain: evolution during disease progression. *Proceedings of the National Academy of Sciences U.S.A.*, **103**, pp. 9999–10004.

35. Parsonnet, J. (1994). Gastric adenocarcinoma and *Helicobacter-pylori* infection. *Western Journal of Medicine*, **161**, p. 60.

36. Parsonnet, J., Friedman, G. D., Vandersteen, D. P., Chang, Y., Vogelman, J. H., Orentreich, N. and Sibley, R. K. (1991). *Helicobacter-pylori* infection and the risk of gastric lymphoma. *New England Journal of Medicine*, **325**, pp. 1127–1131.

37. Parsonnet, J., Replogle, M., Yang, S. and Hiatt, R. (1997). Seroprevalence of CagA-positive strains among *Helicobacter pylori*-infected, healthy young adults. *Journal of Infectious Diseases*, **175**, pp. 1240–1242.

38. Pisani, P., Parkin, D. M., Bray, F. and Ferlay, J. (1999). Estimates of the worldwide mortality from 25 cancers in 1990. *International Journal of Cancer*, **83**, pp. 18–29.

39. Rain, J. C., Selig, L., De Reuse, H., Battaglia, V., Reverdy, C., Simon, S., Lenzen, G., Petel, F., Wojcik, J., Schachter, V., Chemama, Y., Labigne, A. S. and Legrain, P. (2001). The protein-protein interaction map of *Helicobacter pylori. Nature,* **409**, pp. 211–215.

40. Rohde, M., Puls, J., Buhrdorf, R., Fischer, W. and Haas, R. (2003). A novel sheathed surface organelle of the *Helicobacter pylori* cag type IV secretion system. *Molecular Microbiology,* **49**, pp. 219–234.

41. Savvides, S. N., Yeo, H. J., Beck, M. R., Blaesing, F., Lurz, R., Lanka, E., Buhrdorf, R., Fischer, W., Haas, R. and Waksman, G. (2003). VirB11 ATPases are dynamic hexameric assemblies: new insights into bacterial type IV secretion. *EMBO Journal,* **22**, pp. 1969–1980.

42. Schulein, R., Guye, P., Rhomberg, T. A., Schmid, M. C., Schroder, G., Vergunst, A. C., Carena, I. and Dehio C. (2005). A bipartite signal mediates the transfer of type IV secretion substrates of *Bartonella henselae* into human cells. *Proceedings of the National Academy of Sciences U.S.A.,* **102**, pp. 856–861.

43. Smith, T. G., Lim, J. M., Weinberg, M. V., Wells, L. and Hoover, T. R. (2007). Direct analysis of the extracellular proteome from two strains of *Helicobacter pylori. Proteomics,* **7**, pp. 2240–2245.

44. Sozzi, M., Crosatti, M., Kim, S. K., Romero, J. and Blaser, M. J. (2001). Heterogeneity of *Helicobacter pylori* cag genotypes in experimentally infected mice. *FEMS Microbiology Letters,* **203**, pp. 109–114.

45. Suerbaum, S. and Josenhans, C. (2007). *Helicobacter pylori* evolution and phenotypic diversification in a changing host. *Nature Reviews Microbiology,* **5**, pp. 441–452.

46. Tanaka, J., Suzuki, T., Mimuro, H. and Sasakawa, C. (2003). Structural definition on the surface of *Helicobacter pylori* type IV secretion apparatus. *Cellular Microbiology,* **5**, pp. 395–404.

47. Taylor, N. S., Fox, J. G., Akopyants, N. S., Berg, D. E., Thompson, N., Shames, B., Yan, L., Fontham, E., Janney, F. and Hunter, F. M. (1995). Long-term colonization with single and multiple strains of *Helicobacter-pylori* assessed by DNA-fingerprinting. *Journal of Clinical Microbiology,* **33**, pp. 918–923.

48. Terradot, L., Bayliss, R., Oomen, C., Leonard, G. A., Baron, C. and Waksman, G. (2005). Structures of two core subunits of the bacterial type IV secretion system, VirB8 from Brucella suis and ComB10 from *Helicobacter pylori. Proceedings of the National Academy of Sciences U.S.A.,* **102**, pp. 4596–4601.

49. Terradot, L., Durnell, N., Li, M., Li, M., Ory, J., Labigne, A., Legrain, P., Colland, F. and Waksman, G. (2004). Biochemical characterization of protein complexes from the *Helicobacter pylori* protein interaction map — strategies for complex formation and evidence for novel interactions within type IV secretion systems. *Molecular & Cellular Proteomics*, **3**, pp. 809–819.

50. Tomb, J. F., White, O., Kerlavage, A. R., Clayton, R. A., Sutton, G. G., Fleischmann, R. D., Ketchum, K. A., Klenk, H. P., Gill, S., Dougherty, B. A., Nelson, K., Quackenbush, J., Zhou, L. X., Kirkness, E. F., Peterson, S., Loftus, B., Richardson, D., Dodson, R., Khalak, H. G., Glodek, A., McKenney, K., Fitzegerald, L. M., Lee N., Adams, M. D., Hickey, E. K., Berg, D. E., Gocayne, J. D., Utterback, T. R., Peterson, J. D., Kelley, J. M., Cotton, M. D., Weldman, J. M., Fujii, C., Bowman, C., Watthey, L., Wallin, E., Hayes, W. S., Weidman, J. M., Fujii, C., Borodovsky, M., Karp, P. D., Smith, H. O., Fraser, C. M. and Venter, J. C. (1997). The complete genome sequence of the gastric pathogen *Helicobacter pylori*. *Nature*, **388**, pp. 539–547.

51. Tummuru, M. K., Cover, T. L. and Blaser, M. J. (1994). Mutation of the cytotoxin-associated caga gene does not affect the vacuolating cytotoxin activity of *Helicobacter-pylori*. *Infection and Immunology*, **62**, pp. 2609–2613.

52. Vergunst, A. C., van Lier, M. C. M., den Dulk-Ras, A., Stuve, T. A. G., Ouwehand, A. and Hooykaas, P. J. J. (2005). Positive charge is an important feature of the C-terminal transport signal of the VirB/D4-translocated proteins of Agrobacterium. *Proceedings of the National Academy of Sciences U.S.A.*, **102**, pp. 832–837.

53. Viala, J., Chaput, C., Boneca, I. G., Cardona, A., Girardin, S. E., Moran, A. P., Athman, R., Memet, S., Huerre, M. R., Coyle, A. J., DiStefano, P. S., Sansonetti, P. J., Labigne, A., Bertin, J., Philpott, D. J. and Ferrero, R. L. (2004). Nod1 responds to peptidoglycan delivered by the *Helicobacter pylori* cag pathogenicity island. *Nature Immunology*, **5**, pp. 1166–1174.

54. Yeo, H. J., Savvides, S. N., Herr, A. B., Lanka, E. and Waksman, G. (2000). Crystal structure of the hexameric traffic ATPase of the *Helicobacter pylori* type IV secretion system. *Molecular Cell*, **6**, pp. 1461–1472.

55. Yeo, H. J. and Waksman, G. (2004). Unveiling molecular scaffolds of the type IV secretion system. *Journal of Bacteriology*, **186**, pp. 1919–1926.

56. Yeo, H. J., Yuan, Q., Beck, M. R., Baron, C. and Waksman G. (2003). Structural and functional characterization of the VirB5 protein from the type IV secretion system encoded by the conjugative plasmid pKM101. *Proceedings of the National Academy of Sciences U.S.A.*, **100**, pp. 15947–15952.

Chapter 11

THERMOPHILIC ENZYMES OF POTENTIAL INDUSTRIAL USE: STRUCTURE AND FUNCTION

Giuseppe Perugino, Marco Moracci, and Mosè Rossi

[a]*Institute of Protein Biochemistry – Consiglio Nazionale delle Ricerche, Via P. Castellino 111, 80131, Naples, Italy*

[b]*Dipartimento di Biologia Strutturale e Funzionale, Università di Napoli "Federico II," Complesso Universitario di Monte S. Angelo, Via Cinthia 4, 80126 Naples, Italy*

m.rossi@ibp.cnr.it

The exploitation of natural catalysts as whole cells or as purified enzymes is the essence of biotechnology, which was developed in an effort of obtaining industrial processes environmentally more friendly and economically convenient. Nevertheless, several industrial processes are denaturing for most "conventional" organisms and biocatalysts. This may explain why, from their discovery since the 1950s, organisms living under extreme environmental conditions, and the molecules extracted therein, appeared very promising tools for biotechnology. These expectations were motivated by the unusual, for human beings, physical conditions required by extremophiles for growth. As a consequence, biotechnologists can screen extremophiles for the most suitable biocatalyst to apply in a particular industrial field. In this chapter we

Synchrotron Radiation and Structural Proteomics
Edited by Eugenia Pechkova and Christian Riekel

www.panstanford.com

will describe some applications in biotechnology of *extremozymes*, pointing out that most of the applications described here are common to both extremophilic *Bacteria* and *Archaea*. In particular, we will show in detail the potential advantages of mutated glycoside hydrolases from thermophilic organisms, in comparison with their mesophilic counterparts, as biotechnological tools in the oligosaccharide synthesis.

11.1 INTRODUCTION

Biotechnology means any technological application that uses biological systems, living organisms, or derivatives thereof, to make or modify products or processes for specific use. In this sense, we have been constantly involved in making biotechnology, from the ancient civilizations (the use of microbial fermentation for bread-, beer-, and wine-making) to the modern era (the recombinant DNA technology ought to the genetic engineering of animals and plants). In parallel, *biocatalysis* can be defined as the study and the utilization of natural catalysts, called enzymes, to perform chemical transformations on organic compounds. Both isolated enzymes or enzymes still residing inside living cells are employed for this task. The important role of biotechnology and biocatalysis is to transfer the knowledge acquired from biochemistry, molecular biology, microbiology, and genetics to the industrial applications in several fields (human nutrition, mining, pharmaceuticals, chemicals, pulp and paper production, detergents, animal feed, etc.) in an effort of obtaining industrial processes environmentally more friendly and economically convenient (Moracci *et al.*, 2007).

The rising interest in biocatalysts is due to the great advantages that these molecules offer in terms of: (i) increase of rate of chemical reaction, which can be as high as 10^{23}-fold in comparison with the uncatalyzed processes; (ii) transformation under mild conditions; (iii) selectivity, regio- and stereo-specificity, intrinsic properties of the enzymes, which are composed by an unique combination of 19 chiral amino acids which make an asymmetric microenvironment for the substrate; (iv) possibility to promote thousands of different reactions, considering that enzymes are clustered in metabolic cascades involving hundreds of intermediate reactions in the cell; (v) biodiversity, since more than 10,000 species of living organisms are known on Earth, each possessing its own specific enzymatic collection. These characteristics can be exploited in a lot of industrial processes. For example, enzymes can decrease the viscosity, promote liquid mining (hydrolysis of proteins and cellulose) and molecule bioconversion (e.g., the conversion of glucose into fructose for the high fructose corn syrup production), or increase the cleaning power of detergents (removal of stains by enzymatic hydrolysis).

Despite all the benefits mentioned above, the industrial use of enzyme in many cases is still questionable because of the high costs resulting from the purification of enzymes and the adaptation of these biocatalysts to old productive processes. In addition, it is often arduous finding the appropriate enzyme for a specific purpose. Moreover, the main drawback in industrial enzymes is the denaturation and instability of most "conventional" biocatalysts at the harsh reaction conditions of temperature, pH, and ionic strength of several industrial processes.

11.2 THERMOZYMES IN BIOTECHNOLOGY

Since their discovery the 1950s, organisms living under environmental conditions of high temperature, and the molecules extracted therein, appeared very promising tools for biotechnology, and the resulting interests for them were motivated by the unusual, from the human point of view of life, physical conditions required by them to grow. The huge body of data present in the literature on proteins and enzymes from these organisms clearly showed their intrinsic resistance to high temperatures and to other common denaturing agents (such as detergents, proteases, and high concentration of organics) (Danson *et al.*, 1996).

The intrinsic stability of proteins from thermo- and hyperthermophilic organisms resides mainly in their primary structure, but it involves all the structures of higher order (Jaenicke, 2000). A common old-fashioned opinion ascribed the stability of thermophilic proteins to a higher rigidity with respect to their mesophilic counterparts, to compensate the increased fluctuations at high temperatures. However, this theory did not take into account the flexibility which an enzyme needs to accommodate its substrate to perform catalysis. In 1978, Somero demonstrated that every enzyme is in a "corresponding state" at the related optimal temperatures.

This means that all the enzymes show similar behaviors of flexibility/rigidity at their (optimal) physiological temperature.

The comparison of crystallographic data among homologous proteins from psicro-, meso-, and (hyper)thermophilic organisms revealed, mostly, a common overall three-dimensional (3D) structure. With few exceptions, the idea that proteins from (hyper) thermophilic organisms are stabilized at high temperatures by an unique combination of small local modifications is nowadays prevalent. The latter has a cumulative effect on the global structure, including: (i) the substitution of thermolable amino acids as glutamine, asparagines, methionine, and triptophan with more stable residues (Cambillau and Claverie, 2000); (ii) the increase of hydrogen bonds; (iii) the increase of hydrophobic interactions in the core; (iv) the decrease of the number of solvent exposed loops; (v) the tendency to oligomerization (Aguilar *et al.*, 1997); (vi) the presence of large salt bridges networks (Yip *et al.*, 1995; Aguilar *et al.*, 1997; Elcock, 1998; Hough and Danson, 1999). Extensive studies on the structure/function relationship in *thermozymes* will lead to disclose the molecular mechanisms of stabilization of these biocatalysts and to produce stabilized mutants of conventional enzymes (for a review, see Hough and Danson, 1999).

Conduction biocatalysis and biotransformation processes at high temperatures may imply a series of advantages: (i) a reduced viscosity of the medium; (ii) an increased solubility and diffusion of substrates/products in organic compounds; (iii) a reduced risk of contamination by common mesophilic organisms. These features increase both the reaction rates and the process yields by favoring the extraction of volatile organic compounds, the accessibility of recalcitrant substrates to hydrolysis, and the equilibrium displacement in endothermic reactions (Mozhaev, 1993; Krahe *et al.*, 1996). In addition, their peculiar characteristics allow easier purification procedures for the recombinant enzymes (i.e., simple heating steps eliminating the unstable proteins of the host) resulting in a more convenient downstream processing (Moracci *et al.*, 2007). These properties of the thermocatalysts allowed the development of new processes (i.e., the polymerase chain reaction [PCR]) or could be adopted as an alternative to mesophilic enzymes, since their prolonged life span reduces the costs of the continuous addition of fresh biocatalyst (Moracci *et al.*, 2007).

Thermophiles are mainly prokaryotes, with the majority from *Archaea*, but they populate the three living domains since an eukaryotic thermophile has also been recorded (Cary *et al.*, 1998). Biotechnologists can screen thermophiles for the most suitable biocatalyst to apply in a particular industrial field, taking their chance by the biodiversity of these organisms, a key feature for the isolation of organisms with peculiar or unexpected metabolic pathways and of their enzymes with different characteristics in terms of substrate specificity, selectivity, and reaction mechanism. Despite these premises and the extensive research on their peculiar properties, the number of extant applications of extremophiles is still limited. The reasons for this stalling is that their commercial use has to satisfy a number of other different criteria, including technology integration, regulatory compliance, intellectual properties, and security of supply. In other words, microorganisms or biomolecules can be successfully exploited in novel industrial processes if they introduce such an innovation to outcompete the existing plants; in this, thermophiles are no exceptions (Moracci *et al.*, 2007).

11.3 APPLICATION OF THERMOZYMES

The estimated value of the worldwide use of industrial enzymes grew from \$1 billion in 1995 to \$2 billion in 2004 and, depending on different estimates, is expected to rise in the range \$2.4–5.1 billion in 2009 (Business Communication Company, http://www.bccresearch.com; Freedonia group, http://www.freedoniagroup.com/world.html) (Moracci *et al.*, 2007). However, as mentioned above, thermophilic enzymes are interesting alternatives to mesophilic counterparts for the harsh conditions adopted in industry (Demirjian *et al.*, 2001; Kirk *et al.*, 2002). The increasing number of current and promising applications of thermozymes will be briefly described here, emphasizing the cases in which their origin is an added value for biotechnological use. A short list of enzymes from thermophiles is shown in Table 11.1; owing to the limited number of details, the reader is recommended to consult the papers cited in this review and a number of excellent reviews on these biocatalysts that are available in the recent literature (Hough and Danson, 1999; Sellek and Chaudhuri, 1999; Demirjian *et al.*, 2001; Vieille and Zeikus, 2001; Haki and Rakshit, 2003; Atomi, 2005; Moracci *et al.*, 2007).

Table 11.1 Current and potential application of (hyper) thermophilic enzymes (*see next page*) (adapted with the permission from Moracci *et al.* Biotechnology. In *Archaea: Molecular and Cellular Biology* (ed. R. Cavicchioli), pp. 478–495, © 2007, ASM Press).

Enzyme	Organism	Type	Representative applications	T_{opt} (°C)	Remarks
DNA Polymerases	*Pyrococcus furiosus*	*Pfu* pol	In PCR reaction for DNA fragment up to 4 kb	72	Low processivity; uracil stalling
	Thermococcus litoralis	*Vent*® pol		72	Low processivity; uracil stalling
	Pyrococcus sp. GB-D	*Deep Vent*® pol			Low processivity; uracil stalling
	Thermococcus gorgoniarus	*Tgo* pol	In PCR reaction for DNA fragment up to 3.5 kb	72	Low processivity; uracil stalling
	Pyrococcus furiosus	*PfuUltra*™ pol	*Pfu* mutant with improved fidelity; combined with dUTPase (ArchaeMaxx®; patent n. US2005003401) for PCR product up to 17 kb genomic	72	Low processivity; uracil stalling
	Thermococcus kodakaraensis	*Platinum*® Pfx pol	In PCR reaction for DNA fragment up to 12 kb genomic and 20 kb vector	68	High processivity; uracil stalling
DNA ligaseas	*Thermococcus kodakaraensis*			70–80	First DNA ligase from an archaeon
	Pyrococcus furiosus		Stratagene (patent n. US6,280,998)	80	Higher melting temperature for LCR techniques
Alcohol dehydrogenases	*Sulfolobus solfataricus*		Stereochemistry	85	Recombinant; extreme enantiostereoselective
	Aeropyrum pernix			95	Recombinant

Enzyme	Organism	Type	Representative applications	T_{opt} (°C)	Remarks
Proteases	*Pyrococcus furiosus*	Lysine amino-peptidases (KAP)	Baking, brewing, detergents, leather industry	100	First KAP from an archaeon
	Thermococcus kodakaraensis	lon proteases		70	Recombinant; membrane bound; ATP-independent lon protease on unfolded protein substrates
	Sulfolobus solfataricus	Cysteine-protease		70	Intracellular
	Aeropyrum pernix K1	Serine protease		90	Extracellular; recombinant
	Thermococcus stetteri			85	Extracellular; 1% SDS-resistant
	Staphylothermus marinus			90	Membrane associated; resistant after a 135 °C treatment
Esterases/ lipases	*Sulfolobus solfataricus*	Carboxyleterase	Organic synthesis in industrial processes	95	Recombinant
	Pyrobaculum calidifontis			90	Recombinant
	Archeoglobus fulgidus			80	Recombinant
	Thermococcus litoralis	L-aminoacylase	Production of L-amino acids from racemic solutions of N-acetyl-amino acids	85	Recombinant; enantiospecific for L-amino acids
Glycosyl-hydrolases	*Pyrococcus furiosus*	α-amylases	Starch hydrolysis, brewing, baking, detergents	100	Extracellular, recombinant
	Pyrococcus furiosus			100	Extracellular, recombinant
	Desulfurococcus mucosus	Pullulanases type II	Production of linear small sugars	85	Recombinant
	Pyrococcus woesei			100	Recombinant

(*Contd.*)

Table 11.1 (*Contd.*)

Enzyme	Organism	Type	Representative applications	T_{opt} (°C)	Remarks
	Thermococcus aggregans	Pullulanases type III		100	Recombinant
	Picrophilus oshimae	Glucoamylases	Production of glucose	90	
	Picrophilus torridus			90	
	Thermoplasma acidophilum			90	
	Sulfolobus solfataricus	α-glucosidases	Final step of starch degradation	120	Recombinant
	Pyrococcus horikoshii	Endo-β-glucanase	Cotton products biopolishing	85	Recombinant
	Sulfolobus solfataricus		Degradation of cellulose and production of cellooligomers	80	Recombinant
	Sulfolobus solfataricus	Xylanases	Paper bleaching	90	Membrane associated
	Sulfolobus solfataricus	α-xylosidases		90	Recombinant
	Thermococcus chitonophagus	Chitinases	Food, cosmetics, pharmaceuticals, agrochemicals	80	Extracellular part of multi-component enzymatic apparatus
	Pyrococcus horikoshii	β-glycosidases		90	Membrane bound; recombinant
	Sulfolobus solfataricus			85	Recombinant; nucleophile mutant able to synthesize oligosaccharides

11.3.1 Archaeal Enzymes in Molecular Biology

11.3.1.1 DNA polymerases

The amplification of DNA fragments by the well-defined technique of the PCR has led to a burst in the knowledge of life sciences, comprising molecular and cellular biology, genetic engineering, and biotechnology. The advent in the scientific scenario of a novel polymerase from the thermophilic bacterium *Thermus aquaticus* bypassed the problem of the addition of a mesophilic enzyme at each cycle after the denaturation step (Erlich *et al.*, 1988), leading to a rapid development of the PCR technique as we know it today (Chien *et al.*, 1976; Kaledin *et al.*, 1980; Moracci *et al.*, 2007).

The high processivity of *Taq* DNA polymerase still makes this enzyme useful for a wide number of PCR applications; however, the lacking of the 3′–5′ exonuclease activity (*proofreading* activity) leads to a scarce fidelity of the final PCR products, hampering its use in all the cases in which an high-fidelity amplification procedure is required, such as the analysis of particular sequences like allelic stages of single cells, allelic polymorphisms, or rare mutations in human cells.

Significant time and effort has been spent to improve the performance of *Taq* DNA polymerase (Eckert and Kunkel, 1991), but in the last decades, the raising knowledge about archaeal enzymes caused the introduction in the market of a number of DNA polymerases for PCR applications (Table 11.1). Interestingly, all archaeal DNA polymerases (B-type) (Perler *et al.*, 1996) possess proofreading activity and lack of an associated 5′–3′ exonuclease activity, contributing to the great development of PCR techniques. However, most of archaeal DNA polymerases differ from the bacterial *Taq* polymerase also because (i) they exhibit limited processivity *in vitro* (the polymerization rate is in the range 9–25 s^{-1}, compared to that of *Taq* pol which is 47–61 s^{-1}) and (ii) they detect uracil residues (dU), formed by the temperature-dependent deamination of dCTP during the PCR reaction, which causes the stalling in the synthesis and the decrease of their performance (Greagg *et al.*, 1999). These characteristics preclude amplifications of DNA fragments longer than 3–4 kb, since prolonged extension times (1–2 min per kb at 72 °C) promote dUTP formation (Moracci *et al.*, 2007).

The isolation of an archaeal DNA polymerase from *Thermococcus kodakaraensis*, which showed a high extension rate (106–138 s^{-1}) and a low error rate (Takagi *et al.*, 1997), overcame the problem of the low processivity, allowing to perform longer PCR reactions similar to *Taq* DNA polymerase. Moreover, the introduction of a new class of archaeal enzymes, dUTPases, reduced the limitation of the dUTP formation during the PCR reaction. These thermostable biocatalysts from *Pyrococcus* sp. prevent the dUTP incorporation in the PCR products transforming dUTP to dUMP and consequently increasing the PCR final yields (Hogrefe *et al.*, 2002).

Today most commercially available products for the amplification of DNA are composed by an opportune mixture of several thermostable DNA polymerases, including archaeal and bacterial, to guarantee a low error rates and high processivity. This is the case, for instance, of the Herculase™ and PicoMaxx™ (Stratagene, USA), including also the dUTPase (ArchaeMaxx® Polymerase Enhancing Factor), to minimize uracil poisoning, resulting in amplifications as long as 19 kb (Hogrefe *et al.*, 1997; Borns and Hogrefe, 2000; Hogrefe *et al.*, 2002; Moracci *et al.*, 2007).

11.3.1.2 DNA ligases

A variant of the well-established PCR is the ligase chain reaction (LCR). LCR differs from PCR because it amplifies the probe molecule rather than producing amplicon through polymerization of nucleotides. Two probes are used per each DNA strand and are ligated together to form a single probe. LCR uses both a DNA polymerase enzyme and a DNA ligase enzyme to drive the reaction. Like PCR, LCR requires a thermal cycler to drive the reaction, and each cycle results in a doubling of the target nucleic acid molecule. In this way, LCR can have greater specificity than PCR. The advent of this technique has been possible after the isolation of the first thermostable DNA ligase from *T. kodakaraensis* (Nakatani *et al.*, 2000, 2002) and, later on, that isolated (Table 11.1) from *Pyrococcus furiosus*, whose activity is suitable to perform LCR at higher melting temperature.

11.3.2 Alcohol Dehydrogenases

Alcohol dehydrogenases (ADH) from hyperthermophilic Archaea are also worth of mention (Radianingtyas and Wright, 2003). The ADH from *Sulfolobus solfataricus*, for which the 3D structure at high resolution is also available (Esposito *et al.*, 2003), is zinc-containing, NAD(H)-dependent, and it is able to give almost 100% stereoselectivity in the production of (*S*)-3-methylbutan-2-ol from 3-methylbutan-2-one (Raia *et al.*, 2001). The same enzyme from

Aeropyrum pernix shows an enantiomeric excess of about 90% with 2-octanone, 2-nonanone, and 2-decanone substrates (Hirakawa *et al.*, 2004); instead, the ADH from *P. furiosus*, though having wide substrate specificity, presents only slight preference for (*S*)-2-butanol (van der Oost *et al.*, 2001).

11.3.3 Hydrolases

11.3.3.1 Proteases

This class of hydrolases are the major industrial enzymes covering more than 25% of the world market (Business Communication Company, http://www.bccresearch.com). They are extensively used in food, pharmaceutical, leather, and textile industries (Moracci *et al.*, 2007). The stability of the thermophilic proteases to the components (in particular, detergents and alkaline pH) is the main reason for this interest (Kirk *et al.*, 2002). (Hyper)thermophilic proteases are important components of the detergent formulations as they improve the cleaning ability of the detergent on protein-based stains (Eriksen, 1996). Several Archaea harbor an abundant number of intracellular and extracellular proteases, such as *P. furiosus*, which shows more than a dozen of such enzymes (Connaris *et al.*, 1991), *S. solfataricus* (Guagliardi *et al.*, 2002), several thermococcales (Fukui *et al.*, 2002), for which also a patent has been filed in the United States (Stetter, 1998), and *A. pernix* (Catara *et al.*, 2003). The reader can find a detailed description of archaeal proteases in Eichler (2001) and in the consecutive sections of *Methods in Enzymology*, vol. 330, 2001.

11.3.3.2 Esterases/lipases

Esterases and lipases, which differ on the basis of the length of the alkyl chain of the substrate hydrolyzed, are some of the most versatile industrial enzymes being exploited in food industry, in lubricants and cosmetic formulations, and in pulp and paper industry (Tolan, 1996; West, 1996), but mainly in organic synthesis in which they are the most widely used biocatalysts (Demirjian *et al.*, 2001).

The range of reactions in which these enzymes are employed has been extended after the discovery of their thermophilic versions; for instance, thermostable lipases enhanced the physical refining of seed oils, which includes pH 5.0 and separations of the lysophosphatide from the oil at 75 °C (Haki and Rakshit, 2003). Recently, several esterases/lipases have been described from Archaea; the enantioselectivity and the catalytic efficiency in organic solvents of the esterase from *S. solfataricus* have been described in detail (Sehgal and Kelly, 2002; Sehgal *et al.*, 2001, 2002), and, more recently, three open reading frames producing enzymes with esterase activity and an additional phosphotriesterase have been cloned from the same organism (Kim and Lee, 2004; Merone *et al.*, 2005). Other enzymes have been identified in *Pyrobaculum calidifontis* (Hotta *et al.*, 2002), *P. furiousus* (Ikeda *et al.*, 1998), and *Archaeoglobus fulgidus* (Manco *et al.*, 2000) for which the 3D structure is also available (De Simone *et al.*, 2001). Patents of esterase enzymes from various archaeal genera (*Pyrodictium, Archaeoglobus, Thermococcus,* and *Sulfolobus*) have also been filed (Robertson *et al.*, 1999), but no extant application of the listed enzymes is currently known.

11.3.3.3 Glycosidases

Industrial glycoside hydrolases are widely used for the hydrolysis of starch and β-polysaccharides. Every year more than one billion of tons of starch is enzymatically hydrolized for food industry as sugar source (Kaper *et al.*, 2004). This is important in many industrial application, as the production of glucose/fructose syrups for soft drinks, or that of isomalto-oligosaccharides for anti-cariogenic sugars (Kaper *et al.*, 2004), anti-stalling agents in baking (Godfrey, 1996), and cyclic glucans (cyclodextrines). The latter has found applications as carriers of small molecules, artificial protein chaperones, and thermoreversible starch gels (Kaper *et al.*, 2004).

Starch is composed of α-glucose units linked by α-1,4- and α-1,6-glucosidic bonds; the linear polymer amylose consists of α-1,4-linked glucopyranose residues, while amylopectin shows also α-1,6-linked branch points every 17–26 glucose units of the linear polymer.

Liquefaction (gelatinization at 105–110 °C, pH 5.8–6.0) and saccharification (production of maltodextrines by an α-amylase at 95 °C for 2–3 h) are the principal processes for the enzymatic hydrolysis of starch. Glucose and maltose syrups are then produced by the combined action of pullulanase with glucoamylases or β-amylases, respectively. All these processes require a strict control of temperature and pH: gelatinization below 105 °C produces incomplete filtration, while above 105 °C inactivates the enzymes; on the other hand, pHs higher or lower than 5.5 lead to by-products and color formation (Vieille and Zeikus, 2001). To avoid these problems, pH adjustments are required before and after the liquefaction step increasing the chemical costs. Termamyl™ and Fungamyl™ are commercially available enzymes from non-archaeal moderate thermophilic organisms used in the degradation of starch; however, their activity depends on calcium that, forming calcium oxalate, can damage the industrial plans (Haki and Rakshit, 2003). More recently, an engineered version of one of the previous enzymes (Termamyl LC™) drastically reduced the amount of calcium required (Haki and Rakshit, 2003). Making starch conversion more economical would be possible exploiting extremophilic amylases active and stable at 100 °C, pH 4.0–5.0, which do not require calcium in the reaction. In addition, thermostable pullulanases, β-amylases, and glucoamylases could be used in a "one pot" strategy during starch biodegradation.

Amylolytic enzymes produced by (hyper)thermophilic organisms, including Archaea, whose function *in vivo* is, presumably, the hydrolysis of glycogen, include α-amylases, glucoamylases, α-glucosidases, and pullulanases (for a review see Bertoldo and Antranikian, 2002). Enzymes that could find potential applications are α-amylase from *Pyrococcus woesei*, which does not require calcium for activity and stability (Frillingos *et al.*, 2000), and a novel pullulanase from *Thermococcus aggregans*, that was classified as pullulanase of type III. This enzyme attacks simultaneously α-1,4- and α-1,6-glycosidic linkages in pullulan producing a mixture of maltotriose, panose, maltose, and glucose (Niehaus *et al.*, 2000). Finally, glucoamylases belonging to the genera *Picrophilus* and *Thermoplasma* active at 70–100 °C and pH 0.7–3.0 are worth of nothing (Bertoldo and Antranikian, 2002). These activities are interesting candidates for the starch degradation, once methods for their large-scale recombinant expression will be developed.

Cellulose, a linear polymer of β-1,4-linked glucose units, is the most abundant natural polymer on Earth; cellulolytic enzymes are exploited in a variety of industrial fields (for a review, see Bayer *et al.*, 1994). The use of thermostable cellulases would be conceivable, because the high temperature may overcome the problem of the crystalline structure of cellulose, which makes it a recalcitrant substrate for hydrolysis. This condition would allow the loosening of the wood fibers and the access of the hydrolytic enzyme. Therefore, extremophilic cellulases would be very useful in the production of fermentable sugars for fuel ethanol (Wheals *et al.*, 1999), in the biostoning of denim, and in detergent formulas to improve the color brightening and softening by the biopolishing of the cotton fabrics, where temperatures close to 100 °C are needed (Ando *et al.*, 2002), while commercial enzymes are active only at 50–55 °C.

Interesting reviews on hyperthermophilic cellulases and hemicellulases are available in the literature (Bergquist *et al.*, 1999; Sunna *et al.*, 1997). These enzymes are quite uncommon in Archaea; cellulases have been found in *P. furiosus* (Andersen *et al.*, 1998; Cady *et al.*, 2001), in *Pyrococcus horikoshii* (Bergquist *et al.*, 1999), in the archae on AEPII1a (Lam and Mathur, 1998), and in the thermoacidophile *S. solfataricus* (Limauro *et al.*, 2001; Huang *et al.*, 2005). Xylanase activities have been shown in *Thermococcus zilligii* (Uhl and Daniel, 1999), though its unequivocal isolation is still disputed (Rolland *et al.*, 2002), and in *Pyrodictium abyssi* (Andrade *et al.*, 2001), *Halorhabdus utahensis* (Waino and Ingvorsen, 2003), and *S. solfataricus* (Cannio *et al.*, 2004). The xylanases from *P. abyssi* and *S. solfataricus* are highly thermostable and active up to 110 °C and 90 °C, respectively, while the enzyme from *H. utahensis* is, quite surprisingly, active over a broad NaCl concentration range (0–30%).

At present, the main drawback for the exploitation of archaeal cellulolytic enzymes is the difficult expression in heterologous hosts (Huang *et al.*, 2002), making their application to the industrial scale still problematic.

Another class of glycosidases that is worth of mention here are chitinases. Chitin is a waste polymer produced from the exoskeleton of crabs and shrimps and it is used to produce chitosan, applicable in medical fields, cosmetics, paper, and photographic products (Haki and Rakshit, 2003). Extremophilic chitinases would be attractive alternative to the current use of corrosive solutions of 40% sodium hydroxide (Haki and Rakshit, 2003). Interestingly, three enzymes encoded by genes whose function was previously unknown have been recently identified in *T. kodakaraensis*. They constitute a new pathway for the degradation of chitin (Tanaka *et al.*, 2001, 2003, 2004). All the enzymes mentioned above give an idea of how important is the biodiversity of thermophilic organisms as potential tool for novel biotechnological applications.

11.4 OLIGOSACCHARIDE SYNTHESIS BY MUTATED HYPERTHERMOPHILIC GLYCOSIDASES

Nearly all the recent reviews on carbohydrates and glycoenzymes start with a sentence on the importance of these biomolecules in the living organisms. The key informative role of oligosaccharides in biology is becoming more obvious every day. Oligosaccharides, in form of glycoconjugates, play a range of important roles in biological systems including fertilization, embryogenesis, neuronal development, cell proliferation, and metastasis (Varki, 1993; Dwek, 1996; Sears and Wong, 1996), and therefore have considerable potential as therapeutic agents (Zopf and Roth, 1996). However, this potential is limited because the complex structure of the oligosaccharides makes difficult the classical chemical synthesis. In particular, the many protecting group manipulations that are needed to control the stereospecificity (i.e., the formation of only one anomer) and the regiospecificity (i.e., the formation of 1–2 vs. 1–3 vs. 1–4 vs. 1–6 bonds) of the products are demanding and hamper efficient production of oligosaccharides, which is needed for biological testing (Crout and Vic, 1998).

Enzymatic synthesis give the opportunity to produce oligosaccharides in large-scale because the regio- and stereospecificity of the reaction can be controlled (Crout and Vic, 1998). The approaches available so far are based on two major classes of enzymes: glycosyltransferases and glycosidases. The approach involving the use of the formers and nucleotide phosphosugar donors suffers from the scarcity of these enzymes and of the relatively high costs of their substrates. However, the availability of glycosyltransferase from the genomes of several organisms and of methods for the preparation of nucleotide-activated sugars by enzymatic or biological methods is slowly overcoming these problems (Henrissat and Davies, 2000; Sears and Wong, 2001).

11.4.1 Glycosyl Hydrolases in the Oligosaccharide Synthesis

In comparison to glycosyltransferases, glycosidases are particularly abundant and use inexpensive substrates. Despite their hydrolytic nature, glycosidases are attractive enzymes also for the synthesis of oligosaccharides in reactions by reverse hydrolysis (equilibrium-controlled synthesis) or by a typical reaction defined transglycosylation (kinetically controlled process).

Glycoside hydrolases follow two major mechanisms, one giving the inversion and the other giving the overall retention of the anomeric configuration of the substrate (Fig. 11.1).

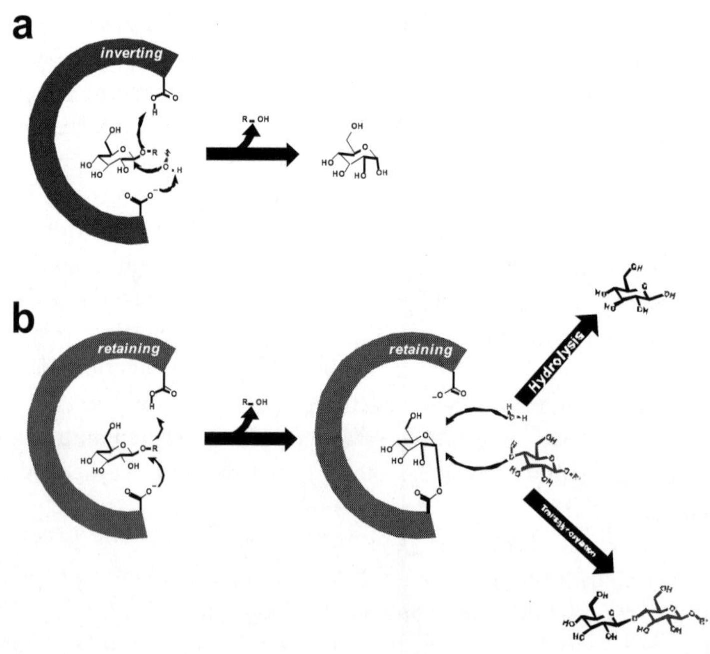

Figure 11.1 Glycoside hydrolases reaction mechanisms. See also Color Insert.

Both mechanisms involve two carboxylic groups in the active site highly conserved in each family (Koshland, 1953). In *inverting* enzymes, these residues function as an acid and a base catalyst, respectively, and operate with a single displacement of the leaving group (Fig. 11.1a) (McCarter and Withers, 1994). Instead, *retaining* enzymes follow a two-step mechanism with the formation of a covalent glycosyl-enzyme intermediate. The carboxyl group in the active center functions as a general acid/base catalyst and the carboxylate as the nucleophile of the reaction (White and Rose, 1997). Enzymatic synthesis by transglycosylation occurs when the glycone is transferred to an acceptor rather than to water (Fig. 11.1b).

The major drawback of this approach is that the reaction products are themselves targets of the hydrolytic activity of the enzymes, thus reducing the yields of the process (10–40%) and precluding their exploitation in large-scale synthesis.

11.4.2 From Glycosidases to Glycosynthases

In 1998, Withers and colleagues described for the first time that a β-glycosidase mutated in the catalytic nucleophile, devoid of hydrolytic activity, produced oligosaccharides with yields >80%. Typically, this totally inactive mutant employed, as donors, glycosyl-fluoride substrates that, possessing the opposite anomeric configuration to that of the normal substrate, mimic the glycosyl-enzyme intermediate (Fig. 11.2a). Under these conditions, several *retaining* glycosidases act as *inverting* enzymes (Wang *et al.*, 1994; MacLeod *et al.*, 1996; Viladot *et al.*, 1998; Shallom *et al.*, 2002), and the reactivation then occurs *via* transglycosylation producing oligosaccharides that the enzyme cannot hydrolyze, thus the desired oligosaccharides accumulate (Mackenzie *et al.*, 1998). The synthesis of the products occurred with a mechanism similar to the second step of the *retaining* hydrolysis reaction, in which the carboxylate of the active site acts as a general base, promoting the attack of the acceptor to the α-glycosyl-fluoride (compare Figs. 11.1b and 11.2a). In the first case reported, Glu358 of the β-glucosidase from *Agrobacterium* sp. (Abg) was replaced by an alanine residue (Mackenzie *et al.*, 1998). This new class of enzymes was named *glycosynthases* (Mackenzie *et al.*, 1998) and since this first report, the "*glycosynthase technology*" was actively developed in other laboratories in the world leading to a variety of novel activities including other exo- and endo-*glycosynthases*, α-*glycosynthases* (Okuyama *et al.*, 2002), mannosynthases, mannansynthases, and galactosynthases (for a review on *glycosynthases*, see Ly and Withers, 1999, and Mackenzie *et al.*, 1998).

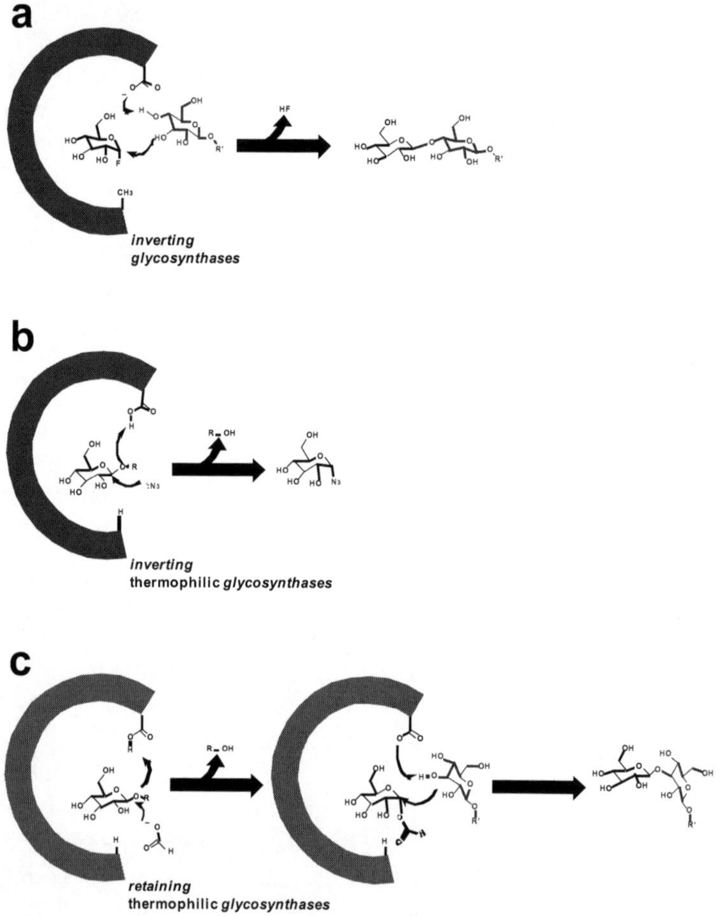

Figure 11.2 Proposed reaction mechanisms of *glycosynthases*. See also Color Insert.

The residue acting as the catalytic nucleophile in *retaining* α- and β-glycosidases (Fig. 11.1b) can be identified using a variety of methods that are described in detail in several excellent reviews (Ly and Withers, 1999; Williams and Withers, 2000). The approaches include mechanism-based inhibitors, inspection of the 3D structure of the enzyme, and site-directed mutagenesis and detailed enzymatic characterization of the mutant. Once the active site residues and the reaction mechanism of a particular enzyme have been experimentally determined, they can be easily extended to all the homologous enzymes by following the classification in families of the more than 2500 glycoside hydrolases (http://www.cazy.org/) (Henrissat, 1991; Henrissat and Bairoch, 1993).

To the best of our knowledge, 17 glycoside hydrolases, belonging to 10 different GH families from all the three living domains, have been modified as efficient *glycosynthases* (Mackenzie *et al.*, 1998; Malet and Planas, 1998; Fort *et al.*, 2000; Mayer *et al.*, 2000; Niehaus *et al.*, 2000; Trincone *et al.*, 2000; Moracci *et al.*, 2001; Nashiru *et al.*, 2001; Fairweather *et al.*, 2002, 2003; Hrmova *et al.*, 2002; Jakeman and Winthers, 2002; Jahn *et al.*, 2003a,b; Perugino *et al.*, 2004; and for an overview, see Hancock *et al.*, 2006). These enzymes are all expressed in recombinant form and can be obtained at amounts as high as 150–250 mg of pure protein per liter of culture (Mayer *et al.*, 2000). Almost every different *glycosynthase* produces a specific type of product; therefore, their biodiversity can be exploited to synthesize oligosaccharides of interest.

11.4.3 Hyperthermophilic Glycosynthases

In a different approach, a completely inactive mutant of the thermophilic β-glycosidase from *S. solfataricus* (*Ss*β-gly) in which the nucleophile of the reaction was changed in a non-nucleophilic residue (Glu387Ala/Gly) could be reactivated by the addition of high amount of external nucleophiles (Table 11.2) and the reaction mechanism depended on the type of nucleophile used (Fig. 11.2) (Moracci *et al.*, 1998).

In particular, the most active Glu387Gly mutant in the presence of sodium azide leaded to α-glucosyl azide as the sole product identified by NMR spectroscopy, indicating that it functions as an *inverting* enzyme (Fig. 11.2b). More interestingly, when 2 M sodium formate is added as nucleophile, the mutant followed the same double-displacement mechanism of *retaining* glycosidases, since it produced 3-O-β-linked disaccharide derivatives of the substrates (compare the transglycosylation reaction of Fig. 11.1b and the *glycosynthase* reaction of Fig. 11.2c). These findings indicated that the *Ss*β-gly Glu387Gly formed the formyl-glucoside intermediate *in situ* (Moracci *et al.*, 1998; Viladot *et al.*, 1998), assigning to formate a biomimetic role restoring the function of the natural nucleophile Glu387 (Perugino *et al.*, 2004). The accumulation of this product during the reaction caused a competition with the original substrate as acceptor molecule in the second step of the reaction.

Table 11.2 Steady-state kinetic constants at 65 °C of *Ss*β-gly wild-type and Glu387Ala/Gly mutants. Kinetic constants were measured in 50 mM sodium phosphate buffer, pH 6.5. Values are corrected for the spontaneous rate observed in the absence of enzyme (reprinted with the permission from Moracci *et al.* Restoration of the activity of active-site mutants of the hyperthermophilic β-glycosidase from *Sulfolobus solfataricus*: dependence of the mechanism on the action of external nucleophiles. *Biochemistry*, 37, pp. 17262–17270, © 1998, American Chemical Society).

	2-Np-β-D-glucopyranoside			**2,4-DNp-β-D-glucopyranoside**		
Enzyme	k_{cat} (s^{-1})	K_M (mM)	k_{cat}/K_M (s^{-1} mM^{-1})	k_{cat} (s^{-1})	K_M (mM)	k_{cat}/K_M (s^{-1} mM^{-1})
Wild type	538 ± 11	1.01 ± 0.24	533	275 ± 16	0.17 ± 0.04	1617
Wild type + 2 M azide	480 ± 13	0.98 ± 0.10	490	428 ± 20	1.64 ± 0.20	261
Wild type + 2 M formate	425 ± 11	0.50 ± 0.07	850	367 ± 12	0.61 ± 0.06	602
Glu387Ala	ND[a]	ND	—	ND	ND	—
Glu387Ala + 2 M azide	ND	ND	—	9.6 ± 0.3	1.21 ± 0.12	8
Glu387Gly	ND	ND	—	ND	ND	—
Glu387Gly + 2 M azide	15 ± 0.5	0.80 ± 0.11	19	110 ± 5	0.18 ± 0.03	611
Glu387Gly + 2 M formate	53 ± 1.2	1.17 ± 0.12	45	42 ± 1	0.13 ± 0.02	323

[a]ND, not detectable.

This competition among the substrate and the products had as result a sort of "transglycosylation chain reaction," producing branched oligosaccharides derivatives of different structure as long as four glycosidic residues containing branched functionalizations of β-1,3 and β-1,6 bonds (Fig. 11.3) (Trincone *et al.*, 2003, 2005).

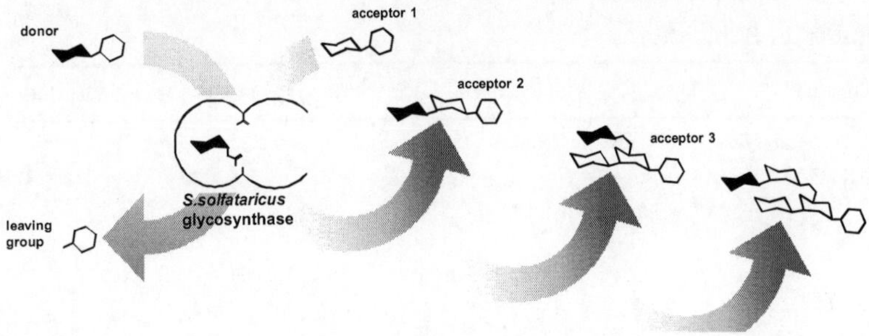

Figure 11.3 The "transglycosylation chain reaction" performed by thermophilic *glycosynthases* in the presence of formate ion as external nucleophile. See also Color Insert.

Among these, of particular interest are β-1,3/1,6-linked tetrasaccharides composing the building blocks of the β-1,3-1,6 glucans that are recognized as elicitors of defense response against pathogens in plants and invertebrates (Ayers *et al.*, 1976; Skriver *et al.*, 1991; Duvic and Soderhall, 1992; Cotè and Hahn, 1994). Several *retaining* glycosidases mutated in the nucleophile residue produced α-glycosyl azide and the products of hydrolysis, in the presence of azide and formate, respectively (Wang *et al.*, 1994; MacLeod *et al.*, 1996; Viladot *et al.*, 1998; Shallom *et al.*, 2002). However, it is worth noting that only mutant glycosidases from hyperthermophiles act efficiently, in the presence of formate, as *retaining glycosynthases* (Table 11.3) (Moracci *et al.*, 1998, 2001; Trincone *et al.*, 2000). Presumably, the intermediate formyl-glycoside, which has been identified in one case (Viladot *et al.*, 2001), does not react efficiently with sugar acceptors in mesophilic enzymes.

Table 11.3 Specificity constants of glycoside hydrolases and related nucleophile mutants reactivated with external nucleophiles (adapted from Moracci *et al.*, 2001, and Perugino *et al.*, 2003).

Enzyme		Substrate	k_{cat}/K_M (s^{-1} mM^{-1})	% reactivation[a]
Agrobacterium β-glucosidase	wt	DNPG[b]	2871	—
	E358A	DNPG	7.1×10^{-5}	—
		DNPG + 2 M azide	0.289	0.01
		DNPG + 4 M formate	2.727	0.10

(*Contd.*)

Table 11.3 (*Contd.*)

Enzyme		Substrate	k_{cat}/K_M (s^{-1} mM^{-1})	% reactivation[a]
Cellulomonas fimi β-glucanase	wt	DNPC	116.7	—
	E233A	DNPC	—	—
		DNPC + 4 M azide	0.09	0.08
		DNPC + 4 M formate	1.73	1.49
Bacillus licheniformis β-glucanase	wt	DNPG4G3G	3015	—
	E134A	DNPG4G3G	4.4×10^{-3}	1.46×10^{-4}
		DNPG4G3G + 3.3 M azide	5.1×10^{-2}	1.69×10^{-3}
		DNPG4G3G + 4 M formate	6	0.2
Sulfolobus solfataricus β-glycosidase	wt	DNPG	1617	—
		ONPG	533	—
	E387A	DNPG	—	—
		DNPG + 2 M azide	8	0.49
	E387G	DNPG	—	—
		DNPG + 2 M azide	611	37.79
		DNPG + 2 M formate	323	19.97
		ONPG	—	—
		ONPG + 2 M azide	19	3.56
		ONPG + 2 M formate	45	8.44
Thermosphaera aggregans β-glycosidase	wt	ONPG	1576	—
	E386G	ONPG	—	—
		ONPG + 2 M azide	16	1.01
		ONPG + 2 M formate	73	4.63

Enzyme		Substrate	k_{cat}/K_M (s^{-1} mM^{-1})	% reactivation[a]
Pyrococcus furiosus β-glycosidase	wt	ONPG	6480	—
	E372A	ONPG	—	—
		ONPG + 2 M azide	0.11	1.69×10^{-3}
		ONPG + 2 M formate	2.6	4.00×10^{-2}

[a]The % of reactivation is the ratio between the specificity constant values of the nucleophile mutant enzyme and their wild-type counterpart (% k_{cat}/K_M mut/k_{cat}/K_M wt)
[b]DNPG, 2,4-dinitro-phenyl-β-D-glucopyranoside; DNPC, 2,4-dinitro-phenyl-β-D-cellobioside; DNPG4G3G, 2,4-dinitro-phenyl-β-D-cellobiosyl-glucopyranoside; ONPG, 2-nitro-phenyl-β-D-glucopyranoside.
[c]Unpublished results.

11.4.4 Strategies for the Improvement of the Glycosynthase Activity

Although *glycosynthase*-catalyzed reactions produce excellent yields since the manipulation of the enzymes precludes hydrolysis of the products, the reactions are sometimes slow and often require substantial quantities of mutant enzyme and/or extended incubation times. This leads to the continuous search to improve enzyme performances by using site-directed and random mutagenesis (Viladot *et al.*, 2001; Kim *et al.*, 2004; Lin *et al.*, 2004). We addressed this issue on the above-mentioned *Ss*β-glyGlu387Gly *glycosynthase* and two other mutants of hyperthermophilic β-glycosidases from *Thermosphaera aggregans* (*Ta*β-glyGlu386Gly) and *P. furiosus* (CelBGlu372Ala).

The reaction mechanism of *retaining glycosynthases* involves a glycosylation step in which the general acid/base catalyst and the formate ion cooperate (Fig. 11.2b). Here, the leaving ability of the group in the substrate and the protonated form of the carboxylic group in the enzyme are essential to perform efficiently the first step of the reaction. Removal of the catalytic nucleophile in *retaining* glycosidases causes a downward shift in the pK_a of the acid/base catalyst (McIntosh *et al.*, 1996; Moracci *et al.*, 1998). Consequently, at pH close to neutrality, this group is converted to its ionized form, hence performing the first step of the reaction less efficiently. By following this line of approach, we tested the activity of hyperthermophilic *glycosynthases* in acidic conditions, assuming that below pH 6.0 the acid/base catalyst would be converted to the protonated form necessary to complete efficiently the first step of the glycosynthetic reaction. As expected, in the presence of a small amount of sodium formate (50 mM) at pH 3.0 and 2-Np-β-Glc substrate, *Ss*β-glyGlu387Gly, *Ta*β-glyGlu386Gly, and CelBGlu372Ala *glycosynthases* showed an increment of the k_{cat} of 43-, 83-, and 105-fold, respectively, if compared to their activity measured at pH 6.0 (Table 11.4). The increment of the activity was of 17-, 13-, and 12-fold, respectively, if compared to that observed in sodium phosphate 50 mM, pH 6.5 and 2 M sodium formate (Perugino *et al.*, 2003). The acidic environment of the reaction mixture not only allowed a noticeable improvement of the efficiency in the synthetic reaction, halving the amount of the catalyst used and the time of the reaction, but also enhanced the synthetic repertoire of these hyperthermophilic *glycosynthases* (Perugino *et al.*, 2003).

Table 11.4 Steady-state kinetic constant of thermophilic mutants for 2-Np-β-D-glc hydrolysis at 65 °C in 50 mM formate buffer (adapted with the permission from Perugino *et al.* Activity of hyperthermophilic glycosynthases is significantly enhanced at acidic pH. *Biochemistry*, 42, pp. 8484–8493, © 2003, American Chemical Society).

	K_M (mM)	k_{cat} (s^{-1})	k_{cat}/K_M (s^{-1} mM^{-1})	K_M (mM)	k_{cat} (s^{-1})	k_{cat}/K_M (s^{-1} mM^{-1})
pH 3.0	6.6 ± 0.6	970 ± 40	147.0	16.4 ±1.6	901 ± 33	54.9
pH 4.0	3.3 ± 0.5	305 ± 14	92.4	4.9 ± 0.5	322 ± 12	65.7
pH 5.0	3.1 ± 0.4	54 ± 2	17.4	2.2 ± 0.5	53 ± 3	24.1
pH 6.0	2.8 ± 0.5	12 ± 1	4.3	1.1 ± 0.4	21 ± 2	19.1
pH 6.5 in formate 2 M	1.0 ± 0.1	73 ± 2	73	1.2 ± 0.1	53 ± 1	44.2

	K_M (mM)	k_{cat} (s^{-1})	k_{cat}/K_M (s^{-1} mM^{-1})
pH 3.0	4.3 ± 0.2	48 ± 1	11.2
pH 4.0	2.1 ± 0.4	7.0 ± 0.4	3.3
pH 5.0	1.3 ± 0.1	0.4 ± 0.0	0.3
pH 6.0	ND[a]	ND	—
pH 6.5 in formate 2 M	1.5 ± 0.1	3.9 ± 0.1	2.6

[a] ND, not detectable.

Finally, it is worth noting that the reactivation of the *Ss*β-glyGlu387Gly *glycosynthase*, if compared to the wild type assayed at optimal conditions (Table 11.2), is the highest among available *glycosynthases* (Table 11.4). This is a clear example of how the unique characteristics of stability of the glycosidases from hyperthermophilic Archaea to high temperatures, acidic pHs, and high concentration of organics allowed the development of a novel methodology opening up new strategies for their exploitation in oligosaccharide synthesis.

11.5 PERSPECTIVE: THE NEXT FIVE YEARS

Thermophilic and, in general, extremophilic organisms, which populate a border line between chemistry and biology, have been always considered interesting source for novel biotechnological tools and methods, placing themselves as an alternative to labile mesophilic organisms and biomolecules. However, products exploiting these organisms and their molecules are a minority of the catalogues of biotechnological industries. One of the most important of them, Genencor, though dealing with enzymes from extremophiles, has no products, either in development or on the market, that are derived from Archaea. At present, Archaezyme Ltd (Jerusalem, Israel) and ArchaeaSolutions Inc. (Tyrone, GA, USA) are the only two companies fully committed with the exploitation of Archaea and their biomolecules.

Fortunately, the scenario is rapidly evolving: just in the last year several exciting findings have been reported — just two among the others, the genomic sequence of the extreme acidophile *Picrophilus torridus* (Futterer *et al.*, 2004) and a new mechanism of gene expression in *Nanoarchaeum equitans* (Randau *et al.*, 2001) — justifying great expectation for the future.

In conclusion, still it is a long way to the current exploitation of these organisms, depending not only on the scientific advances, but also on the market rules and on the propensity of industries and governments in making investments in this field (Moracci *et al.*, 2007).

Acknowledgments

We would like to thank the Agenzia Spaziale Italiana and MIUR for support of this work (I/R/365/02 and "Folding di proteine: l'altra metà del codice genetico").

References

1. Aguilar, C. F., Sanderson, J., Moracci, M., Ciaramella, M., Nucci, R., Rossi, M. and Pearl, L. H. (1997). Crystal structure of the β-glucosidase from the hyperthermophilic archeon *Sulfolobus solfataricus*: resilience as a key factor in thermostability. *Journal of Molecular Biology*, **271**, pp. 789–802.

2. Andersen, L. N., Bjoernvad, M. E. and Schuelein N. (1998). *A thermostable endo-β-1,4-glucanase*. Patent number WO9833895.

3. Ando, S., Ishida, H., Kosugi, Y. and Ishikawa, K. (2002). Hyperthermostable endoglucanase from *Pyrococcus horikoshii*. *Applied and Environmental Microbiology*, **68**, pp. 430–433.

4. Andrade, C. M., Aguiar, W. B. and Antranikian, G. (2001). Physiological aspects involved in production of xylanolytic enzymes by deep-sea hyperthermophilic archaeon *Pyrodictium abyssi*. *Applied Biochemistry and Biotechnology*, **91–93**, pp. 655–669.

5. Atomi, H. (2005). Recent progress towards the application of hyperthermophiles and their enzymes. *Current Opinion in Chemical Biology*, **9**, pp. 166–173.

6. Ayers, A. R., Ebel, J., Finelli, F., Berger, N. and Albersheim, P. (1976). Host-pathogen interactions. IX. Quantitative essays of elictor activity and characteristics of the elicitor present in the extracellular medium of cultures of *Phytophthora megasperma* var. sojae. *Plant Physiology*, **57**, pp. 751–759.

7. Bayer, E. A., Morag, E. and Lamed, R. (1994). The cellulosome-a treasure-trove for biotechnology. *Trends in Biotechnology*, **12**, pp. 379–386.

8. Bergquist, P. L., Gibbs, M. D., Morris, D. D., Te'o, V. S., Saul, D. J. and Moran, H. W. (1999). Molecular diversity of thermophilic cellulolytic and hemicellulolytic bacteria. *FEMS Microbiology Ecology*, **28**, pp. 99–110.

9. Bertoldo, C. and Antranikian, G. (2002). Starch-hydrolyzing enzymes from thermophilic archaea and bacteria. *Current Opinion in Chemical Biology*, **6**, pp. 151–160.

10. Borns, M. and Hogrefe, H. (2000). Unique enhanced DNA polymerase delivers high fidelity and great PCR performance. *Strategies*, **13**, pp. 76–79.

11. Cady, S. G., Bauer, M. W., Callen, W., Snead, M. A., Mathur, E. J., Short, J. M. and Kelly, R. M. (2001). β-Endoglucanase from *Pyrococcus furiosus*. *Methods in Enzymology*, **330**, pp. 346–354.

12. Cambillau, C. and Claverie. J. (2000). Structural and genomic correlates of hyperthermostability. *Journal of Biological Chemistry*, **275**, pp. 32383–32386.

13. Cannio, R., Di Prizito, N., Rossi, M. and Morana, A. (2004). A xylan-degrading strain of *Sulfolobus solfataricus*: isolation and characterization of the xylanase activity. *Extremophiles*, **8**, pp. 117–124.

14. Cary, S. C., Shank, T. and Stein, J. (1998). Worms bask in extreme temperatures. *Nature*, **391**, pp. 545–546.

15. Catara, G., Ruggiero, G., La Cara, F., Digilio, F. A., Capasso, A. and Rossi, M. (2003). A novel extracellular subtilisin-like protease from the hyperthermophile *Aeropyrum pernix* K1: biochemical properties, cloning, and expression. *Extremophiles*, **7**, pp. 391–399.

16. Chien, A., Edgar, D. and Trela, J. (1976). Deoxyribonucleic acid polymerase from the extreme thermophilile *Thermus aquaticus*. *Journal of Bacteriology*, **127**, pp. 1550–1557.

17. Connaris, H., Cowan, D. A. and Sharp, R. J. (1991). Heterogeneity of proteinases from the hyperthermophilic archaeobacterium *Pyrococcus furiosus*. *Journal of General Microbiology*, **137**, pp. 1193–1199.

18. Cotè, F. and Hahn, M. G. (1994). Oligosaccharins: structures and signal transduction. *Plant Molecular Biology*, **26**, pp. 1379–1411.

19. Crout, D. H. and Vic, G. (1998). Glycosidases and glycosyl transferases in glycoside and oligosaccharide synthesis. *Current Opinion in Chemical Biology*, **2**, pp. 98–111.

20. Danson, M. J., Hough, D. W., Russel, R. J. M., Taylor, G. L. and Pearl, L. H. (1996). Enzyme thermostability and thermoactivity. *Protein Engineering*, **9**, pp. 629–630.

21. De Simone, G., Menchise, V., Manco, G., Mandrich, L., Sorrentino, N., Lang, D., Rossi, M. and Pedone, C. (2001). The crystal structure of a hyper-thermophilic carboxylesterase from the archaeon *Archaeoglobus fulgidus*. *Journal of Molecular Biology*, **314**, pp. 507–518.

22. Demirjian, D. C., Moris-Varas, F. and Cassidy, C. S. (2001). Enzymes from extremophiles. *Current Opinion in Chemical Biology*, **5**, pp. 144–151.

23. Duvic, B. and Soderhall, K. (1992). Purification and partial characterization of a β-1,3-glucan-binding-protein membrane receptor from blood cells of the crayfish *Pacifastacus leniusculus*. *European Journal of Biochemistry*, **207**, pp. 223–228.

24. Dwek, R. A. (1996). Glycobiology: toward understanding the function of sugars. *Chemical Reviews*, **96**, pp. 683–720.

25. Eckert, K. A. and Kunkel, T. A. (1991). DNA polymerase fidelity and the polymerase chain reaction. *PCR Methods and Applications*, **1**, pp. 17–24.

26. Eichler, J. (2001). Biotechnological uses of archaeal extremozymes. *Biotechnology Advances*, **19**, pp. 261–278.

27. Elcock, A. H. (1998). The stability of salt bridges at high temperatures: implications for hyperthermophilic proteins. *Journal of Molecular Biology*, **284**, 489–502.

28. Eriksen, N. (1996). Detergents. In *Industrial Enzymes* (eds. T. Godfrey and S. West), 2nd ed., pp.187–200. The Macmillan Press Ltd., Basingstoke, UK.

29. Erlich, H., Gelfand, D. and Saiki, R. (1988). Specific DNA amplification product review. *Nature*, **33**, pp. 461–462.

30. Esposito, L., Bruno, I., Sica, F., Raia, C. A., Giordano, A., Rossi, M., Mazzarella, L. and Zagari, A. (2003). Crystal structure of a ternary complex of the alcohol dehydrogenase from *Sulfolobus solfataricus*. *Biochemistry*, **42**, pp. 14397–14407.

31. Fairweather, J. K., Faijes, M., Driguez, H. and Planas, A. (2002). Specificity studies of bacillus 1,3-1,4-β-glucanases and application to *glycosynthase*-catalyzed transglycosylation. *ChemBioChem*, **3**, pp. 866–873.

32. Fairweather, J. K., Hrmova, M., Rutten, S. G., Fincher, J. B. and Driguez, H. (2003). Synthesis of complex oligosaccharides by using a mutated (1,3)-β-D-glucan endohydrolase from barley. *Chemistry*, **9**, pp. 2603–2610.

33. Fort, S., Boyer, V., Greffe, L., Davies, G. J., Moroz, O., Christiansen, L., Schülein, M., Cottaz, S. and Driguez, H. (2000). Highly efficient synthesis of β(1-4)-oligo- and — polysaccarides using a mutant cellulase. *Journal of the American Chemical Society*, **122**, pp. 5429–5437.

34. Frillingos, S., Linden, A., Niehaus, F., Vargas, C., Nieto, J. J., Ventosa, A., Antranikian, G. and Drainas, C. (2000). Cloning and expression of alpha-amylase from the hyperthermophilic archaeon *Pyrococcus weosei* in the moderately halophilic bacterium *Halomonas elongata*. *Journal of Applied Microbiology*, **88**, pp. 495–503.

35. Fukui, T., Eguchi, T., Atomi, H. and Imanaka, T. (2002). A membrane-bound archaeal Lon protease displays ATP-independent proteolytic activity towards unfolded proteins and ATP-dependent activity for folded proteins. *Journal of Bacteriology*, **184**, pp. 3689–3698.

36. Futterer, O., Angelov, A., Liesegang, H., Gottschalk, G., Schleper, C., Schepers, B., Dock, C., Antranikian, G. and Liebl, W. (2004). Genome sequence of *Picrophilus torridus* and its implications for life around pH 0. *Proceeding of the National Academy of Sciences USA*, **101**, pp. 9091–9096.

37. Godfrey, T. (1996). Baking. In *Industrial Enzymes* (eds. T. Godfrey and S. West), 2nd ed., pp. 87–101. The Macmillan Press Ltd., Basingstoke, UK.

38. Greagg, M. A., Fogg, M. J., Panayotou, G., Evans, S. J., Connolly, B. A. and Pearl, L. H. (1999). A read-head function in archaeal DNA polymerases detects promutagenic template-strand uracil. *Proceeding of the National Academy of Sciences USA*, **96**, pp. 9045–9050.

39. Guagliardi, A., Cerchia, L. and Rossi, M. (2002). An intracellular protease of the crenarchaeon *Sulfolobus solfataricus*, which has sequence similarity to eukaryotic peptidases of the CD clan. *Biochemical Journal*, **368**, pp. 357–363.

40. Haki, G. D. and Rakshit, S. K. (2003). Developments in industrially important thermostable enzymes: a review. *Bioresource Technology*, **89**, pp. 17–34.

41. Hancock, S. M., Vaughan, M. D. and Withers, S. G. (2006). Engineering of glycosidases and glycosyltransferases. *Current Opinion in Chemical Biology*, **10**, pp. 509–519.

42. Henrissat, B. (1991). A classification of glycosyl hydrolases based on amino acid sequence similarities. *Biochemical Journal*, **280**, pp. 309–316.

43. Henrissat, B. and Bairoch, A. (1993). New families in the classification of glycosyl hydrolases based on amino acid sequence similarities. *Biochemical Journal*, **293**, pp. 781–788.

44. Henrissat, B. and Davies, G. J. (2000). Glycoside hydrolases and glycosyltransferases. Families, modules, and implications for genomics. *Plant Physiology*, **124**, pp. 1515–1519.

45. Hirakawa, H., Kamiya, N., Kawarabayashi, Y. and Nagamune, T. (2004). Properties of an alcohol dehydrogenase from the hyperthermophilic archaeon *Aeropyrum pernix* K1. *Journal of Bioscience and Bioengineering*, **97**, pp. 202–206.

46. Hogrefe, H. H., Hansen, C. J., Scott, B. R. and Nielson KB. (2002). Archaeal dUTPase enhances PCR amplifications with archaeal DNA polymerase by preventing dUTP incorporation. *Proceeding of the National Academy of Sciences USA*, **99**, pp. 596–601.

47. Hogrefe, H. H., Scott, B. R., Nielson, K. B., Hedden, V., Hansen, C. J., Cline, J., Bai, F., Amberg, J., Allen, R. and Madden, M. (1997). Novel PCR enhancing factor improves performance of *Pfu* DNA polymerase. *Strategies*, **10**, pp. 93–96.

48. Hotta, Y., Ezaki, S., Atomi, H. and Imanaka, T. (2002). Extremely stable and versatile carboxylesterase from a hyperthermophilic archaeon. *Applied and Environmental Microbiology*, **68**, pp. 3925–3931.

49. Hough, D. W. and Danson, M. J. (1999). Extremozymes. *Current Opinion in Chemical Biology*, **3**, pp. 39–46.

50. Hrmova, M., Imai, T., Rutten, S. J., Fairweather, J. K., Pelosi, L., Bulone, V., Driguez, H. and Fincher J. B. (2002). Mutated barley (1,3)-β-D-glucan endohydrolases synthesize crystalline (1,3)-β-D-glucans. *Journal of Biological Chemistry*, **277**, pp. 30102–30111.

51. Huang, L. N., Zhou, H., Chen, Y. Q., Luo, S., Lan, C. Y. and Qu, L. H. (2002). Diversity and structure of the archaeal community in the leachate of a full-scale recirculating landfill as examined by direct 16S rRNA gene sequence retrieval. *FEMS Microbiology Letters*, **214**, pp. 235–240.

52. Huang, Y., Krauss, G., Cottaz, S., Driguez, H. and Lipps, G. (2005). A highly acid-stable and thermostable endo-beta-glucanase from the thermoacidophilic archaeon *Sulfolobus solfataricus*. *Biochemical Journal*, **385**, pp. 581–588.

53. Ikeda, M. and Clark, D. S. (1998). Molecular cloning of extremely thermostable esterase gene from hyperthermophilic archaeon

Pyrococcus furiosus in *Escherichia coli*. *Biotechnology and Bioengineering*, **57**, pp. 624–629.

54. Jaenicke, R. (2000). Do ultrastable proteins from hyperthermophiles have high or low conformational rigidity? *Proceeding of the National Academy of Sciences USA*, **97**, pp. 2962–2964.

55. Jahn, M., Marles, J., Warren, R. A. J. and Withers, S. G. (2003a). Thioglycoligases: mutant glycosidases for thioglycoside synthesis. *Angewandte Chemie-International Edition*, **42**, pp. 352–354.

56. Jahn, M., Stoll, D., Warren, R. A. J., Szabó, L., Singh, P., Gilbert, H. J., Ducros, V. M., Davies, G. J. and Withers, S. G. (2003b). Expansion of the *glycosynthase* repertoire to produce defined manno-oligosaccharides. *Chemical Communications*, **12**, pp. 1327–1329.

57. Jakeman, D. L. and Withers, S. G. (2002). On expanding the repertoire of *glycosynthases*: mutant β-galactosidases forming β-(1,6)-linkages. *Canadian Journal of Chemistry*, **80**, pp. 866–870.

58. Kaledin, A., Sliusarenko, A. and Gorodetski, S. (1980). Isolation and properties of DNA polymerase from extreme thermophilic bacteria *Thermus aquaticus* YT-1. *Biochimiya*, **45**, pp. 644–651.

59. Kaper, T., van der Maarel, M. J., Euverink, G. J. and Dijkhuizen, L. (2004). Exploring and exploiting starch-modifying amylomaltases from thermophiles. *Biochemical Society Transactions*, **32**, pp. 279–282.

60. Kim, S. and Lee, S. B. (2004). Thermostable esterase from a thermoacidophilic archaeon: purification and characterization for enzymatic resolution of a chiral compound. *Bioscience Biotechnology and Biochemistry*, **68**, pp. 2289–2298.

61. Kim, Y. W., Lee, S. S., Warren, R. A. and Withers, S. G. (2004). Directed evolution of a *glycosynthase* from *Agrobacterium* sp. increases its catalytic activity dramatically and expands its substrate repertoire. *Journal of Biological Chemistry*, **279**, pp. 42787–42793.

62. Kirk, O., Borchert, T. V. and Fuglsang, C. C. (2002). Industrial enzyme applications. *Current Opinion in Biotechnology*, **13**, pp. 345–351.

63. Koshland, D. E. (1953). Stereochemistry and the mechanism of enzymic reactions. *Biological Reviews of the Cambridge Philosophical Society*, **28**, pp. 416–436.

64. Krahe, M., Antranikian, G. and Markl, H. (1996). Fermentation of extremophilic microorganisms. *FEMS Microbiology Reviews*, **18**, pp. 271–285.

65. Lam, D. E. and Mathur, E. J. (1998). *Endoglucanases.* US Patent number 5: 789, 228.

66. Limauro, D., Cannio, R., Fiorentino, G., Rossi, M. and Bartolucci, S. (2001). Identification and molecular characterization of an endoglucanase gene, *celS*, from the extremely thermophilic archaeon *Sulfolobus solfataricus. Extremophiles*, **5**, pp. 213–219.

67. Lin, H., Tao, H. and Cornish, V. W. (2004). Directed evolution of a *glycosynthase* via chemical complementation. *Journal of the American Chemical Society*, **126**, pp. 15051–15059.

68. Ly, H. D. and Withers, S. G. (1999). Mutagenesis of glycosidases. *Annual Review of Biochemistry*, **68**, pp. 487–522.

69. Mackenzie, L. F., Wang, Q., Warren, R. A. J. and Withers, S. G. (1998). *Glycosynthases*: mutant glycosidases for oligosaccharide synthesis. *Journal of the American Chemical Society*, **120**, pp. 5583–5584.

70. MacLeod, A. M., Tull, D., Rupitz, K., Warren, R. A. J. and Withers, S. G. (1996). Mechanistic consequences of mutation of active site carboxylates in a *retaining* β-1,4-glycanase from *Cellulomonas fimi. Biochemistry*, **35**, pp. 13165–13172.

71. Malet, C. and Planas, A. (1998). From β-glucanase to β-glucansynthase: glycosyl transfer to α-glycosyl fluorides catalyzed by a mutant endoglucanase lacking its catalytic nucleophile. *FEBS Letters*, **440**, pp. 208–212.

72. Manco, G., Giosuè, E., D'Auria, S., Herman, P., Carrea, G. and Rossi, M. (2000). Cloning, overexpression, and properties of a new thermophilic and thermostable esterase with sequence similarity to hormone-sensitive lipase subfamily from the archaeon *Archaeoglobus fulgidus. Archives of Biochemistry and Biophysics*, **373**, pp. 182–192.

73. Mayer, C., Zechel, D. L., Reid, S. P., Warren, R. A. J. and Withers, S. G. (2000). The E358S mutant of *Agrobacterium* sp. β-glucosidase is a greatly improved *glycosynthase. FEBS Letters*, **466**, pp. 40–44.

74. McCarter, J. D. and Withers, S. G. (1994). Mechanisms of enzymatic glycoside hydrolysis. *Current Opinion in Structural Biology*, **4**, pp. 885–892.

75. McIntosh, L. P., Hand, G., Johnson, P. E., Joshi, M. D., Korner, M., Plesniak, L. A., Ziser, L., Wakarchuk, W. W. and Withers, S. G. (1996). The pK_a of the general acid/base carboxyl group of a glycosidase cycles during catalysis. A ^{13}C-NMR study of *Bacillus circulans* xylanase. *Biochemistry*, **35**, pp. 9958–9966.

76. Merone, L., Mandrich, L., Rossi, M. and Manco, G. (2005). A thermostable phosphotriesterase from the archaeon *Sulfolobus solfataricus*: cloning, overexpression and properties. *Extremophiles,* **9**, pp. 297–305.

77. Moracci, M., Cobucci-Ponzano, B., Perugino, G. and Rossi, M. (2007). Biotechnology. In *Archaea: Molecular and Cellular Biology* (ed. R. Cavicchioli), pp. 478–495. ASM Press, Washington, DC.

78. Moracci, M., Trincone, A., Cobucci-Ponzano, B., Perugino, G., Ciaramella, M. and Rossi, M. (2001). Enzymatic synthesis of oligosaccharides by two glycosyl hydrolasesof *Sulfolobus solfataricus*. *Extremophiles,* **5**, pp. 145–152.

79. Moracci, M., Trincone, A., Perugino, G., Ciaramella, M. and Rossi, M. (1998). Restoration of the activity of active-site mutants of the hyperthermophilic β-glycosidase from *Sulfolobus solfataricus*: dependence of the mechanism on the action of external nucleophiles. *Biochemistry,* **37**, pp. 17262–17270.

80. Mozhaev, V. V. (1993). Mechanism-based strategies for protein thermostabilization. *Trends in Biotechnology,* **11**, pp. 88–95.

81. Nakatani, M., Ezaki, S., Atomi, H. and Imanaka, T. (2000). A DNA ligase from a hyperthermophilic archaeon with unique cofactor specificity. *Journal of Bacteriology,* **182**, pp. 6424–6433.

82. Nakatani, M., Ezaki, S., Atomi, H. and Imanaka, T. (2002). Substrate recognition and fidelity of strand joining by an archaeal DNA ligase. *European Journal of Biochemistry,* **269**, pp. 650–656.

83. Nashiru, O., Zechel, D. L., Stoll, D., Mohammadzadeh, T., Warren, R. A. J. and Withers, S. G. (2001). β-Mannosynthase: synthesis of β-mannosides with a mutant β-mannosidase. *Angewandte Chemie-International Edition,* **40**, pp. 417–420.

84. Niehaus, F., Peters, A., Groudieva, T. and Antranikian, G. (2000). Cloning, expression and biochemical characterisation of a unique thermostable pullulan-hydrolysing enzyme from the hyperthermophilic archaeon *Thermococcus aggregans. FEMS Microbiology Letters,* **190**, pp. 223–229.

85. Okuyama, M., Mori, H., Watanabe, K., Kimura, A. and Chiba, S. (2002). α-glucosidase mutant catalyzes "*α-glycosynthase*"-type reaction. *Bioscience Biotechnology and Biochemistry,* **66**, pp. 928–933.

86. Perler, F. B., Kumar, S. and Kong H. (1996). Thermostable DNA polymerases. *Advances in Protein Chemistry,* **48**, pp. 377–435.

87. Perugino, G., Trincone, A., Giordano, A., van der Oost, J., Kaper, T., Rossi, M. and Moracci, M. (2003). Activity of hyperthermophilic *glycosynthases* is significantly enhanced at acidic pH. *Biochemistry*, **42**, pp. 8484–8493.

88. Perugino, G., Trincone, A., Rossi, M. and Moracci, M. (2004). Olisaccharide synthesis by *glycosynthases*. *Trends in Biotechnology*, **1**, pp. 31–37.

89. Radianingtyas, H. and Wright, P. C. (2003). Alcohol dehydrogenases from thermophilic and hyperthermophilic archaea and bacteria. *FEMS Microbiology Reviews*, **27**, 593–616.

90. Raia, C. A., Giordano, A. and Rossi, M. (2001). Alcohol dehydrogenase from *Sulfolobus solfataricus*. *Methods in Enzymology*, **331**, pp. 176–195.

91. Randau, L., Munch, R., Hohn, M. J., Jahn, D. and Soll D. (2005). *Nanoarchaeum equitans* creates functional tRNAs from separate genes for their 5'- and 3'-halves. *Nature*, **433**, pp. 537–541.

92. Robertson, D. E., Murphy, D., Reid, J., Maffia, A. M., Link, S., Swanson, R. V., Warren, P. V. and Kosmotka A. (1999). *Esterases*. US Patent number 5,942,430.

93. Rolland, J. L., Gueguen, Y., Flament, D., Pouliquen, Y., Street, P. F. and Dietrich J. (2002). Comment on "The first description of an archaeal hemicellulase: the xylanase from *Thermococcus zilligii* strain AN1": evidence that the unique N-terminal sequence proposed comes from a maltodextrin phosphorylase. *Extremophiles*, **6**, pp. 349–350.

94. Sears, P. and Wong, C. H. (1996). Intervention of carbohydrate recognition by proteins and nucleic acids. *Proceeding of the National Academy of Sciences USA*, **93**, pp. 12086–12093.

95. Sears, P. and Wong, C. H. (2001). Toward automated synthesis of oligosaccharides and glycoproteins. *Science*, **291**, pp. 2344–2350.

96. Sehgal, A. C., Callen, W., Mathur, E. J., Short, J. M. and Kelly, R. M. (2001). Carboxylesterase from *Sulfolobus solfataricus* P1. *Methods in Enzymology*, **330**, pp. 461–471.

97. Sehgal, A. C. and Kelly, R. M. (2002). Enantiomeric resolution of 2-aryl propionic esters with hyperthermophilic and mesophilic esterases: contrasting thermodynamic mechanisms. *Journal of the American Chemical Society*, **124**, pp. 8190–8191.

98. Sehgal, A. C., Tompson, R., Cavanagh, J. and Kelly, R. M. (2002). Structural and catalytic response to temperature and cosolvents of

carboxylesterase EST1 from the extremely thermoacidophilic archaeon *Sulfolobus solfataricus* P1. *Biotechnology and Bioengineering*, **80**, pp. 784–793.

99. Sellek, G. A. and Chaudhuri, J. B. (1999). Biocatalysis in organic media using enzymes from extremophiles. *Enzyme and Microbial Technology*, **25**, pp. 471–482.

100. Shallom, D., Belakhov, V., Solomon, D., Shoham, G., Baasov, T. and Shoham, Y. (2002). Detailed kinetic analysis and identification of the nucleophile in α-L-arabinofuranosidase from *Geobacillus stearothermophilus* T-6, a family 51 glycoside hydrolase. *Journal of Biological Chemistry*, **277**, pp. 43667–43673.

101. Skriver, K., Olsen, F. L., Rogers, J. C. and Mundy, J. (1991). Cis-acting DNA elements responsive to gibberellin and its antagonist abscisic acid. *Proceedings of the National Academy of Sciences USA*, **88**, pp. 7266–7270.

102. Stetter, K. O. (1998). *Thermococcus AV4 and enzymes produced by the same.* US Patent number 5,714,373.

103. Sunna, A., Moracci, M., Rossi, M. and Antranikian, G. (1997). Glycosyl hydrolases from hyperthermophiles. *Extremophiles*, **1**, pp. 2–13.

104. Takagi, M., Nishioka, M., Kakihara, H., Kitabayashi, M., Inoue, H., Kawakami, B., Oka, M. and Imanaka, T. (1997). Characterization of DNA polymerase from *Pyrococcus* sp. Strain KOD1 and its application to PCR. *Applied and Environmental Microbiology*, **63**, pp. 4504–4510.

105. Tanaka, T., Fukui, T. and Imanaka, T. (2001). Different cleavage specificities of the dual catalytic domains in chitinase from the hyperthermophilic archaeon *Thermococcus kodakaraensis* KOD1. *Journal of Biological Chemistry*, **276**, pp. 35629–35635.

106. Tanaka, T., Fukui, T., Fujiwara, S., Atomi, H. and Imanaka, T. (2003). Characterization of an exo-beta-D-glucosaminidase involved in a novel chitinolytic pathway from the hyperthermophilic archaeon *Thermococcus kodakaraensis* KOD1. *Journal of Bacteriology*, **185**, pp. 5175–5181.

107. Tanaka, T., Fukui, T., Fujiwara, S., Atomi, H. and Imanaka, T. (2004). Concerted action of diacetylchitobiose deacetylase and exo-beta-D-glucosaminidase in a novel chitinolytic pathway in the hyperthermophilic archaeon *Thermococcus kodakaraensis* KOD1. *Journal of Biological Chemistry*, **279**, pp. 30021–30027.

108. Tolan, J. S. (1996). Pulp and paper. In *Industrial Enzymes* (eds. T. Godfrey and S. West), 2nd ed., pp. 327–338. The Macmillan Press Ltd., Basingstoke, UK.

109. Trincone, A., Giordano, A., Perugino, G., Rossi, M. and Moracci, M. (2003). *Glycosynthase*-catalysed syntheses at pH below neutrality. *Bioorganic & Medicinal Chemistry Letters*, **13**, pp. 4039–4042.

110. Trincone, A., Giordano, A., Perugino, G., Rossi, M. and Moracci, M. (2005). Highly productive autocondensation and transglycosylation reactions with *Sulfolobus solfataricus glycosynthase*. *ChemBioChem*, **6**, pp. 1431–1437.

111. Trincone, A., Perugino, G., Rossi, M. and Moracci, M. (2000). A novel thermophilic *glycosynthase* that effects branching glycosylation. *Bioorganic & Medicinal Chemistry Letters*, **4**, pp. 365–368.

112. Uhl, A. M. and Daniel, R. M. (1999). The first description of an archaeal hemicellulase: the xylanase from *Thermococcus zilligii* strain AN1. *Extremophiles*, **3**, pp. 263–267.

113. van der Oost, J., Voorhorst, J. W., Kengen, S. W., Geerling, A. C., Wittenhorst, V., Gueguen, Y. and de Vos, W. M. (2001). Genetic and biochemical characterization of a short-chain alcohol dehydrogenase from the hyperthermophilic archaeon *Pyrococcus furiosus*. *European Journal of Biochemistry*, **268**, pp. 3062–3068.

114. Varki, A. (1993). Biological roles of oligosaccharides: all of the theories are correct. *Glycobiology*, **2**, pp. 97–130.

115. Vieille, C. and Zeikus, G. J. (2001). Hyperthermophilic enzymes: sources, uses, and molecular mechanisms for thermostability. *Microbiology and Molecular Biology Reviews*, **65**, pp. 1–43.

116. Viladot, J. L., Canals, F., Batllori, X. and Planas, A. (2001). Long-lived glycosyl-enzyme intermediate mimic produced by formate re-activation of a mutant endoglucanase lacking its catalytic nucleophile. *Biochemical Journal*, **355**, pp. 79–86.

117. Viladot, J. L., de Ramon, E., Durany, O. and Planas, A. (1998). Probing the mechanism of *Bacillus* 1,3-1,4-β-D-glucan 4-glucanohydrolases by chemical rescue of inactive mutants at catalytically essential residues. *Biochemistry*, **37**, pp. 11332–11342.

118. Waino, M. and Ingvorsen, K. (2003). Production of beta-xylanase and beta-xylosidase by the extremely halophilic archaeon *Halorhabdus utahensis*. *Extremophiles*, **7**, pp. 87–93.

119. Wang, Q., Graham, R. W., Trimbur, D., Warren, R. A. J. and Withers, S. G. (1994). Changing enzymatic reaction mechanism by mutagenesis: conversion of a *retaining* glucosidase to an *inverting* enzyme. *Journal of the American Chemical Society*, **116**, pp. 11594–11595.

120. West, S. (1996). Flavour production with enzymes. In *Industrial Enzymes* (eds. T. Godfrey and S. West), 2nd ed., pp. 209–224. The Macmillan Press Ltd., Basingstoke, UK.

121. Wheals, A. E., Basso, L. C., Alves, D. M. and Amorim, H. V. (1999). Fuel ethanol after 25 years. *Trends in Biotechnology*, **17**, pp. 482–487.

122. White, A. and Rose, D. R. (1997). Mechanism of catalysis by *retaining* β-glycosyl hydrolases. *Current Opinion in Structural Biology*, **7**, pp. 645–651.

123. Williams, S. J. and Withers, S. G. (2000). Glycosyl fluorides in enzymatic reactions. *Carbohydrate Research*, **327**, pp. 27–46.

124. Yip, K. S. P., Stillman, T. J., Britton, K. L., Artymiuk, P. J., Baker, P. J., Sedelnikova, S. E., Engel, P. C., Pasquo, A., Chiaraluce, R., Consalvi, V., Scandurra, R. and Rice, D. W. (1995). The structure of *Pyrococcus furiosus* glutamate dehydrogenase reveals a key role for ion-pair network in mantaining enzyme stability at extreme temperatures. *Structure*, **3**, pp. 1147–1158.

125. Zopf, D. and Roth, S. (1996). Oligosaccharide anti-infective agents. *Lancet*, **347**, pp. 1017–1021.

Chapter 12

USING X-RAY SCATTERING TO STUDY THE STRUCTURES OF MEMBRANE-ASSOCIATED PROTEINS

Lin Yang[a] **and Masafumi Fukuto**[b]

[a]*National Synchrotron Light Source, Brookhaven National Laboratory, Upton, NY 11973, USA*

lyang@bnl.gov

[b]*Condensed Matter Physics and Materials Science Department, Brookhaven National Laboratory, Upton, NY 11973, USA*

fukuto@bnl.gov

Structural determination of membrane proteins remains a grand challenge in structural biology. Studying the structures of membrane proteins in two-dimensional (2D) membranes that resemble their native environment is a promising alternative to the prevalent method of X-ray crystallography using crystals produced from detergent-extracted membrane proteins in bulk solutions. Here, we explore the possibility of using X-ray scattering to study the structure of membrane-associated proteins in 2D solutions of fluid, single-layered planar lipid membranes that contain these proteins. To illustrate the feasibility of this approach, we review recent results using tobacco mosaic virus adsorbed to a substrate-supported lipid membrane as a model protein. We discuss the information that can be extracted from these data and the prospect of applying these methods to actual membrane proteins.

Synchrotron Radiation and Structural Proteomics
Edited by Eugenia Pechkova and Christian Riekel

www.panstanford.com

12.1 INTRODUCTION

X-ray scattering and diffraction are fundamental tools to help structural biologists understand how proteins function by learning about their structures. Synchrotron-based protein crystallography and solution scattering (commonly known as small angle X-ray scattering, or SAXS) have contributed tremendously to structural determination of soluble proteins. These techniques, however, have made little impact in the studies of membrane proteins. This is perhaps not surprising since the native environment of membrane proteins, namely, the cell and organelle membranes, are fundamentally different than that of soluble proteins, i.e., the cytoplasm. Solution scattering measurements are carried out in bulk aqueous solutions that resemble the cytoplasm, and protein crystallography requires samples (crystals) derived from such solutions. Membrane proteins, on the other hand, do not dissolve in water.

The prevailing method to make the traditional X-ray scattering and diffraction methods applicable to membrane proteins is to extract the membrane proteins out of the membranes by detergent solubilization and deal with the soluble detergent–protein complex as a whole. Unfortunately, the presence of the detergents greatly hinders the effort to grow high-quality crystals that produce high-resolution X-ray diffraction data. Interpretation of solution scattering data is also never straightforward since the structure of the detergent micelle surrounding the membrane protein is generally not known. More importantly, extracting the membrane proteins out of the membrane necessarily eliminates the possibility to examine how membrane proteins and the membrane interact and how the membrane might regulate the membrane proteins' function. It is therefore logical to instead study membrane proteins in lipid membranes that resemble their native environment. But doing so will require a set of tools other than traditional protein crystallography and solution scattering.

Structural biologists and electron microscopists have already begun to explore structural determination using two-dimensional (2D) membrane protein crystals by reconstituting the membrane proteins back into a lipid bilayer (Hasler *et al.*, 1998; Werten *et al.*, 2002). The obtained 2D crystal is then used in electron diffraction experiments to determine the structure of the membrane protein. This method is somewhat analogous to X-ray crystallography for soluble protein crystals, in the sense that both permit diffraction-based structural determination using crystals produced in a near-native environment for the proteins, i.e., soluble proteins in aqueous solutions and membrane proteins in lipid bilayers.

For soluble proteins, solution scattering is now being used increasingly frequently to deal with proteins that cannot be easily crystallized and to study the structures of proteins in *in situ* settings. Extending the parallel between soluble protein in bulk solution and membrane proteins in membranes, we will explore using X-ray scattering to characterize the structures of membrane-associated proteins in a solution state as well, with the lipid membrane playing the role of a 2D buffer solution. Like X-ray scattering from bulk solutions, 2D solution scattering will not produce atomic resolution structure. However, it will be able to provide lower resolution molecular envelope and can be used to monitor overall structural changes in membrane proteins *in situ*, for instance as they bind to ligands. This method can also potentially provide information on the structure of the lipid membrane, as we will discuss below. 2D solution scattering may also be utilized to characterize the interaction between membrane-bound proteins to help optimizing the 2D crystallization process.

12.2 EXISTING STUDIES

2D solution scattering measurements require appropriately prepared membrane samples. The existing studies of membrane structures commonly employ two sample configurations: substrate-supported lipid multi-bilayers and Langmuir lipid monolayers. In this section, we review studies relevant to 2D solution scattering and discuss the pros and cons for these sample configurations. We then turn to substrate-supported single layers in the next section, which are rarely used for X-ray studies and but have some important advantages over the other two configurations for the application of structural characterization by X-ray scattering.

A well-documented example of X-ray scattering studies of non-crystal structures of membrane-associated proteins is that on Gramicidin ion-channel and small pore-forming peptides in substrate-supported, *multilayered* model membranes (see review by Huang and Yang, 2009). The ion-channels or the transmembrane pores induced by pore-forming peptides collectively behave as 2D liquids confined in each of the lipid bilayers within the multilayered stack. X-ray scattering from Gramicidin channels can be observed using lab-based tube X-ray source (He *et al.*, 1993; Yang *et al.*, 1999). Due to the poor scattering contrast, studies of the transmembrane pores usually resorted to neutron scattering with deuterium-hydrogen substitution to enhance scattering contrast. Synchrotron-based experiments have also showed similar scattering patterns from 2D liquids of transmembrane pores (Huang and Yang, 2009). Furthermore, as the sample is dehydrated, the positional correlations between pores located in neighboring bilayers start to develop and the pores can form 3D periodic structures that diffract like a crystal. The structures of the pores formed by alamethicin (Qian *et al.*, 2008a) and a Bax-α5 (Qian *et al.*, 2008b) in Br-containing lipid bilayers have recently been solved using the multi-wavelength anomalous dispersion method at Br K-edge.

A major drawback of structural studies utilizing lipid multi-bilayers is that the samples are measured in a partially dehydrated state in a humidified environment, where the water vapor surrounding the sample maintain a water layer that is ~1 nm thick between bilayers. The chemical condition in the inter-bilayer water is not well defined and cannot be easily characterized or controlled. This method therefore is not suitable for generic membrane-associated proteins, which may require specific chemical environment to function. Small peptides can be introduced into the supported lipid multilayer by pre-mixing the organic solutions of lipids and peptides when preparing the multilayer. This again may not be feasible for membrane-associated proteins in general.

For easier introduction of proteins and control of the chemical environment around the proteins, Langmuir monolayers are often utilized. Langmuir monolayers are single-leaflet lipid membranes spread at the air–water interface. Proteins can be introduced into the aqueous subphase and interact with the lipid membrane. Most X-ray scattering studies of proteins adsorbed to a Langmuir monolayer focused on specular reflectivity (Vaknin *et al.*, 1993;

Haas and Moehwald, 1994; Weygand *et al.*, 1999; Zheng *et al.*, 2001), which provides information only on the average structure along the direction normal to the monolayer. Clearly, more information can be extracted from off-specular scattering due to the in-plane structure. For instance, in a recent study of poorly ordered crystals of cholera toxin (Miller *et al.*, 2008), the q_z-dependence of the diffraction intensity along the off-specular quasi-Bragg rods (q_z is the component of the scattering vector normal to the membrane) was analyzed to infer the molecular orientation of the bound proteins with respect to the membrane.

Under appropriate condition in the subphase, the proteins adsorbed to the Langmuir monolayer can form 2D crystals (Kornberg and Darst, 1991; Dietrich and Vénien-Bryan, 2005). Grazing incidence X-ray diffraction from such 2D protein crystals has been demonstrated (Haas *et al.*, 1995; Lenne *et al.*, 2000). While structural reconstruction has not been reported using diffraction data obtained from these 2D crystals, Lenne *et al.* were able to collect from 2D streptavidin crystals diffraction data corresponding to 10 Å in-plane resolution (Lenne *et al.*, 2000). It should be noted that 2D protein crystals are usually very small (for instance, lateral dimension of streptavidin 2D crystals can be as large as tens of microns). Therefore, in these grazing incidence diffraction measurements X-ray diffraction comes from an ensemble of crystals with random in-plane orientations within the illuminated area, which is usually centimeters long along the beam direction. This could pose a problem for structural determination since Bragg rods at identical or similar q_r positions may not be resolved (q_r is magnitude of the component of the scattering vector parallel to the membrane).

12.3 SUBSTRATE-SUPPORTED, SINGLE-LAYERED LIPID MEMBRANES

Substrate-supported *single-layered* lipid membranes submerged under a buffer solution present a unique opportunity for X-ray scattering studies of the structure of membrane-associated proteins. These membranes can be prepared in different configurations to accommodate various types of membrane-associated proteins

(Fig. 12.1a–c). Like Langmuir monolayers, these lipid membranes are in direct contact with the bulk aqueous solution, which provides a means to control the chemical environment and incorporate proteins into the membranes. FRAP measurements have shown that the lipid molecules in these supported membranes remain mobile (Boxer, 2000; Yang *et al.*, 2009), as in native biological membranes. The mobility of lipids, which is essential for the biological function of the proteins, is also required to create 2D crystals of lipid-bound proteins (Kornberg and Darst, 1991; Mosser and Brisson, 1991; Weisenhorn *et al.*, 1992; Dietrich and Vénien-Bryan, 2005). For instance, 2D streptavidin crystals formed on supported lipid bilayers

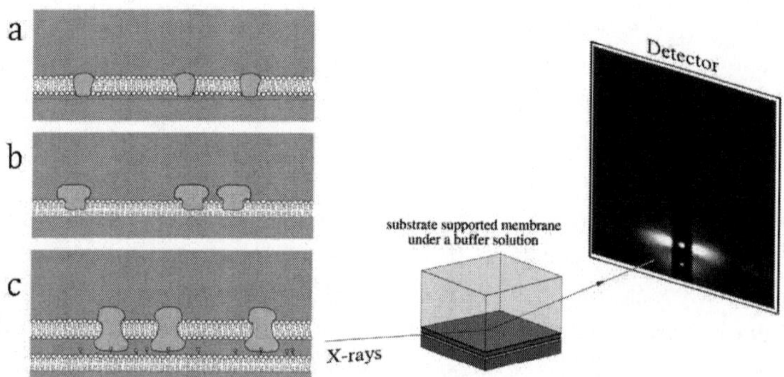

Figure 12.1 Schematic representation of X-ray scattering from 2D solution of membrane-associated proteins. The proteins are embedded in a substrate-supported lipid membrane, which in turn is submerged in a buffer solution that provides a suitable chemical environment for the protein to function. The lipid membrane and the proteins can be associated in several different configurations. (a) A substrate-supported bilayer is suitable for transmembrane proteins that do not extend beyond the thickness of the lipid bilayer. (b) Peripheral proteins that only penetrate into one leaflet of the bilayer can be studied in a lipid monolayer supported by a hydrophobic substrate (e.g., Plant, 1999; Scheuring *et al.*, 1999). (c) In the most general case, a complete bilayer is assembled above the substrate-supported monolayer. Transmembrane proteins that are embedded in the lipid bilayer are constrained to the monolayer by Ni-His tag linkers. This process has been used with Langmuir lipid monolayers to create 2D membrane protein crystals (e.g., Hasler *et al.*, 1998; Lévy *et al.*, 1999). His tags are commonly present in recombinant membrane proteins due to the requirement for the purification process and Ni-chelating lipids are commercially available. See also Color Insert.

and monolayers have been reported (Calvert and Leckband, 1997; Scheuring *et al.*, 1999; Lou *et al.*, 2007). The solid substrate may be further utilized to control the orientation of the crystalline domains. As is being explored in other self-assembled structures (Chai *et al.*, 2007; Bita *et al.*, 2008), physical or chemical patterns on substrate can be used to influence the process of self-assembly and extend the range of structural order. Crystal domains aligned by the patterned substrate would behave as part of a large single crystal. The aforementioned problem of unresolved diffraction peaks therefore can be avoided.

These supported membranes also have several advantages over Langmuir monolayers in grazing incidence X-ray scattering (GIXS) measurements. Since the membrane is formed on a solid support, the incident angle of the X-ray can be adjusted simply by tilting the substrate. In contrast, the X-ray beam must be deflected downward in measurements of Langmuir monolayers because the water surface always remains horizontal. The amount of proteins required for preparing a substrate-supported sample is also relatively small. The total volume of the bulk solution to which proteins are introduced can be much less than 1 ml, compare to ~20 ml (Lenne *et al.*, 2000) and 170 ml (Haas *et al.*, 1995), respectively, in the streptavidin studies cited above. Therefore, at the same protein concentration, substrate-supported membrane samples consume less protein than Langmuir monolayer samples. They are therefore more practical for the studies of membrane proteins that are often available only in very limited quantities. The small sample size and mechanical stability also makes the substrate-supported sample geometry well suited for high throughput-type of approach due to easier sample handling, and for *in situ* measurements to explore the behavior of the protein under a variety of different conditions.

12.4 CASE STUDY WITH TOBACCO MOSAIC VIRUSES AS MODEL PROTEINS

X-ray scattering from proteins in a bulk solution is isotropic. In contrast, the GIXS data from supported 2D membranes contains both in-plane and out-of-plane information due to the planar geometry of the structure. Following the description of bulk solution scattering, we will write the scattering intensity as the product of the form factor F, which describes the structure of individual proteins, and the structure factor S, which describes the positional correlations between proteins:

$$I(q_r, q_z) \sim \langle |F(q_x, q_y, q_z)|^2 S(q_x, q_y) \rangle . \quad (12.1)$$

Here, x and y are the in-plane coordinates in real space. The bracket $\langle\rangle$ denotes the average over all possible in-plane orientations since in general the organization of the membrane-associated proteins within the plane of the membrane is isotropic (liquid-like or contain many small, randomly oriented 2D crystalline domains) and the measured scattering intensity only depends on the magnitude of the in-plane scattering vector $\mathbf{q}_r = (q_x, q_y)$. A full analysis of the scattering data should include q_r– and q_z– dependence of the scattering intensity. However, because the positional correlations between membrane-bound proteins are confined to the plane of the membrane, the structure factor is independent of q_z. Therefore, a simplified analysis can be performed on the in-plane scattering intensity near $q_z = 0$:

$$I(q_r, q_z = 0) \sim \langle |F(q_x, q_y, 0)|^2 S(q_x, q_y) \rangle \quad (12.2)$$

This equation still contains the complete structure factor and the value of the form factor at $q_z = 0$, which corresponds to the structure of the protein projected onto the membrane plane. Below we will examine the content of structural information that can be extracted from Eq. 12.2 using the example of a recent study (Yang *et al.*, 2009) on tobacco mosaic viruses (TMVs) organized on a substrate-supported lipid monolayer (Fig. 12.2).

TMV is a rod-like virus and has a diameter of 18 nm and a length of 300 nm. At physiological pH, TMVs are negatively charged and can bind to a substrate-supported monolayer containing cationic lipids. The lipid-bound TMVs are oriented such that their long axes lie parallel to the membrane as demonstrated in Fig. 12.2a. The positional correlation between TMVs is only significant within a domain and depends only on the lateral distance between the rods. The structure factor therefore is a function of q_x only (we choose x to be the direction perpendicular to the long axis of the TMV) and the scattering intensity is further simplified as:

$$I(q_r) \sim \langle |F(q_x, q_y)|^2 S(q_x) \rangle = \frac{2}{\pi} \int_0^{\pi/2} |F(q_r \sin\vartheta, q_r \cos\vartheta)|^2 S(q_r \sin\vartheta) d\vartheta, \tag{12.3}$$

where ϑ is the angle between the TMV long axis and the in-plane scattering vector $\mathbf{q}_r$.

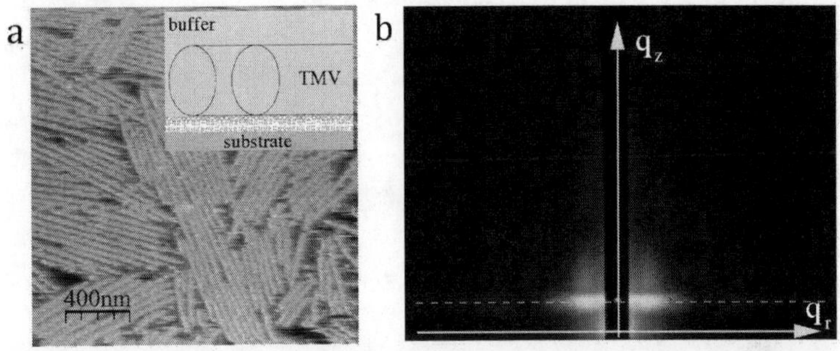

Figure 12.2 (a) Atomic force microscope image of TMVs organized on a substrate-supported lipid monolayer. Due to their long rod-like shape, TMVs lie parallel to the lipid membrane and also approximately parallel to each other within each loosely organized domain, as depicted in the inset of schematic side view. The sample configuration is similar to that in Fig. 12.1b. (b) GIXS data obtained from the same structure. A slight intensity maximum at low q_r suggests position correlation between the TMV rods. The in-plane intensity along the dashed line is analyzed in Fig. 12.3 (adapted from Yang *et al.* Structure and interaction in 2D assemblies of tobacco mosaic viruses. *Soft Matter*, 5, pp. 4951–4961, © 2009 by permission of The Royal Society of Chemistry). See also Color Insert.

This in-plane scattering occurs along the dashed line in the 2D GIXS pattern in Fig. 12.2b, and the corresponding intensity profile is plotted as symbols in Fig. 12.3a. For the purpose of analysis of the structure of *individual* proteins, solution scattering studies should be performed at low protein concentration, so that the structure factor is identically unity. On the other hand, the scattering data shown in Figs. 12.2 and 12.3 contain a maximum at $q_r \sim 0.02$ Å^{-1}, suggesting a non-negligible contribution from the structure factor, i.e., positional correlations between neighboring TMVs. This is similar to neutron scattering from transmembrane pores the multilayer. In the analysis of such neutron scattering data, the pore structure is usually assumed to be a simple cylinder, defined by the D_2O enhanced water lumen through the transmembrane pore (Ludtke, *et al.*, 1996; Yang *et al.*, 1999). The size of the pore is then determined as one of the fitting parameters in the data analysis. In contrast, the radial electron density distribution of TMV can be obtained by modeling scattering data from TMVs in a bulk solution. The 2D form factor can be then calculated and used in the analysis of the structure factor to extract information on the electrostatic interaction between TMVs. The details of this analysis are described in Yang *et al.* (2009). Briefly, an approximate experimental structure factor (symbols in the inset of Fig. 12.3a) is obtained by dividing the data by the calculated form factor. The structure factor is then calculated from results of Monte Carlo simulations based on adjustable parameters that describes the TMV–TMV interaction to reproduce the experimental structure factor. The scattering data can be well described by a structure factor based on screened Coulomb repulsion due to ~−90e carried by each TMV rod (magenta line in the inset of Fig. 12.3a).

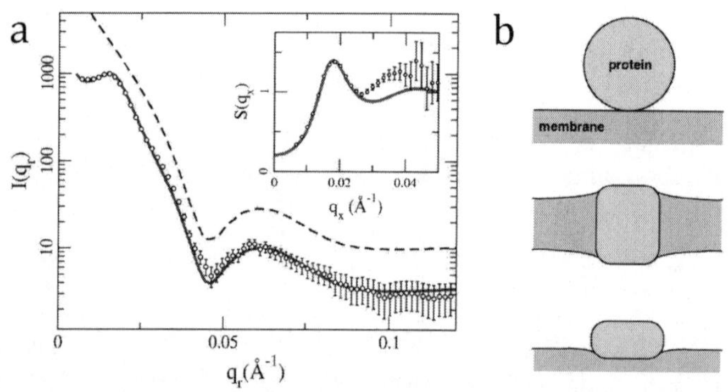

Figure 12.3 (a) Analysis of the experimental in-plane GIXS data (symbols) shown in Fig. 12.2. The fit (red line) includes contributions from the form factor (blue dashed line, offset for clarity) calculated from bulk solution scattering and the structure factor (magenta line in the inset) calculated based on screened Coulomb repulsion between two charged rods. The symbols in the inset show the experimental structure factor obtained by dividing the experimental data by the form factor. (adapted from Yang *et al.* Structure and interaction in 2D assemblies of tobacco mosaic viruses. *Soft Matter*, 5, pp. 4951–4961, © 2009 by permission of The Royal Society of Chemistry). (b) Different scenarios that may affect the calculation of the form factor. The form factor used here is calculated based on the assumption that the membrane does not contribute to the form factor (top). However, membrane distortions, such as those produced by hydrophobic matching (middle) or thinning (bottom) due to expansion of lipid headgroup, must be included in the calculation of the form factor for small proteins. See also Color Insert.

This method of using structure factor analysis to elucidate inter-particle interaction is analogous to those employed in the studies of protein crystallization in bulk solutions using light scattering and X-ray scattering (George and Wilson, 1994; Finet *et al.*, 2004). The nature of the protein–protein interaction ultimately determines whether the proteins would form crystals, aggregates, or would not nucleate at all. Clearly, analyses of this kind could also be useful in searching for the optimal conditions for 2D crystallization of membrane proteins.

The form factor F in Eq. 12.1 is determined by the electron density contrast of the protein against its environment. In the analysis of the TMV scattering data, we have assumed that TMVs are simply adsorbed to the membrane without disturbing the membrane structure. The in-plane electron density contrast is therefore due to electron density of TMV against that of the buffer solution, as in bulk solution scattering measurements.

For actual membrane proteins, however, the form factor may contain a contribution from the membrane itself. It is well known that protein–membrane interaction tends to distort the profile of the lipid bilayer. For integral membrane proteins, the lipid bilayer thickness must match the hydrophobic portion of the membrane protein. This "hydrophobic matching" usually creates a non-uniform bilayer thickness profile centered at the membrane protein (Harroun *et al.*, 1999). Such distortion in bilayer thickness profile can also be induced by peripheral membrane proteins and membrane active peptides that preferentially interact with the lipid headgroups, therefore effectively expand the headgroups relative to the lipid chains and create "thinning" where the membrane protein and lipid membrane make contact (Huang, 2000). Distortions in lipid bilayer structure (Fig. 12.3b) alter the electron density contrast around the protein and therefore the form factor. The resulted change in the form factor will be most pronounced where the protein form factor is near zero, and therefore sensitive to small perturbations.

Due to the large volume of the TMV, membrane distortion due to TMV adsorption is unlikely to cause the form factor to significantly deviate from that calculated from bulk solution scattering. However, such perturbation may have to be taken into account when analyzing scattering data from smaller membrane-associated proteins. While we only examined the in-plane scattering data for TMVs, a complete data analysis for membrane-associated proteins should include the complete 2D GIXS data. Such analysis can also incorporate the low-resolution molecular envelope or atomic resolution rigid body modeling methods (e.g., see review by Svergun and Koch, 2002) that are widely used to analyze bulk protein scattering data today.

12.5 DISCUSSION AND OUTLOOK

The above example of TMV adsorbed on a substrate-supported lipid monolayer provides a glimpse into the rich structural information that can be potentially retrieved from the GIXS data from membranes with associated proteins. Both the structure of the individual protein and the interactions between proteins as well as between the membrane and the protein can be probed. The challenge, however, is to collect scattering data of high quality so that model fitting can provide reliable information. The volume of TMV ($\sim 7.3 \times 10^4$ nm^3) is much larger than typical membrane proteins. For instance, the volume of the KcsA potassium channel is approximately 6 nm high and 4 nm in diameter, with its overall volume smaller than that of TMV by a factor of ~1000. On the other hand, membrane proteins have higher electron density contrast against the membrane compared to that of TMV against water (the average electron density is 0.33 e/Å^3 for water, ~0.25 e/Å^3 for lipid chains, as shown in Wang *et al.*, 2008, and ~0.42 e/Å^3 for proteins). Membrane proteins should have comparable electron density as the TMV shell. Membrane proteins can also be bound to the membrane at a higher number density, and therefore more membrane proteins can contribute to the scattering intensity. Considering all these factors, the scattering intensity from membrane-associated proteins is expected to be lower by a factor of $\sim 10^2$–10^3 than that observed in the TMV data presented here. The background scattering from the buffer solution therefore must be reduced by the same factor to obtain GIXS data of comparable signal-to-background ratio. This may be achievable by reducing the thickness of the bulk solution covering the membrane and using focused X-ray beam to reduce the volume of the bulk solvent that is illuminated by X-rays. In situations where the use of a solid support is not crucial, Langmuir monolayer may be a viable option for performing 2D solution scattering, as the incident X-rays do not penetrate into the bulk water at grazing incident angles, resulting in low background scattering. For instance, Langmuir monolayers may be used for exploring the optimal condition for 2D crystallization (Fukuto *et al.*, 2010); 2D crystals formed under such condition and aligned on the substrate then may be used to collect diffraction data for structural determination.

Another challenge for GIXS measurements on single-layered membrane structures in general is to mitigate radiation damage. Due to the small sample volume, the sample must be exposed to X-rays longer and receive high dose per unit volume to accumulate enough scattering intensity comparable to that from bulk samples. Cryocooling has been used effectively in protein crystallography to mitigate radiation damage. The same general principle should also apply for planar membrane structures. However, the process of initially freezing planar membrane samples and maintaining sample temperature will likely be different from those for bulk protein crystals and have to be developed.

Acknowledgments

The measurements on TMVs were carried out at the National Synchrotron Light Source of Brookhaven National Laboratory, supported by US Department of Energy, Office of Basic Energy Sciences, under Contract No. DE-AC02-98CH10886. We thank Prof. Qian Wang's group at University of South Carolina for providing the supply of TMVs used in these measurements. Parts of Figs. 12.2 and 12.3 are adapted from Yang *et al.*, 2009, with permission of The Royal Society of Chemistry.

References

1. Bita, I., Yang, J. K. W., Jung, Y. S., Ross, C. A., Thomas, E. L. and Berggren, K. K. (2008). Graphoepitaxy of self-assembled block copolymers on two-dimensional periodic patterned templates. *Science*, **321**, pp. 939–943.

2. Boxer, S. G. (2000). Molecular transport and organization in supported lipid membranes. *Current Opinion in Chemical Biology*, **4**, pp. 704–709.

3. Calvert, T. L. and Leckband, D. (1997). Two-dimensional protein crystallization at solid-liquid interfaces. *Langmuir*, **13**, pp. 6737–6745.

4. Chai, J., Wang, D., Fan, X. and Buriak, J. M. (2007). Assembly of aligned linear metallic patterns on silicon. *Nature Nanotechnology*, **2**, pp. 500–506.

5. Dietrich, J. and Vénien-Bryan, C. (2005). *Strategies for Two-Dimensional Crystallization of Proteins Using Lipid Monolayers.* Imperial College Press, pp. 1–156.

6. Finet, S., Skouri-Panet, F., Casselyn, M., Bonneté, F. and Tardieu, A. (2004). The Hofmeister effect as seen by SAXS in protein solution. *Current Opinion in Colloid & Interface Science*, **9**, pp. 112–116.

7. Fukuto, M., Wang, S. T., Lohr, M. A., Kewalramani, S. and Yang, L. (2010). Effects of surface ligand density on lipid-monolayer-mediated 2D assembly of proteins. *Soft Matter*, **6**, pp. 1513–1519.

8. George, A. and Wilson, W. W. (1994). Predicting protein crystallization from a dilute-solution property. *Acta Crystallographica Section D — Biological Crystallography*, **50**, pp. 361–365.

9. Haas, H., Brezesinski, G. and Möhwald, H. (1995). X-ray diffraction of a protein crystal anchored at the air-water-interface. *Biophysical Journal*, **68**, pp. 312–314.

10. Haas, H. and Moehwald, H. (1994). Ordered protein arrays as mesophases. *Langmuir*, **10**, pp. 363–366.

11. Harroun, T. A., Heller, W. T., Weiss, T. M., Yang, L. and Huang, H. W. (1999). Theoretical analysis of hydrophobic matching and membrane-mediated interactions in lipid bilayers containing gramicidin. *Biophysical Journal*, **76**, pp. 3176–3185.

12. Hasler, L., Heymann, J. B., Engel, A., Kistler, J. and Walz, T. (1998). 2D crystallization of membrane proteins: rationales and examples. *Journal of Structural Biology*, **121**, pp. 162–171.

13. He, K., Ludtke, S. J., Wu, Y. L. and Huang, H. W. (1993). X-ray-scattering with momentum-transfer in the plane of membrane — application to gramicidin organization. *Biophysical Journal*, **64**, pp. 157–162.

14. Huang, H. W. (2000). Action of antimicrobial peptides: two-state model. *Biochemistry*, **39**, pp. 8347–8352.

15. Huang, H. W. and Yang, L. (2009). In *Handbook of Molecular Biophysics: Methods and Applications.* Wiley-VCH, Berlin, pp. 457–502.

16. Kornberg, R. D. and Darst, S. A. (1991). Two-dimensional crystals of proteins on lipid layers. *Current Opinion in Structural Biology*, **1**, pp. 642–646.

17. Lenne, P. F., Berge, B., Renault, A., Zakri, C., Venien-Bryan, C., Courty, S., Balavoine, F., Bergsma-Schutter, W., Brisson, A., Grubel, G., Boudet, N., Konovalov, O. and Legrand, J. F. (2000). Synchrotron radiation diffraction from two-dimensional protein crystals at the air/water interface. *Biophysical Journal*, **79**, pp. 496–500.

18. Lévy, D., Mosser, G., Lambert, O., Moeck, G. S., Bald, D. and Rigaud, J. L. (1999). Two-dimensional crystallization on lipid layer: a successful approach for membrane proteins. *Journal of Structural Biology*, **127**, pp. 44–52.

19. Lou, C., Wang, Z. and Wang, S. W. (2007). Two-dimensional protein crystals on a solid substrate: effect of surface ligand concentration. *Langmuir*, **23**, pp. 9752–9759.

20. Ludtke, S. J., He, K., Heller, W. T., Harroun, T. A., Yang, L. and Huang, H. W. (1996). Membrane pores induced by magainin. *Biochemistry*, **35**, pp. 13723–13728.

21. Miller, C. E., Majewski, J., Watkins, E. B., Weygand, M. and Kuhl, T. L. (2008). Part II. Diffraction from two-dimensional cholera toxin crystals bound to their receptors in a lipid monolayer. *Biophysical Journal*, **95**, pp. 641–647.

22. Mosser, G. and Brisson, A. (1991). Conditions of 2-dimensional crystallization of cholera-toxin B-subunit on lipid films containing ganglioside GM1. *Journal of Structural Biology*, **106**, pp. 191–198.

23. Plant, A. L. (1999). Supported hybrid bilayer membranes as rugged cell membrane mimics. *Langmuir*, **15**, pp. 5128–5135.

24. Qian, S., Wang, W. C., Yang, L. and Huang, H. W. (2008a). Structure of the alamethicin pore reconstructed by x-ray diffraction analysis. *Biophysical Journal*, **94**, pp. 3512–3522.

25. Qian, S., Wang, W., Yang, L. and Huang, H. W. (2008b). Structure of transmembrane pore induced by Bax-derived peptide: evidence for lipidic pores. *Proceedings of the National Academy of Sciences USA*, **105**, pp. 17379–17383.

26. Scheuring, S., Muller, D. J., Ringler, P., Heymann, J. B. and Engel, A. (1999). Imaging streptavidin 2D crystals on biotinylated lipid monolayers at high resolution with the atomic force microscope. *Journal of Microscopy-Oxford*, **193**, pp. 28–35.

27. Svergun, D. I. and Koch, M. H. J. (2002). Advances in structure analysis using small-angle scattering in solution. *Current Opinion in Structural Biology*, **12**, pp. 654–660.

28. Vaknin, D., Kjaer, K., Ringsdorf, H., Blankenburg, R., Piepenstock, M., Diederich, A. and Losche, M. (1993). X-ray and neutron reflectivity studies of a protein monolayer adsorbed to a functionalized aqueous surface. *Langmuir*, **9**, pp. 1171–1174.

29. Wang, S., Fukuto, M. and Yang, L. (2008). In situ x-ray reflectivity studies on the formation of substrate-supported phospholipid bilayers and monolayers. *Physical Review E*, **77**, art. no. 031909.

30. Weisenhorn, A. L., Schmitt, F. J., Knoll, W. and Hansma, P. K. (1992). Streptavidin binding observed with an atomic force microscope. *Ultramicroscopy*, **42** (Part B), pp. 1125–1132.

31. Werten, P. J. L., Remigy, H. W., de Groot, B. L., Fotiadis, D., Philippsen, A., Stahlberg, H., Grubmuller, H. and Engel, A. (2002). Progress in the analysis of membrane protein structure and function. *FEBS Letters*, **529**, pp. 65–72.

32. Weygand, M., Wetzer, B., Pum, D., Sleytr, U. B., Cuvillier, N., Kjaer, K., Howes, P. B. and Lösche, M. (1999). Bacterial S-layer protein coupling to lipids: x-ray reflectivity and grazing incidence diffraction studies. *Biophysical Journal*, **76**, pp. 458–468.

33. Yang, L., Wang, S. T., Fukuto, M., Checco, A., Niu, Z. W. and Wang, Q. (2009). Structure and interaction in 2D assemblies of tobacco mosaic viruses. *Soft Matter*, **5**, pp. 4951–4961.

34. Yang, L., Weiss, T. M., Harroun, T. A., Heller, W. T. and Huang, H. W. (1999). Supramolecular structures of peptide assemblies in membranes by neutron off-plane scattering: method of analysis. *Biophysical Journal*, **77**, pp. 2648–2656.

35. Zheng, S., Strzalka, J., Ma, C., Opella, S. J., Ocko, B. M. and Blasie, J. K. (2001). Structural studies of the HIV-1 accessory protein Vpu in Langmuir monolayers: synchrotron X-ray reflectivity. *Biophysical Journal*, **80**, pp. 1837–1850.

Chapter 13

STRUCTURAL ANALYSIS OF THE β-SUBUNIT OF THE TRANSLATION INITIATION FACTOR aIF2 FROM DIFFERENT SPECIES: ROLE OF ZN IONS

Francesca Vasile, Eugenia Pechkova, and Claudio Nicolini

Nanoworld Institute, Fondazione EL.B.A., University of Genoa Medical School, Italy

manuscript@ibf.unige.it

The translation initiation factor aIF2 belongs to the aIF-2-beta/aIF-5 family which is active in the early steps of protein synthesis by forming a ternary complex with GTP and the initiator tRNA. It is involved in the delivery of Met-$tRNA_i^{Met}$ to the 40S ribosomal subunit. The solution structure of the intact β-subunit of aIF2 from *Sulfolobus solfataricus* has been solved by ^1H NMR and results composed of an unfolded N-terminus and a folded core domain with four α-helices and three β-strands. The comparison of this structure with the protein of the same family, that we reported here, suggested that that Zn ions can be useful for the correct folding of the C-terminus portion. Starting from this evidence, here we also report a structural homology calculation. The comparison between the Zn-free and Zn-bound forms suggests a possible structure of the terminal region β-subunit of aIF2 from *S. solfataricus* in presence of Zn.

Synchrotron Radiation and Structural Proteomics
Edited by Eugenia Pechkova and Christian Riekel

www.panstanford.com

13.1 INTRODUCTION

The translation initiation factor aIF2 is a heterotrimeric protein, consisting of α-, β-, and γ-subunits, with high sequence similarity among proteins of the same family.

It plays a critical role in the initiation of protein synthesis by forming a ternary complex with GTP (guanosine-5'-triphosphate) and the aminoacylated initiator methionyl-tRNA (Met-$tRNA_i^{Met}$). This complex binds to the small ribosomal subunit (Bell and Jackson, 1998), and with the aid of other translation factors scans mRNA from the 5' end. Upon recognition of the initiation codon, GTP is hydrolyzed and the eIF2-GDP (guanosine diphosphate) complex is released. This leads to the assembly of the 80S ribosome at the initiation codon and to the start of protein elongation. The recycling of eIF2 between successive rounds of translation requires an additional protein factor, the guanine nucleotide exchange factor IF2B, which catalyzes the exchange of GDP bound to eIF2 for GTP (Kimball, 1999; Pestova and Hellen, 2000).

Distinct functions have been observed for each subunit of eIF2. The β-subunit is a global regulator of protein synthesis in eukaryotes. Phosphorylation of eIF2α regulates the exchange rate of GDP to GTP in a/eIF2, altering its availability to initiate the translation through the inhibition of Met-$tRNA_i^{Met}$ binding (Pain, 1996). The β-subunit is responsible for GTP binding, and its similarity to EF-Tu (~27% identity, ~50% similarity) allowed the identification of the Met-$tRNA_i^{Met}$ binding region (Schmitt *et al.*, 2002). The β-subunit of eIF2 is involved in a variety of interactions with other translation factors. For example, its N-terminus binds to eIF5, the GTPase activating factor for eIF2, and to the β-subunit of the exchange factor eIF2β (Asano *et al.*, 1999). This region has also been shown to bind RNA *in vitro* through three lysine repeats (Laurino *et al.*, 1999). The C-terminal region of eIF2β contains another potential RNA binding motif. Mutations in this C2–C2 zinc finger result in a spontaneous GTPase activity and alter the correct recognition of the AUG codon (Huang *et al.*, 1997). The β-subunit has also been involved in the binding to the β-subunit of eIF2B and to crosslink GTP and Met-$tRNA_i^{Met}$ (Bommer and Kurzchalia, 1989; Gaspar *et al.*, 1994).

The genome of *Sulfolobus solfataricus*, as well as those of the other archaea sequenced so far, contains homologues of all three subunits of the eukaryal factor (She *et al.*, 2001). In *S. solfataricus*, the largest subunit is the γ-homologue, which comprises 415 amino acids and contains a recognizable G-domain. The α-subunit homologue contains 266 amino acids. The β-subunit is the smallest polypeptide of a/eIF2 (139 amino acids in *S. solfataricus*) and has experienced the most extensive evolutionary drift with respect to the eukaryotic protein. Indeed, eukaryotic β-subunits are about twice the size of archaeal ones, and include domains involved in the interaction with two other proteins essential for eIF2 function: eIF2β, a factor required for GDP/GTP exchange (Kimball *et al.*, 1987; Das *et al.*, 1997; Asano *et al.*, 1999) and eIF5, necessary for the hydrolysis of eIF2-bound GTP (Das *et al.*, 1997; Asano *et al.*, 1999; Das and Maitra, 2000). Neither eIF2B nor eIF5 have homologues in archaeal genomes. The conserved region of the β-subunit in archaea and eukaryotes includes a domain containing a zinc-bound β-ribbon motif, which has been implicated in controlling the accuracy of initiation codon recognition (Huang *et al.*, 1997).

The γ-subunit has both in eukaryotic and in archeal some intrinsic ability to interact with Met-$tRNA_i$, to bind to the small ribosomal subunits, and to promote the interaction between Met-$tRNA_i$ and the ribosome. The α-subunit is clearly involved in stabilizing Met-$tRNA_i$ binding, while the action of the small β-subunit remains elusive at the moment (Pedullà *et al.*, 2005).

We published the solution structure of the intact translation initiation factor β-subunit from *S. solfataricus* obtained by ^{1}H NMR (Vasile *et al.*, 2008). We compare this structure with the homologous proteins with known structure to investigate the importance of Zn ions for the correct folding of the C-terminal part (Kim *et al.*, 1998; Gutiérrez *et al.*, 2002, 2004). We calculated a new structure for the β-aIF2 from *S. solfataricus* by homology modeling starting from our previous NMR structure and using the aIF2 β-subunit from *Methanobacterium thermoautotrophicum* as model (Gutiérrez *et al.*, 2002, 2004). The *apo* and *holo* forms stability was investigated by molecular dynamic simulations.

13.2 MATERIALS AND METHODS

13.2.1 NMR and Molecular Dynamics Analysis

The structure determination was performed at pH 7.0 and 25 °C. Homonuclear ^{1}H NMR experiments were recorded at 500 MHz on a Bruker AMX spectrometer. The probe temperature was maintained at 298 K, and the water suppression was carried out using the watergate scheme for the COSY (COrrelation SpectroscopY) and the DPFGSE (double pulsed field gradient spin echo) scheme in the case of TOCSY (Total COrrelation SpectroscopY) and NOESY (Nuclear Overhauser Effect SpectroscopY) spectra. In TOCSY experiments, a mixing time of 80 ms was applied to obtain remote scalar connectivities. NOESY spectra were recorded with mixing time of 150 and 300 ms for spin system and sequential assignments. The spectral width was about 16 ppm. The complete sequence-specific ^{1}H resonance assignment was obtained from standard ^{1}H 2D experiments (DQF [double quantum filtering], TOCSY, NOESY) following standard procedures (Wüthrich, 1986). Spectra were analyzed with the program XEASY (Bartels *et al.*, 1995), while the programs DYANA (dynamics algorithm for NMR applications) (Güntert *et al.*, 1997) were used for structure calculation. A total of 100 structures were generated with the program DYANA, using a torsion angle dynamics with a simulated annealing. The following analyses were carried out on the 10 structures with the lowest target function. The selected structures were energy minimized with GROMACS (GROningen MAchine for Chemical Simulations) (Berendsen *et al.*, 1995) through cycles of steepest descent and conjugated gradient optimization. The quality of results was evaluated with the molecular graphic program, MOLMOL (Koradi *et al.*, 1996).

13.2.2 Modeling and Molecular Dynamic Simulations

The *holo* form has been obtained by modeling the sequence of the *apo*-2NXU (Protein Data Bank [PDB] ID 2NXU, Vasile *et al.*, 2008) on the *holo*-1NEE.pdb (PDB ID 1NEE, Gutiérrez *et al.*, 2004), which displays 30% sequence identity and root mean square deviation (RMSD) (calculated over CA) of 3.6 Å. The model has been obtained using the backbone of 1NEE.pdb and adjusting the side chains with the SCWRL algorithm (Canutescu *et al.*, 2003). The insertion of K78 in the model has been carried out using SPDBV (Swiss PDB Viewer) (Guex and Peitsch, 1997), modifying locally the corresponding loop so as to mimic the dihedrals of 2NXU.

Molecular dynamics simulations have been performed with the GROMACS package (Berendsen *et al.*, 1995; Van der Spoel *et al.*, 2005; Hess *et al.*, 2008) with the Gromos96 force field. The simulations are carried out at 300 K for 1500 ps with a time step of 2 fs in a box with ~10,000 water molecules and Cl-ions to neutralize the total charge of the system. The box is dodecahedral with primitive vectors of length 4.9 nm, 5.0 nm, 3.3 nm; periodic boundary conditions apply. Electrostatics is treated by particle mesh Ewalds and the system is coupled to a thermal bath by means of a Berendsen thermostat. The solvent is equilibrated before the simulation through a linear quenching of 500 ps starting at 350 K.

13.3 CONCLUSIONS

The NMR solution structures of β-aIF2 from *S. solfataricus* show an unfolded N- and C-terminal, while the core of the protein (the region between residues 30–110) is composed by three antiparallel beta strands packed against four alpha helices.

The three antiparallel β-strands forms two β-sheets, β_A (37–39), β_B (42–45), β_C (82–85), and four alpha helices were formed by residues 48–55 (α1), 59–68 (α2), 93–99 (α3), 103–106 (α4). The atomic coordinate of β-aIF2 from *S. solfataricus* was deposited on the RCSB PDB (Berman *et al.*, 2000, 2003) and the PDB code is 2NXU (Fig. 13.1). The structure of the β-aIF2 from *S. solfataricus* was compared with the structures of aIF2 β-subunit from *M. thermoautotrophicum* and from *Methanococcus jannaschii* (Kim *et al.*, 1998). The protein solved by Gutiérrez *et al.* (2004) appears to have a core domain (residues 30–98) composed of three α-helices and three antiparallel β-strands. The N-terminus is unfolded but shows some secondary chemical shift typical of helix structures. The C-terminus is a zinc finger domain (residues 99–135) composed of three antiparallel β-strand. In Gutiérrez *et al.* (2002, 2004), it is also shown that zinc is required for the stability of this domain. The sequence alignment for the three analyzed protein is reported in Fig. 13.2, with the indication of the secondary structure elements found for our protein.

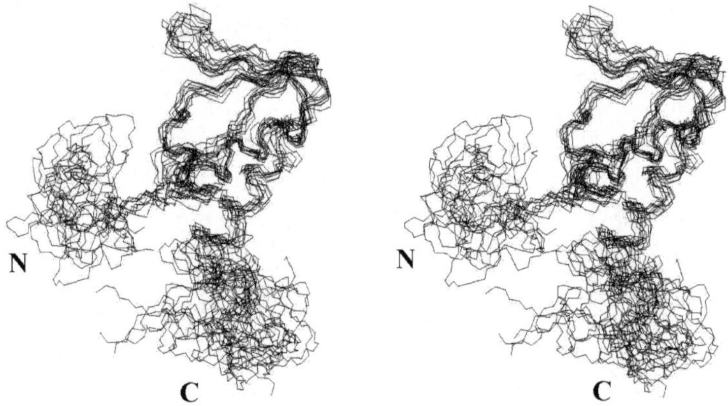

Figure 13.1 Stereoview of the best 10 structures obtained for the β-subunit of aIF2 from *S. solfataricus*. The N- and C-terminal domains result quite unstructured. (Reprinted with the permission from Vasile *et al.* Solution structure of the β-subunit of the translation initiation factor aIF2 from archaebacteria *Sulfolobus solfataricus. Proteins: Structure, Function, and Bioinformatics*, 70, pp. 1112–1115, © 2008 Wiley-Liss, Inc., A Wiley Company.) See also Color Insert.

```
                 1                                                     55
IF2B_SULSO   -----MSSEKEYVEMLDRLYSKLPEKGRKEGTQSLPNMIILNIGNTTIIRNFAEYCDRIR
IF2B_METJA   MSDLENIDYYDYKALLKRARSQIPDYVFQKDRFELPEIEILIEGNRTIIRNFRELAKAVN
IF2B_METTH   --------MDDYEKLLERAIDQLPPEVFETKRFEVPKAYSVIQGNRTFIQNFREVADALN
                .*  .* *   ..*           .*    .  ** * *.** *  .  .

             56                                                        115
IF2B_SULSO   REDKICMKYLLKELAAPGNVDDKGELVIQGKFSSQVINTLMERFLKAYVECSTCKSLDTI
IF2B_METJA   RDEEFFAKYLLKETGSAGNLEG-GRLILQRRISPELLKSRINDFLREYVICRECGKPDTK
IF2B_METTH   RDPQHLLKFLLRELGTAGNLEG-GRAILQGKFTHFLINERIEDYVNKFVICHECNRPDTR
             *.     *.**.*  . **..  *  ..* . .  ..   .  ..  .* *  *   **

            116                     139
IF2B_SULSO   LKKEKKSWYIVCLACGAQTPVKPL
IF2B_METJA   IIKEGRVHLLKCMACGAIRPIRMI
IF2B_METTH   IIREGRISLLKCEACGAKAPLKNV
             . .* .   . * ****  *.. .
```

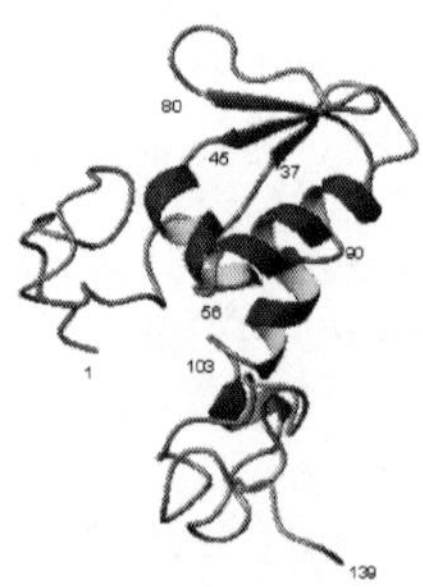

Figure 13.2 (Left) Alignment of the β-aIF2 from *S. solfataricus* (IF2B_SULSO) with its homologues from *M. thermoautotrophicum* (IF2B_METTH) and *M. Jannaschii* (IF2B_METJA). Above the IF2B_SULSO sequence, the secondary structure elements are indicated. (Right) The more representative structure (with lowest energy) was evidenced to show the secondary structure elements (the four alpha helices and the three beta strands). (Reprinted with the permission from Vasile *et al.* Solution structure of the β-subunit of the translation initiation factor aIF2 from archaebacteria *Sulfolobus solfataricus. Proteins: Structure, Function, and Bioinformatics*, 70, pp. 1112–1115, © 2008 Wiley-Liss, Inc., A Wiley Company.) See also Color Insert.

The sequence identity is 35% with *M. jannaschii* and 32% with *M. thermoautotrophicum* protein. The main folding of the core domain is well conserved in the three structures, but we have no evidence of secondary structure elements in both N- and C-terminal regions.

For the C-terminus, the lack of folding can be due to the fact that we have studied the protein in the absence of Zn, while both *M. thermoautotrophicum* and *M. jannaschii* were solved in presence of Zn ions. On the other hand, the unfolding of N-terminus was also reported for the two compared structures. Since the similarity of the NMR structure of the β-subunit of initiation factor from *S. solfataricus* with its homologous proteins from *M. thermoautotrophicum* and *M. jannaschii* appears high validating the previously inferred implications of the observed tertiary and secondary structure for the initiation of translation. The unfolded N-terminus (equivalent to the eukaryotic central portion of IF2β) may be the part necessary for the interaction with γ-subunit with its subsequent folding possibly occurring with the binding. The C-terminus domain (apparently a zinc finger domain) is normally associated with the recognition of sequence-specific nucleic acid. We performed a new analysis of the β-aIF2 from *S. solfataricus*, obtaining a new structure by homology modeling with the homologous protein from *M. thermoautotrophicum*. In particular, starting from our NMR structure we modeled the C-terminal part to investigate the effect of the Zn ion on the folding of the proteins. The model is displayed with a dark-blue cartoon, the *apo*-2NXU structure with a light-blue cartoon in Fig. 13.3. Molecular dynamics simulations have been performed with the *apo* and *holo* forms, starting from the NMR and from the model structure, respectively. At 300 K both of them display large fluctuations, reaching an RMSD of 5 A with respect to the initial structure (see Fig. 13.4, the black curve is the *apo* form, the red curve is the *holo* form).

It is noticeable that the *holo* form does not seem more stable than the *apo* form. The reason could be associated with the fact that the cysteines which bind the zinc atom are quite close along the chain, and consequently are not able to exert an overall stabilizing effect.

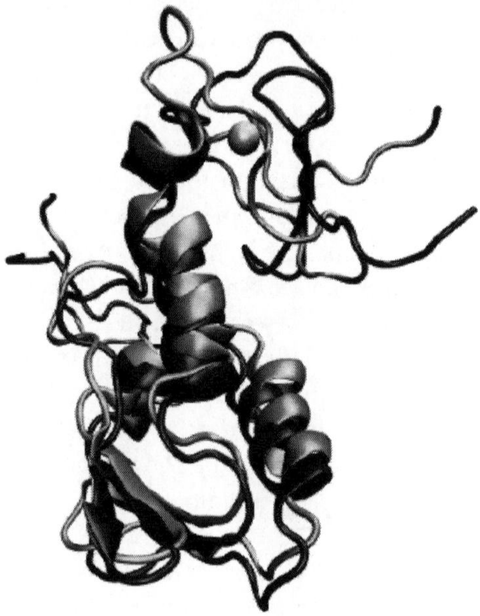

Figure 13.3 Superimposition of the NMR structure of β-aIF2 from *Sulfolobus solfataricus* (light blue) and the model obtained for the same protein with Zn ion. See also Color Insert.

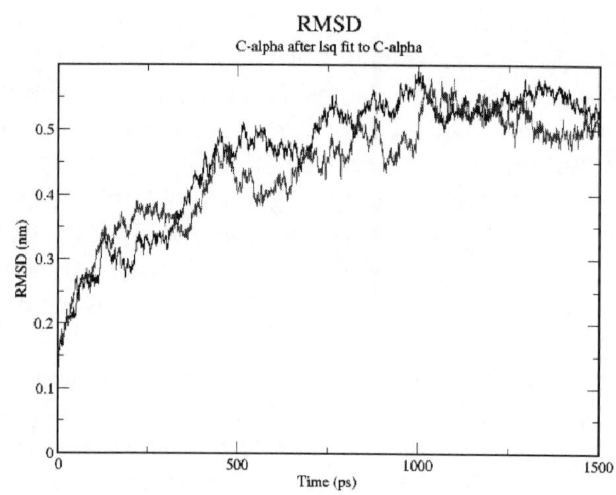

Figure 13.4 RMSD analysis of the *apo* (black curve) and *holo* form (red curve) with respect to their initial structures obtained by molecular dynamic simulations of 1500 ps. See also Color Insert.

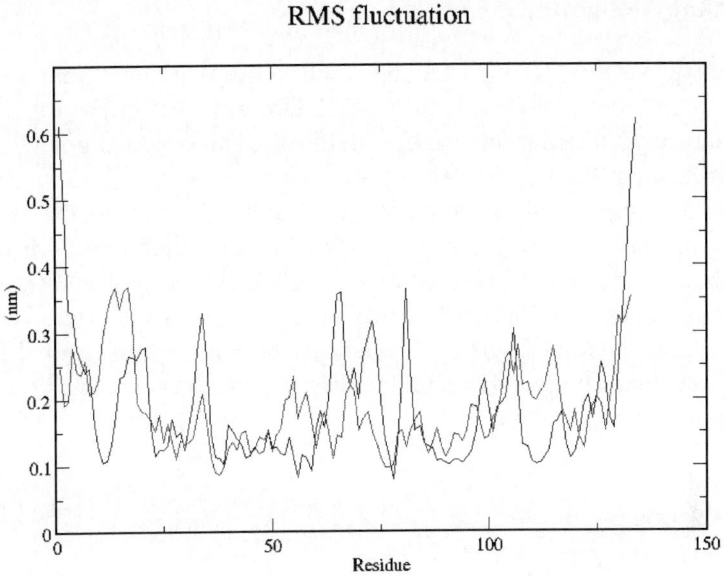

Figure 13.5 RMS fluctuation per residues of the *apo* (black curve) and *holo* form (red curve) obtained by molecular dynamic simulations of 1500 ps. See also Color Insert.

The difference between the two forms of the protein is mainly in how the fluctuations involve the different residues. In Fig. 13.5 it is displayed the RMS fluctuations per residue. In the case of the *apo* form (black curve) fluctuations are more concentrated towards the terminals and the region 60–80. In the case of the *holo* form, they are distributed more evenly over the chain.

The flexibility of the C-terminal part of this protein seems to be quite conserved and probably important for the biological activity of the translation initiation factor.

This conclusion seems to be confirmed from the analysis of the crystal structure of the full-sized heterotrimeric aIF2 from *S. solfataricus* (Stolboushkina *et al.*, 2008) in the nucleotide-free form, that has been determined at 2.8 Å resolution.

The full-sized aIF2 consists of a rigid central part (formed by the γ-subunit, domain 3 of the α-subunit, and the N-terminal α-helix of the β-subunit) and two mobile "wings" (formed by domains 1 and 2 of the α-subunit, the central part and the zinc-binding domain of the β-subunit).

Acknowledgments

This project was supported by a FIRB International Grant on Proteomics and Cell Cycle (RBIN04RXHS) from MIUR to CIRSDNNOB-Nanoworld Institute of the University of Genova and a grant to Fondazione EL.B.A. by MIUR for "Funzionamento." This work was also supported by the grant FIRB RBPR05JH2P from MIUR to Claudio Nicolini of University of Genova. We are grateful to Fabrizio Nozza of the Nanoworld Institute for his technical assistance, to Marina Garber and Elena Stolboushkina at the Institute of Protein Research, Russian Academy of Sciences, Pushchino, Russia, for the protein being provided, and to Udo Bläsi (Vienna Biocenter, Austria) for providing the clone.

References

1. Asano, K., Krishnamoorthy, T., Phan, L., Pavitt, G. D. and Hinnebusch, A. G. (1999). Conserved bipartite motifs in yeast eIF5 and eIF2B epsilon, GTPaseactivating and GDP-GTP exchange factors in translation initiation, mediate binding to their common substrate eIF2. *EMBO Journal*, **18**, pp. 1673–1688.

2. Bartels, C., Xia, T., Billeter, M., Güntert, P. and Wüthrich, K. (1995). The program XEASY for computer-supported NMR spectral analysis of biological macromolecules. *Journal of Biomolecular NMR*, **5**, pp. 1–10.

3. Bell, S. D. and Jackson, S. P. (1998). Transcription and translation in Archaea: a mosaic of eukaryal and bacterial features. *Trends in Microbiology*, **6**, pp. 222–228.

4. Berendsen, H. J. C., Van der Spoel, D. and Vandrunen, R. (1995). GROMACS: a message passing parallel molecular dynamics implementation. *Computer Physics Communications*, **91**, pp. 43–56.

5. Berman, H. M., Henrick, K. and Nakamura, H. (2003). Announcing the worldwide Protein Data Bank. *Nature Structural Biology*, **10** (12), pp. 980.

6. Berman, H. M., Westbrook, J., Feng, Z., Gilliland, G., Bhat, T. N., Weissig, H., Shindyalov, I. N. and Bourne, P. E. (2000). The Protein Data Bank. *Nucleic Acids Research*, **28**, pp. 235–242.

7. Bommer, U. A. and Kurzchalia, T. V. (1989). GTP interacts through its ribose and phosphate moieties with different subunits of the eukaryotic initiation factor eIF-2. *FEBS Letters*, **244**, pp. 323–327.

8. Canutescu, A. A., Shelenkov, A. A. and Dunbrack, R. L. (2003). A graph theory algorithm for protein side-chain prediction. *Protein Science*, **12**, pp. 2001–2014.

9. Das, S., Maiti, T., Das, K. and Maitra, U. (1997). Specific interaction of eukaryotic translation initiation factor 5 (eIF5) with the β-subunit of eIF2. *Journal of Biological Chemistry*, **272**, pp. 31712–31718.

10. Das, S. and Maitra, U. (2000). Mutational analysis of mammalian translation initiation factor 5 (eIF5): role of the interaction between the beta subunit of eIF2 and eIF5 in eIF5 function *in vivo* and *in vitro*. *Molecular and Cellular Biology*, **11**, pp. 3942–3950.

11. Gaspar, N. J., Kinzy, T. G., Scherer, B. J., Humbelin, M., Hershey, J. W. and Merrick, W. C. (1994). Translation initiation factor eIF-2. Cloning and expression of the human cDNA encoding the β-subunit. *Journal of Biological Chemistry*, **269**, pp. 3415–3422.

12. Guex, N. and Peitsch, M. C. (1997). SWISS-MODEL and the Swiss-PdbViewer: an environment for comparative protein modeling. *Electrophoresis*, **18**, pp. 2714–2723.

13. Güntert, P., Mumenthaler, C. and Wüthrich, K. (1997). Torsion angle dynamics for NMR structure calculation with the new program DYANA. *Journal of Molecular Biology*, **273**, pp. 283–298.

14. Gutiérrez, P., Coillet-Matillon, S., Arrowsmith, C. and Gehring, K. (2002). Zinc is required for structural stability of the C terminus of archaeal translation initiation factor aIF2β. *FEBS Letters*, **517**, pp. 155–158.

15. Gutiérrez, P., Osborne, M. J., Siddiqui, N., Trempe, J. F., Arrowsmith, J. and Gehring, K. (2004). Structure of the archaeal translation initiation factor aIF2β from *Methanobacterium thermoautotrophicum*: implications for translation initiation. *Protein Science*, **13**, pp. 659–667.

16. Hess, B., Kutzner, C., Van Der Spoel, D. and Lindahl, E. (2008). GROMACS 4: algorithms for highly efficient, load-balanced, and scalable molecular simulation. *Journal of Chemical Theory and Computation*, **4**, pp. 435–447.

17. Huang, H. K., Yoon, H., Hannig, E. M. and Donahue, T. F. (1997). GTP hydrolysis controls stringent selection of the AUG start codon

during translation initiation in *Saccharomyces cerevisiae*. *Genes & Development*, **11**, pp. 2396–2413.

18. Kim, K. K., Hung, L. W., Yokota, H., Kim, R. and Kim, S. H. (1998). Crystal structures of eukaryotic translation initiation factor 5A from *Methanococcus jannaschii* at 1.8 Å resolution. *Proceedings of the National Academy of Sciences U.S.A.*, **95**, pp. 10419–10424.
19. Kimball, S. R. (1999). Eukaryotic initiation factor eIF2. *International Journal of Biochemistry & Cell Biology*, **31**, pp. 25–29.
20. Kimball, S. R., Everson, W. V., Myers, L. M. and Jefferson, L. S. (1987). Purification and characterization of eukaryotic initiation factor 2 and a guanine nucleotide exchange factor from rat liver. *Journal of Biological Chemistry*, **262**, pp. 2220–2227.
21. Koradi, R., Billeter, M. and Wutrich, K. (1996). MOLMOL: a program for display and analysis of macromolecular structures. *Journal of Molecular Graphics*, **14**, pp. 51–55.
22. Laurino, J. P., Thompson, G. M., Pacheco, E. and Castilho, B. A. (1999). The b subunit of eukaryotic translation initiation factor 2 binds mRNA through the lysine repeats and a region comprising the C2–C2 motif. *Molecular and Cellular Biology*, **19**, pp. 173–181.
23. Pain, V. M. (1996). Initiation of protein synthesis in eukaryotic cells. *European Journal of Biochemistry*, **236**, pp. 747–771.
24. Pedullà, N., Palermo, R., Hasenöhrl, D., Bläsi, U., Cammarano, P. and Londei, P. (2005). The archaeal eIF2 homologue: functional properties of an ancient translation initiation factor. *Nucleic Acids Research*, **33**, pp. 1804–1812.
25. Pestova, T. V. and Hellen, C. U. (2000). The structure and function of initiation factors in eukaryotic protein synthesis. *Cellular and Molecular Life Sciences*, **57**, pp. 651–674.
26. Schmitt, E., Blanquet, S. and Mechulam, Y. (2002). The large subunit of initiation factor aIF2 is a close structural homologue of elongation factors. *EMBO Journal*, **21**, pp. 1821–1832.
27. She, Q., Singh, R. K., Confalonieri, F., Zivanovic, Y., Allard, G., Awayez, M. J., Chan-Weiher, C. C. Y., Clausen, I. G., Curtis, B. A., De Moors, A., Erauso, G., Fletcher, C., Gordon, P. M. K., Heikamp-de Jong, I., Jeffries, A. C., Kozera, C. J., Medina, N., Peng, X., Thi-Ngoc, H. P., Redder, P., Schenk, M. E., Theriault, C., Tolstrup, N., Charlebois, R. L., Doolittle, W. F., Duguet, M., Gaasterland, T., Garrett, R. A., Ragan, M. A., Sensen, C. W. and Van der Oost, J. (2001). The complete genome of the

crenarchaeon *Sulfolobus solfataricus* P2. *Procedings of the National Academy of Sciences U.S.A.*, **98**, pp. 7835–7840.

28. Stolboushkina, E., Nikonov, S., Nikulin, A., Bläsi, U., Manstein, D. J., Fedorov, R., Garber, M. and Nikonov, O. (2008). Crystal structure of the intact archaea translation initiation factor 2 demonstrates very high conformational flexibility in the alpha- and beta-subunits. *Journal of Molecular Biology*, **382**, pp. 680–691.

29. Van Der Spoel, D., Lindahl, E., Hess, B., Groenhof, G., Mark, A. E. and Berendsen, H. J. (2005). GROMACS: fast, flexible, and free. *Journal of Computational Chemistry*, **26**, pp. 1701–1718.

30. Vasile, F., Pechkova, E. and Nicolini C. (2008). Solution structure of the beta-subunit of the translation initiation factor aIF2 from archaebacteria *Sulfolobus solfataricus*. *Proteins: Structure, Function and Bioinformatics*, **70**, pp. 1112–1115.

31. Wüthrich, K. (1986). *NMR of Proteins and Nucleic Acids*. Wiley-Interscience, New York, USA.

Chapter 14

CRYSTAL QUALITY: A QUEST FOR STRUCTURAL PROTEOMICS

Vivian Stojanoff
Brookhaven National Laboratory, National Synchrotron Light Source Building 725D, Upton NY 11973, USA
stojanoff@bnl.gov

In spite of recent progress in the understanding of the nucleation process and the exploitation of new crystallization methods the understanding of the crystallization process of proteins remains a trial and error process. Different approaches to the crystallization of these molecules and the impact on the diffraction quality and consequently on the structure determination of these molecules will be discussed in light of crystallization and characterization methods in specific high-resolution diffraction imaging methods.

14.1 INTRODUCTION

Traditional crystallization methods found new life with the advent of the so-called "omics" initiatives, such as structural genomics and proteomics. Structural proteomics adopts a large-scale approach to describe the three-dimensional (3D) structure–function relation of proteins. Potentially the results will lead to the design of new

Synchrotron Radiation and Structural Proteomics
Edited by Eugenia Pechkova and Christian Riekel

www.panstanford.com

drugs, vaccines, targets for new therapies, or the engineering of new proteins for industrial applications. There are currently two methods to determine the molecular structure of proteins: X-ray crystallography and nuclear magnetic resonance; approximately 86.7% of the 3D molecular structures deposited in the Protein Data Bank to date (Berman *et al.*, 2003) are determined by X-ray diffraction methods. The determination of the 3D structure of a protein involves several steps. Once a protein is isolated, cloned, expressed, solubilized, and purified comes the difficult task of crystallization. One can be lucky and find the crystallization conditions in a few hours, but it can take weeks and even years to find and optimize any crystallization condition to produce good quality diffracting crystals. Once a crystal is obtained, it is tested for its diffraction properties at a home source and/or taken to a synchrotron radiation facility. If the crystal diffracts up to a reasonable resolution diffraction data are collected and an electron density determined. Solving the molecular structure is to fit the full sequence of amino acids of the molecule under consideration into this electron density. A final 3D model for the macromolecule is obtained after several fitting and refinement cycles. A summary of these steps is illustrated in Fig. 14.1.

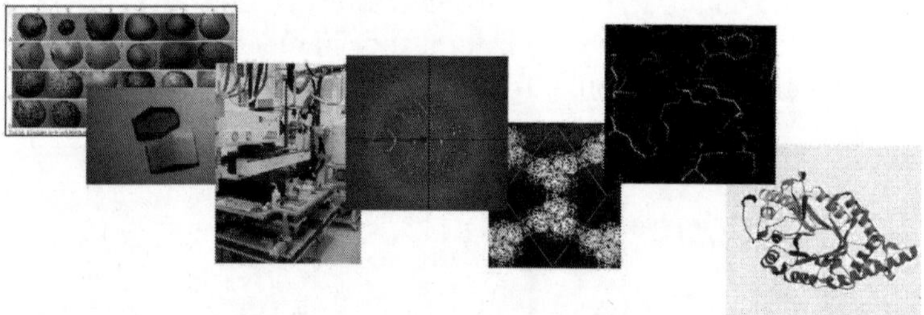

Figure 14.1 From crystal to structure. See also Color Insert.

Major legacy of the structural "omics" initiatives are the development of automated mass screening tools for crystallization conditions using very small amounts of protein. Combined with emerging databases and genetic engineering protocols, it is now possible to choose the best screening procedures for a specific protein. In spite of the great progress observed and significant efforts in characterizing the crystallization process and in understanding the physical process behind it, obtaining crystals of high diffraction quality continues to be the bottleneck for the structural "-omics" community.

The quality of the molecular structure depends on the quality of the diffraction pattern, which in turn is related to the quality of the crystal. Figure 14.2 illustrates this point. The patterns are from three lysozyme crystals grown by the temperature control method (Rosenberger *et al.*, 1988).

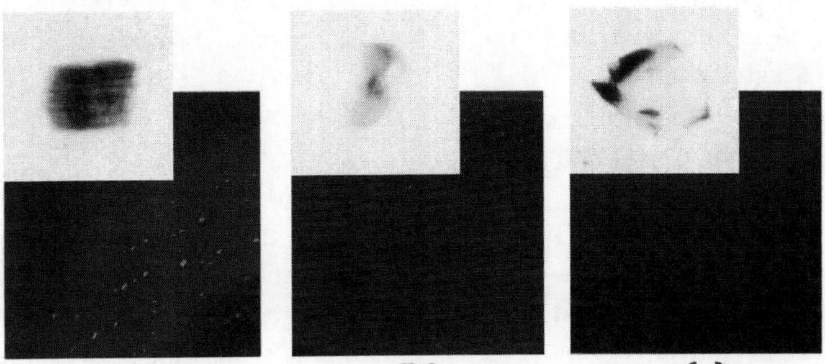

(a) (b) (c)

Figure 14.2 The quality of the 3D structure depends on the quality of the crystals. The diffraction pattern and corresponding X-ray diffraction topographs of three hen egg white lysozyme crystals grown by the temperature control method (Rosenberger *et al.*, 1988). (a) Temperature was cycled at 1 °C/day between 11 and 14 °C, crystal diffracted only to 2.5 Å. (b) Constant temperature set at 14 °C, crystal diffracted only to 1.8 Å. (c) Constant temperature 11 °C, crystal diffracted to 1.2 Å. See also Color Insert.

Crystals presenting poor quality diffraction (split diffraction spots and low-resolution diffraction limits) also show high density of defects (Fig. 14.2a). Conversely for crystals that present very sharp and well-defined diffraction spots and diffract to high resolution, the topographic image shows a very low concentration of defects (Fig. 14.2c). As it so happens many structures have been solved using diffraction patterns of much worse quality than that shown in Fig. 14.2a. For most protein crystallographers a single crystal will suffice for structure determination. The same is not the case for biomolecular crystal growers who seek the optimization of the crystal growth process for all proteins.

Several approaches are being pursued in the attempt to find a unique method that would truly allow the high-throughput determination of the 3D molecular structure. These can be basically divided into three major fronts: crystallization, characterization methods, and simulations. Below we attempt to discuss some thoughts on the current status of each of these fronts and how they can and are contributing to the development of a high-throughput 3D structure determination. These are not limited to the proteomics projects but are equally important for all "-omics" projects in need of 3D structures.

14.2 CRYSTALLIZATION

The crystallization of biological molecules is a multi-parametric process whose properties are influenced by a large variety of physical, chemical and biological parameters (Giege and Ducruix, 1992). Contrary to inorganic crystallography the phase diagram (Fig. 14.3), which allows the quantitative description of these different parameters, is not known except for a few cases. Recently Talreja *et al.* (2010) proposed to map the phase space using a modified vapor diffusion platform that allows the use of minimal material quantities.

Protein crystal growth methods, such as batch techniques, dialysis, vapor diffusion, were developed early on to study protein crystallization; for a concise history see McPherson (1991, 1999). Of these the vapor diffusion method is by far the most popular (Hampel *et al.*, 1968; Kim *et al.*, 1971). It consists in mixing the protein and the precipitating agent in a drop and letting it equilibrate against a reservoir. Crystals may appear after a few hours or may take month to grow. Very sensitive to any changes in the environment or containment it is the least controllable of the growth methods. Further improvement is achieved by the use of seeding techniques (Bergfors, 2003).

In an attempt to improve the crystal quality and the diffraction resolution crystallization in a microgravity environment was proposed early on. The thoughts behind were to minimize sedimentation and convective currents. Littke and John (1984, 1986) proposed that in a reduced gravity environment the simultaneous absence of convection and sedimentation would minimize the probability of nucleation and favor isotropic growth, thus favoring the formation of a reduced number of larger crystals. The liquid-liquid diffusion technique they used to grow β-galactosidase and lysozyme crystals, on board of the sounding rocket TEXUS3, showed for the first time the perfect and continuous interpenetration of two liquids in low gravity conditions. The sharp interface observed between the two liquids a consequence of diffusion alone, free of perturbations produced by convective mixing. The β-galactosidase and lysozyme crystals obtained by Littke and John although far from perfect were proof that crystals of the same or better quality than those grown in 1 g can be grown in low gravity environments.

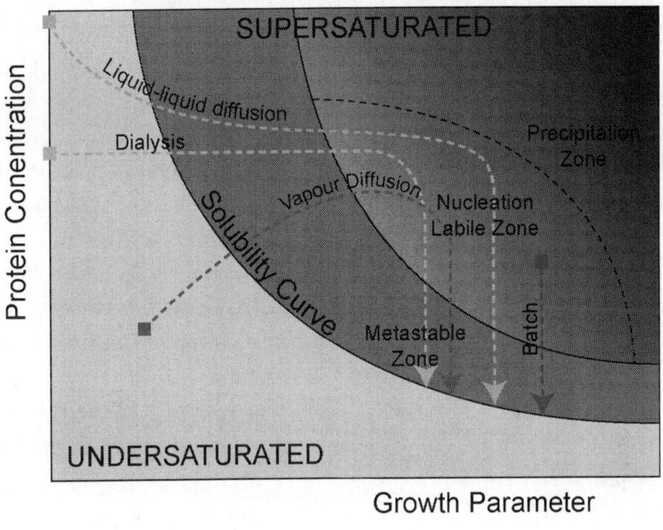

Figure 14.3 Phase diagram (cartoon). The horizontal axis can represent any number of parameters such as precipitant concentration, pH, temperature, etc. In the figure different crystallization methods are represented assuming precipitant concentration as a variable. Along the different pathways physical and chemical processes such as nucleation and growth occur (adapted from Chayen, 1998). See also Color Insert.

In a reduced gravity environment sedimentation and buoyancy driven convection are in principle suppressed, as there is no convective mixing of the solution. Under these conditions nutrient transport is dominated entirely by diffusion mass transport, which is extremely slow for protein molecules. The first attempt to characterize the solution flow surrounding a developing crystal was reported by

Shlichta (1986). Following Schlichta's report Rosenberger and co-workers observed the formation of a nutrient depleted zone at the protein crystal-solution interface (Monaco and Rosenberger, 1993; Rosenberger *et al.* 1993; Vekilov *et al.*, 1999). Because in reduced gravity environments the diffusion is very slow the depletion zone is quasi-stable and a local decrease in supersaturation is created. The net effect is the creation of a zone where optimal growth may be expected to occur equivalent to the metastable region in the phase diagram, and consequently an improvement in the crystal quality can be obtained. An uncontrolled growth process in a 1 g environment can therefore become self-regulating in reduced gravity environments. As was pointed out by McPherson (1993) if this argument is correct the greatest improvements should be observed for those proteins that show the highest growth rates or those which nucleate at high supersaturation concentrations.

Significant improvement could be expected if the convective driven flow could be eliminated or restricted. The depletion zone around the growing crystal could then be controlled. This has been achieved in part by growing crystals in gel media or in highly viscous media (Pietras *et al.*, 2010, and references therein). Higher supersaturation as that attained by centrifugal increase of concentration allows for a diminished nucleation rate and improved crystal growth (see, for example, Nanev *et al.*, 2006; Lorber, 2008). Significant damping of convective flow has been achieved through levitation techniques, using immiscible fluids (Chayen, 1996; Lorber and Giege, 1996), or acoustic fields (Chung and Trinh, 1998). Preferential orientation and consequent improvement in crystal quality has been observed by the application of external electric (Hammadi and Veesler, 2009, and references therein) and magnetic fields (Ataka *et al.*, 1997; Sazaki *et al.*, 1997; Lübbert *et al.*, 2004; Moreno *et al.*, 2009). Convection in this method is decreased due to damping by increased viscosity of the medium and reduced diffusion. Sedimentation, however, will not recede unless some stirring is applied.

In the quest to grow high-quality crystals of diffraction quality two approaches can be distinguished. The first relies on the assessment of a wide range of conditions. Vapor diffusion and microbatch methods have been adapted to automated environments, miniaturization allowing thousands of conditions to be probed at once. This has been the approach pursued by most structural proteomics projects which use robotic nanoliter dispensers adapted with imaging capabilities and software that allows the ranking of successful conditions. Robotic screening of protein crystal growth conditions have been also made available to the academic community, organizations such as Hauptman Woodward Institute, HWI (Luft *et al.*, 2003), providing screening of 1000+ conditions using the microbatch methods.

The Microfluidic Large-Scale Integration technology pioneered by Quake's group (Thorsen *et al.*, 2002; Ng *et al.*, 2008, and references therein; Perry *et al.*, 2009) (Fig. 14.4) enabled a new approach to the crystallization process allowing complex fluid manipulations at the nanoliter scale. Based on free interface (liquid-liquid) diffusion phenomena microfluidic devices allow multiple sampling of the phase space. Microfluidic chips are being developed to allow crystals obtained along the length of the microchannel chambers to be systematically analyzed by *in situ* X-ray analysis. Available commercially microfluidic chips are being further developed to allow for full automation of the screening process and simultaneous characterization by X-ray diffraction and structure solution.

The use of mineral substrates to induce nucleation under reduced saturation conditions is not new (McPherson and Shlichta, 1988). In an effort to better understand and control the crystallization process, different substrates such as porous silicon (Chayen *et al.*, 2001), microporous synthetic zeolites (Sugahara *et al.*, 2008), gold particles (Hodzhaoglu *et al.*, 2008), carbon-nanotubes, and gel-glasses (Chayen *et al.*, 2006; Asanith *et al.*, 2009) have been proposed. Mesopourous materials (Chayen *et al.*, 2006) with pore size between 2 and 50 nm have been suggested to be the best candidates for heterogeneous nucleants. Following this line of thought agarose gel media preferentially employed in counter diffusion methods (Garcia-Ruiz and Moreno, 1994) is a perfect candidate as different pore sizes can be obtained by varying the agar concentration (Narayanan *et al.*, 2006). In turn it should be possible to "taylor" the size of the pores to match the molecules to be crystallized. The same is true for lipid cubic phase matrices (Landau and Rosenbusch, 1996; Caffrey, 2000, and references therein; Nollert *et al.*, 2001; Caffrey and Cherezov, 2009). One aspect to be kept in mind when using these media for crystallization essays is the approximate size of the molecules being crystallized.

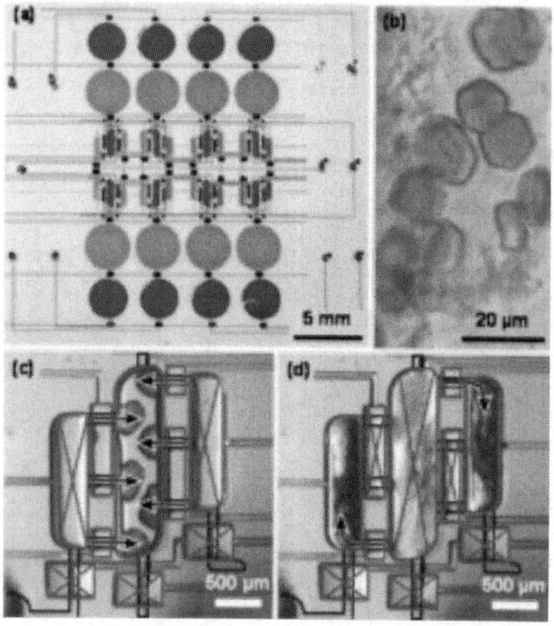

Figure 14.4 Microfluidic chip: (a) An optical micrograph of microfluidic chip containing a 4 × 2 array of devices capable of preparing crystallization trials in parallel using the lipidic cubic phase method at a volume of > 20 nl. Preparation of the cubic mesophase is done in the three chambered mixing portion of the device (blue and purple). Crystallization occurs in a separate crystallization chamber (light green) where the protein-containing mesophase is contacted with a precipitant solution that is introduced from a separate circular chamber (dark green). A single set of valve lines is used to control all of the mixing and crystallization units simultaneously. (b) Optical micrograph of bacteriorhodopsin crystals grown on chip via the lipidic cubic phase method. (c–d) Optical micrographs on an aqueous 13.5 mg/ml bacteriorhodopsin solution being mixed with the lipid monoolein. The blue lines delineate the edges of the fluidic channels containing the protein and lipid solutions while green lines highlight the location of microfluidic pneumatic valves. (c) Straight-line injection of the protein solution from the side chambers into the lipid-containing center chamber. (d) Chamber-to-chamber injection of the fluid mixture through different sets of inlets to create a net circulatory motion. (Courtesy: Sarah Perry [Perry *et al.*, 2009]). See also Color Insert.

The concept of nucleation was taken a step further by Nicolini and collaborators. These researchers showed that it is possible to introduce proteins in Langmuir–Blodgett (LB) films without considerable denaturation of the protein (Berzina *et al.*, 1996).

LB films for each protein of interest are deposited on siliconized cover slips and used in vapor diffusion hanging drop experiments (Pechkova *et al.*, 2008) leading to nearly homogeneous nucleation. For a review see Nicolini and Pechkova in this issue and references therein (Nicolini *et al.*, 2010). Micron-sized crystals of several proteins could be successfully be grown on such substrate-specific templates leading the authors to coin the expression "nanocrystallography," as high diffraction quality crystals were obtained by true nanotechnology methods.

14.3 CRYSTAL QUALITY

The 3D molecular structure determination for structural proteomics is fully automated from data collection to structure refinement. Assessment of crystal quality, however, remains rather indirect. The quality of a crystal is generally assessed from parameters derived from the data collection process by the X-ray oscillation method, Fig. 14.5.

Crystals that diffract to higher resolutions, that is, for which the smallest dimensions in the reciprocal space can be determined, with smaller mosaicities, high signal-to-noise ratios, are thought to be of better quality than those crystals which diffract to low resolutions and present high mosaicities and low signal-to-noise. Thus, a crystal diffracting to 0.09 nm (0.9 Å) is considered of superior quality than one diffracting to 0.3 nm (3 Å), or a crystal with 0.3° mosaicity is worse than a crystal 0.1° mosaicity. These values, however, are average dimensions and represent a convolution of contributing factors allowing therefore only for qualitative information.

In the search for a quantitative description, several methods were developed or adapted. Probably the most popular characterization techniques rely on microscopic and interferometric methods. These methods allow to correlate supersaturation and crystal growth within a semi-quantitative frame. Michelson interferometry has been used extensively to study depletion zones around the growing crystals (Monaco and Rosenberger, 1993, Vekilov *et al.*, 1999). Mach-Zehnder interferometers were adapted to the hardware used on board of several space missions and in conjunction with an optical microscope allowed the observation of crystal formation (Garcia-Ruiz and Novella, 2000).

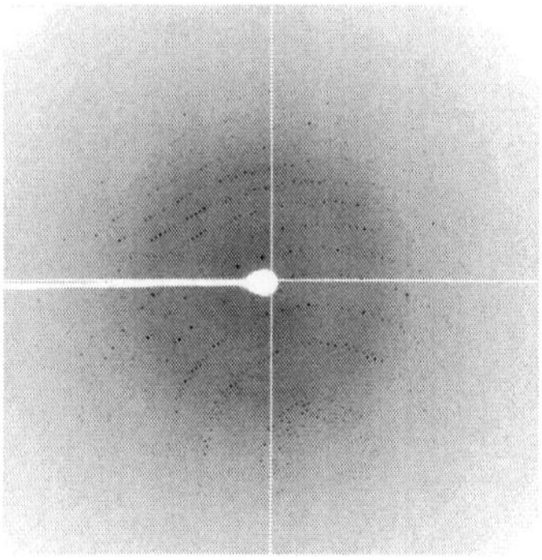

Resolution range (A)	20.00 - 0.88
Mosaicity (HKL2000)	0.14
I/σI	61(2.8)
Rmerge(%)	4.6 (63)
Completeness(%)	95 (75)

Figure 14.5 The oscillation method is commonly used to determine the 3D molecular structure of proteins. Crystal quality is usually assessed from parameters such as resolution range, mosaicity, signal-to-noise ratio, and data completeness. Tetragonal hen egg white lysozyme diffraction pattern collected at beamline X6A, National Synchrotron Light Source.

Further characterization of the protein can be obtained by dynamic or quasi-elastic light scattering (Lomakin *et al.*, 2005; Dierks *et al.*, 2008, and references therein). These methods allow to determine particle size and dispersion in solution. To crystallize the protein should be fairly pure in other words present a monodispersive distribution in solution. This method is nicely complemented by small-angle X-ray (SAXS) or neutron scattering (SANS). While SANS has been widely used to characterize complex mixtures, SAXS has been recently exploited to study the on-set of crystallization, i.e., to study early stages of nucleation (Robinson *et al.*, 2004). The particular advantage of this technique is the ability to tune in on just one scattering object in a complex mixture. Therefore, it is a very suitable technique to study membrane proteins. SAXS is also a powerful technique to predict the envelop of complex molecules, such as molecular machines and viruses (see Svergun and Koch, 2003, Petoukhov and Svergun, 2007). Grazing incidence small angle X-ray scattering (GISAXS) has been widely employed to assess surface quality and regularity and is very sensitive to phase-transitions proving and ideal tool to probe for nucleation. In combination with a micrometer-sized X-ray beam, so-called micro-beam GISAXS, μGISAXS (Riekel, 2000), Pechkova and

Nicolini investigated the effect of temperature on the reorganization of nanostructured protein films. Controlled *in situ* studies of nanostructured Langmuir–Schaeffer template induced nucleation and crystallization allowed to confirm the participation of protein molecules from the template in the nucleation process (Nicolini *et al.*, 2010, and references therein).

Atomic force microscopy (AFM) is by far the most popular method to study defects and growth mechanisms in protein crystals (see Malkin and Thorne, 2004, and references therein). Screw dislocations and step growth functions have been pointed out as possible growth mechanisms for crystals of biological molecules (McPherson, 1999, and references therein). Several reviews have discussed the different mechanisms as a function of entropy and energetic considerations (Vekilov, 2008; DeYoreo *et al.*, 2009). So far there has been no single model upon the whole scientific community would agree upon. Furthermore, as was observed by X-ray diffraction imaging methods (Capelle *et al.*, 2004) screw dislocations are unlikely to form in biomolecular crystals as the energetic considerations are not favorable for the formation of these defects. As discussed by Capelle and co-workers, edge dislocations could present a screw dislocation pattern on the crystal surface (Strunk, 1996) and therefore cause AFM to mistakenly classify these defects as screw dislocations as this is a surface technique.

Being a surface method AFM does not allow the observation of defect formation in the crystal bulk. In an attempt to visualize *in situ* defect formation Sazaki (Sazaki *et al.*, 2005) combined laser confocal microscopy with differential interference contrast microscopy and used phase contrast microscopy to study defect formations in {1 1 0} planes of tetragonal hen egg white lysozyme. *In-situ* observation of the growing crystals allowed the identification of large inclusions ranging from 60 to 300 μm. The presence of dislocations and other defects was confirmed by a combination of surface etching and microscopy images. Still these studies are limited to a restricted area.

High-resolution X-ray diffraction imaging techniques (Stojanoff and Siddons, 1996; Dobrianov *et al.*, 1999; Volz and Matyi, 1999; Boggon *et al.*, 2000; Hu *et al.*, 2001) allow the non-destructive study of defect formations in the bulk of the crystal (Bowen and Tanner, 1998; Authier, 2001). Monochromatic coherent X-rays diffracted by the still or rotating crystal are recorded on film, to produce an image, or on a simple scintillation counter, to provide a distribution of the diffracting power. The image recorded on photographic film of the diffracted intensity as a function of position within the crystal is known as diffraction topograph (Fig. 14.6, left panel). The intensity distribution of the diffracting power as the crystal rotates through the Bragg position, rocking curve (Fig. 14.6, central panel), of a specific reflection provides information on the mosaicity of the sample in a specific crystallographic direction (Stojanoff *et al.*, 2004). In an attempt to obtain quantitative information on the quality of the crystals Lovelace and Borgstahl (2003) proposed to record the diffracted intensity from several crystallographic planes simultaneously on a CCD detector. By rotating the crystals in very small increments, typically of the order of milidegrees, in the oscillation method, and integrating the diffracted intensity for each reflection over this angular distance it is possible to build a distribution of intensities as a function of angular position, i.e., a rocking curve, for several reflections simultaneously. This method allows the correlation of the effect of specific parameters on the order–disorder in a crystal. Temperature cycling was employed to improve the mosaicity of glycerol kinase crystals and allow the structure determination to higher resolution limits (Vahedi-Faridi *et al.*, 2005). Juers *et al.* (2007) used this method to study the effect of cryocooling on the mosaicity as crystals were cycled from room temperature to liquid nitrogen temperatures. Recent studies of the effect of strong magnetic fields on the quality of solution grown tetragonal hen egg lysozyme crystals revealed that certain crystallographic directions are more susceptible to order–disorder than others. For example, the mosaicity measured for [0 k l] directions improved significantly with magnetic field (Moreno *et al.*, 2009). Reciprocal space mapping (Fig. 14.6, right panel) has been employed by several groups (Volz and Matyi, 1999; Boggon *et al.*, 2000; Hu *et al.*, 2001) to further ascertain the nature of the predominant defects in a crystal. The two-dimensional (2D) map that is obtained makes it possible to distinguish between the mosaic degree and strain in specific crystallographic directions.

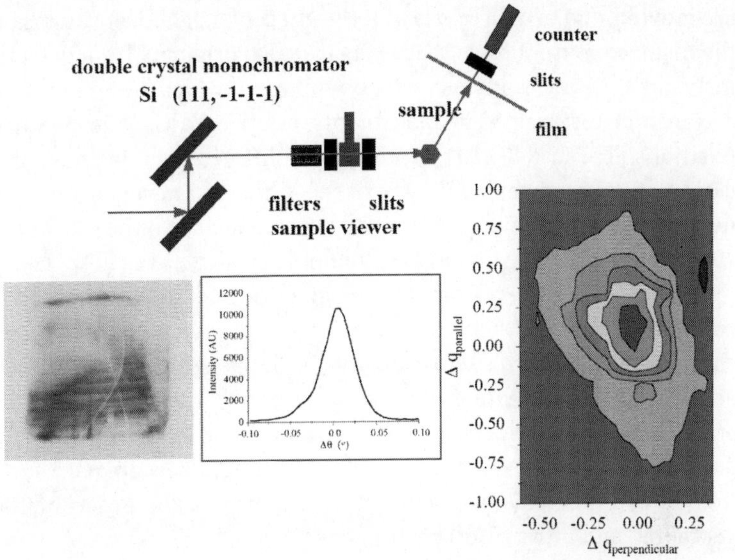

Figure 14.6 High-resolution diffraction imaging. Top: Scheme of the experimental layout at ID19 at the European Synchrotron Radiation Facility; after the identification of the reflection spot by the oscillation method, the 2D detector is removed and a film is set at the proper angle to record the X-ray topography image. The film is then removed to allow the measurement of the rocking curve and reciprocal space mapping (courtesy of J. Härtwig). Bottom: High-resolution diffraction image of reflection (0 14 –22) from a thaumatin crystal grown by the batch method in gel media at room temperature. From left to right: X-ray diffraction topography, dislocations (vertical lines at the bottom) and growth sectors are present in the image; rocking curve showing extended shoulders for angles smaller than the Bragg peak; the rather small angular misorientations (mosaicity) is also observed in the reciprocal space map ($\Delta q_{\text{perpendicular}}$) the extensive variation observed along the scattering vector ($\Delta q_{\text{parallel}}$) are a reflection of the strain field, and consequent interplanar (d-spacing) spacing variations result of defect formations such as the dislocations observed at the bottom of the X-ray topography image. See also Color Insert.

Dynamic conformational changes of the proteins at the molecular level give rise to diffracted intensities that cannot be predicted by Braggs law. It is well known that the static atomic temperature factor determined by standard structural crystallography is high for highly mobile regions of the protein. Analysis of this diffuse scattering (Lonsdale, 1942) provides an overall description of the molecular motion allowing to determine how much and in what direction atoms are moving, and to a certain extent the effect of neighboring atoms on the displacement. Diffuse scattering can also originate from defects: inclusions, dislocations, growth boundaries, which ultimately are a reflection of atomic displacements relative to the ideal lattice positions. These defects give origin to diffracted intensities very close to the Bragg peak. Analysis of the diffuse scattering allows the determination of the predominant type of defects and the size of the strain or stress field surrounding them.

High-resolution diffraction imaging (Fig. 14.6) methods are routinely employed to study the effect of several physical and chemical parameters, such as purity (Caylor *et al.*, 1999; Yoshizakia *et al.*, 2006), pH, temperature, protein concentration (Dobrianov *et al.*, 1998), and electrical, magnetic (Moreno *et al.*, 2009), and gravitational fields (Boggon *et al.*, 2000). Studies are currently limited by film, detector resolution (grain, pixel size), experimental geometry, and intrinsic diffraction effects.

Visualization of individual defects in protein crystals is not easy. The diffraction power for a dislocation in a protein crystal is expected to be very weak (Stojanoff *et al.*, 2004). Simulations have shown that compared to inorganic crystals such as silicon, individual dislocation in proteins will present nearly no contrast on an empty background, pretty much like the image observed in Fig. 14.2c. Furthermore, the width of the image will be largest at the Bragg peak decreasing in width for higher or lower angles. The X-ray topography image shown in Fig. 14.6 was obtained in the so-called weak beam condition, i.e., away several tenth of a degree from the Bragg peak. Using weak beam geometry and monochromatic beam, Capelle and co-workers (2004) showed the formation of individual dislocations in tetragonal lysozyme along the [1 1 0] direction not far from the normal to the faces. The dislocations with Burgers vector along the [0 0 1] direction were identified as mixed dislocations mostly edge dislocations in character, as any other direction would be energetically unfavorable. Using monochromatic weak beam conditions and protein crystals with thickness of the order of the extinction length for individual reflections Koishi and colleagues (2007) obtained clear topographic images of dislocations arranged in bundles in tetragonal, orthorhombic, and monoclinic hen egg white lysozyme crystals grown by the vapor diffusion method and levitation between immiscible fluids (Chayen, 1996; Adachi, *et al.*,

2003). Still it is possible to study defect formation and obtain clear images from small protein crystals (of the order of 100 μm in size) in the weak beam geometry at third-generation synchrotron beam lines, such as ID19 at the ESRF.

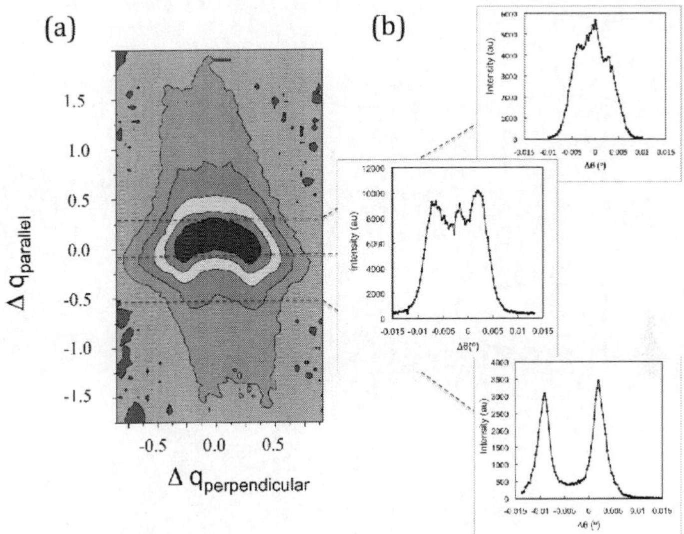

Figure 14.7 High-resolution diffraction images of reflection (15 15 0) of a hen egg white lysozyme crystal grown by the batch method in gel media at room temperature in supersaturation conditions. (a) Reciprocal space map, both interplanar spacing variations ($\Delta q_{\text{parallel}}$, i.e., d-spacing variations), and orientational changes $\Delta q_{\text{perpendicular}}$, i.e., $\Delta\theta$) are observed. (b) The rocking curves were collected at different positions along the scattering vector showing the differentiated angular misorientation in the crystal volume. See also Color Insert.

Figure 14.7 shows the effect of supersaturation on the quality of a lysozyme crystal grown by the batch method in 10% agarose media at room temperature. The reciprocal map (Fig. 14.7a) of the (15 15 0) reflection recorded with the *c*-axis parallel to the horizontal rotation axis and perpendicular to the beam direction shows rather extended variations in interplanar spacing ($\Delta q_{\text{parallel}}$). As the saturation decreases during growth, lattice angular distortions increase and so does the lattice parameter in the [0 0 1] direction. The rocking curves in Fig. 14.7b were recorded at different positions along the scattering vector. The contribution perpendicular to the scattering vector ($\Delta q_{\text{perpendicular}}$) represents the angular misorientation ($\Delta\theta$) between different regions in the crystal. The rocking curves shown in Fig. 14.7b show important differences that taken individually could be misleading; therefore, rocking curve measurements should always be combined with reciprocal space maps and X-ray topography images.

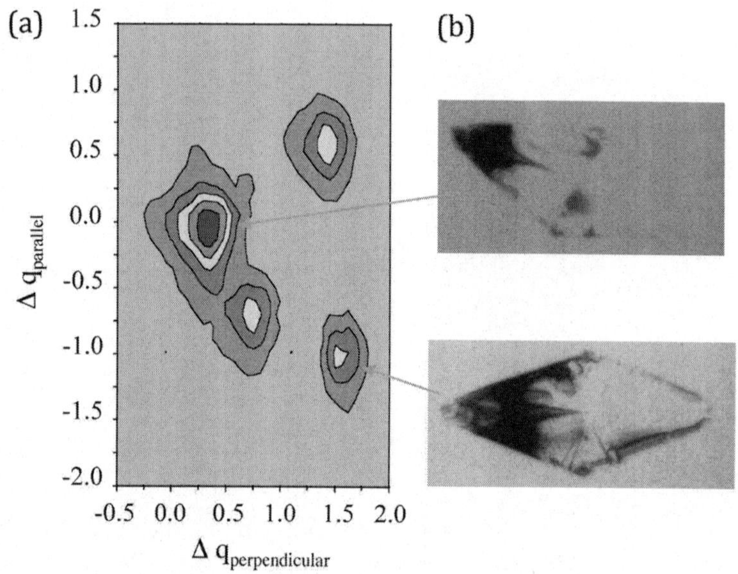

Figure 14.8 High-resolution diffraction image from a thaumatin crystal grown by the batch method at room temperature in the presence of agarose gel media. The *c*-axis is horizontal and parallel to the plane of the figure so is the *a*-axis. (a) Reciprocal space map of reflection (17–1 0), the four distinct peaks are result of differentiated lattice spacing ($\Delta q_{\text{parallel}}$) and orientational, angular ($\Delta q_{\text{perpendicular}}$) variations found in the crystal. (b) The rhombohedral outline of the diffraction topography images correspond to the projection of the crystal form in the scattering direction. The two images were recorded at different positions in reciprocal space. It is worth noting the straight lines at the bottom of the images moving outwards from the centre of the crystal toward the crystal faces (Stojanoff *et al.*, 2003). See also Color Insert.

Recent studies of thaumatin crystals grown by the batch method in gel media showed the formation of line dislocations. Figure 14.8 shows the reciprocal map of the (17–1 0) reflection, the *a*- and *c*-axis are in the plane of the figure with the *c*-axis in the horizontal (Stojanoff *et al.*, 2002, 2003). Four distinct regions are observed in the reciprocal space, indicative of differentiated variations in the lattice spacing, strain "fields," in the crystal. The straight lines observed at the bottom are dislocation lines that move out, from a central nuclei, toward the crystal faces as the crystal grows. The darker regions in the figure are most probably due to unresolved bundles of dislocations.

14.4 CONCLUDING REMARKS

Here we discussed some of the most commonly used methods in the characterization of the nucleation and growth process of proteins. As in the real world there is very little connection between structural biologists and protein crystal growers. However, several technological developments required by high-throughput structural biology projects such as structural proteomics brought to the front once again the need for a better understanding of the crystallization process. Brute force procedures using robotics have being combined with "intelligent" observations (see for example Chayen, 2004). "Intelligent" automated screening allows the use of very small quantities of proteins and with some rational input improves significantly the degree of success. Crystals obtained through these methods are usually very small with sizes of the order of few microns. This is generally a problem for standard protein crystallography beamlines. With the advent of third-generation synchrotron sources, flux and beam size are no longer a problem. Beam sizes that match crystal size are commonly achievable. Pioneered by Riekel, the ID13 beamline was the first beamline to achieve micron-sized beams (Riekel, 2000). There are currently a few beamlines that allow for micro beams as well, such is the case of GMCA CAT at the APS. Micron-sized beams have the advantage that it allows to match crystal size with beam sizes. Further advantage is the possibility of beam translation along long but thin needle-shaped crystals. This has two advantages: one that if there is radiation damage, it is possible to pick a fresh part of the crystal. Second, if the quality of the crystal changes significantly along the crystal it will be possible to avoid highly disordered regions and select a higher quality section within the crystal, thus being able to determine the molecular structure. An example of the first is the structure of cross-β spine of amyloid-like fibrils (Nelson *et al.*, 2005) that could not be determined till crystals were exposed on the ID13 beamline at the ESRF (Riekel *et al.*, 2005). The crystals were very radiation sensitive, to obtain the 3D structure, data from several different regions from a single crystal and several others had to be combined. An example of the second situation was the observation by Riekel and colleagues (Riekel, personal communication) that as micron-sized crystals grown by the LB template method were exposed to a micron-sized beam very large variations in the mosaicity could be observed.

There is no doubt that significant progress has been observed in recent years both toward the understanding of the protein crystallization process and application and development of the characterization methods; however, the need remains for systematic controlled studies as the very fundamental questions remain unanswered.

Acknowledgments

To the many friends and colleagues who throughout the years showed that science can be fun, thanks. The author thanks S. Perry (University of Illinois at Urbana-Champaing), who kindly provided Fig. 14.4. The X6A beamline at the National Synchrotron Light Source (DE-AC02-98CH10886) is part of the NIGMS-ECSBF supported by the National Institute of General Medical Sciences, National Institute of Health under agreement GM-0080.

References

1. Adachi, H., Takano, K., Yoshimura, M., Mori, Y. and Sasaki, T. (2003). Application of a stirring method to microscale and vapor diffusion protein crystallization. *Japanese Journal of Applied Physics Part 2-Letters*, **42**, pp. L314–L315.

2. Asanithi, P., Saridakis, E., Govada, L., Jurewicz, I., Brunner, E. W., Ponnusamy, R., Cleaver, J. A. S., Dalton, A. B., Chayen, N. E. and Sear, R. P. (2009). Carbon-nanotube-based materials for protein crystallization. *ACS Applied Materials & Interfaces*, **1**, pp. 1203–1210.

3. Ataka, M., Katoh, E. and Wakayama, N. I. (1997). Magnetic orientation as a tool to study the initial stage of crystallization of lysozyme. *Journal of Crystal Growth*, **173**, pp. 592–596.

4. Authier, A. (2001). *Dynamical Theory of X-ray Diffraction*. Oxford University Press, Oxford.

5. Bergfors, T. (2003). Seeds to crystals. *Journal of Structural Biology*, **142**, pp. 66–76.

6. Berman, H. M., Henrick, K. and Nakamura, H. (2003). Announcing the worldwide Protein Data Bank. *Nature Structural Biology*, **10**, pp. 980. http://www.pdb.org/

7. Berzina, T. S., Troitsky, V. I., Petrigliano, A., Alliata, D., Tronin, A. Y. U. and Nicolini, C. (1996). Langmuir-Blodgett films composed of monolayers of amphiphilic molecules and adsorbed soluble proteins. *Thin Solid Films*, **284–285**, pp. 757–761.

8. Boggon, T., Helliwell, J. R., Judge, R. A., Olczak, A., Siddons, D. P., Snell, E. H. and Stojanoff, V. (2000). Synchrotron X-ray reciprocal-space mapping, topography and diffraction resolution studies of macromolecular crystal quality. *Acta Crystallographica D*, **56**, pp. 868–880.

9. Bowen, D. K. and Tanner, B. K. (1998). *High Resolution X-ray Diffractometry and Topography*. Taylor and Francis, London.

10. Caffrey, M. (2000). A lipid's eye view of membrane protein crystallization in mesophases. *Current Opinion in Structural Biology*, **10**, pp. 486–497.

11. Caffrey, M. and Cherezov, V. (2009). Crystallizing membrane proteins using lipidic mesophases. *Nature Protocols*, **4**, pp. 706–731.

12. Capelle, B., Epelboin, Y., Härtwig, J., Moraleda, A. B., Otálora, F. and Stojanoff, V. (2004). Characterization of dislocations in protein crystals by means of synchrotron double crystal topography. *J. Appl. Crystallography*, **37**, pp. 67–71.

13. Chayen, N. E. (1996). A novel technique for containerless protein crystallization. *Protein Engineering*, **9**, pp. 927–929.

14. Chayen, N. E. (1998). Comparative studies of protein crystallization by vapour-diffusion and microbatch techniques. *Acta Crystallographica D*, **54**, pp. 8–15.

15. Chayen, N. E. (2004). Turning protein crystallisation from an art into a science. *Current Opinion in Structural Biology*, **14**, pp. 577–583.

16. Chayen, N. E., Saridakis, E., El-Bahar, R. and Nemirovsky, Y. (2001). Porous silicon: an effective nucleation-inducing material for protein crystallization. *Journal of Molecular Biology*, **312**, pp. 591–595.

17. Chayen, N. E., Saridakis, E. and Sear R. (2006). Experiment and theory for heterogeneous nucleation of protein crystals in a porous medium. *Proceedings of the National Academy of Sciences USA*, **103**, pp. 597–601.

18. Caylor, C. L., Dobrianov, I., Lemay, S. G., Kimmer, C., Kriminski, S., Finkelstein, K. D., Zipfel, W., Webb, W. W., Thomas, B. R., Chernov, A. A. and Thorn, R. E. (1999). Macromolecular impurities and disorder in protein crystals. *Proteins: Structure, Function, and Genetics*, **36**, pp. 270–281.

19. Chung, S. K. and Trinh, E. H. (1998). Containerless protein crystal growth in rotating levitated drops. *Journal of Crystal Growth*, **194**, pp. 384–397.

20. DeYoreo, J. J., Zepeda-Ruiz, L. A., Friddle, R. W., Qiu, S. R., Wasylenki, L. E., Chernov, A. A., Gilmer, G. H. and Dove, P. M. (2009). Rethinking classical crystal growth models through molecular scale insights: consequences of kink-limited kinetics. *Crystal Growth & Design*, **9**, pp. 5135–5144.

21. Dierks, K., Meyer, A., Einspahr, H. and Betzel, C. (2008). Dynamic light scattering in protein crystallization droplets: adaptations for analysis and optimization of crystallization processes. *Crystal Growth & Design*, **8**, pp. 1628–1634.

22. Dobrianov, I., Caylor, C., Lemay, S. G., Finkelstein, K. D. and Thorne, R. E. (1999). X-ray diffraction studies of protein crystal disorder. *Journal of Crystal Growth*, **196**, pp. 511–523.

23. Dobrianov, I., Finkelstein, K. D., Lemay, S. G. and Thorne, R. E. (1998). X-ray topographic studies of protein crystal perfection. *Acta Crystallographica D*, **54**, pp. 922–937.

24. Garcia-Ruiz, J. M. and Moreno, A. (1994). Investigations on protein crystal growth by the gel acupuncture method. *Acta Crystallographica D*, **50**, pp. 484–490.

25. Garcia-Ruiz, J. M. and Novella, M. L. (2000). 1st International Microgravity Conference, Sorrento.

26. Giege, R. and Ducruix, D. (1992). *An Introduction to the Crystallogenesis of Biological Macromolecules. Crystallization of Nucleic Acids and Proteins: A Practical Approach* (eds. D. Rickwood and B. D. Hames), pp. 1–18. Oxford University Press. ISBN-10: 0199632464.

27. Hammadi, Z. and Veesler, S. (2009). New approaches on crystallization under electric fields. *Progress in Biophysics and Molecular Biology*, **101**, pp. 38–44.

28. Hampel, A., Labanauskas, M., Connors, P. G., Kirkegard, L., RajBhandary, U. L., Sigler, P. B. and Bock, R. M. (1968). Single crystal transfer RNA from formylmethionine and phenylalanine transfer RNA's. *Science*, **162**, pp. 1384–1387.

29. Hodzhaoglu, F., Kurniawan, F., Mirsky, V. and Nanev, C. (2008). Gold nanoparticles induce protein crystallization. *Crystal Research and Technology*, **43**, pp. 588–593.

30. Hu, Z. W., Lai, B., Chu, Y. S., Cai, Z., Mancini, D. C., Thomas, B. R. and Chernov, A. A. (2001). Phase sensitive X-ray diffraction imaging of defects in biological macromolecular crystals. *Physical Review Letters*, **87**, pp. 1481011–14810114.

31. Juers, D. H., Lovelace, J., Bellamy, H. D., Snell, E. H., Matthews, B. W. and Borgstahl, G. E. (2007). Changes to crystals of *Escherichia coli* beta-galactosidase during room-temperature/low-temperature cycling and their relation to cryo-annealing. *Acta Crystallographica D*, **63**, pp. 1139–1153.

32. Kim, S. H., Quigley, G., Suddath, F. L. and Rich A. (1971). High-resolution x-ray diffraction patterns of crystalline transfer RNA that show helical regions. *Proceedings of the National Academy of Sciences USA*, **68**, pp. 841–845.

33. Koishi, M., Ohya, N., Mubayashi, Y., Koizumi, H., Kojima, K. and Tachibana, M. (2007). Observation of clear images of dislocations in protein crystals by synchrotron monochromatic-beam X-ray topography. *Crystal Growth & Design*, **7**, pp. 2182–2186.

34. Landau, E. M. and Rosenbusch, J. P. (1996). Lipidic cubic phases: a novel concept for the crystallization of membrane proteins. *Proceedings of the National Academy of Sciences USA*, **93**, pp. 14532–14535.

35. Littke, W. and John, C. (1984). Protein single crystal growth under microgravity. *5th European Symposium on Materials Sciences under Microgravity, Schloss Elmau, Germany, ESA SP-222*, pp. 185–188.

36. Littke, W. and John, C. (1986). Protein single crystal growth under microgravity. *Journal of Crystal Growth*, **76**, pp. 663–672.

37. Lomakin, A., Teplow, D. B. and Benedek, G. B. (2005). Quasielastic light scattering for protein assembly study. In *Methods in Molecular Biology* (ed. E. M. Sigurdsson), **299**, pp. 153–174. Humana Press Inc., Totowa, NJ.

38. Lonsdale, K. (1942). X-ray study of crystal dynamics: an historical and critical survey of experiment and theory. *Proceedings of the Physical Society*, **54**, pp. 314–353.

39. Lorber, B. (2008). Virus and protein crystallization under hypergravity. *Crystal Growth & Design*, **8**, pp. 2964–2969.

40. Lorber, B. and Giege, R. (1996). Containerless protein crystallization in floating drops: application to crystal growth monitoring under reduced nucleation conditions. *Journal of Crystal Growth*, **168**, pp. 204–215.

41. Lovelace, J. and Borgstahl, G. E. O. (2003). BEAM-ish2.0: a graphical user interface for the physical characterization of macromolecular crystals. *Journal of Applied Crystallography*, **36**, pp. 1101–1102.

42. Lübbert, D., Meents, A. and Weckert, E. (2004). Accurate rocking-curve measurements on protein crystals grown in a homogeneous magnetic field of 2.4 T. *Acta Crystallographica D*, **60**, pp. 987–998.

43. Luft, J. R., Collins, R. J., Fehrman, N. A., Lauricella, A. M., Veatch, C. K. and DeTitta, G. T. (2003). A deliberate approach to screening for initial crystallization conditions of biological macromolecules. *Journal of Structural Biology*, **142**, pp. 170–179.

44. Malkin, A. J. and Thorne, R. E. (2004). Growth and disorder of macromolecular crystals: insights from atomic force microscopy and X-ray diffraction studies. *Methods*, **34**, pp. 273–299.

45. Moreno, A., Yokaishia, F., Dimasi, E. and Stojanoff, V. (2009). Growth and characterization of high-quality protein crystals for X-ray crystallography. *Annals of the New York Academy of Sciences*, **1161**, pp. 429–436.

46. McPherson, A. (1991). A brief history of protein crystal growth. *Journal of Crystal Growth*, **110**, pp. 1–10.

47. McPherson, A. (1993). Effects of a microgravity environment on the crystallization of biological macromolecules. *Microgravity Science and Technology*, **6**, pp. 101–109.

48. McPherson, A. (1999). *Crystallization of Biological Macromolecules*. Cold Spring Harbor Press, Cold Spring Harbor, NY.

49. McPherson, A. and Shlichta, P. (1988). Heterogeneous and epitaxial nucleation of protein crystals on mineral surfaces. *Science*, **239**, pp. 385–387.

50. Monaco, L. A. and Rosenberger, F. (1993). Growth and etching kinetics of tetragonal lysozyme. *Journal of Crystal Growth*, **129**, pp. 465–484.

51. Nanev, C. N., Dimitrov, I. and Hodjaoglu, F. (2006). Hypergravity as a crystallization tool. *Annals of the New York Academy of Sciences*, **1077**, pp. 172–183.

52. Narayanan, J., Xiong, J.-Y. and Liu, X.-Y. (2006). Determination of agarose gel pore size: absorbance measurements vis a vis other techniques. *Journal of Physics: Conference Series*, **28**, pp. 83–86.

53. Nelson, R., Sawaya, M. R., Balbirnie, M., Madsen A. O., Riekel, C., Grothe, R. and Eisenberg, D. (2005). Structure of the cross-b spine of amyloid-like fibrils. *Nature*, **435**, pp. 773–778.

54. Ng, J. D., Clark, P. J., Stevens, R. C. and Kuhn, P. (2008). In situ X-ray analysis of protein crystals in low-birefringent and X-ray transmissive plastic microchannels. *Acta Crystallographica D*, **64**, pp. 189–197.

55. Nicolini, C., Riekel, C. and Pechkova, E. (2010) Growth and organization of LB protein crystals via in situ GISAXS, laser, nanodiffraction, Raman and AFM, this volume.

56. Nollert, P., Qiu, H., Caffrey, M., Rosenbusch, J. P. and Landau, E. M. (2001). Molecular mechanism for the crystallization of bacteriorhodopsin in lipidic cubic phases. *FEBS Letters*, **504**, pp. 179–186.

57. Pechkova, E., Vasile, F., Spera, R. and Nicolini, C. (2008). Crystallization of alpha and beta subunits of IF2 translation initiation factor from Archabacteria *Sulfolobus solfataricus*. *Journal of Crystal Growth*, **310**, pp. 3767–3770.

58. Perry, S. L., Roberts, G. W., Tice, J. D., Gennis, R. B. and Kenis, P. J. A. (2009). Microfluidic generation of lipidic mesophases for membrane protein crystallization. *Crystal Growth & Design*, **9**, pp. 2566–2569.

59. Petoukhov, M. V. and Svergun, D.I. (2007). Analysis of X-ray and neutron scattering from biomacromolecular solutions. *Current Opinion in Structural Biology*, **17**, pp. 562–571.

60. Pietras, Z., Lin, H.-T., Surade, S., Luisi, B., Slattery, O., Pos, K. M. and Moreno, A. (2010). The use of novel organic gels and hydrogels in protein crystallization. *Journal of Applied Crystallography*, **43**, pp. 58–63.

61. Riekel, C. (2000). New avenues in X-ray microbeam experiments. *Reports on Progress in Physics*, **63**, pp. 233–262.

62. Riekel, C., Burghammer, M. and Schertler, G. (2005). Protein crystallography microdiffraction. *Current Opinion in Structural Biology*, **15**, pp. 556–562.

63. Robinson, I., Bouet, S., Lin, Y. and Stojanoff, V. (2004). Detection of early nucleation stages of HEWL by SAXS, unpublished results.

64. Rosenberger, F., Howard, S. B., Sowers, J. W. and Nyce, T. A. (1993). Temperature dependence of protein solubility — determination and

application to crystallization in X-ray capillaries. *Journal of Crystal Growth*, **129**, pp. 1–2.

65. Rosenberger, F. and Meehan, E. J. (1988). Control of nucleation and growth in protein crystal growth. *Journal of Crystal Growth*, **90**, pp. 74–78.

66. Sazaki, G., Yoshida, E., Komatsu, H., Nakada, T., Miyashita, S. and Watanabe, K. (1997). Effects of a magnetic field on the nucleation and growth of protein crystals. *Journal of Crystal Growth*, **173**, pp. 231–234.

67. Sazaki, G., Tsukamoto, K., Satomi Yai, S., Okada, M. and Nakajima, K. (2005). In situ observation of dislocations in protein crystals during growth by advanced optical microscopy. *Crystal Growth & Design*, **5**, pp. 1729–1735.

68. Sazaki, G., Yoshida, E., Komatsu, H., Nakada, T., Miyashita, S. and Watanabe, K. (1997). Effects of a magnetic field on the nucleation and growth of protein crystals. *Journal of Crystal Growth*, **173**, pp. 231–234.

69. Shlichta, P. J. (1986). Feasibility of mapping solution properties during the growth of protein crystals. *Journal of Crystal Growth*, **76**, pp. 656–662.

70. Stojanoff, V., Cappelle, B., Epelboin, Y., Hartwig, J., Moradela, A. B. and Otalora, F. (2002). LS1860 European Synchrotron Radiation Facility, data analysis in progress.

71. Stojanoff, V., Cappelle, B., Epelboin, Y., Hartwig, J., Moradela, A. B. and Otalora, F. (2003). SC1220 European Synchrotron Radiation Facility, data analysis in progress.

72. Stojanoff, V., Cappelle, B., Epelboin, Y., Hartwig, J., Moradela, A. B. and Otalora, F. (2004). *Annals New York Academy of Sciences*, **1027**, pp. 48–55.

73. Stojanoff, V. and Siddons, D. P. (1996). X-ray topography of a lysozyme crystal. *Acta Crystallographica A*, **52**, pp. 498–499.

74. Strunk, H. P. (1996). Edge dislocations may cause growth spirals. *Journal of Crystal Growth*, **160**, pp. 184–185.

75. Sugahara, M., Asada, Y., Morikawa, Y., Kageyama, Y. and Kunishima, N. (2008). Nucleant-mediated protein crystallization with the application of microporous synthetic zeolites. *Acta Crystallographica D*, **64**, pp. 686–695.

76. Svergun, D. I. and Koch, M. H. J. (2003). Small-angle scattering studies of biological macromolecules in solution. *Reports on Progress in Physics*, **66**, pp. 1735–1782.

77. Talreja, S., Perry, S. L., Guha S., Zukoski, C. F. and Keni, P. J. A. (2010). Determination of the phase diagram for soluble and membrane proteins. *The Journal of Physical Chemistry B*, **114**, pp. 4432–4441. Thorsen, T., Maerkl, S. J. and Quake, S. R. (2002). Microfluidic large-scale integration. *Science*, **298**, pp. 580–584.

78. Vahedi-Faridi, A., Stojanoff, V. and Yeh, J. I. (2005). The effects of flash-annealing on glycerol kinase crystals. *Acta Crystallographica D*, **61**, pp. 982–989.

79. Vekilov, P. G. (2008). Kinetics and mechanisms of protein crystallization at the molecular level. *Methods in Molecular Biology, Protein Nanotechnology*, **300**, pp. 15–52.

80. Vekilov, P. G., Rosenberger, F., Lin, H. and Thomas, B. R. (1999). Nonlinear dynamics of layer growth and consequences for protein crystal perfection. *Journal of Crystal Growth*, **196**, pp. 261–275.

81. Volz, H. M. and Matyi, R. J. (1999). High resolution X-ray diffraction analysis of protein crystals. *Philosophical Transactions of the Royal Society A*, **357**, pp. 2789–2799.

82. Yoshizakia, I., Fukuyamab, S., Koizumic, H., Tachibanac, M., Kojimac, K., Matsuurad, Y., Tanakae, M., Igarashie, N., Kadowakif, A., Ronga, L., Adachia, S., Yodaa, S., Komatsu, H., Yoshizakia, I., Fukuyamab, S., Koizumic, H., Tachibanac, M., Kojimac, K., Matsuurad, Y., Tanakae, M., Igarashie, N., Kadowakif, A., Ronga, L., Adachia, S., Yodaa, S. and Komatsu, H. (2006). Impurity-induced defect and its effect on protein crystal perfection. *Journal of Crystal Growth*, **290**, pp. 185–191.

Chapter 15

GROWTH AND ORGANIZATION OF LANGMUIR–BLODGETT PROTEIN CRYSTALS VIA *IN SITU* GISAXS, LASER-MICRODISSECTION, NANODIFFRACTION, RAMAN SPECTROSCOPY, AND ATOMIC FORCE MICROSCOPY

Claudio Nicolini[a], Christian Riekel[b], and Eugenia Pechkova[a]

[a]*Nanoworld Institute, Fondazione EL.B.A., University of Genoa Medical School, Italy*

[b]*European Synchrotron Radiation Facility, B.P.220, F38043 Grenoble Cedex, France*

Recent atomic scale characterization of Langmuir–Blodgett (LB)-based protein crystals here reviewed, namely via laser-microdissection, Raman spectroscopy, *in situ* grazing-incidence small-angle X-ray scattering, atomic force microscopy, and synchrotron radiation diffraction down to the nanofocus scale, provides new and unexpected insights on the role of LB nanotemplate on protein crystal domain organization, nucleation, and growth. These insights are quite superior to the already

Synchrotron Radiation and Structural Proteomics
Edited by Eugenia Pechkova and Christian Riekel

www.panstanford.com

significant properties discovered five years ago which led to proteins being uniquely crystallizable only with LB nanotemplate (CK2α kinase and P450scc). They complement the recently discovered unique radiation resistance for crystals of six model proteins reported in this volume. The major challenges here successfully addressed with LB nanotemplate proved to be the lack of crystallization (for the first time with IF2α and β, ribosomal protein, phage GroEL), and the domains identification and characterization.

15.1 INTRODUCTION

This review provides an overview on largely unpublished results recently obtained by Langmuir–Blodgett (LB) nanotemplate (Pechkova and Nicolini, 2002, 2003). Our aim is to show that significant progress in understanding the unique properties of crystals grown on LB nanotemplate can be obtained by a combination of advanced structural and spectroscopic techniques.

In situ micro grazing-incidence small-angle X-ray scattering (microGISAXS) of enzyme crystal growth with and without homologous nanotemplate added new insights to early events in the nucleation and crystallization processes originally discovered by *ex situ* GISAXS. The major challenges successfully addressed with LB nanotemplate using microprotein crystallography (microPX) at the ESRF-ID13 and ESRF-ID23 beamlines proved to be radiation damage and lack of crystallization (Pechkova and Nicolini, 2004a), but also the domain identification and characterization (Nicolini and Pechkova, 2010). Single crystals of all proteins obtained for the first time by LB nanotemplate[a] confirm earlier findings with CK2α human kinase (Pechkova *et al.*, 2004b; Pechkova and Nicolini, 2004b) pointing to a new generation of bionanomaterials of unique structure–function relationship capable to overcome the difficulties of classical vapor diffusion method to yield protein crystals with optimal diffraction quality, order, and radiation resistance.

[a] Bovine cytochrome P450scc; IF2α and β ribosomal protein subunits from archeabacteria *Sulfolobus solfataricus*; phage GroEL over-expressed in *Escherichia coli.*

The translation initiation factor aIF2 belongs to the aIF-2-β/aIF-5 family, which is active in the early steps of protein synthesis by forming a ternary complex with GTP and the initiator tRNA (Bell and Jackson, 1998; Pestova and Hellen, 2000). It is a heterotrimeric protein, consisting of α-, β-, and γ-subunits, displaying high sequence similarity with the other proteins of its family and is involved in the delivery of Met-tRNA$_i^{Met}$ to the 40S ribosomal subunit (Das and Maitra, 2000). Since the molecular weight (MW) of the α- and β-subunits is relatively low (SsIF2α — MW 32 KDa, 266 amino acids; SsIF2β — MW 15 KDa, 139 amino acids), it is possible to study these subunits separately by NMR and crystallography for the structure comparison in the crystal (Pechkova *et al.*, 2008b) and in the solution state (Vasile et al., 2008).

The solution structure of the intact β-subunit of aIF2 from *Sulfolobus solfataricus* (Pedulla *et al.*, 2000; She *et al.*, 2001) (IF2B_SULSO) (PF01873) was recently solved by ^{1}H NMR (PDB code 2NXU) (Vasile *et al.*, 2008), but this type of protein has great difficulties to be crystallized by classical crystallization method and represents a good system to apply the recently developed LB nanotemplate crystallization method (Pechkova and Nicolini, 2004a). Similarly for phage growth λ E large (GroEL) in the *E. coli* cytoplasm, member of HSP60 chaperone family (Cheng *et al.*, 1989; Martin *et al.*, 1992; Soltys and Gupta, 1996), recognized as a chromosomally encoded product having a role in protein assembly (Horwich and Willison, 1993). Indeed, despite the fact that the protein was widely studied, in some conditions the crystals possess low diffraction quality and appear to be quite disordered which could be improved by LB nanotemplate.

MicroPX using monochromatic beams has become a routine tool at third-generation synchrotron radiation sources (Cusack *et al.*, 1998; Perrakis *et al.*, 1999; Riekel, 2004; Riekel *et al.*, 2005) with illuminated crystal volumes per single exposure of less than 100 μm^3, as shown for the case of bovine rhodopsin (Li *et al.*, 2004), which was refined to a resolution of 2.6 Å and an amyloid fiber structure (Nelson *et al.*, 2005) (eight amino acids).

It has recently been suggested that a reduction of the dimension of a protein crystal in the horizontal scattering plane to a few micrometers or less should result in a reduction in secondary radiation damage owing to the high escape probability of the photoelectrons (Nave and Hill, 2005; Cowan and Nave, 2008). The use of ultrasmall crystals as obtainable by laser-microdissection techniques described in this article is therefore an interesting and desirable avenue to explore the lower limit of accessible crystal volumes, but has several consequences for beamline instrumentation.

Indeed, the beam size has to be reduced by focusing to the micrometer or better submicrometer scale to increase the flux density at the sample. The sphere of confusion of the goniometer rotation axis should be less than the beam size to keep the sample in the beam during data collection. Owing to the radiation dose limit of protein crystals (Henderson, 1990), one also has to be able to distribute the overall dose by multiple exposures using a combination of scanning and rotation across a crystal or a number of small crystals differing in orientation (Riekel, 2004).

We exploited the radiation stability of LB crystals (Pechkova *et al.*, 2009) in an experiment of a laser-cut LB lysozyme crystal using a focused submicrometer-sized beam in combination with an optimized microcollimating system developed for scanning micro-SAXS/WAXS experiments (Riekel, 2000) to demonstrate for the first time that it is possible to determine a high-resolution structure using diffraction patterns obtained from less than 1 μm^3 protein crystal volume.

This review intends in summary to characterize the growth characteristics and the organization down to the atomic scale of the protein LB crystal and their domains by a combination of atomic force microscopy (AFM), laser-cutter, nanodiffraction, and Raman spectroscopy. Single-crystal diffraction does not indeed provide all answers and the complementary techniques here introduced prove quite useful.

15.2 NEW PROTEINS CRYSTALLIZED BY LB NANOTEMPLATE

15.2.1 LB Nanotemplate Primer

In the thin film LB nanotemplate method (Pechkova and Nicolini, 2004a) the droplet with protein solution and the crystallizing agent are placed on the thin protein film (1 or 2 monolayers), deposited on the siliconized glass cover slide and equilibrated against the reservoir solution containing the crystallizing agent with the concentration twice as much as in the droplet.

For these experiments, thin protein films of ribosomal protein were prepared by Teflon-trough LB technology (Nicolini and Pechkova, 2006a) which has recently shown to be capable of improving the crystal quality by applying the LB nanotemplate method for their growth in several new protein systems (Pechkova *et al.*, 2008a,b).

The crystallization of the ribosomal proteins SsIF2α- and SsIF2β-subunits (Pechkova *et al.*, 2008b) and of phage GroEL (Pechkova *et al.*, 2008a) was indeed obtained (Fig. 15.1) toward the independent determination of their crystal structure, as earlier carried out for human protein kinase Ck2α (Fig. 15.2a,b) (Pechkova and Nicolini, 2003, 2004b; Pechkova *et al.*, 2004b) and of cytochrome P450scc (Fig. 15.2c,d), fully crystallizable only by LB nanotemplate applied to the hanging drop vapor diffusion (Pechkova and Nicolini, 2004a).

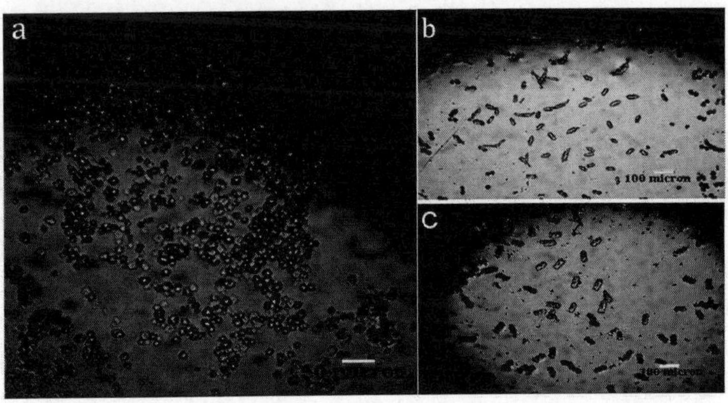

Figure 15.1 (a) aIFα microcrystals from archaebacteria Sulfolobus solfataricus obtained by LB nanotemplate; (b) aIF2β microcrystals from same source; (c) Groel microcrystals from *E. Coli* (polarized light microscope images). (Reprinted from Pechkova, E., Vasile, F., Spera, R. and Nicolini, C. (2008a). Crystallization of alpha and beta subunits of IF2 translation initiation factor from archabacteria *Sulfolobus solfataricus. Journal of Crystal Growth*, 310, pp. 3767–3770, © 2008 with the permission from Elsevier.) See also Color Insert.

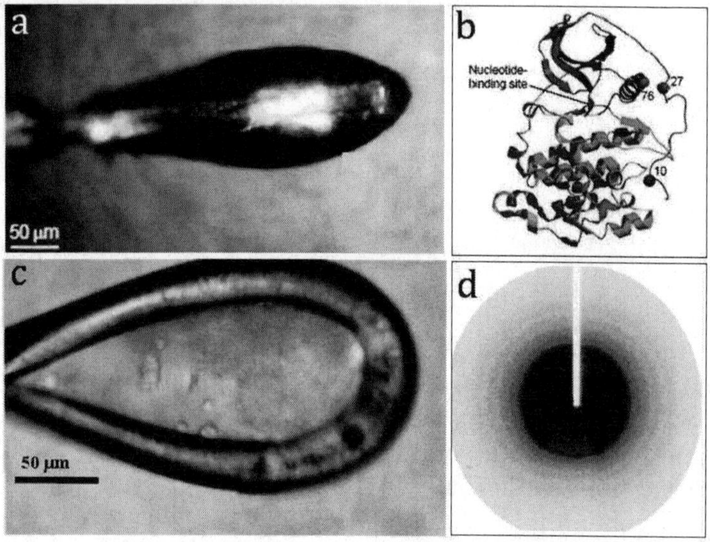

Figure 15.2 (a) Single crystal of human kinase CK2α in cryoloop. (b) Atomic structure of human kinase CK2α. (c) Microcrystals of P450scc protein. (d) P450scc powder pattern. See also Color Insert.

The summary of all proteins so far crystallized by LB nanotemplate is shown in Table 15.1.

Table 15.1 Proteins uniquely crystallized by LB nanotemplate with three-dimensional atomic structures parameters determined by microPX.

Proteins	**Average size in µm**
Cytochrome P450scc (Nicolini and Pechkova, 2006a)	5 × 5 × 5
Human kinase CK2α(Pechova and Nicolini, 2003)	50 × 10 × 10
IF2 αI (Pechkova *et al.*, 2008b)	20 × 10 × 10
IF2 β Pechkova *et al.*, 2008b)	50 × 10 × 10
GroEL (Pechkova *et al.*, 2008a)	60 × 10 × 20

It is sometimes difficult and/or time-consuming to obtain crystals of suitable size for single-crystal X-ray diffraction measurements. However, with the LB method, microcrystals of three different ribosomal proteins were also obtained and their diffraction analysis has become possible using microPX at the ID13-ESRF microfocus beamline (Riekel, 2000) and powder diffraction at the ESRF-ID11 beamline (Margiolaki *et al.*, 2005).

Since many useful proteins are difficult to crystallize and do not optimally diffract, powder diffraction techniques serve indeed as an important tool to predict the space group of protein crystals and can give a range of complementary information, which is difficult to get from single-crystal X-ray diffraction techniques (Dreele, 1999; Margiolaki *et al.*, 2005). Diffraction peaks obtained in powder diffraction analysis depend on the microstructure of materials, and thus accurate unit cell information can be obtained even from poorly diffracting ribosomal proteins crystals (Pechkova *et al.*, 2008a,b).

15.2.2 Protein Expression and Crystallization

The phage GroEL protein was overproduced at *Protein Research Institute, Puchino* (Russia), in bacterial expression system. *E. coli* BL21(DE3) transformed with the plasmid carrying GroEL gene was grown at 37 °C (Pechkova *et al.*, 2008a). Expression of GroEL was induced by IPTG adding (final concentration 1 mM), followed by 3 h culture growth at 37 °C.

The protein was eluted by buffer A containing 200 mM NaCl and 25% ethanol. Fractions containing GroEL were combined, concentrated up to 20–30 mg/ml using "Vivaspin" concentrator 100 kDa, and subjected to crystallization (Pechkova *et al.*, 2008a). For LB crystallization, the sample of GroEL with concentration 15 mg/ml was prepared in the following buffer: 50 mM Tris-HCl pH 8, 200 mM NaCl (buffer A). Screening for crystallization conditions was carried out using classical hanging drop vapor diffusion technique.

Initially, crystalline precipitation were observed in solution containing 100 mM Hepes, pH 7.5, 10% PEG 8000, 8% ethylene glycol (Hampton research Screen II No37) (M. Garber, personal communication). The various trials were performed to optimize these conditions.

For LB crystallization, the prepared sample of aIF2α (Pechkova *et al.*, 2008b; Vasile *et al.*, 2008) was concentrated up to 50 mg/ml in 50 mM Tris-HCl pH 8 and 200 mM NaCl, while the aIF2β sample was concentrated up to 12.5 mg/ml in 50 mM Tris-HCl pH 8, 200 mM NaCl, 10 M mercaptoethanol. Droplets of the aIFα (6.25 mg/ml) with various concentration of PEG 4000 (9–14% (m/v)) with an amount of lithium sulphate from 0.1–0.2 M were placed on the nanofilm template and equilibrated at 20 °C against 1000 ml reservoir solution with PEG 4000 concentration of 18–28% (m/v) and lithium sulphate concentration from 0.2–0.5 M (Pechkova *et al.*, 2008a).

The genes encoding α- and β-subunits of *S. solfataricus* aIF2 were cloned into pET22b and pET28b, correspondingly; and the resulting vectors were introduced in *E. coli* BL21 (DE3) cells.

15.2.3 Protein Characterization by Mass Spectrometry

We used matrix-assisted laser desorption ionization time of flight mass spectrometry (MALDI-TOF MS, Bruker) to monitor the purity of proteins solutions of crystallographic interest. These solutions had been preventively dialyzed and/or the proteins had been precipitated in a solution of tricloroacetic acid to eliminate any trace of glycerol, salt, and detergent that prevent the protein ionization.

We confirmed by MALDI-TOF MS the identification of the corresponding dissolved crystals as formed by ssIF2 α, β and GroEL proteins (Fig. 15.3).

The protein samples were diluted in a 0.1% (v/v) TFA solution. The matrix used for the mass spectrometric analysis was a saturated solution of acid (α-cyano-4-hydroxycinnamic acid for light proteins and sinapinic acid for heavy proteins, Bruker Daltonics) dissolved in 2/3 of 0.1% (v/v) TFA and 1/3 of acetonitrile. 1.5 μl of matrix solution was mixed with 1.5 μl of sample, then 1 μl of this mixture is spotted onto a suitable aluminum plate and air-dried.

MALDI-MS spectra were acquired in positive ion linear mode using an Autoflex mass spectrometer (Bruker Daltonics) externally calibrated using a solution of protein of known masses resulting in a mass accuracy <100 ppm for intact proteins (Pechkova *et al.*, 2008b). A trypsin digestion was occasionally performed for fingerprint analysis, and the results are consistent and satisfactory, confirming the high purity level of each solution.

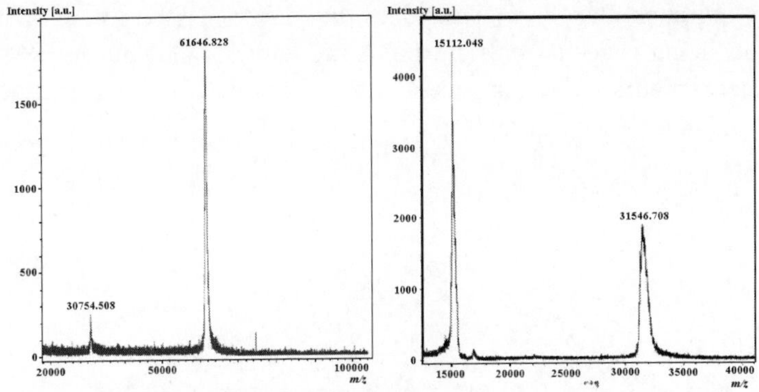

Figure 15.3 Matrix-assisted laser desorption ionization time of flight mass spectrometry (MALDI-TOF MS) of the dissolved crystals of GroEL (left) and aiFα and β (right). (Left: Reprinted from Pechkova, E., Tripathi, S., Spera, R. and Nicolini, C. Groel crystal growth and characterization. *Biosystems*, 94, pp. 223–227, © 2008 with the permission from Elsevier. Right: Reprinted from Pechkova, E., Vasile, F., Spera, R. and Nicolini, C. Crystallization of alpha and beta subunits of IF2 translation initiation factor from archabacteria *Sulfolobus solfataricus*. *Journal of Crystal Growth*, 310, pp. 3767–3770, © 2008 with the permission from Elsevier.)

15.3 DOMAIN ORGANIZATION OF LB PROTEIN CRYSTALS BY AFM

15.3.1 Domains Organization Revealed by AFM

We used an MFP-3D™ stand-alone AFM (Asylum Research; Santa Barbara, USA[b]) in the stand-alone mode. This low noise head AFM provides optimal imaging of protein crystals in solution (Fig. 15.4).

The AFM was tested on crystals (LB) films of lysozyme proteins enabling the periodicity and morphology of crystals anchored to the mica surface in their mother liquid, thereby preserving the native periodic protein crystal structure and making apparent the domain structure, which is typically destroyed by drying. Moreover (Pechkova and Nicolini, 2010), the instrument appears to distinguish protein crystals from the types of preparation either classical (Fig. 15.4) or LB, which under the optical microscope

[b] www.asylumresearch.com/

are quite similar in morphology, the difference between them often being revealed only during X-ray analysis and by this AFM measurements revealing quite larger domain (2 μm rather than 0.2 μm in the classical crystal).

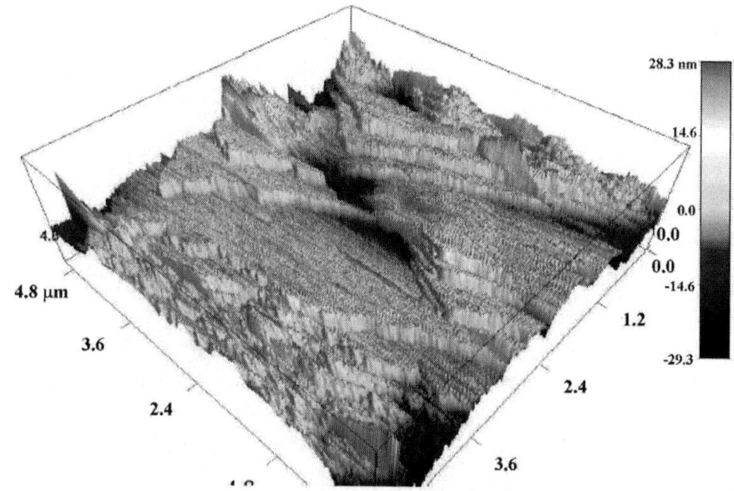

Figure 15.4 AFM image of lysozyme crystal in solution prepared according to standard crystallization techniques at the 5 μm scale (Nicolini and Pechkova, 2010). See also Color Insert.

15.4 DOMAINS ORGANIZATION OF LB PROTEIN CRYSTALS BY LASER-MICRODISSECTION AND LIGHT MICROSCOPY

Laser-microdissection allows preparing sections of biological materials for synchrotron radiation experiments without the necessity of embedding (Seidel *et al.*, 2008; Davies *et al.*, 2008). We applied a PALM MicroBeam laser-microdissection system[c] to two different types of samples, namely, classical protein crystals and LB proteins which were stored in a glycerol solution (Fig. 15.5).

The samples could be electrostatically fixed to a mica sheet while they were found to be floating on glass slides. The immobilization made a rather precise cut possible and led to the following

[c] www.zeiss.de/mikro

observation: the crystal indeed appeared to split along the cutting surfaces immediately after being hit by a laser pulse, with the cutting requiring a continuous laser beam of approximately 90 s in the LB and respectively 20 s in the classical crystal.

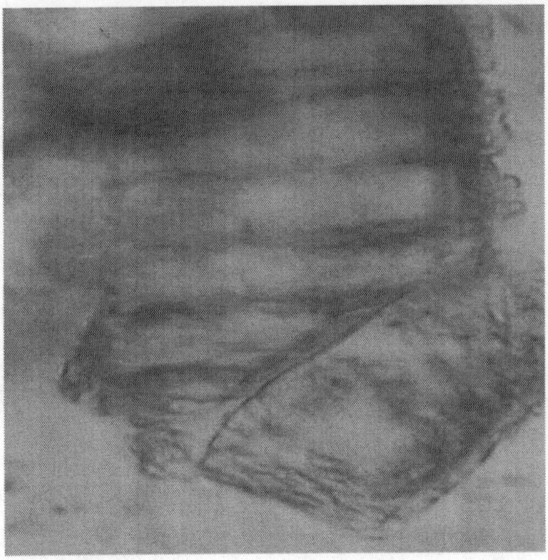

Figure 15.5 Domains in LB crystals induced by laser-microdissection (Nicolini *et al.*, unpublished). See also Color Insert.

The protein is stored in a very viscous solution causing black bubbles to develop after a laser hit. Figure 15.5 shows a crystal after only two single, short laser pulses focused on just one point of the domains structures apparent in LB crystal quite similarly to what apparent by AFM analysis, but displaying large differences between classical and LB crystals. By subsequent processing, a crystal size of approximately 5 × 5 μm^2 could be achieved. Within the first minutes after the beginning of the laser-microdissection experiment, the typical break along domain lines could not be observed. It occurs indeed only after some time depending on the type of the crystal; the proteins started to dissolve and, in doing so, showed their domain structure (small and very quickly in the classical crystal and large and very slowly in the LB crystal). When hit at one of these domain lines in the LB crystal (Fig. 15.5), they easily split. This dissolution took approximately 30 min for the classical crystal and more than 45 min for the LB crystals (Fig. 15.5). Striking and reproducible are the differences in domains apparently visible in crystal produced by classical hanging drop technique with respect to LB crystal produced by nanotemplate hanging drop technique. Images are taken with crystal maintained under same buffer conditions but at different times, due to the higher stability of LB crystal.

The solution in which the protein crystals are stored has apparently an important effect on their behavior. In an aqueous solution, they tend to float around and therefore cannot be handled very precisely. Additionally, the water is evaporating very quickly, so there is a limited time in which the protein crystals have to be manipulated, due to the increasing percentage of salt in the solution (~15 min). By contrast, when being stored in a viscous solution like glycerol, the pieces of a crystal cannot be separated after the cutting and at the same time, the evaporation is not a problem. The solution utilized by us fits the best to the required needs (manipulation time, kinetics, etc.).

15.5 NANODIFFRACTION OF LASER-MICRODISSECTED LB CRYSTALS

In this study, we have used a 0.5 μm beam for microPX at the ESRF-ID13 beamline (Riekel *et al.*, 2009) and have determined the structure of a well-diffracting crystal of lysozyme (Fig. 15.6) to a resolution of 2.0 Å.[d] The proteins crystal being here studied is the lysozyme prepared by LB nanotemplate technique. The data have been acquired after a series of laser irradiations of the lysozyme crystal shown in Fig. 15.5. The size of the thin protein crystal after laser irradiation is of submicron range and cannot be seen in the cryoimmobilized crystal within the nylon loop using light microscopy. The diffraction pattern from such an ultrasmall crystal was identifiable only by a blind eyes processing (Fig. 15.7).

The final electron-density maps after proper refinement (Table 15.2) showed excellent densities for all residues, with the exception of a few that are not ordered and are located on the protein surface (Pechkova et al., 2010).

[d] Experimental conditions: λ = 0.1 nm, beam size ~0.5 × 0.5 μm^2; exposure time/frame: 1 s, flux at sample: ~7 × 10^{10} photons/s corresponding to a flux density of ~3 × 10^{11} ph/s/μm^2.

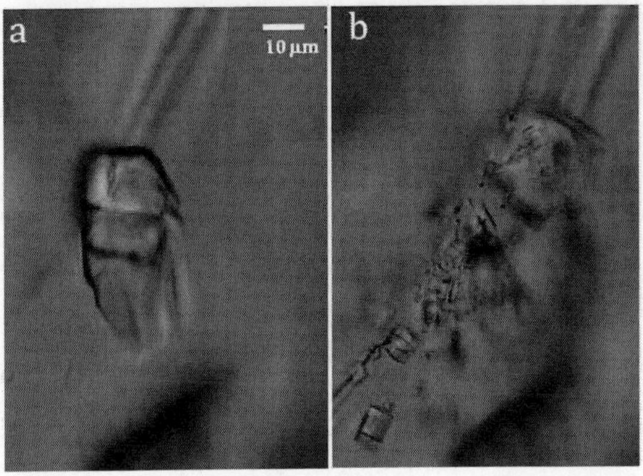

Figure 15.6 LB lysozyme crystal in a nylon cryoloop before (a) and after (b) laser irradiation. See also Color Insert.

It has been shown that increasing irradiation of protein crystals results in structural damage of disulfide bridges, glutamates and aspartates decarboxylate, tyrosine residues losing their hydroxyl group, and methionines losing their C–S bond (Burmeister, 2000; Pechkova *et al.*, 2009). For the current LB nanotemplate prepared lysozyme submicron crystal there was, however, no indication of specific radiation damage from the electron-density maps for these residues, confirming early recent data on LB crystal of proteinase K (Pechkova *et al.*, 2009). Very similar B-factors are observed for glutamates and aspartates in the larger crystal.

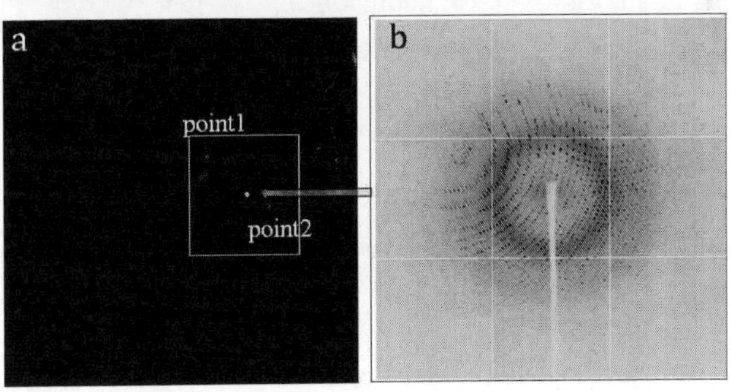

Figure 15.7 (a) Optical image of laser-microdissected lysozyme crystal shown in Fig. 15.6b after cryofreezing in nylon loop. A two-dimensional raster scan was performed in the yellow rectangle with a 0.5 μm beam around point 2. (b) Typical diffraction pattern recorded at point 1. See also Color Insert.

No radiation damage is indicated for tyrosines and methionine which agrees to an enhanced radiation stability reported for a xylanase II crystal studied with a 1 μm beam (Moukhametzianov *et al.*, 2008).

Table 15.2 Submicron-sized lysozyme crystal structural parameters.

Parameters	LB5	LB4
Resolution (Å)	2.0	2.0
Unit cell a,c (Å) ($a = b; \alpha = \beta = \gamma = 90°$)[e]	80.1(4), 37.4(2)	78.9(3), 37.1(2)
Mosaicity (°)	1.3	2.6
Rmerge (%)	23.7	9.8
Individual data set Rfree	25.7	24.4
Unit cell volume (nm^3)	239.9587(3)	230.9553(3)
I/Sigma(I) (%)	6.0 (2.4)[f]	10.9 (6.7)[g]
Completeness, %	77.9 (81.6)[g]	99.8 (100)[g]
Total number of reflections/ independent reflections	33084/6736	114126/8322
Mean B-value	15.30	15.65

Sets of diffraction patterns collected from different-sized crystals were shown to comprise data of good quality, which allowed a 2.0 Å resolution lysozyme structure to be obtained (similar to what has been obtained for a considerably larger lysozyme crystal; Pechkova *et al.*, 2004a). We have therefore demonstrated for the first time for a laser-microdissected LB lysozyme crystal that a ~0.5 μm X-ray beam allows obtaining a high-resolution structure of an ultrasmall protein crystal of <1 μm^3 crystal volume, corresponding to <4 × 10^7 unit cells.[g] Despite the high irradiation dose, it was possible to obtain an excellent high-resolution electron-density map and it could be concluded from the individual atomic B-factors that there was no evidence of significant radiation damage in this ultrasmall LB crystal. Set aside an enhanced radiation damage resistance of LB crystals, the photoelectron escape from a

[e] Spacegroup: P43212.
[f] Value in brackets correspond to the outer resolution shell 2.1–2.0.
[g] Experiments were performed at 100 K using a MAR CCD; λ = 0.9611 Å.

narrow diffraction channel is a possible explanation for reduced radiation damage (Nave and Hill, 2005; Cowan and Nave, 2008). These results open many new opportunities in scanning microPX and make random data collection from microcrystals a real possibility, therefore enabling structures to be solved from much smaller crystals than previously anticipated.

15.6 *IN SITU* MICROGISAXS OF LB CRYSTAL GROWTH

As discussed in the previous chapter, protein crystals grown on nanotemplate surfaces have particular features such as an enhanced radiation resistance. To understand such features one has to study not only the final product but also the growth process of such crystals. The growth of lysozyme films on lysozyme crystal surfaces has been studied by AFM techniques (Li *et al.*, 1999) but microscopic details on the crystal growth process under standard crystal growth geometries, such as the hanging drop technique, are largely unknown. We have therefore used GISAXS, which is an advanced scattering techniques developed to investigate large-scale structures in thin films (Müller-Buschbaum, 2003).

A microGISAXS setup is available at the ESRF-ID13 beamline (Müller-Buschbaum *et al.*, 2007; Gebhardt et al., 2009) and has been used for studying enzyme surface gradients (Gebhardt *et al.*, 2010a) or nanoparticles at confined surfaces (Roth *et al.*, 2007).

The potential for studying thin protein films by microGISAXS during nanotemplate-assisted crystallization experiments was recently demonstrated in several *ex situ* experiments (Pechkova *et al.*, 2005; Pechkova and Nicolini, 2006; Nicolini and Pechkova, 2006b). *Ex situ* data obtained with both cytochrome P450scc and lysozyme were extremely complex to interpret and needed to be studied *in* parallel with microscopy characterization, using either classical or nanotemplate hanging drop crystallization technique. The formation of thaumatin crystals by LB nanotemplate was recently studied by *in situ* microGISAXS (Pechkova and Nicolini, 2010) using the hanging drop technique in a flow-through cell (Fig. 15.8a). The thaumatin sample was glued onto the inner cylinder of the crystallization unit. For rapid buffer exchange the reservoirs were connected via Teflon tubes with two Harvard syringe pumps (Pechkova and Nicolini, 2010).

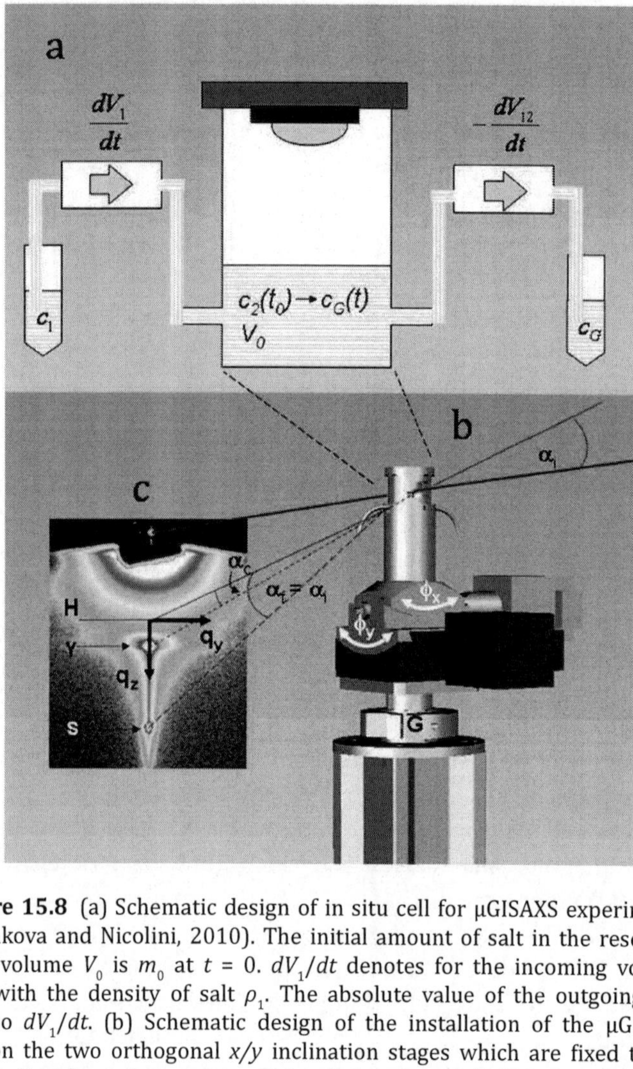

Figure 15.8 (a) Schematic design of in situ cell for μGISAXS experiments (Pechkova and Nicolini, 2010). The initial amount of salt in the reservoir with volume V_0 is m_0 at $t = 0$. dV_1/dt denotes for the incoming volume flux with the density of salt ρ_1. The absolute value of the outgoing flux is also dV_1/dt. (b) Schematic design of the installation of the μGISAXS cell on the two orthogonal *x*/*y* inclination stages which are fixed to the supporting element via a clamp G. An *x*/*y*/*z* stage attached to the supporting element (not shown) allows positioning of a specific point of the cell in the beam or provides raster-capabilities (Pechkova and Nicolini, 2002). (c) Typical μGISAXS pattern; *H*: height of the direct beam; *Y*: Yoneda peak at the critical angle $\alpha_c = 0.11°$; *S*: specular peak at $\alpha_f = \alpha_i$ with $\alpha_i = 0.71°$. (Reprinted with the permission from Pechkova *et al.* In situ GISAXS: I. Experimental setup for submicron study of protein nucleation and growth, *Biophysical Journal*, in press, doi:10.1016/j.bpj.2010.03.069, © 2010 Cell Press.) See also Color Insert.

Two Kapton windows allowed scattering experiments under GISAXS conditions with an angle of incidence $\alpha_i = 0.71°$ (Fig. 15.8b) (Müller-Buschbaum, 2003).[h] Note that the choice of a larger α_i value than the critical angle $\alpha_c = 0.11°$ allows separating the Yoneda peak from the specular peak (Fig. 15.8c). The critical angles of thaumatin and glass were calculated on the basis of their chemical formulas and densities.

The kinetics of crystallization was measured directly on the interface of the LB film crystallization nanotemplate. The evolution of the microGISAXS patterns suggests that the increase in intensity in the Yoneda region is due to protein incorporation into the LB film. The intensity variation suggests several steps, which were modeled by system dynamics, based on first order differential equations (Gebhardt *et al.*, 2010b). The kinetic data can be described by two processes, which take place on the LB film. A first, fast, process is attributed to the crystal growth and its detachment from the LB film. The second, slower, process is attributed to an unordered association and conversion of protein on the LB film (Gebhardt *et al.*, 2010b).

A typical μGISAXS pattern is shown in Fig. 15.8c. Specular scattering is observed for $Q_x = Q_y = 0$, $Q_z > 0$ and diffuse scattering for Q_x, $Q_y \neq 0$. Correlations vertical to the sample surface can be probed along Q_z at $Q_y = 0$. Characteristic morphological parameters such as shape and distances of the sample can be extracted by analysis of out of plan scans in Q_y direction as discussed elsewhere in detail (Gebhardt *et al.*, 2010b). Detector scans were created from the two-dimensional GISAXS pattern by cutting along the Q_z-axis at $Q_y = 0$. Figure 15.9 shows a projection of the temporal variation of the detector-cuts. The direct beam at $\alpha_f = -0.7°$ and the specular beam at $\alpha_f = 0.7°$ are striking features and do not change in position with time. Both beams generate scattering SAXS signal. This scattering broadens the specular beam and causes scattered intensity between $\alpha_f = -0.2°$ to $-0.5°$ for the direct beam. Variation in intensity with time can be found in the angular range from $\alpha_f = 0.1°$ to $0.6°$. The intensity increases after the first 50 min and finally drops after 400 min (Fig. 15.9).

[h] Experimental conditions: 1 μm incident beam with ~10^{10} photons/s at $\lambda = 0.96$ Å and 200 mA ring current; the beam footprint on the sample is for $\alpha_f = 0.7$ mrad ~87 μm for a sampling depth of about 700 nm; GISAXS patterns were recorded by a MAR165 CCD detector for a sample-to-detector distance of 791 mm, determined by an Ag-behenate standard.

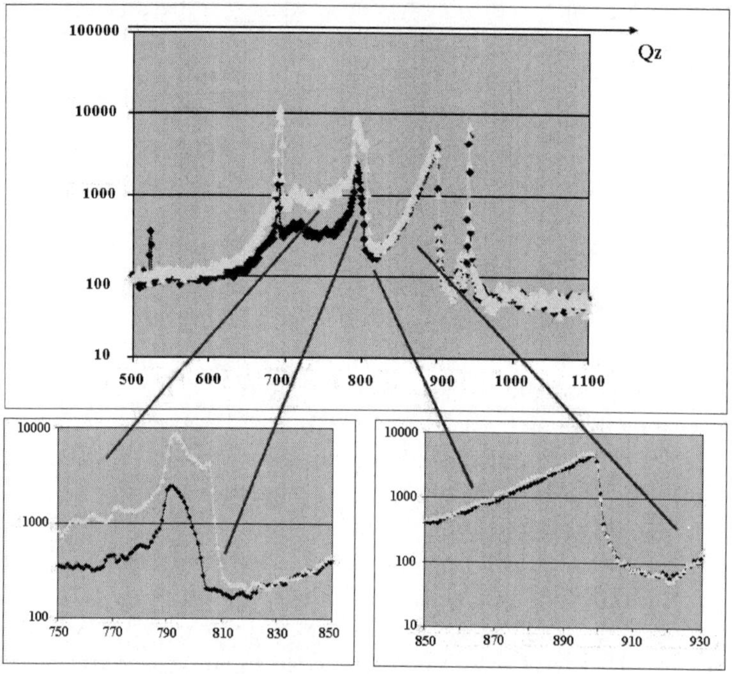

Figure 15.9 Detector-cut of GISAXS of thaumatin LB crystal. See also Color Insert.

The Yoneda region of the background corrected detector scans consists of a contribution from glass (α_c = 0.14°) and a contribution from protein (α_c = 0.11°). Compared to the glass signal the scattering contribution of the protein is weak. After 400 min the maximum intensity of the composed protein/glass Yoneda peak is increased by one order of magnitude compared to the case at the beginning of the kinetics. After this, the intensity drops on approximately one half of the value at the end of the measured kinetics (t = 972 min).

Figure 15.10 depicts the temporal variation of protein in relation to glass, namely the crystal growth and the decay due to radiation damage. The kinetics of the thaumatin μGISAXS pattern was measured at different positions on the LB film. This temporal variation of the thaumatin Yoneda peak intensities, evaluated at the critical angle of the protein (α_c = 0.11°), allows to discriminate among the crystal growth and the crystal radiation damage.

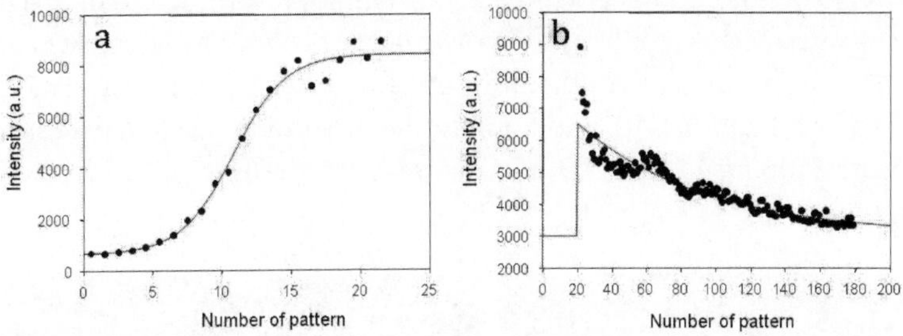

Figure 15.10 (a) The experimental thaumatin growth kinetics (black symbols) is modeled by a sigmoidal curve (in red). (b) The experimental intensity decay (black symbols) is modeled by an exponential decay curve (in red).

15.7 *IN SITU* LB PROTEIN CRYSTAL GROWTH AND CHARACTERIZATION BY RAMAN SPECTROSCOPY

15.7.1 Introduction

Raman spectroscopy probes the vibrational energy levels of chemical bonds even in an aqueous environment (Carey, 1987). Raman spectroscopy has been shown to be an ideal tool for application to aqueous protein solutions, resulting in a powerful non-invasive probe of the changing environment within the solution during crystallization in a vapor diffusion hanging drop experiment in real time (Pechkova and Nicolini, 2004a; Nicolini and Pechkova, 2006a). We have measured Raman spectra to simultaneous evaluate the concentration of lysozyme and the amount of water within the hanging drop with and without LB nanotemplate. Raman spectra have previously been subjected to quantitative infrared partial least-squares (PLS) models with remarkable success (Cooper *et al.*, 1995). The PLS model correlates the spectral region from 2700 to 3600 cm^{-1} with the concentration (g/ml) of lysozyme. This spectral region encompasses vibrations due to the protein CH stretches centered at 2950 cm^{-1} and the water OH stretches centered at 3230 cm^{-1} (Schwartz and Berglund, 1999, 2000). We have reproduced these data using the modified hanging drop method with the LB nanotemplate of the protein deposited on the glass (or quartz) cover slide, brought in the contact with the protein solution drop. Using the lysozyme as the model systems, we have preliminarily confirmed from the Raman spectra the differences between the mechanism of the crystallization with and without the nanotemplate, and estimated the influence of the LB nanotemplate to the protein nucleation and growth.

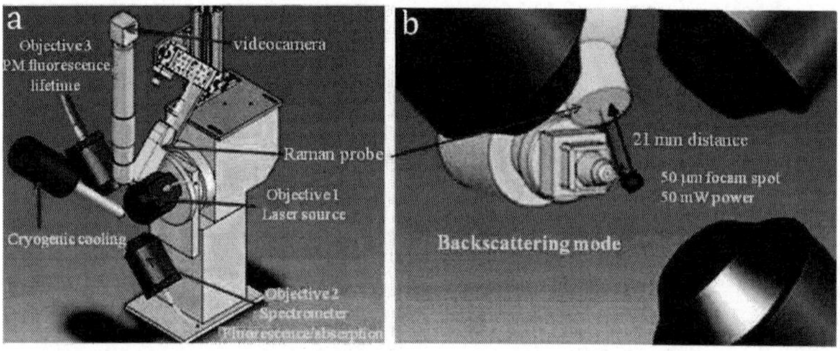

Figure 15.11 (a) Schematic design of IBS cryobench setup at ESRF for microspectroscopy and Raman spectroscopy.[i] (b) Zoom of sample environment showing space available for lysozyme crystal nucleation and growth with flow-through cell. See also Color Insert.

The Raman spectroscopy experiments were performed at the IBS cryobench setup located at ESRF using a $\lambda = 785$ nm laser source.[k] We have utilized an adaptation of our *in situ* chamber previously utilized with μGISAXS at ESRF-ID13 (Fig. 15.11).

15.7.2 Experimental Protocol

A 6 μl protein drop is deposited on a microscope cover glass. Details of the reaction cell and the buffer solution are reported elsewhere. The incident, laser light travels through the cover glass directly into the hanging drop of protein. Since the cover glass does not scatter well at 785 nm, the cover glass is optically transparent to the laser beam permitting the measurement of lysozyme concentration without contact with the solution and provides spectral coverage from 400 to 3650 cm^{-1}. Selected results showing the interest of the Raman probes within the context of the nucleation and growth

[i] www.esrf.eu/UsersAndScience/Experiments/MX/Cryobench/Instruments/raman

of LB crystals is being presented separately (manuscript is in preparation), namely the time sequence of the spectral response of growing lysozyme crystal from their initial induction in an hanging drop vapor diffusion with and without homologous LB nanotemplate.

15.8 CONCLUSIONS

The current review provides a summary of recent results obtained with various techniques on LB nanotemplate crystals and their growth which are still largely pending journal publication. The data conclusively point to the role of LB nanotemplate in determining the morphology, the topology, and the structure of protein crystal to a level unprecedented up to now. Furthermore, while the GroEL protein crystal was grown better with LB nanotemplate but grew even well without LB (even if smaller), the two initiation factor subunits microcrystals were obtained for the first time by LB protein thin film template crystallization method but not by the classical method; all three were proven to be indeed protein crystals by mass spectrometry analysis. Protein powder diffraction on GroEL, aIF2α and β obtained with LB nanotemplate (Pechkova et al., 2008a), has allowed obtaining spacegroup and unit cell information from randomly oriented crystallites through an analysis of the powder rings. The size of the microcrystals obtained suggests, however, that they should be accessible to microPX.

Acknowledgments

This work was supported by an International FIRB-MIUR grant on Functional Proteomics and Cell Cycle to the CIRSDNNOB-Nanoworld Institute of University of Genoa, by a PNR-MIUR grant on Biocatalysis to the University of Genoa and to the Fondazione El.B.A., and by a yearly MIUR Grant for Funzionamento to Fondazione El.B.A. We acknowledge the use of the ESRF Surface Science Lab atomic force microscope and the Raman system of the IBS-ESRF cryobench lab. C.R. acknowledges the attribution of SAXIER-FP6 and HFPS grants for the ESRF-ID13 BL nanobeam R&D, microGISAXS, and computing environment. C. Koenig and I. Snigireva provided support with the use of the ESRF laser-microdissection system, D. Popov with the ID13 microPX experiments, and R. Gebhardt with the ID13 microGISAXS experiments.

References

1. Bell, S. D. and Jackson, S. P. (1998). Transcription and translation in Archaea: a mosaic of eukaryal and bacterial features. *Trends in Microbiology*, **6**, pp. 222–228.

2. Burmeister, W. P. (2000). Structural changes in a cryo-cooled protein crystal owing to radiation damage. *Acta Crystallographica* D — Biological Crystallography, **56**, pp. 328–341.

3. Carey, P. R. (1987). *Raman Spectroscopy*. Academic Press Inc., New York.

4. Cheng, M. Y., Urlich Hartl, F., Martin, J., Pollock, M. A., Kalousek, F., Neuper, W., Hallberg, E. M., Hallber, R. L. and Horwich, A. L. (1989). Mitochondrial heat-shock protein hsp60 is essential for assembly of proteins imported into yeast mitochondria. *Nature*, **337**, pp. 620–625.

5. Cooper, J. B., Wise, K. L., Groves, J. and Welch, W. T. (1995). Determination of octane numbers and reid vapor pressure of commercial petroleum fuels using FT-raman spectroscopy and partial least-squares regression analysis. *Analytical Chemistry*, **67**, pp. 4096–4100.

6. Cowan, J. A. and Nave, C. (2008). The optimum conditions to collect X-ray data from very small samples. *Journal of Synchrotron Radiation*, **15**, pp. 458–462.

7. Cusack, S., Belrhali, H., Bram, A., Burghammer, M., Perrakis, A. and Riekel, C. (1998). Small is beautiful: protein micro-crystallography. *Nature Structural Biology*, **5**, pp. 634–637.

8. Das, S. and Maitra, U. (2000). Mutational analysis of mammalian translation initiation factor 5 (eIF5): role of the interaction between the beta subunit of eIF2 and eIF5 in eIF5 function in vivo and in vitro. *Molecular Cell Biology*, **11**, pp. 3942–3950

9. Davies, R. J., Koenig, C., Burghammer, M. and Riekel, C. (2008). On-axis microbeam wide- and small-angle scattering experiments of a sectioned poly(p-phenylene terephthalamide) fiber. *Applied Physics Letters*, **92**, pp. 101903–101901.

10. Dreele, R. B. V. (1999). Combined Rietveld and stereochemical restraint refinement of a protein crystal structure. *Journal of Applied Crystallography*, **32**, pp. 1084–1089.

11. Gebhardt, R., Burghammer, M., Riekel, C., Kulozik, U. and Müller-Buschbaum, P. (2010a). Surface modification of casein films by rennet-enzyme action as probed with micro-beam grazing incidence small angle X-ray Scattering. *Dairy Science and Technology*, **90**, pp. 75–86

12. Gebhardt, R., Pechkova, E., Riekel, C. and Nicolini, C. (2010b). In situ microGISAXS: II. Thaumatin crystal growth kinetics. *Biophysical Journal*, **99**, pp. 1262–1267.

13. Gebhardt, R., Vendrely, C., Burghammer, M. and Riekel, C. (2009). Characterization of the boundary zone of a cast protein drop: fibroin β-sheet and nanofibril formation. *Langmuir*, **25**, pp. 6307–6311.

14. Henderson, R. (1990). Cryo-protection of protein crystals against radiation damage in electron and X-ray diffraction. *Proceedings of the Royal Society of London Series B — Biological Sciences*, **241**, pp. 6–8.

15. Horwich, A. L. and Willison, K. R. (1993). Protein folding in the cell: functions of two families of molecular chaperone, hsp 60 and TF55-TCP1. *Philosophical Transactions of the Royal Society of London Series B — Biological Sciences*, **339**, pp. 313–326.

16. Li, H. Y., Nadarajah, A. and Pusey, M. L. (1999). Determining the molecular-growth mechanisms of protein crystal faces by atomic force microscopy. *Acta Crystallographica D — Biological Crystallography*, **55**, pp. 1036–1045.

17. Li, J., Edwards, P., Burghammer, M., Villa, C. and Schertler, G. F. X. (2004). Structure of bovine rhodosin in a trigonal crystal form. Journal of Molecular *Biology*, **343**, pp. 1409–1438.

18. Margiolaki, M. I., Wright, J. P., Fitch, A. N., Fox, G. C. and Dreele, R. B. V. (2005). Synchrotron x-ray powder diffraction study of hexagonal turkey egg-white lysozyme. *Acta Crystallographica D — Biological Crystallography*, **61**, pp. 423–432.

19. Martin, J., Horwich, A. L. and Hartl, F. U. A. (1992). Prevention of protein denaturation under heat stress by the chaperonin Hsp60. Science, **258**, pp. 995–998.

20. Moukhametzianov, R., Burghammer, M., Edwards, P. C., Petitdemange, S., Popov, D., Fransen, M., McMullan, G., Schertler, G. F. X. and Riekel, C. (2008). Protein crystallography with a micron-sized

synchrotron radiation beam. *Acta Crystallographica D — Biological Crystallography*, **64**, pp. 158–166.

21. Müller-Buschbaum, P. (2003). Grazing incidence small-angle x-ray scattering — an advanced scattering technique for the investigation of nanostructured polymer films. *Analytical and Bioanalytical Chemistry*, **376**, pp. 3–10.

22. Müller-Buschbaum, P., Gebhardt, R., Roth, S., Metwalli, E. and Doster, W. (2007). Effect of calcium concentration on the Structure of casein micelles in thin films. *Biophysical Journal*, **93**, pp. 960–968.

23. Nave, C. and Hill, M. A. (2005). Will reduced radiation damage occur with very small crystals? *Journal of Synchrotron Radiation*, **12**, pp. 299–303.

24. Nelson, R., Sawaya, M. R., Balbirnie, M., Madsen, A. O., Riekel, C., Grothe, R. and Eisenberg, D. (2005). Structure of the cross-beta spine of amyloid-like fibrils. *Nature*, **435**, pp. 773–778.

25. Nicolini, C. and Pechkova, E. (2006a). Nanostructured biofilms and biocrystals (a review). Journal of Nanoscience and Nanotechnology, **6**, pp. 2209–2236.

26. Nicolini, C. and Pechkova, E. (2006b). Structure and growth of ultrasmall protein microcrystals by synchrotron radiation: I microGISAXS and microdiffraction of P450scc. *Journal of Cellular Biochemistry*, **57**, pp. 544–552.

27. Nicolini, C. and Pechkova, E. (2010). LB protein crystals organization and characterization. *Anticancer Research*, **30**, pp. 2745–2748.

28. Pechkova, E., Gebhardt, R., Riekel, C. and Nicolini C. (2010). In situ GISAXS: I. Experimental setup for submicron study of protein nucleation and growth, *Biophysical Journal*, **99**, pp. 1256–1261

29. Pechkova, E. and Nicolini, C. (2002). From art to science in protein crystallization by means of thin film technology. *Nanotechnology*, **13**, pp. 460–464.

30. Pechkova, E. and Nicolini, C. (2003). *Proteomics and Nanocrystallography*. Kluwer Academic Plenum Publishers, New York, pp. 1–190.

31. Pechkova, E. and Nicolini, C. (2004a). Protein nanocrystallography: a new approach to structural proteomics. Trends in Biotechnology, **22**, pp. 117–122.

32. Pechkova, E. and Nicolini, C. (2004b). Atomic structure of the CK2alpha human kinase by microfocus diffraction of extra-small

microcrystals grown with nanobiofilm template. *Journal of Cellular Biochemistry*, **91**, pp. 1010–1020.

33. Pechkova, E. and Nicolini, C. (2006). Structure and growth of ultrasmall protein microcrystals by synchrotron radiation: II microGISAX and microscopy of lysozyme. *Journal of Cellular Biochemistry*, **97**, pp. 553–560.

34. Pechkova, E. and Nicolini, C. (2010). *In situ* study of nanotemplate-induced growth of lysozyme microcrystals by submicron GISAXS. *Journal of Synchrotron Radiation*, accepted.

35. Pechkova, E., Roth, S. V., Burghammer, M., Fontani, D., Riekel, C. and Nicolini, C. (2005). MicroGISAXS and protein nanotemplate crystallization: methods and instrumentation. *Journal of Synchrotron Radiation*, **12**, pp. 713–716.

36. Pechkova, E., Tripathi, S., Ravelli, R. G., McSweeney, S. and Nicolini, C. (2009). Radiation stability of proteinase K grown by LB nanotemplate method. *Journal of Structural Biology*, **168**, pp. 409–418.

37. Pechkova, E., Tripathi, S., Spera, R. and Nicolini, C. (2008a). Groel crystal growth and characterization. Biosystems, **94**, pp. 223–227.

38. Pechkova, E., Tropiano, G., Riekel, C. and Nicolini, C. (2004a). Radiation stability of protein crystals grown by nanostructured templates: synchrotron microfocus analysis. *Spectrochimica Acta B — Atomic Spectroscopy*, **59**, pp. 1687–1693.

39. Pechkova, E., Vasile, F., Spera, R. and Nicolini, C. (2008b). Crystallization of alpha and beta subunits of IF2 translation initiation factor from archabacteria *Sulfolobus solfataricus*. *Journal of Crystal Growth*, **310**, pp. 3767–3770.

40. Pechkova, E., Zanotti, G. and Nicolini, C. (2004b). Three-dimensional atomic structure of a catalytic subunit mutant of human protein kinase CK2. Acta Crystallographica D — *Biological Crystallography*, **59**, pp. 2133–2139.

41. Pedulla, N., Palermo, R., Hasenohrl, D., Blasi, D., Cammarano, P. and Londei, P. (2000). The archaeal eIF2 homologue: functional properties of an ancient translation initiation factor. *Nucleic Acids Research*, **33**, pp. 1804–1812.

42. Perrakis, A., Cipriani, F., Castagna, J. C., Claustre, L., Burghammer, M., Riekel, C. and Cusack, S. (1999). Protein microcrystals and the design of a diffractometer: current experience and plans at EMBL and ESRF/ID13. *Acta Crystallographica D — Biological Crystallography*, **55**, pp. 1765–1770.

43. Pestova, T. V. and Hellen, C. U. (2000). The structure and function of initiation factors in eukaryotic protein synthesis. *Cellular and Molecular Life Sciences*, **57**, pp. 651–674.

44. Riekel, C. (2000). New avenues in X-ray microbeam experiments. *Reports on Progress in Physics*, **63**, pp. 233–262.

45. Riekel, C. (2004). Recent developments in microdiffraction on protein crystals. *Journal of Synchrotron Radiation*, **11**, pp. 4–6.

46. Riekel, C., Burghammer, M., Davies, R., Gebhardt, R. and Popov, D. (2009). *Fundaments of Soft Condensed Matter Scattering and Diffraction with Microfocus Techniques in Applications of Synchrotron Light to Scattering and Diffraction in Materials.* (eds. T. A. Ezquerra, M. Garcia-Gutierrez, A. Nogales and M. Gomez), Vol. 776, pp. 91–104. Springer, Heidelberg.

47. Riekel, C., Burghammer, M. and Schertler, G. (2005). Protein crystallography microdiffraction. *Current Opinion in Structural Biology*, **15**, pp. 5560–562.

48. Roth, S. V, Autenrieth, T., Grubel, G., Riekel, C., Burghammer, M., Hengstler, R., Schulz, L. and Müller-Buschbaum, P. (2007). *In-situ* observation of nanoparticle ordering at the air-water-substrate boundary in colloidal solutions using x-ray nanobeams. *Applied Physics Letters*, **91**, art. no. 081915.

49. Schwartz, A. M. and Berglund, K. A. (1999). The use of Raman spectroscopy for *in situ* monitoring of lysozyme concentration during crystallization in a hanging drop. *Journal of Crystal Growth*, **203**, pp. 599–603.

50. Schwartz, A. M. and Berglund, K. A. (2000). *In situ* monitoring and control of lysozyme concentration during crystallization in a hanging drop. *Journal of Crystal Growth*, **210**, pp. 753–760.

51. Seidel, R., Gourrier, A., Burghammer, M., Riekel, C., Jeronimidis, G. and Paris, O. (2008). Mapping fibre orientation in complex-shaped biological systems with micrometre resolution by scanning X-ray microdiffraction. *Micron*, **39**, pp. 198–205.

52. She, Q., Singh, R. K., Confalonieri, F., Zivanovic, Y., Allard, G., Awayez, M. J., Chan-Weiher, C. C. Y., Clausen, I. G., Curtis, B. A., De Moors, A., Erauso, G., Fletcher, C., Gordon, P. M. K., Heikamp-de Jong, I., Jeffries, A. C., Kozera, C. J., Medina, N., Peng, X., Thi-Ngoc, H. P., Redder, P., Schenk, M. E., Theriault, C., Tolstrup, N., Charlebois, R. L., Doolittle, W. F., Duguet, M., Gaasterland, T., Garrett, R. A., Ragan, M. A., Sensen, C. W. and Van der Oost, J. (2001). The complete genome of the

crenarchaeon *Sulfolobus solfataricus P2. Proceedings of the National Academy of Sciences USA*, **98**, pp. 7835–7840.

53. Soltys, B. J. and Gupta, R. S. (1996). Immunoelectron microscopic localization of the 60-kDa heat shock chaperonin protein (Hsp60) in mammalian cells. *Experimental Cell Research*, **222**, pp. 16–27.

54. Vasile, F., Pechkova, E. and Nicolini, C. (2008). Atomic structure of the beta-subunit of the translation initiation factor Aif2 from Archaebacteria *Sulfolobus solfataricus*: high resolution NMR in solution. *Proteins: Structure, Function and Bioinformatics*, **70**, pp. 1112–1115.

Index

Color Insert

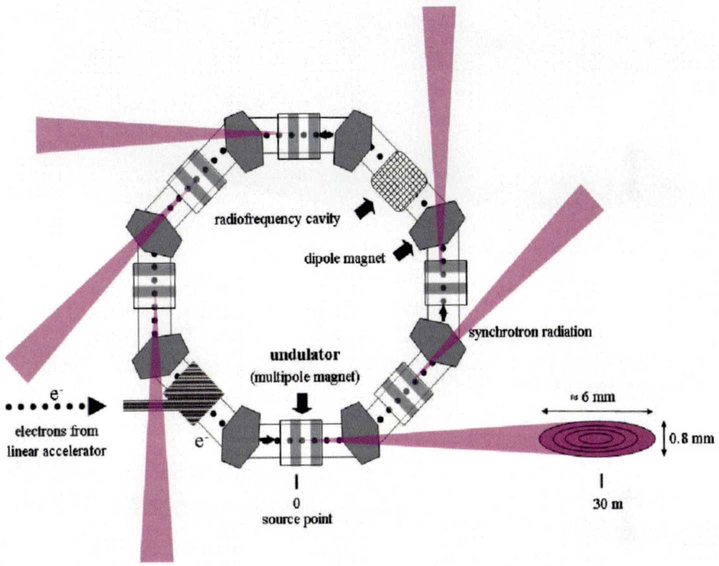

Figure 1.2 Schematic design of a third-generation SR source. The size of the synchrotron radiation beam at 30 m (fwhm) corresponds to an ESRF low-β undulator (Table 1.2). (Reprinted with the permission from Burghammer *et al.* X-ray synchrotron radiation microdiffraction on fibrous biopolymers like cellulose and in particular spider silks. *In Recent Research Developments in Macromolecules.* Vol. 7, pp. 103–125, © 2003, Research Signpost.)

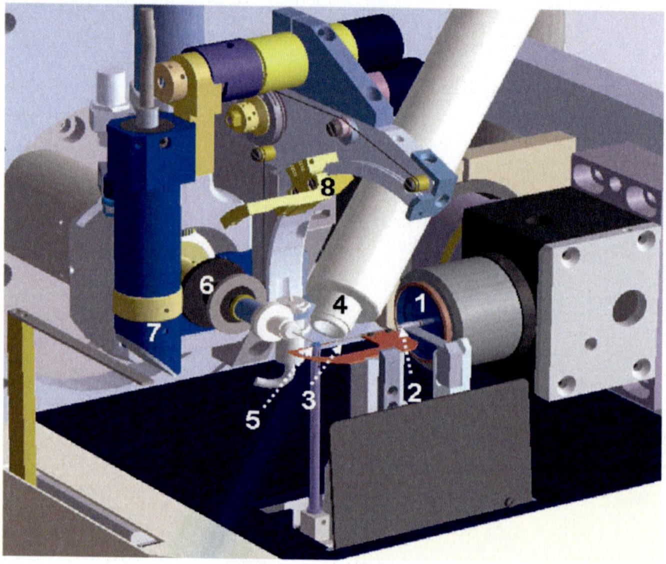

Figure 1.3 Schematic design of μPX goniometer for ≥5 μm beams; 1: pierced zoom objective; 2: beam-defining aperture; 3: guard aperture; 4: cryoflow nozzle; 5: scintillator and photodiode; 6: horizontal rotation axis; 7: illumination; 8: sample transfer system. (Reprinted with the permission from Perrakis *et al.* Protein microcrystals and the design of a diffractometer: current experience and plans at EMBL and ESRF/ID13. Acta Crystallographica D, 55, pp. 1765–1770, © 1999, International Union of Crystallography.)

Figure 1.4 Schematic design of µPX goniometer for 1 µm beam (Volkringer *et al.*, 2007; Moukhametzianov *et al.*, 2008). 1: KB mirror; 2: microionization chamber and aperture; 3: cryoflow system; 4: piezo manipulator; 5: air bearing rotation axis; 6: microscopy objective. The *x/y/z* scanning stage is not visible. (Reprinted with the permission from Volkringer *et al.* A new microdiffraction set-up for nanoporous MOF-type solids. *Nature Materials*, 6, pp. 760–764, © 2007, Nature Publishing Group, division of Macmillan Publishers Limited.)

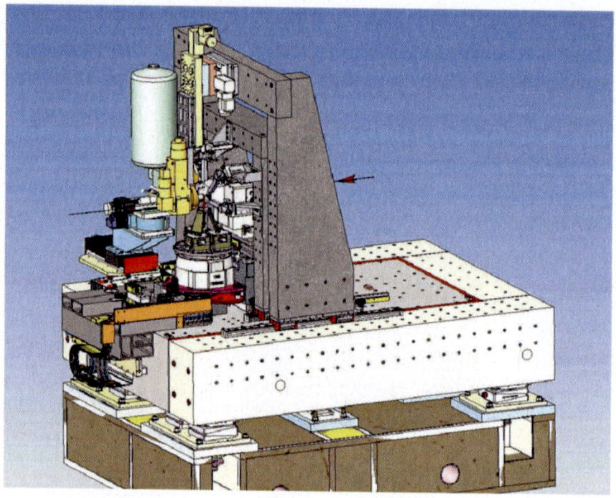

Figure 1.6 Nanogoniometer based on *x/y* air bearing translational stages and a vertical air bearing rotation axis carrying a microhexapod. The mechanical elements requiring the highest stability (e.g., apertures) are located on an inner marble support structure while elements which might introduce vibrations (e.g., cryoflow, fluorescence detector, optical microscope) are installed on a separate outer marble structure.

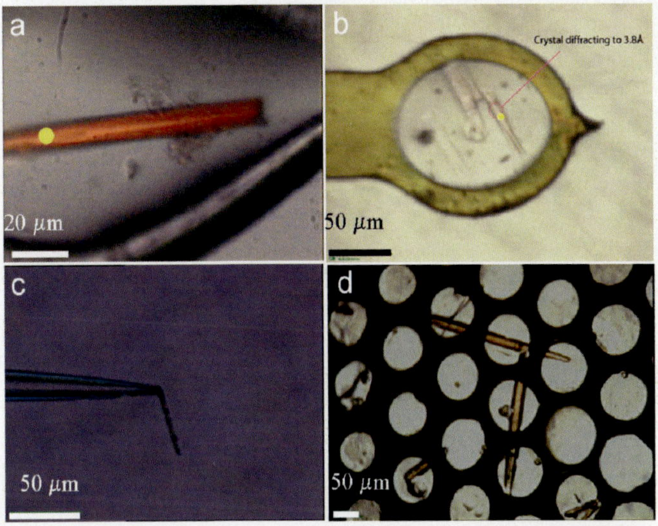

Figure 1.8 (a) Bovine rhodopsin crystal in a cryoloop (Li *et al.*, 2004) (circle corresponds to 5 μm beam). (b) Human β1-adrenergic G-protein-coupled receptor crystals in a Kapton MicroLoopTM (Rasmussen *et al.*, 2007) (circle corresponds to 1 μm beam). (c) Amyloid fibril attached to a glass capillary tip (Nelson *et al.*, 2005). (d) Multiple bovine rhodopsin crystals supported on a metal grid covered with thin carbon foil (Riekel *et al.*, 2005). (Part a: reprinted with permission from Li *et al.* Structure of bovine rhodopsin in a trigonal crystal form. *Journal of Molecular Biology*, 343, pp. 1409–1438, © 2004 Elsevier; Part b: reprinted with permission from Rasmussen *et al.* Crystal structure of the human [bgr]: 2: adrenergic G-protein-coupled receptor. *Nature*, 450, pp. 383–387, © 2007, Nature Publishing Group, division of Macmillan Publishers Limited; Part c: reprinted with permission from Nelson *et al.* Structure of the cross-beta spine of amyloid-like fibrils. *Nature*, 435, pp. 773–778, © 2005, Nature Publishing Group, division of Macmillan Publishers Limited; Part d: reprinted with permission from Riekel *et al.* Protein crystallography microdiffraction. *Current Opinion in Structural Biology*, 15, pp. 556–562, © 2005, Elsevier.)

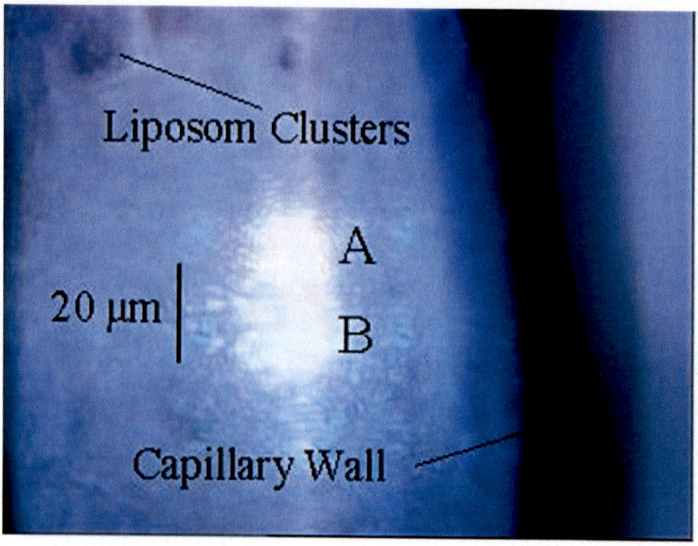

Figure 1.10 Optical microscopy image of two laser-optical trapping spots A, B in a liquid filled capillary, under which two POPE liposome clusters (palmitoyl-oleyl-phosphatidylethanolamine) have been trapped (not seen due to the intense infrared light). A non-trapped cluster is also indicated. (Reprinted with the permission from Cojoc *et al.* Scanning X-ray micro-diffraction of optically manipulated liposomes. *Applied Physics Letters*, 91, art. no. 234107, © 2007, American Institute of Physics.)

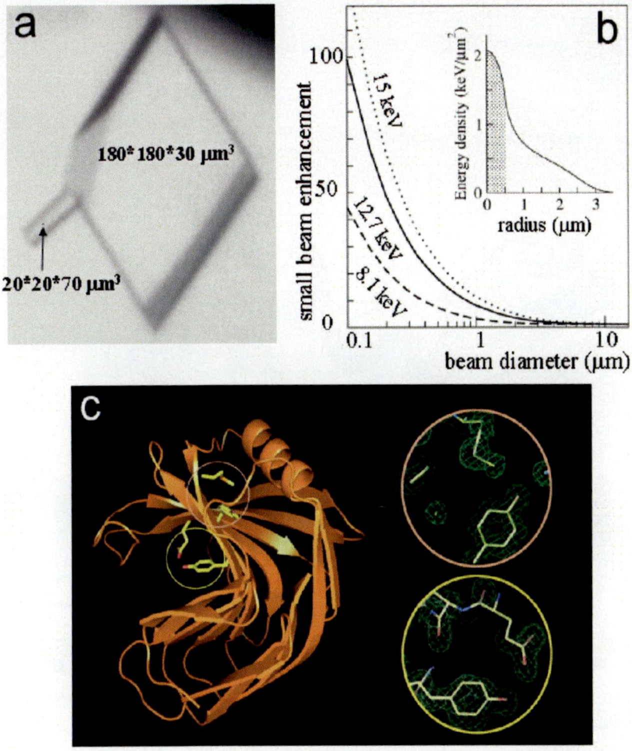

Figure 1.12 (a) The arrow points to the 1 µm beam spot in the smaller part of the Xylanase II crystal. (b) Calculated small beam enhancement (ratio of the total energy deposited in the crystal relative to that actually deposited in the beam volume) as a function of the beam diameter for three energies. Inset: calculated deposited energy density per generated 12.7 keV photoelectron as a function of radius for a 20 µm thick sample illuminated by a 1 µm diameter beam. On average, 12.1 keV is deposited in the whole crystal, but only 1.4 keV of this is contained within the illuminated volume (indicated by the shading). (c) Secondary Xylanase II structure and selected residues shown by electron density map. (Reprinted with the permission from Moukhametzianov *et al.* Protein crystallography with a micron-sized synchrotron radiation beam. *Acta Crystallographica D*, 64, pp. 158–166, © 2008, International Union of Crystallography.)

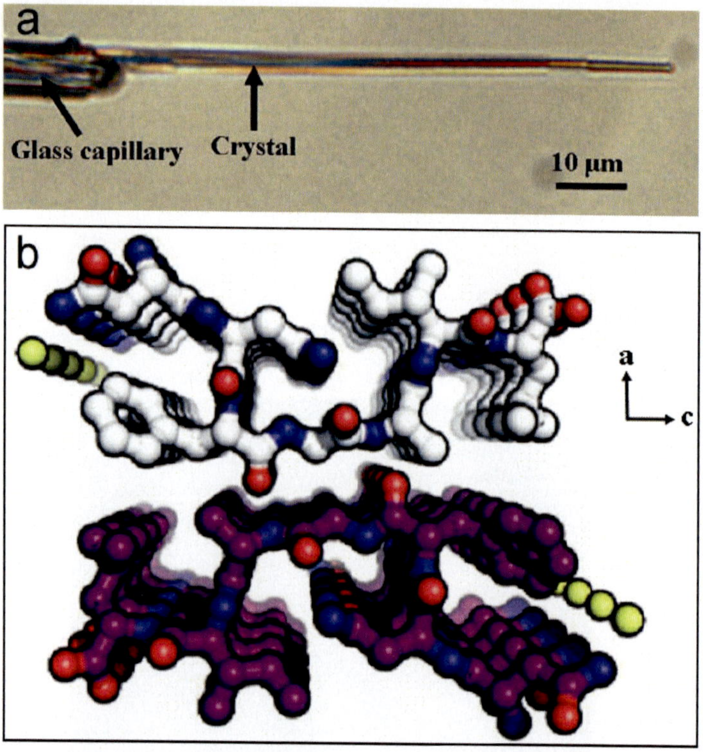

Figure 1.14 (a) Typical crystal of NNFGAIL used to collect intensity data (Wiltzius *et al.*, 2009). (b) The structure of NNFGAIL consists of two close-packed β-sheets, viewed down the sheets. Water molecules are shown as yellow spheres. (Reprinted with the permission from Wiltzius *et al.*, © 2009, The Protein Society.)

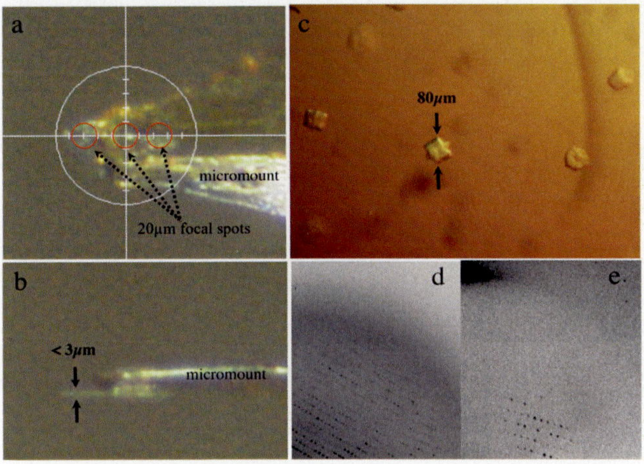

Figure 2.4 Recent uses of capillary microbeam in protein crystallography. Top view (a) and side view (b) of a single serinocylin crystal discovered in the bottom of a sample tube. The thin (<3 μm) plate was exposed at three points (small circles), the central spot yielded high-resolution data. The specimen is shown mounted on a thicker plastic micromount (MiTeGen LLC, NY) with a thin layer of oil. Twinned crystals of the protein CD31 (c), which diffracted with a large P622 unit cell (pattern d), were crushed to yield 20 μm untwined fragments (not shown) with a smaller I222 unit cell (pattern e).

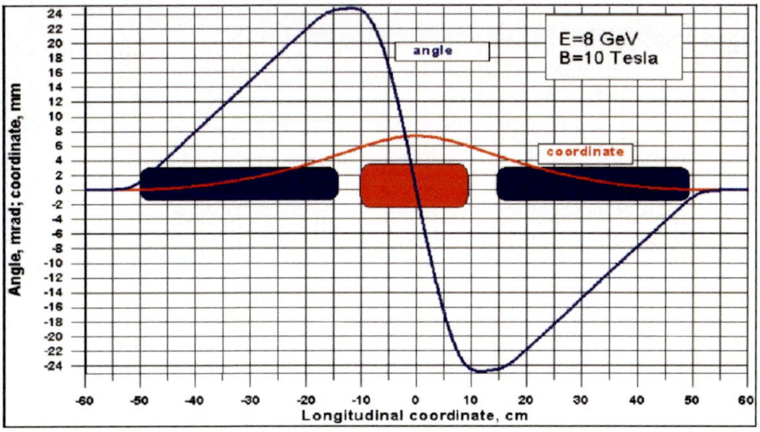

Figure 3.2 Angle deviation and orbit distortion inside the 10 T SC WLS for Spring-8.

Figure 3.4 Photo of the 7 T WLS with fixed radiation point for BESSY-2.

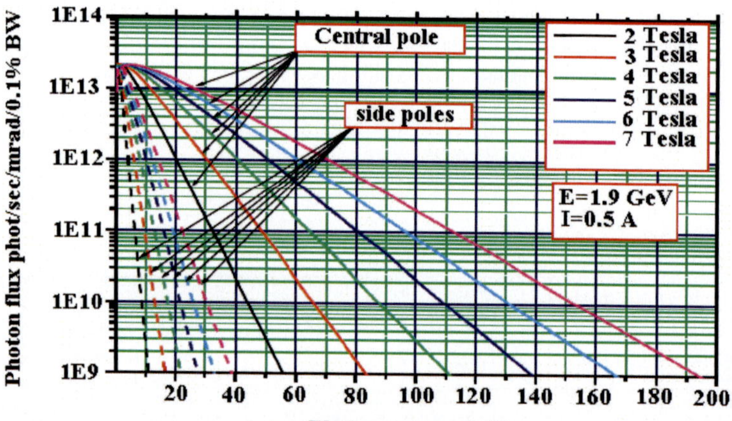

Figure 3.5 Spectral property of SR for some field levels from SC WLS installed on BESSY-2. Spectrum shown from central and side poles for demonstration of difference.

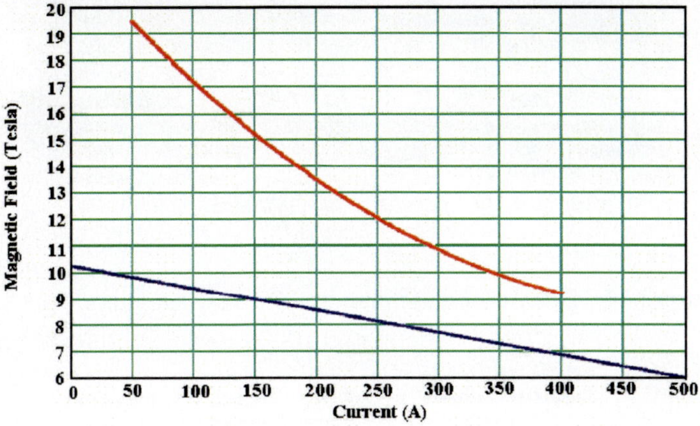

Figure 3.7 Critical curves for Nb_3Sn SC (upper curve) and Nb–Ti SC (lower curve).

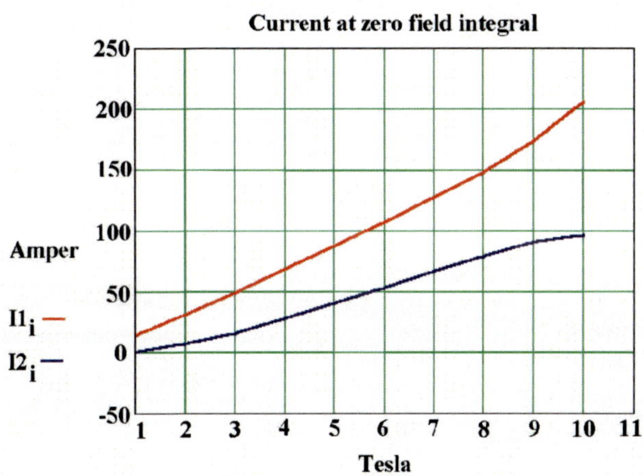

Figure 3.9 Currents in first and second sections depend on maximum magnetic field in median plane of WLS.

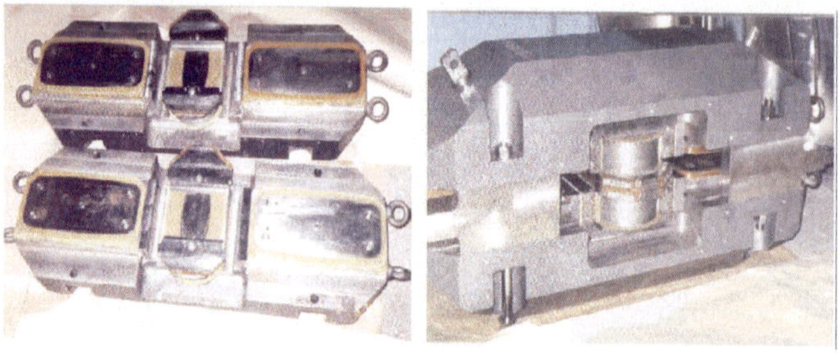

Figure 3.11 Photos of magnet disassembled into two parts (left) and assembled magnet of 10 T WLS.

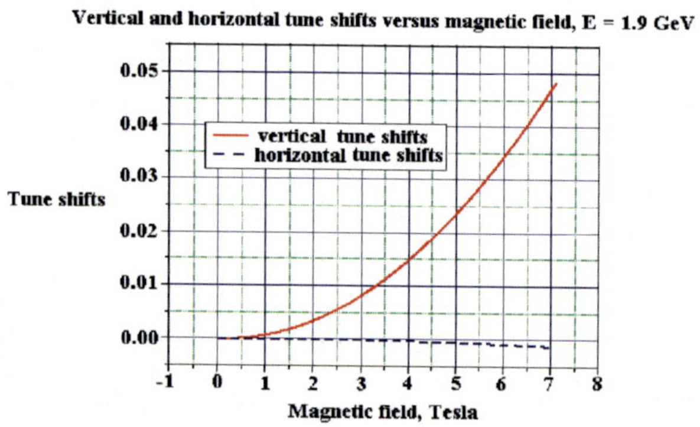

Figure 3.12 Vertical and horizontal betatron tune shifts for BESSY SC 7 T WLS versus magnetic field level.

Figure 3.16 Photo of assembled pole.

Figure 3.18 Photo of assembled magnet.

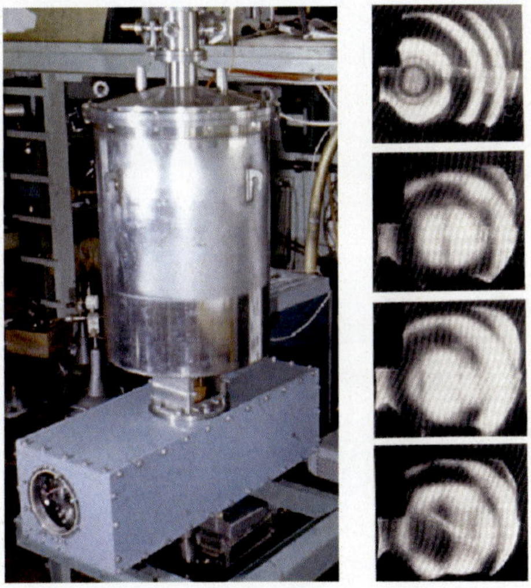

Figure 3.19 (a) Photo of the first assembled SC wiggler before installing on the ring. (b) Photographs of light intensity from the wiggler through red color filter on a screen for different field level: first, second, third, and fourth harmonics of undulator radiation are visible.

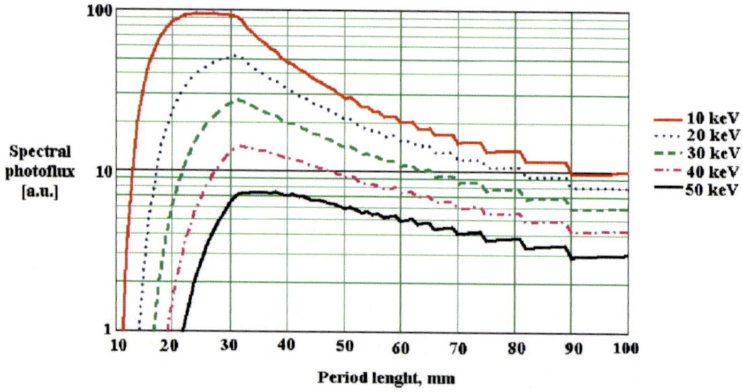

Figure 3.22 Spectral photon flux versus wiggler period length.

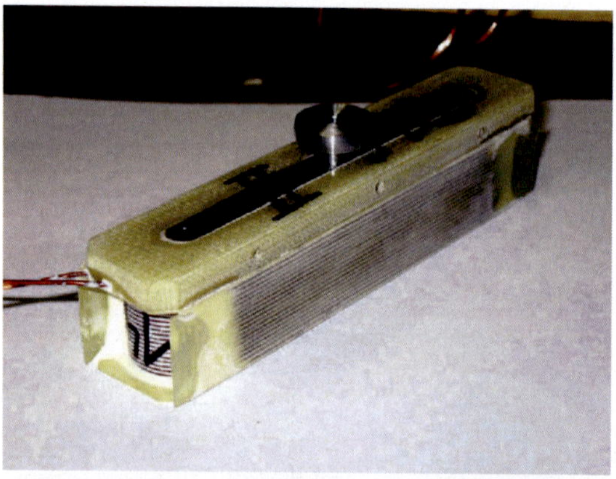

(a)

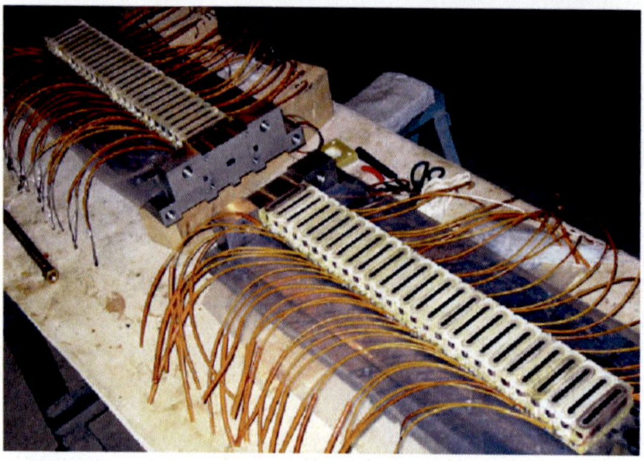

(b)

Figure 3.23 (a) 4.2 T half pole, and (b) two pole arrays of 4.2 T wiggler before commutation and assembling.

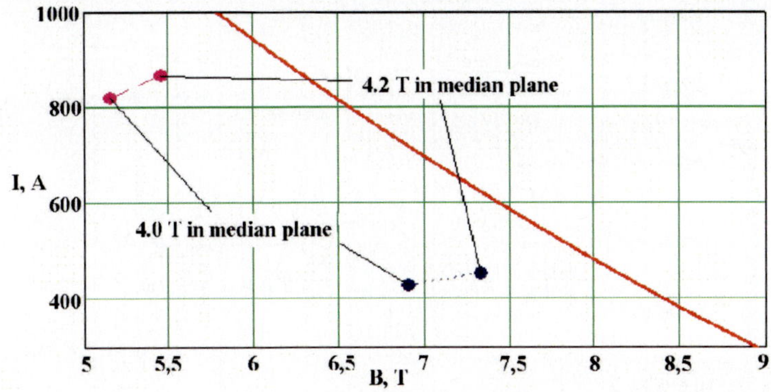

Figure 3.25 Critical curve of SC wire used for medium field wigglers. Blue and magenta points correspond to load lines of first and second coil sections correspondingly relative critical curve at different field level in median plane.

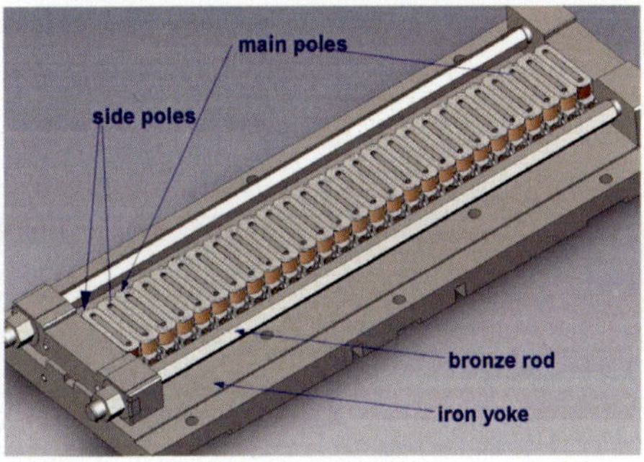

Figure 3.26 Half of assembled wiggler magnet.

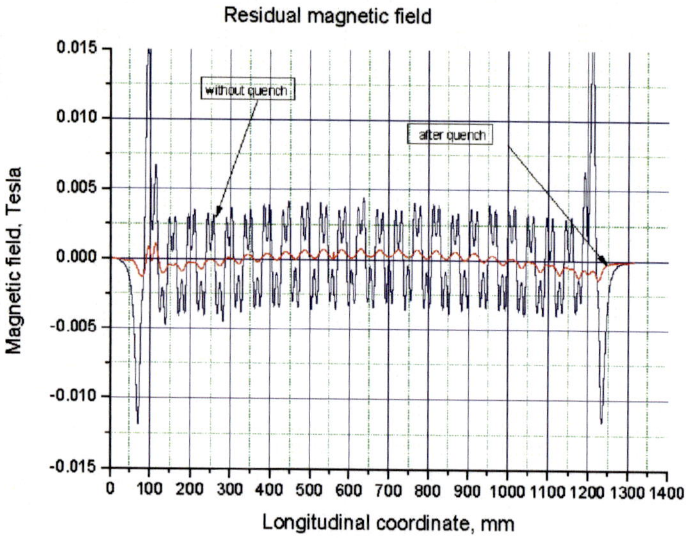

Figure 3.28 Measured residual magnetic field distribution in DLS SC wiggler for zero currents in the wiggler after field ramping down without quench (blue curve) and after quench (red curve).

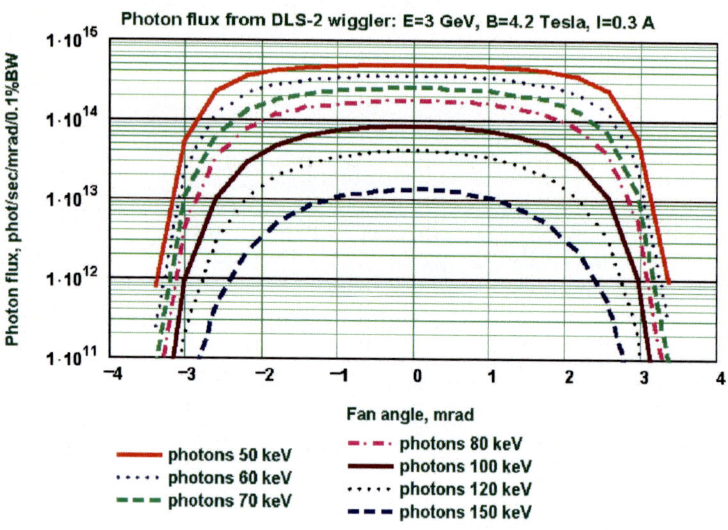

Figure 3.31 Photon flux angle distribution from the wiggler: B = 4.2 T, E = 3 GeV, I = 0.3 A.

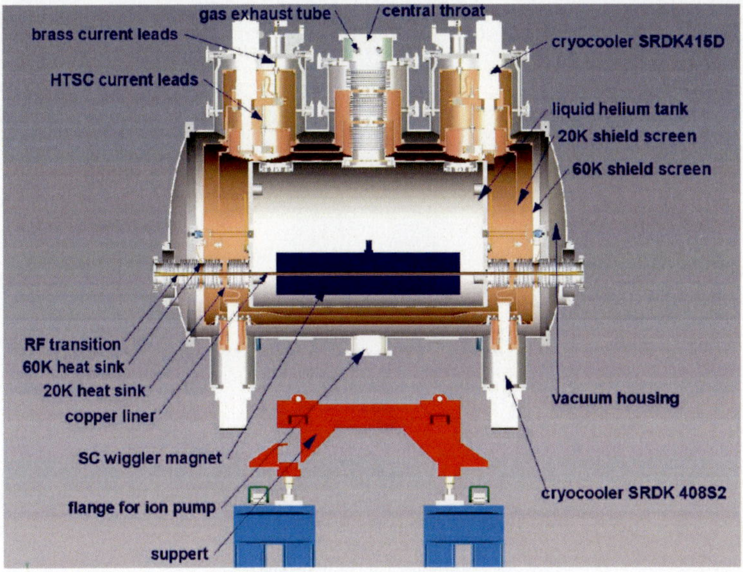

Figure 3.33 Pattern of a cryostat cross section.

Figure 3.34 Photo of SC wiggler installed on CLS storage ring.

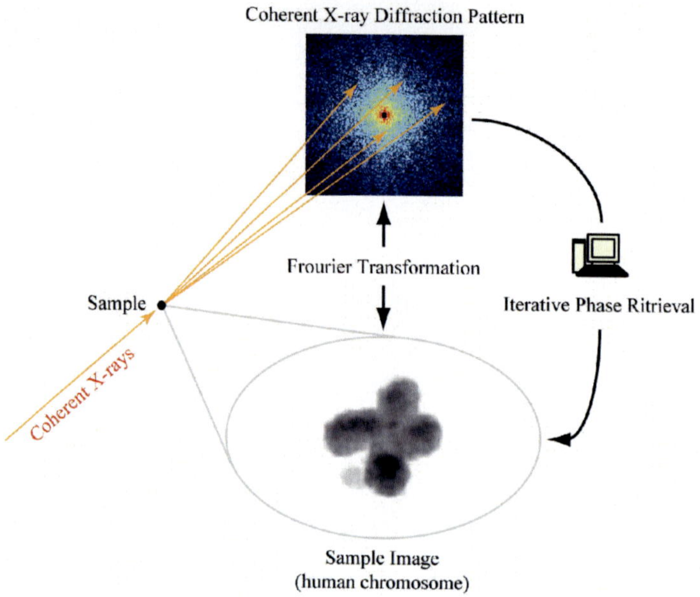

Figure 4.1 Schematic of X-ray diffraction microscopy (XDM). In XDM, a sample is illuminated with coherent X-rays, and the coherent diffraction data are recorded in the far field. The sample structure is reconstructed from the coherent diffraction data by using an iterative phase retrieval algorithm.

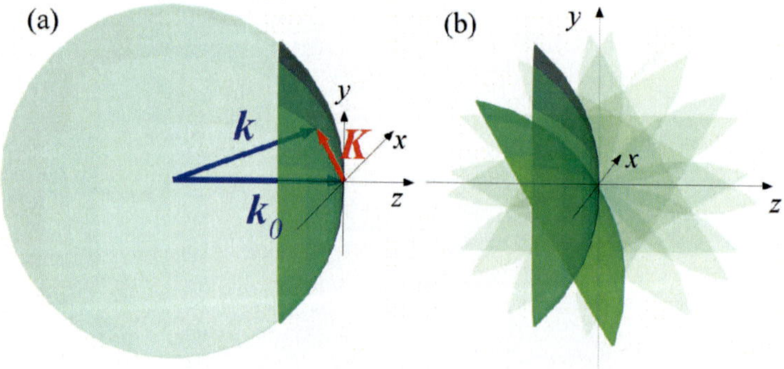

Figure 4.2 Regions in reciprocal space covered by coherent X-ray diffraction experiments. (a) A single diffraction measurement provides diffraction intensities in a near-forward region on the Ewald sphere. In measurements within a sufficiently small range of the solid angle, the measured part of the sphere may be approximated by a plane. (b) Diffraction data sets at a large number of different incident angles can cover nearly an isotropic region around the reciprocal space origin.

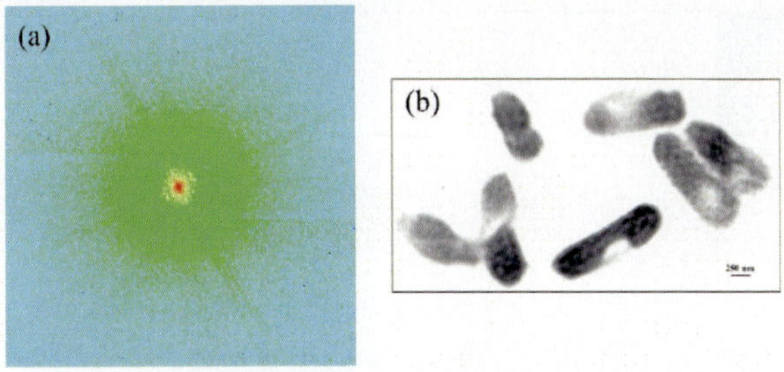

Figure 4.5 Observation of *Escherichia coli* bacteria in XDM. (a) The coherent diffraction pattern and (b) the reconstructed two-dimensional projection image are shown. The higher density regions inside the bacteria are likely the distribution of proteins labeled with $KMnO_4$. (Adapted from Miao *et al.* Imaging whole *E. coli* bacteria by using single-particle x-ray diffraction. *Proceedings of the National Academy of Sciences USA*, 100, pp. 110–112, © 2003, National Academy of Sciences, U.S.A.)

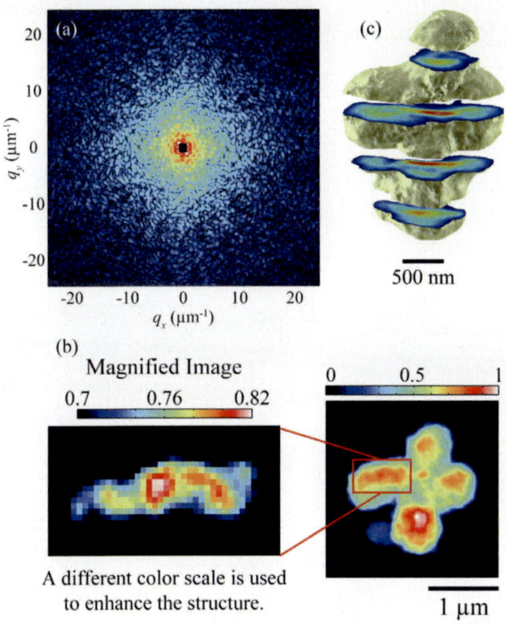

Figure 4.6 Observation of an unstained human chromosome in XDM. (a) The coherent diffraction pattern, (b) the two-dimensional reconstruction, and (c) the three-dimensional reconstruction are shown. In the magnified image of an arm region in (b), an axial wavy structure was observed first in an unstained chromosome. (Adapted from Nishino *et al.* Three-dimensional visualization of a human chromosome using coherent X-ray diffraction. *Physical Review Letters*, 102, art. no. 018101, © 2009, The American Physical Society.)

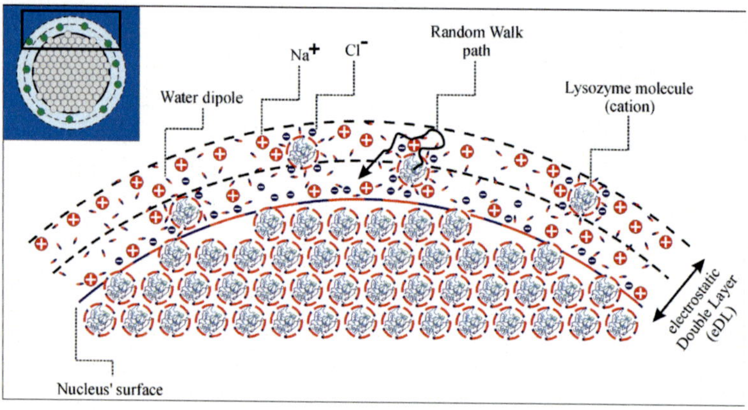

Figure 5.1 A zoomed piece of the spherical nucleus (upper left corner), with the electrical double layer surrounding it, the latter containing the most important electrostatic field emitting components (ions and dipoles); see also text for details.

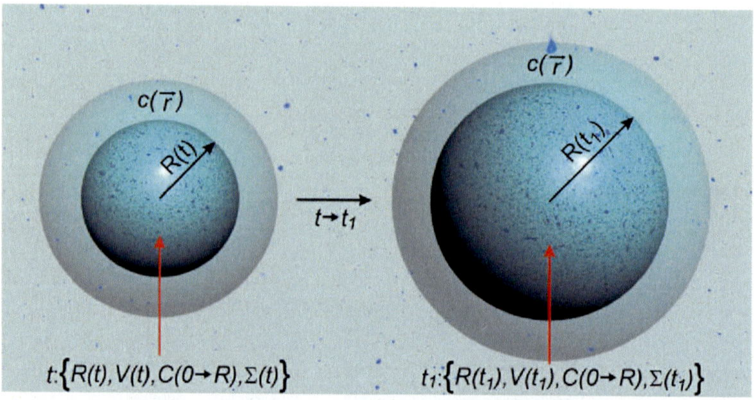

Figure 5.3 A growth of the (simplified) spherical nucleus in two consecutive time instants; for explanations of symbols used, see text, especially a sketch of the derivation of the deterministic growth rule ($0 \rightarrow R$ means: the crystal's radius range – interior of the crystal).

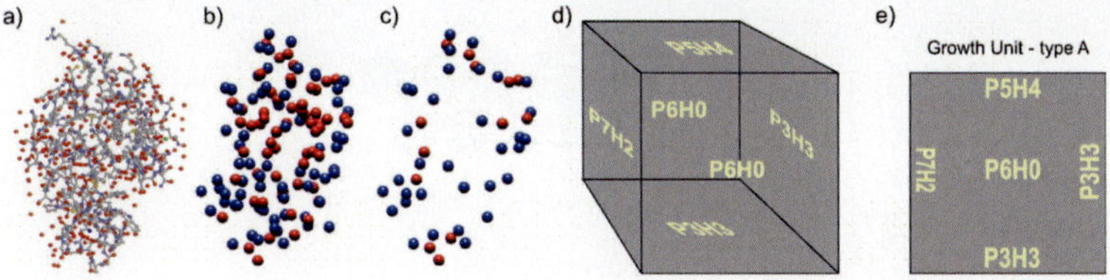

Figure 5.5 From the full atom representation of the lysozyme (PDB ID 193L) to the 2D growth unit: (a) Full atom representation of the lysozyme, (b) HP-amino acidic (hydrophobic–hydrophilic) representation — blue balls represent hydrophilic (P) whereas red balls represent hydrophobic (H) amino acids, (c) exterior part of the macromolecule, (d) 3D cubic representation. The net excess in a number of one of the monomer types determines the type of the wall (side) of the cubic box. (e) A 2D growth unit of A-type (reprinted with the permission from Siódmiak and Gadomski. Growing lysozyme crystals under variety of physicochemical conditions — a computer modelling. *Journal of Non-Crystalline Solids,* 354, pp. 4221–4226, © 2008, Elsevier).

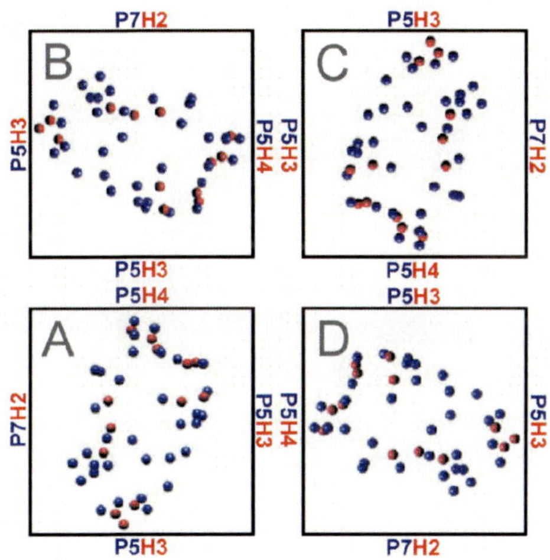

Figure 5.10 Non-Kossel-type (Chernov, 1997) ABCD unit cell (for growth unit type see text and Fig. 5.5c,e; growth units in ABCD unit cell: A — lower left, B — upper left, C — upper right, D — lower right). A hydrophobic core and also hydrophilic surface are visible (reprinted with the permission from Siódmiak and Gadomski. Growing lysozyme crystals under variety of physicochemical conditions — a computer modelling. *Journal of Non-Crystalline Solids,* 354, pp. 4221–4226, © 2008, Elsevier).

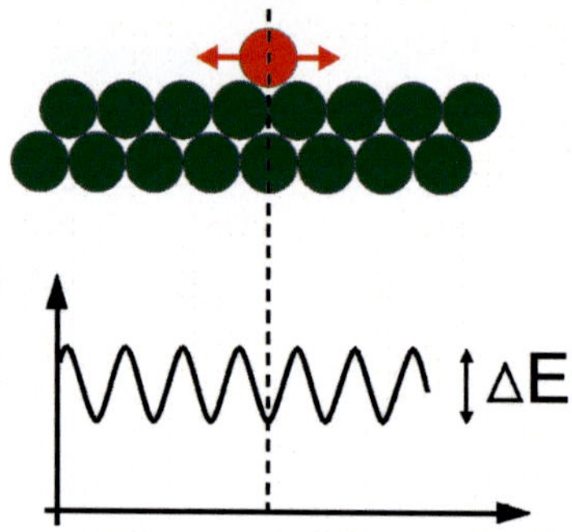

Figure 5.11 Energetic barrier between two neighboring points at the crystal's surface in the Frenkel–Kontorova (solid on solid) models. A commensurability of two neighboring crystal layers is revealed.

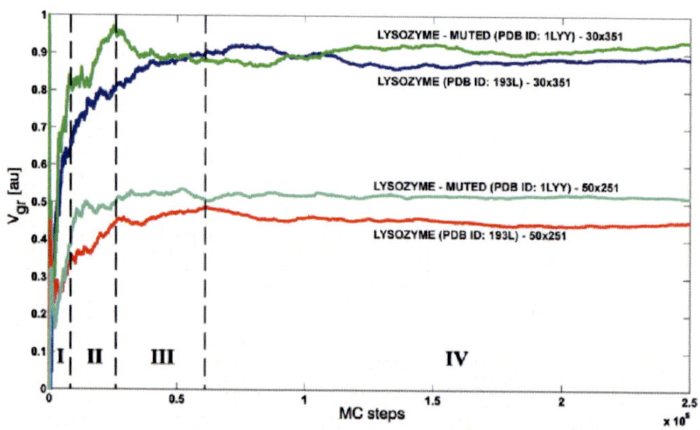

Figure 5.13 Certain most repeatable tendencies observed while examining the rate of the formation of mutant versus non-mutant lysozyme crystal, depending on the lattice size: $30(2R_M) \times 351(2R_M)$ and $50(2R_M) \times 251(2R_M)$ and temperature $T = 310$ K, *cf.* Table 5.1. Four regions are depicted on the plot: I — region of pre-nucleational effects, II — region of nucleus formation, III — region of non-stationary crystal growth, IV — region of stationary, nearly constant-tempo crystal growth. The as yet detected differences are mostly due to a finite lattice size effect. The curves have been obtained for the 6% protein concentration which is noticed to show optimal growing aggregation-oriented, thermodynamic–kinetic trends for both protein forms under study. The final values of the growth rates for the lattice size $30(2R_M) \times 351(2R_M)$ correspond to the appropriate values in Fig. 5.14 (reprinted with the permission from Siódmiak and Gadomski. Computer model of biopolymer crystal growth and aggregation by addition of macromolecular units. A comparative study. *International Journal of Modern Physics C*, 17, pp. 1037–1053, © 2006, World Scientific Publishing Company).

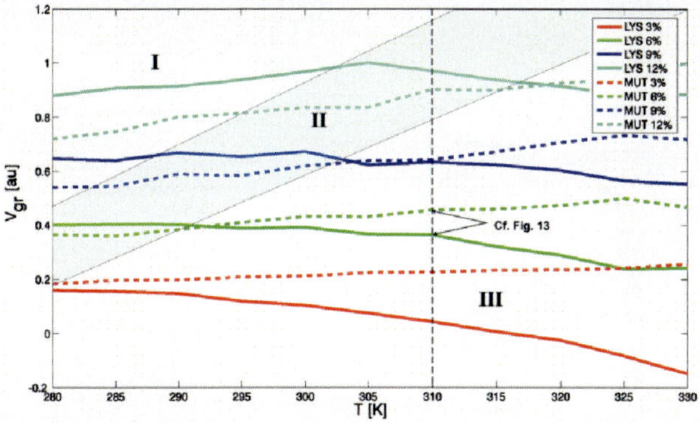

Figure 5.14 A comparative plot, revealing three regions of temperature behavior seen in terms of the tempo of the growing (poly)crystalline aggregates for the set of protein (non-mutant lysozyme) concentrations: 3, 6, 9, 12%. I — region below an optimal temperature for a given lysozyme concentration, II — crystallization window, III — region above an optimal temperature for a given lysozyme concentration. The data points for T = 310 K and concentration 6%, both for non-muted and muted lysozyme, correspond to the final values of the growth rates from Fig. 5.13 for lattice size $30(2R_M) \times 351(2R_M)$ (reprinted with the permission from Siódmiak and Gadomski, Computer model of biopolymer crystal growth and aggregation by addition of macromolecular units. A comparative study. *International Journal of Modern Physics C*, 17, pp. 1037–1053, © 2006, World Scientific Publishing Company).

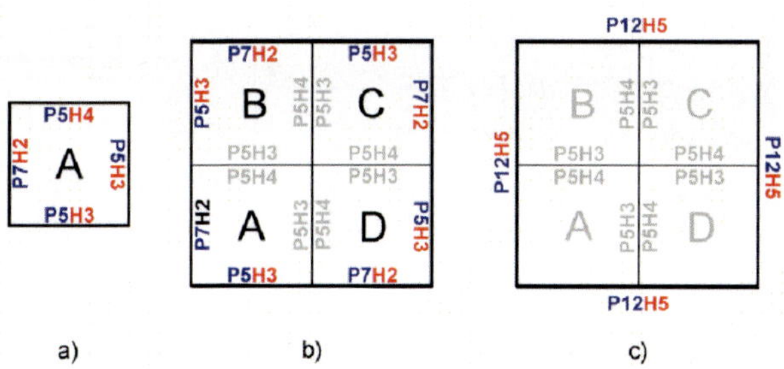

Figure 5.15 Different growth units: (a) monomeric growth unit of A-type, (b) energetically favorable (the lowest energy) configuration of four monomeric growth units, (c) generalized tetrameric growth unit. For details see Siódmiak *et al.* (2006). It can be seen that in the case of tetrameric growth unit (c) in comparison to monomeric growth unit (a) every side look the same what plays significant role in calculation of the movement probability (practically the movement is made without crossing of any energetic barrier because the geometrically smooth surface is also energetically smooth/flat) (Siódmiak *et al.* Computer model of a lysozyme crystal growth with/without nanotemplate — a comparison. *International Journal of Modern Physics C*, 17, pp. 1359–1366, © 2006, World Scientific Publishing Company).

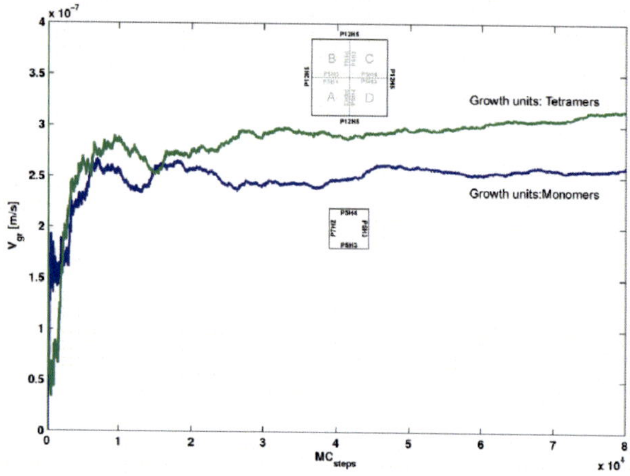

Figure 5.16 Simulated growth rate of lysozyme crystal obtained from the Langmuir–Blodgett–Schaefer-type template motivated technique (tetrameric growth units) — green (higher) line versus classical method (monomeric growth units) — blue (lower) line (Siódmiak *et al.* Computer model of a lysozyme crystal growth with/without nanotemplate — a comparison. *International Journal of Modern Physics C*, 17, pp. 1359–1366, © 2006, World Scientific Publishing Company).

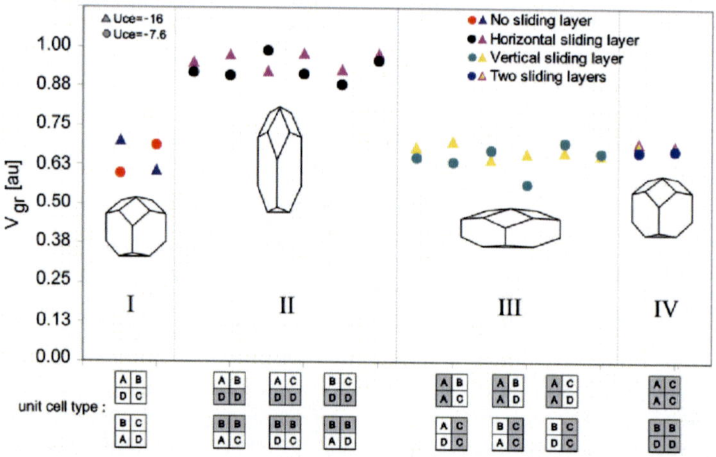

Figure 5.17 A (picturesque) diagram of the anisotropic growth. It can be seen that if the crystal is made up of unit cells which form the horizontal (or vertical) sliding layer (monolayer), group II (or group III), modeled crystal grows faster in a direction perpendicular to this layer and for a long-time period an elongated shape is observed. In the case when no sliding layer is formed, group I, or two horizontal or vertical sliding layers are formed, group IV, no elongated shape is admitted to the system.

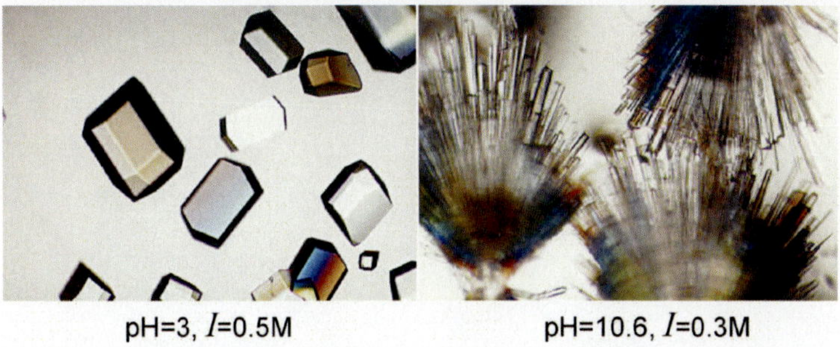

Figure 5.18 Extremely pH value and electrolyte concentration (ionic strength)-dependent shapes effect experimentally observed for the lysozyme crystals. For low values of the pH exactly 3D (rhomboid like) crystals are obtained, whereas for the high values of the pH significantly elongated (needle like) crystals are detected (reprinted with the permission from Velev *et al.* Protein interactions in solution characterized by light and neturon scattering. Comparison of lysozyme and chymotrypsinogen. *Biophysical Journal*, 75, 2682–2697, © 1998, Elsevier).

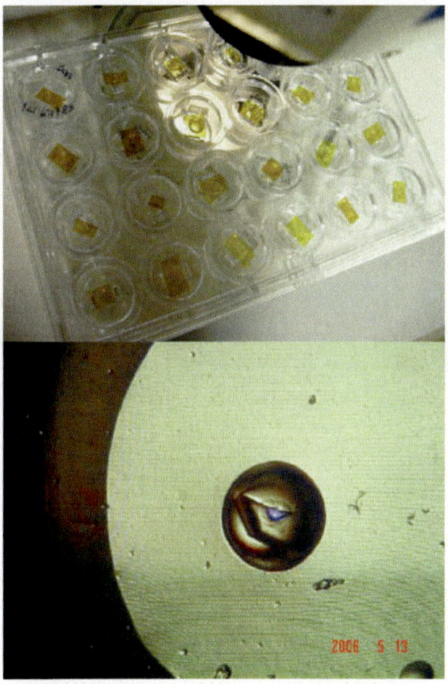

Figure 5.19 Lysozyme crystals grown out from orifices punched in kapton flat and porous membranes (Siwy *et al.*, 2003): top view, a setup consisting of 24 Limbro cells; bottom view, a magnified picture of a single crystal of lysozyme grown from the kapton membrane orifice (see, a paper "Could one effectively grow protein crystals: what comes out from theory and what from experiment?" at http://www.mischer-expo.de/achema/achema-2.htm, presented by A. Gadomski at Achema 2006, in Frankfurt on Main, and coauthored by C. B. Trame).

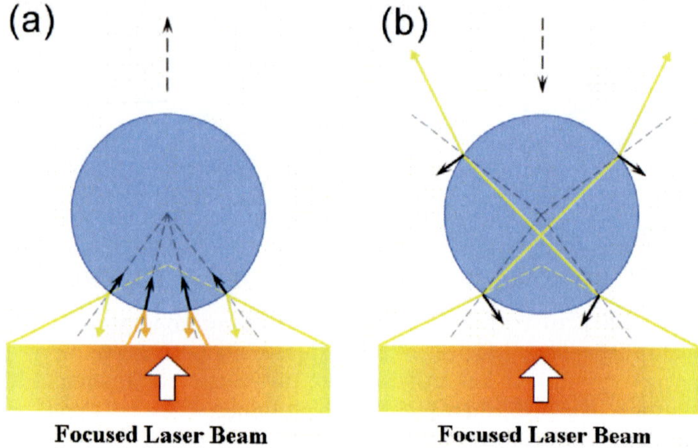

Figure 6.1 Schematic principle of optical trapping. A large particle of high refractive index is illuminated by a TEM_{00} Gaussian intensity profile laser beam focused by a high numerical aperture (NA) microscope objective. (a) The ideal case of a perfectly reflecting particle. Light rays (colored lines) are reflected at the surface. Dashed lines identify the normal directions. Change in light direction implies the transfer of linear momentum (black short arrows) to the particle. Due to the Gaussian profile, the particle is constrained along the laser axis and pushed in the propagation direction (long dashed arrow). (b) The ideal case of a perfectly transmitting particle. At every change of direction, linear momentum is transferred to the sphere (short black arrows). The high NA of the microscope objective makes the total momentum transfer oriented in the opposite verse of the light propagation (long dashed arrow). In practice, when 3D stable trapping is achieved both effects occur and balance one the other.

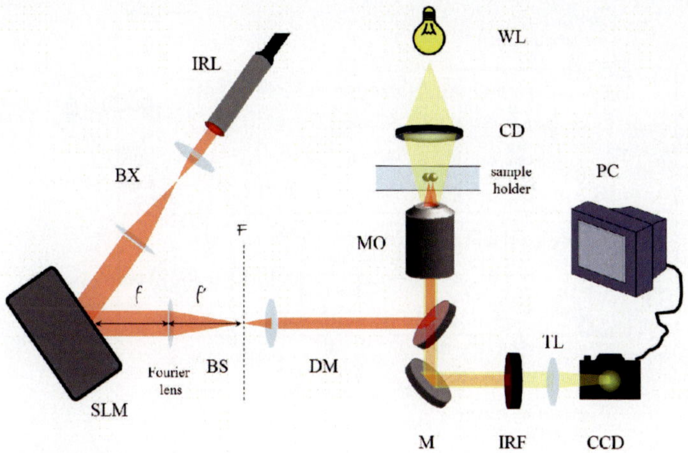

Figure 6.2 Scheme of an OT setup. In red: laser path; in yellow: optical imaging path. **Laser path**. The beam of a collimated infrared laser (IRL) is expanded by a telescopic couple of lenses (BX) and impinges onto a spatial light modulator (SLM). By means of diffractive optical elements, the SLM modulates the phase of the laser wavefront to split the beam and obtain the desired configuration in the Fourier plane (F). The modulated beam is transferred by the second lens of the telescope BS and the microscope objective (MO) in the focal plane of the MO (Dufresne and Grier, 1998). **Imaging path**. White light (WL) illuminates the sample through the condenser, passes through the dichroic mirror (DM), and is reflected by the mirror M onto a CCD through a tube lens (TL). To obtain clean images, a filter (IRF) blocks spurious IR reflections.

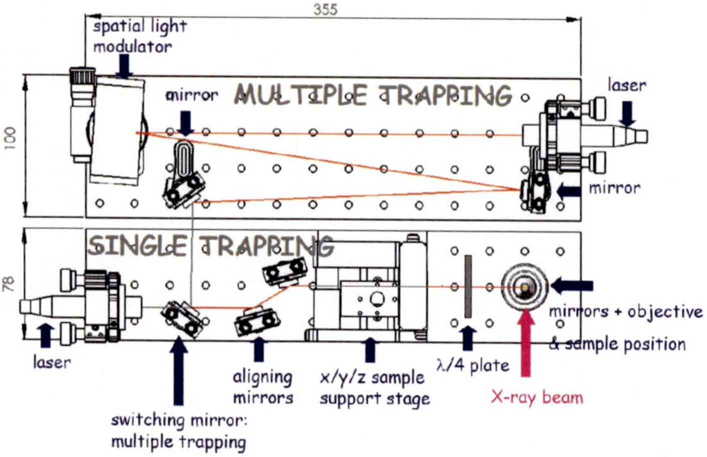

Figure 6.3 Top view of a project of the optical manipulator at the ESRF. Quotes are expressed in millimeters. The setup is organized in two separate modules, one for single trapping and rotation by means of a $\lambda/4$ plate of a single particle at a time, and an additive module for multiple trapping. The pink arrow corresponds to the incoming X-ray beam. The imaging path is not shown in this view.

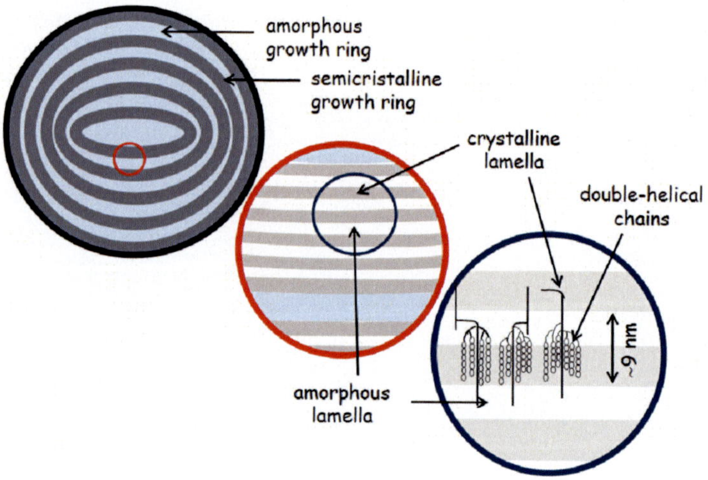

Figure 6.6 Hierarchical organization of starch granule (see text).

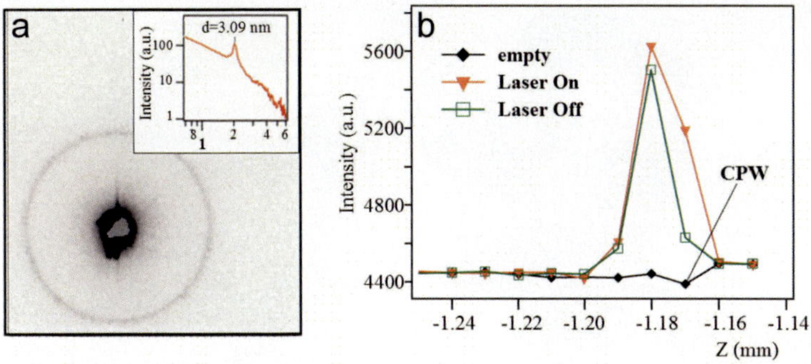

Figure 6.8 Scanning micro-SAXS of optically induced sedimentation of mesoporous aggregates. (a) Diffraction pattern and azimuthally integrated intensity of one aggregate showing the first-order of a hexagonal mesoporous structure. (b) The results of vertical micro-SAXS raster scan along the bottom of the capillary (integrated intensity of the diffraction pattern versus scanning position Z): before the trap was activated (empty), after the laser trap was switched on (laser on), after the trap was deactivated (laser off). CPW indicates roughly the position of the capillary wall.

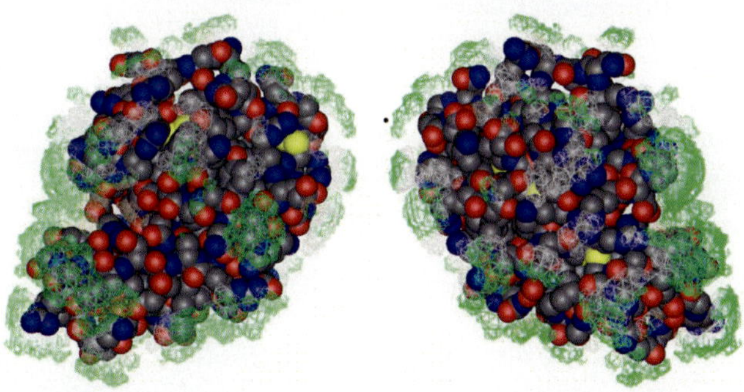

Figure 7.1 Hydrophobic patches surrounding the lysozyme molecule shown in CPK representation. The patches are colored according to electrostatic potential: dark gray, positive; light gray, negative; almost white, neutral. The two images differ by 180° rotation around the vertical axis.

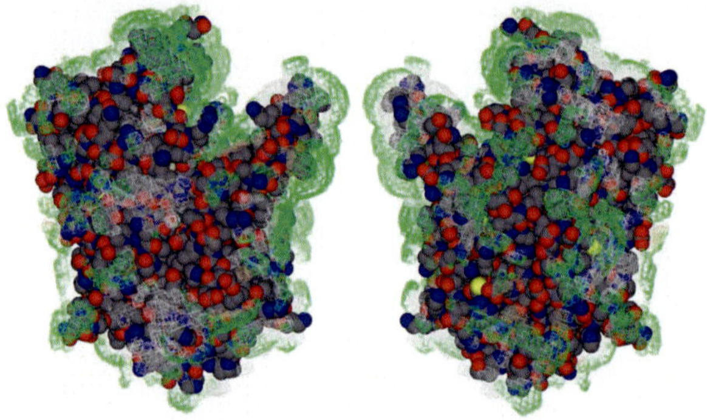

Figure 7.2 Hydrophobic parches around cytochrome P4502B4.

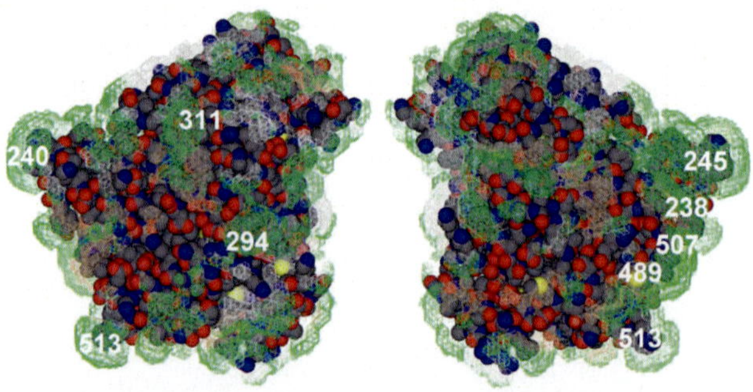

Figure 7.3 Hydrophobic patches around the molecule of cytochrome P450scc, according to the model based on cytochrome P450cam.

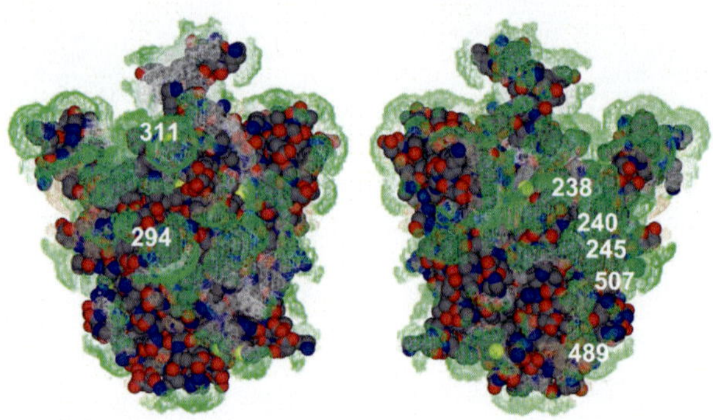

Figure 7.4 Hydrophobic patches around the molecule of cytochrome P450scc, according to the model based on cytochrome P4502B4.

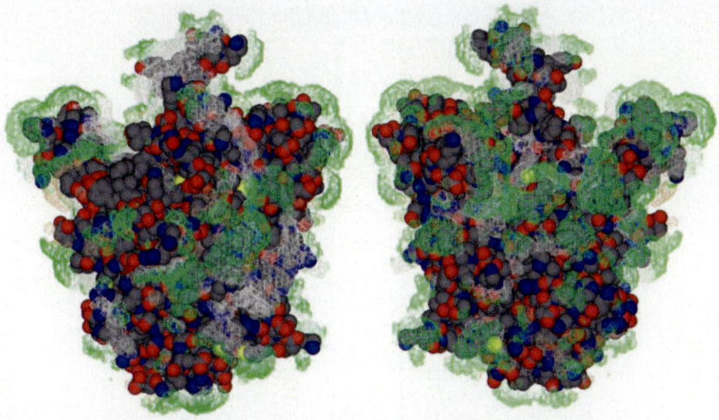

Figure 7.5 The molecule of cytochrome P450scc, according to the model based on cytochrome P4502B4 after suggested mutations.

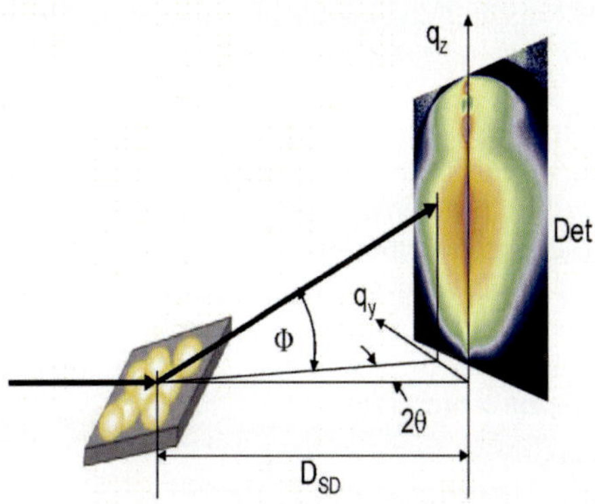

Figure 8.1 Geometry of the GIUSAXS experiment. $\Phi = \alpha_i + \alpha_f$ denotes the sum of incident and exit angle, 2θ is the scattering angle, and q_y, q_z denote the reciprocal space coordinate system. D_{SD} = 12.87 m is the sample-to-detector distance. Det denotes a two-dimensional detector.

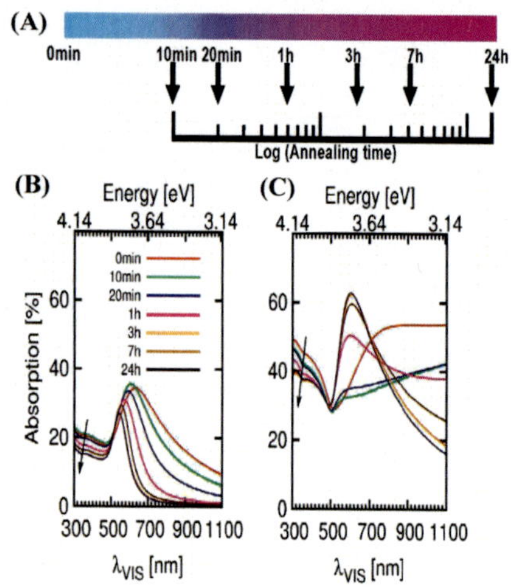

Figure 8.2 (a) Annealing times used (logarithmic scale) and pictorial representation of the color change during annealing. The color bar shows the transition from blue to red in the visible wavelength regime (red-shift) due to the break up of the coalesced gold layer during annealing for a deposited mass thickness of $d_3 = 8$ nm. (b) Optical absorption spectra for gold mass thickness $d_1 = 3$ nm: The spectra indicate a granular island structure. Upon annealing a shift of the maximum is observed, indicating a transition to a monodisperse size distribution. The arrow indicates the increasing annealing time. (c) Optical absorption spectra for gold mass thickness $d_2 =$ 8 nm: A clear transition from a coalescent, smooth gold layer to a granular film is observed. Included is an energy axis to indicate that energetic aspects are correlated with the structures.

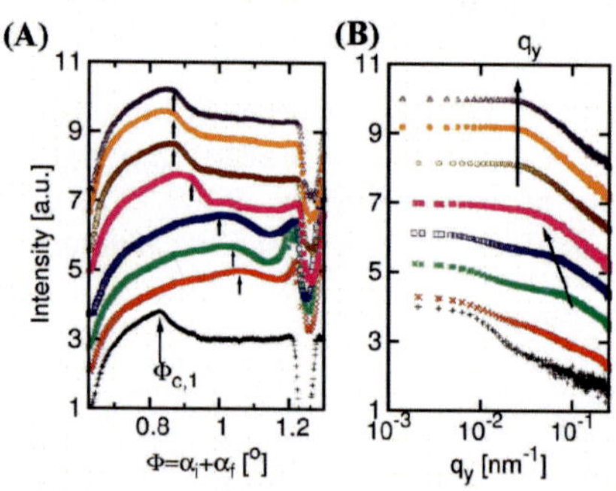

Figure 8.3 GIUSAXS data for $d_3 = 8$ nm; from bottom to top: glass, 0 min, 10 min, 20 min, 1 h, 3 h, 7 h, and 24 h annealing time. The curves are shifted from the bottom to the top along the intensity axis for better visibility. (a) Detector cuts for $q_y = 0$. $\Phi_{c,1}$ is the position of the Yoneda peak of glass. The small arrows indicate the evolution of the Yoneda peak of gold. (b) Out-of-plane cuts for at $\alpha_f = 0.43°$. The arrows indicate the evolution of the side maxima, from which ζ is deduced.

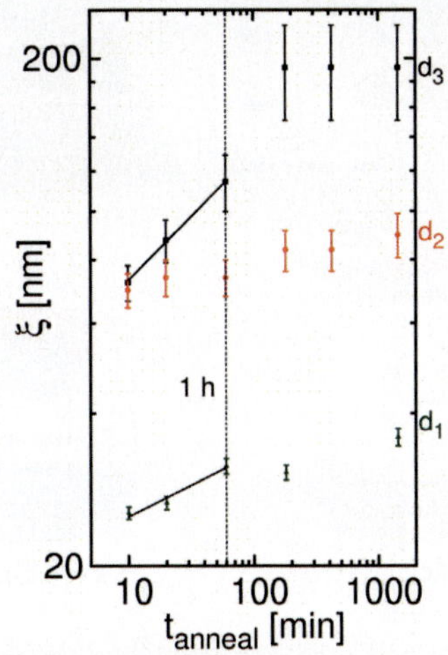

Figure 8.4 Evolution of the most prominent in-plane length ξ as a function of annealing time for different thicknesses d_1 = 3 nm, d_2 = 5 nm, d_3 = 8 nm. The solid lines correspond to power laws $\xi \propto t_{\text{anneal}}^{0.25\pm0.01}$ for d_3 and $\xi \propto t_{\text{anneal}}^{0.12\pm0.02}$ for d_1.

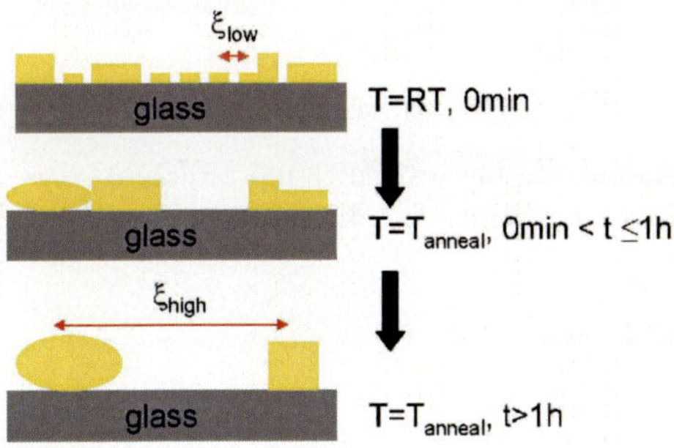

Figure 8.5 Model used for the GIUSAXS data evaluation of the structural evolution for d = 8 nm. ξ_{low} denotes the value for ξ at 0 min (before starting annealing), ξ_{high} the value for t > 1 h annealing.

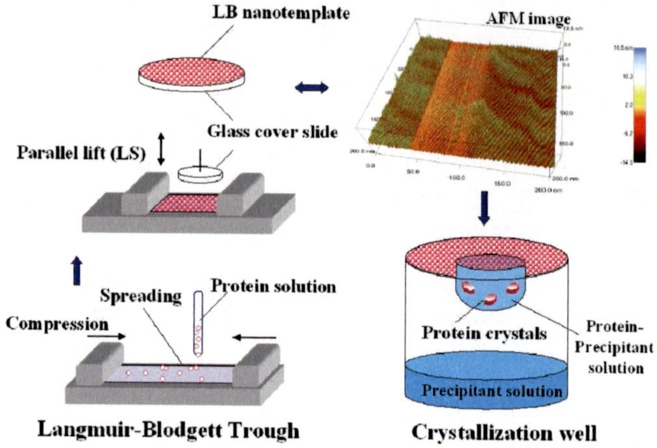

Figure 9.1 LB nanotemplate crystallization method.

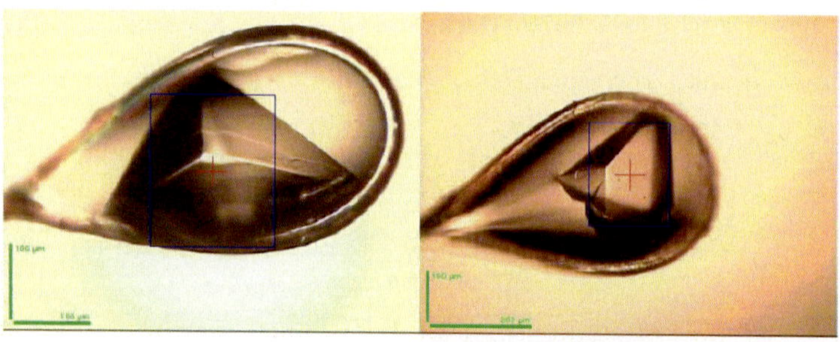

Figure 9.2 Proteinase K classical (left) and LB (right) crystal in the nylon loop in the cryogenic stream. The blue square contour indicates the ID 14-2 beam position.

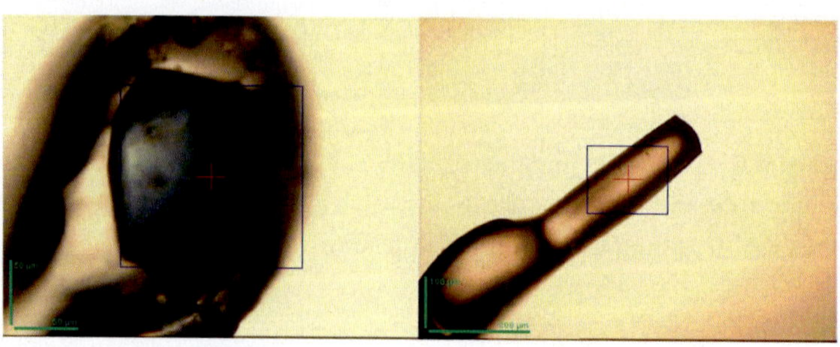

Figure 9.3 LB crystal of ribonuclease A (left) and thermolysin (right) mounted in the nylon loop in the cryogenic stream. The blue square contour indicates the ID14-2 beam position.

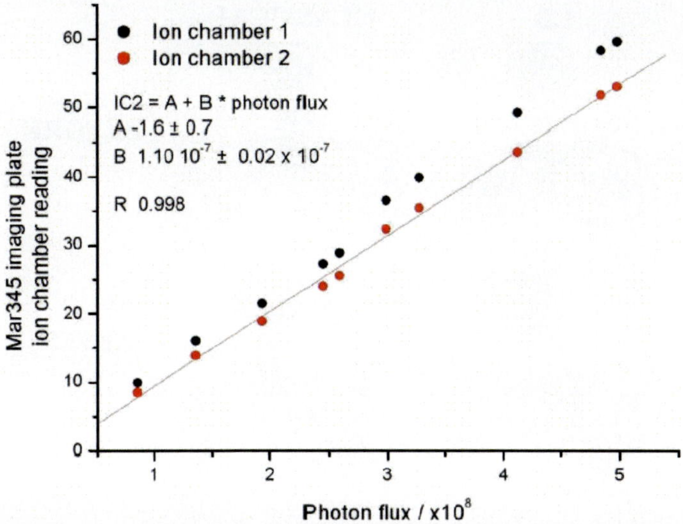

Figure 9.4 Example of ion camber calibration utilized by Garman group. This calibration was produced using a silicon photodiode (Hamamatsu, model S3204-, diode thickness of 500 μm), valid only for the generator used in T5, LMB, Oxford (not anymore in use) (data reproduced with permission of Elspeth Garman from http://lmb.bioch.ox.ac.uk/www/garman/lab_tools.html#guide9).

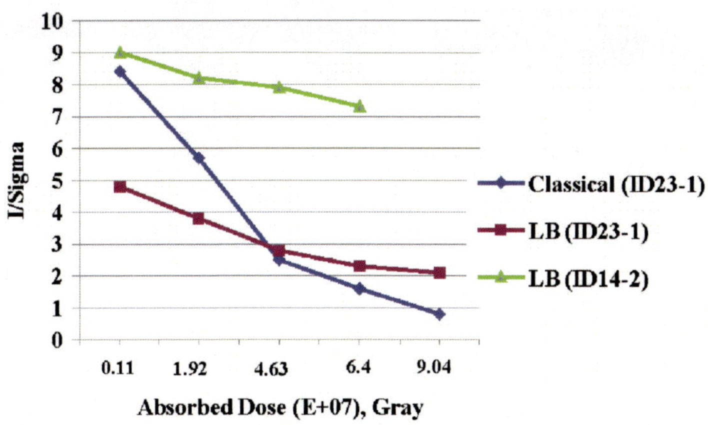

Figure 9.5 Intensity per sigma versus adsorbed dose for proteinase K LB and classical crystals at ID23-1 ESRF beamline.

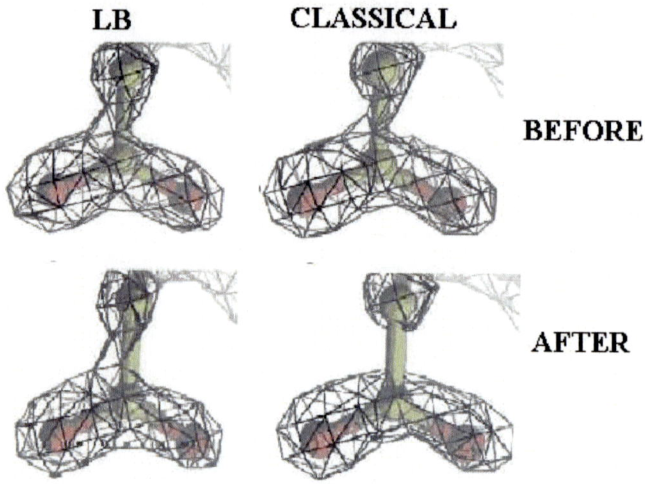

Figure 9.6 Comparison of electron density for carboxylic group of aspartic acid of LB and classical thaumatine crystal before and after high radiation dose exposure at ID 29 ESRF beamline.

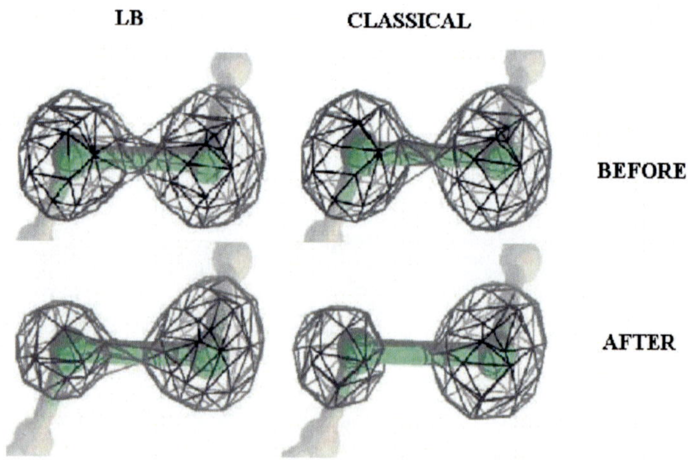

Figure 9.7 Comparison of electron density for the disulphide bond of 71 cysteine residue of LB and classical thaumatine crystal before and after high radiation dose exposure at ID 29 ESRF beamline.

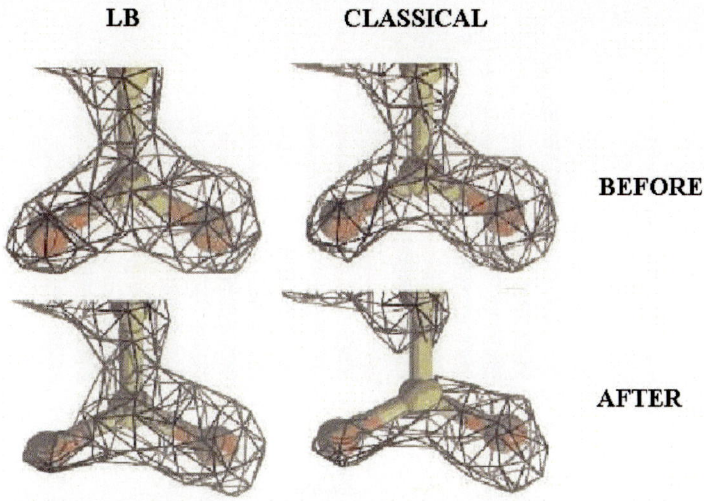

Figure 9.8 Comparison of electron density for carboxylic group of glutamic acid residue at 42 position in amino acid chain of LB and classical thaumatine crystal before and after high radiation dose exposure at ID 29 ESRF beamline.

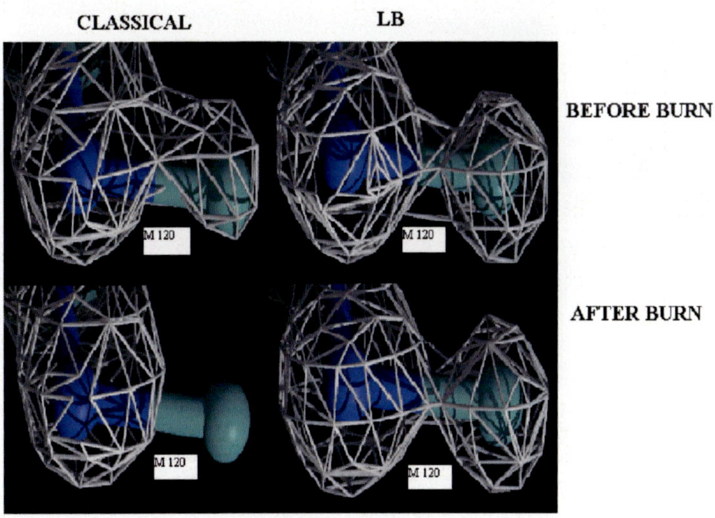

Figure 9.9 Comparison of electron density for the methyl group of methionine residue at 120 position in amino acid chain of LB and classical thermolysine crystal before and after high radiation dose exposure at ID 14-2 ESRF beamline.

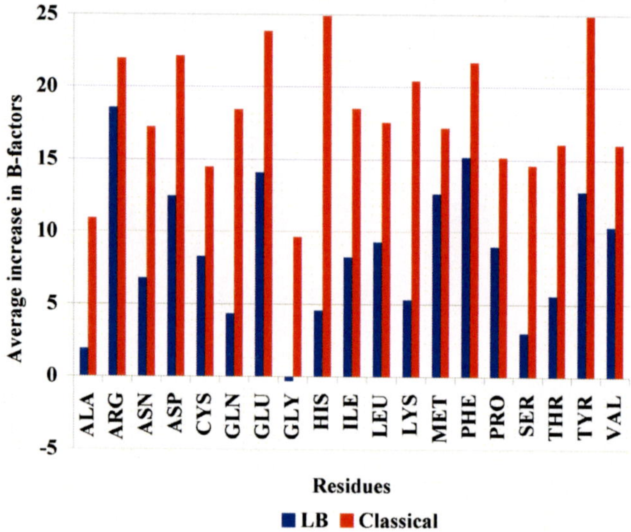

Figure 9.10 Increase in average B factors calculated for side chains of each residue for ribonuclease A. The effect is quite more evident using the relative B factor.

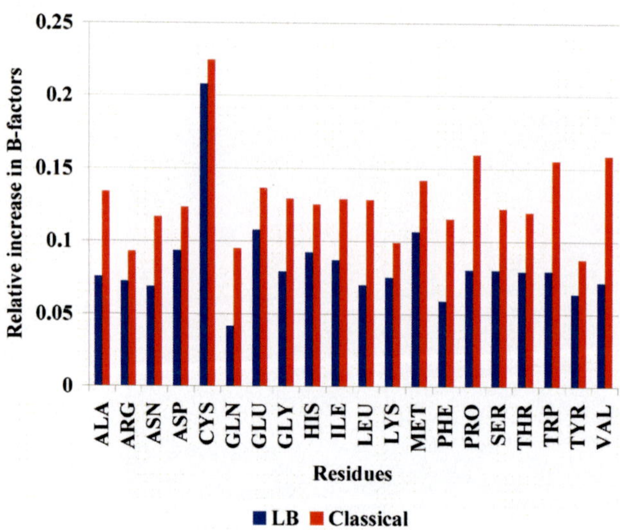

Figure 9.11 Relative B factors associated with side chains of the protein backbone for LB and classical crystal at same dose of 2.43 MGy, for proteinase K protein (ID14). For higher dose the difference is quite more striking (Pechkova *et al.*, 2009).

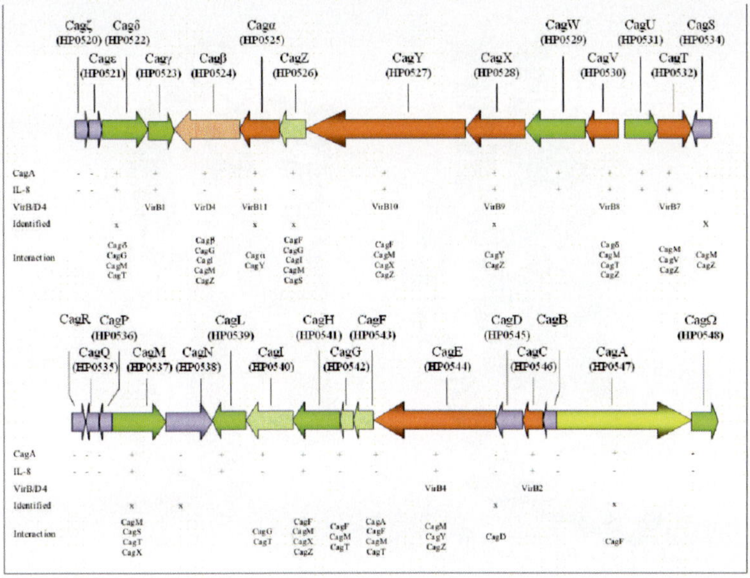

Figure 10.1 Graphical representation of the *cag*-PAI from *H. pylori* strain CCUG 17874 showing relative gene size and orientation. Most of the *cag* genes are probably involved in the assembly of the type IV secretion system that translocates the protein CagA into the gastric epithelial cells. Seven genes (marked in orange) show similarity to components of the VirB/D4 type IV secretion system of the plant pathogen *A. tumefaciens*. Proteins encoded by the island are involved in two major processes, the induction of interleukin-8 (IL-8) production by gastric epithelial cells and the translocation of CagA from the bacterium into host cells. All genes depicted by orange and green arrows are essential for CagA translocation, whereas lighter shades of red and green indicate genes that are not involved in this process but are however important for the IL-8 secretion. The purple arrows indicate genes that are required neither for translocation of CagA nor for IL-8 secretion. Based on mutagenesis study (Fischer *et al.*, 2001) the gene products essential for CagA delivery (Cag^{PY}) and IL-8 induction are indicated with "+," whereas the non-essential genes are indicated with "–." The proteins which have been detected *in vivo* by two-dimensional gel electrophoresis and mass spectrometry (MS) analysis (Busler *et al.*, 2006; Smith *et al.*, 2007) are indicated with "x." Interactions among *cag*-PAI encoded proteins have been detected using a yeast two-hybrid system (Busler *et al.*, 2006) and electron microscopy (Rohde *et al.*, 2003; Tanaka *et al.*, 2003).

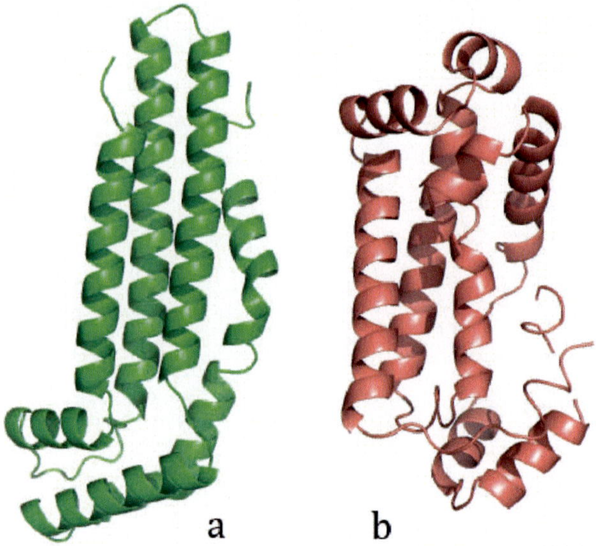

Figure 10.2 Cartoon view of the crystal structure of CagZ (a, left) and CagS (b, right).

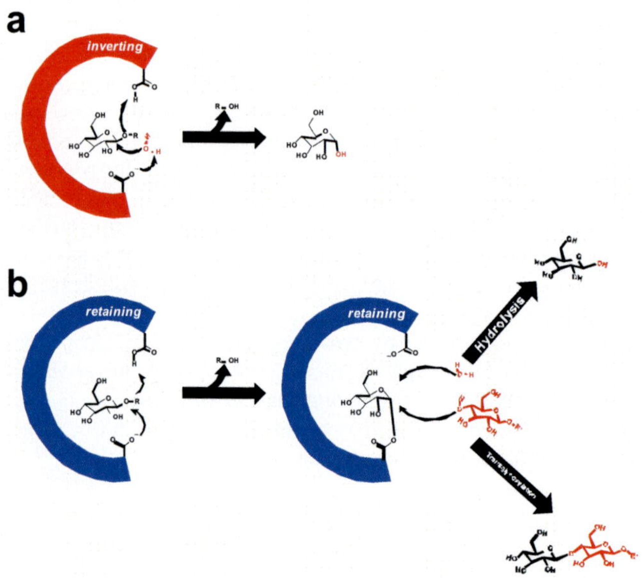

Figure 11.1 Glycoside hydrolases reaction mechanisms.

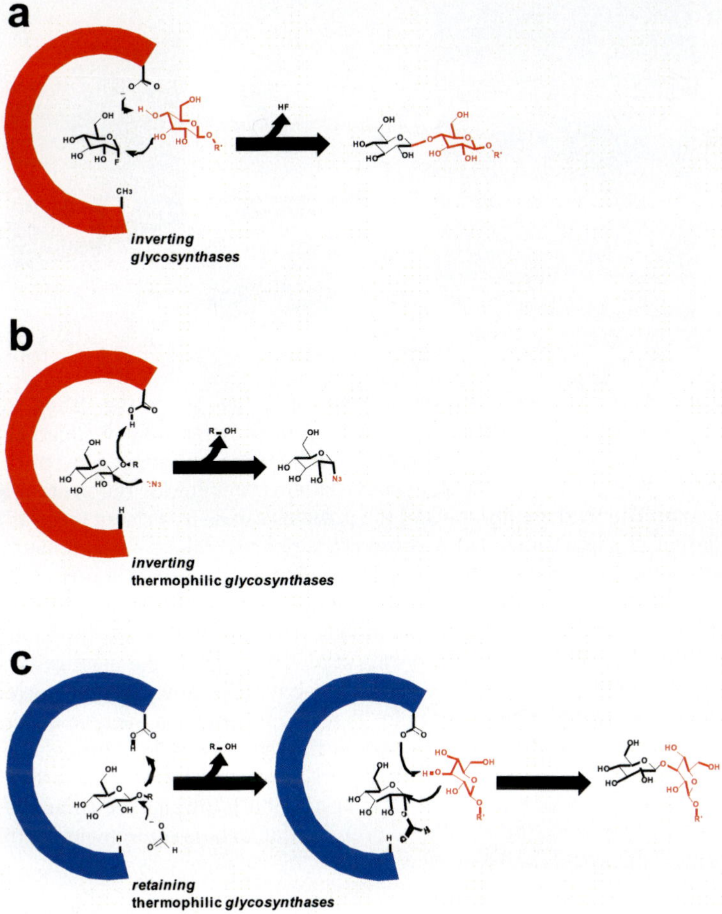

Figure 11.2 Proposed reaction mechanisms of *glycosynthases.*

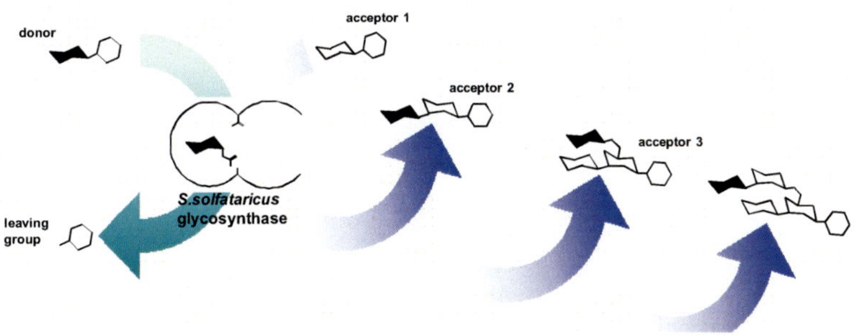

Figure 11.3 The "transglycosylation chain reaction" performed by thermophilic *glycosynthases* in the presence of formate ion as external nucleophile.

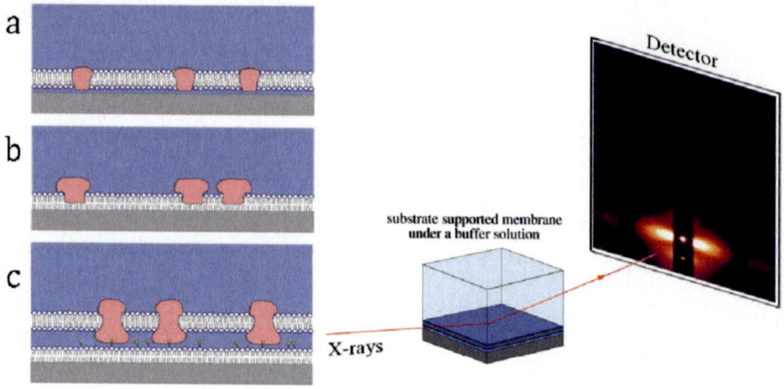

Figure 12.1 Schematic representation of X-ray scattering from 2D solution of membrane-associated proteins. The proteins are embedded in a substrate-supported lipid membrane, which in turn is submerged in a buffer solution that provides a suitable chemical environment for the protein to function. The lipid membrane and the proteins can be associated in several different configurations. (a) A substrate-supported bilayer is suitable for transmembrane proteins that do not extend beyond the thickness of the lipid bilayer. (b) Peripheral proteins that only penetrate into one leaflet of the bilayer can be studied in a lipid monolayer supported by a hydrophobic substrate (e.g., Plant, 1999; Scheuring *et al.*, 1999). (c) In the most general case, a complete bilayer is assembled above the substrate-supported monolayer. Transmembrane proteins that are embedded in the lipid bilayer are constrained to the monolayer by Ni-His tag linkers. This process has been used with Langmuir lipid monolayers to create 2D membrane protein crystals (e.g., Hasler *et al.*, 1998; Lévy *et al.*, 1999). His tags are commonly present in recombinant membrane proteins due to the requirement for the purification process and Ni-chelating lipids are commercially available.

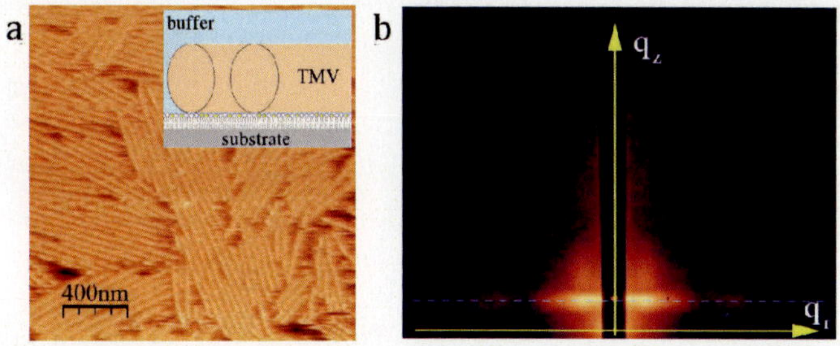

Figure 12.2 (a) Atomic force microscope image of TMVs organized on a substrate-supported lipid monolayer. Due to their long rod-like shape, TMVs lie parallel to the lipid membrane and also approximately parallel to each other within each loosely organized domain, as depicted in the inset of schematic side view. The sample configuration is similar to that in Fig. 12.1b. (b) GIXS data obtained from the same structure. A slight intensity maximum at low q_r suggests position correlation between the TMV rods. The in-plane intensity along the dashed line is analyzed in Fig. 12.3 (adapted from Yang *et al.* Structure and interaction in 2D assemblies of tobacco mosaic viruses. *Soft Matter*, 5, pp. 4951–4961, © 2009 by permission of The Royal Society of Chemistry).

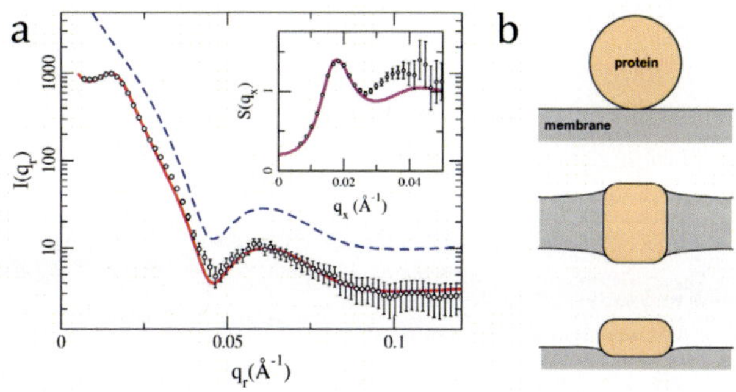

Figure 12.3 (a) Analysis of the experimental in-plane GIXS data (symbols) shown in Fig. 12.2. The fit (red line) includes contributions from the form factor (blue dashed line, offset for clarity) calculated from bulk solution scattering and the structure factor (magenta line in the inset) calculated based on screened Coulomb repulsion between two charged rods. The symbols in the inset show the experimental structure factor obtained by dividing the experimental data by the form factor. (adapted from Yang *et al.* Structure and interaction in 2D assemblies of tobacco mosaic viruses. *Soft Matter*, 5, pp. 4951–4961, © 2009 by permission of The Royal Society of Chemistry). (b) Different scenarios that may affect the calculation of the form factor. The form factor used here is calculated based on the assumption that the membrane does not contribute to the form factor (top). However, membrane distortions, such as those produced by hydrophobic matching (middle) or thinning (bottom) due to expansion of lipid headgroup, must be included in the calculation of the form factor for small proteins.

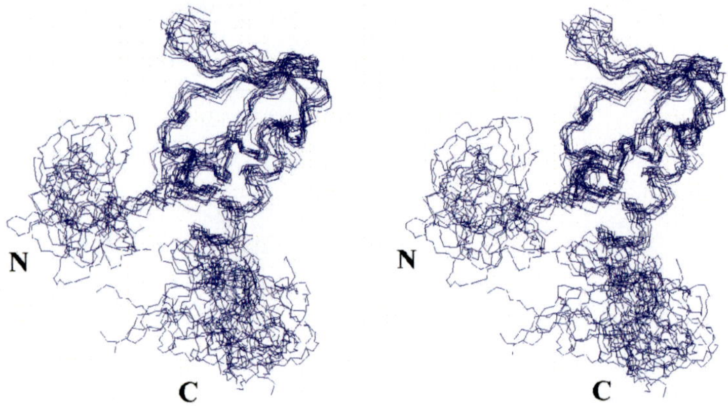

Figure 13.1 Stereoview of the best 10 structures obtained for the β-subunit of aIF2 from *S. solfataricus*. The N- and C-terminal domains result quite unstructured. (Reprinted with the permission from Vasile *et al.* Solution structure of the β-subunit of the translation initiation factor aIF2 from archaebacteria *Sulfolobus solfataricus*. *Proteins: Structure, Function, and Bioinformatics*, 70, pp. 1112–1115, © 2008 Wiley-Liss, Inc., A Wiley Company.)

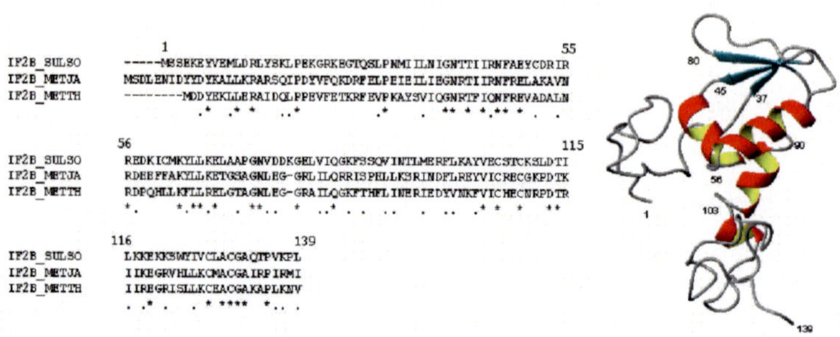

Figure 13.2 (Left) Alignment of the β-aIF2 from *S. solfataricus* (IF2B_SULSO) with its homologues from *M. thermoautotrophicum* (IF2B_METTH) and *M. Jannaschii* (IF2B_METJA). Above the IF2B_SULSO sequence, the secondary structure elements are indicated. (Right) The more representative structure (with lowest energy) was evidenced to show the secondary structure elements (the four alpha helices and the three beta strands). (Reprinted with the permission from Vasile *et al.* Solution structure of the β-subunit of the translation initiation factor aIF2 from archaebacteria *Sulfolobus solfataricus*. *Proteins: Structure, Function, and Bioinformatics*, 70, pp. 1112–1115, © 2008 Wiley-Liss, Inc., A Wiley Company.)

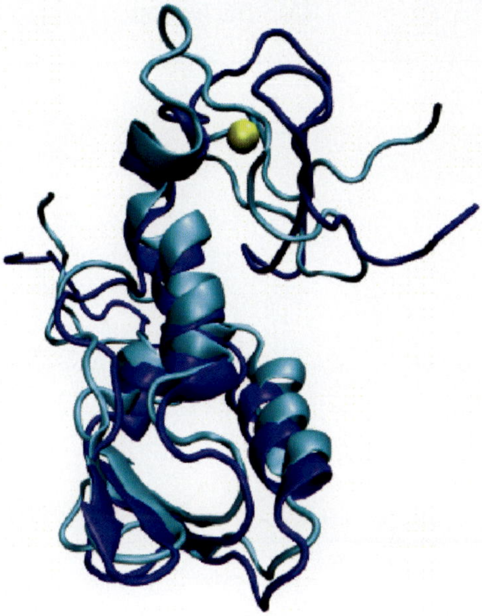

Figure 13.3 Superimposition of the NMR structure of β-aIF2 from *Sulfolobus solfataricus* (light blue) and the model obtained for the same protein with Zn ion.

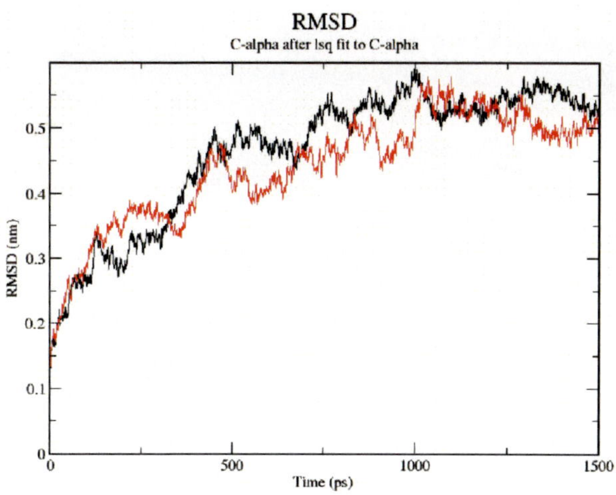

Figure 13.4 RMSD analysis of the *apo* (black curve) and *holo* form (red curve) with respect to their initial structures obtained by molecular dynamic simulations of 1500 ps.

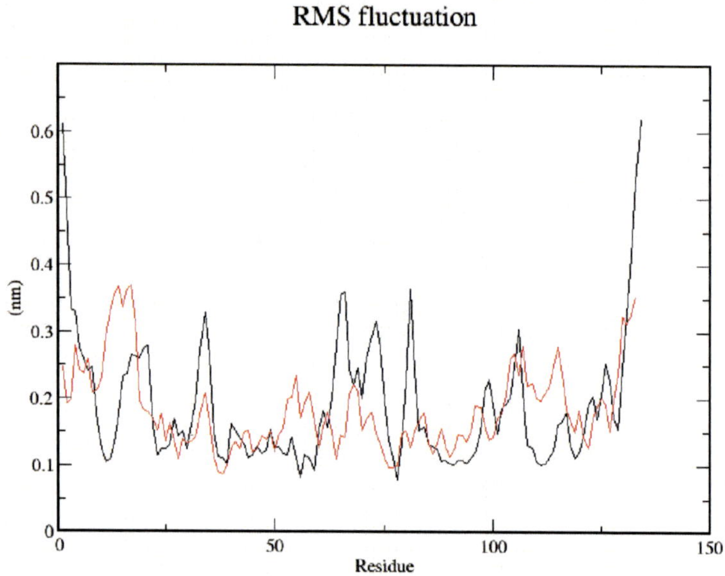

Figure 13.5 RMS fluctuation per residues of the *apo* (black curve) and *holo* form (red curve) obtained by molecular dynamic simulations of 1500 ps.

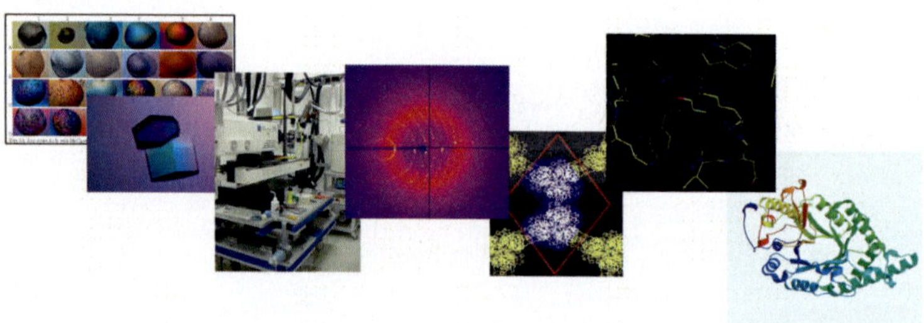

Figure 14.1 From crystal to structure.

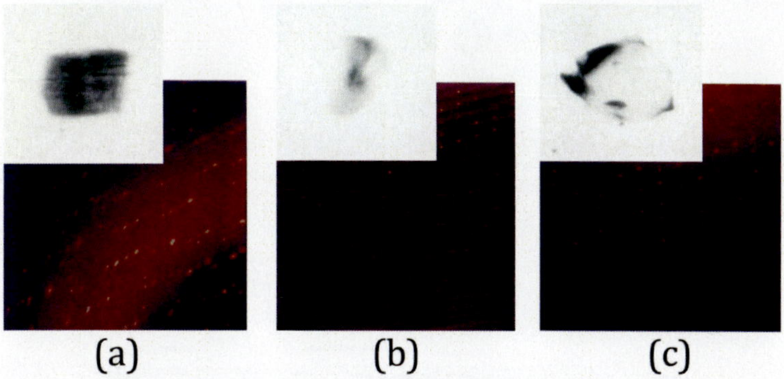

Figure 14.2 The quality of the 3D structure depends on the quality of the crystals. The diffraction pattern and corresponding X-ray diffraction topographs of three hen egg white lysozyme crystals grown by the temperature control method (Rosenberger *et al.*, 1988). (a) Temperature was cycled at 1 °C/day between 11 and 14 °C, crystal diffracted only to 2.5 Å. (b) Constant temperature set at 14 °C, crystal diffracted only to 1.8 Å. (c) Constant temperature 11 °C, crystal diffracted to 1.2 Å.

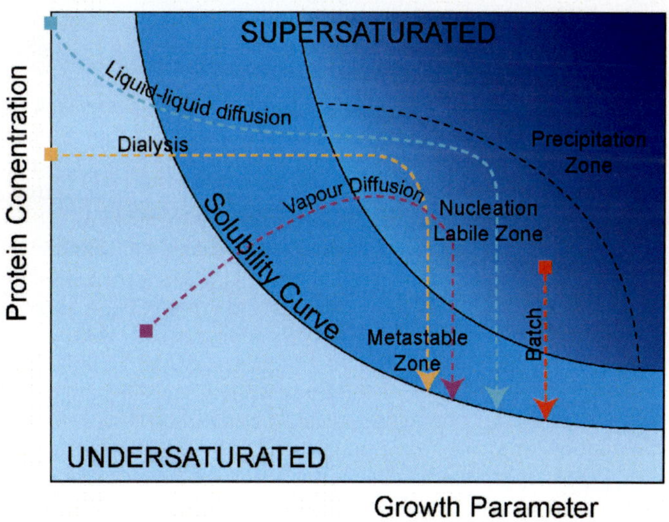

Figure 14.3 Phase diagram (cartoon). The horizontal axis can represent any number of parameters such as precipitant concentration, pH, temperature, etc. In the figure different crystallization methods are represented assuming precipitant concentration as a variable. Along the different pathways physical and chemical processes such as nucleation and growth occur (adapted from Chayen, 1998).

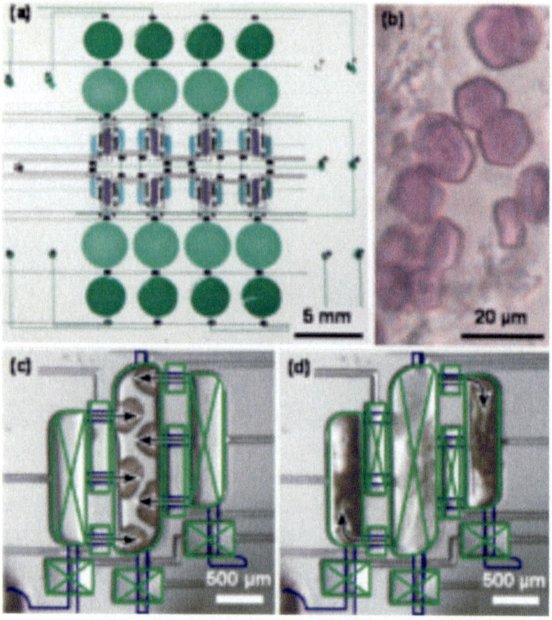

Figure 14.4 Microfluidic chip: (a) An optical micrograph of microfluidic chip containing a 4 × 2 array of devices capable of preparing crystallization trials in parallel using the lipidic cubic phase method at a volume of > 20 nl. Preparation of the cubic mesophase is done in the three chambered mixing portion of the device (blue and purple). Crystallization occurs in a separate crystallization chamber (light green) where the protein-containing mesophase is contacted with a precipitant solution that is introduced from a separate circular chamber (dark green). A single set of valve lines is used to control all of the mixing and crystallization units simultaneously. (b) Optical micrograph of bacteriorhodopsin crystals grown on chip via the lipidic cubic phase method. (c–d) Optical micrographs on an aqueous 13.5 mg/ml bacteriorhodopsin solution being mixed with the lipid monoolein. The blue lines delineate the edges of the fluidic channels containing the protein and lipid solutions while green lines highlight the location of microfluidic pneumatic valves. (c) Straight-line injection of the protein solution from the side chambers into the lipid-containing center chamber. (d) Chamber-to-chamber injection of the fluid mixture through different sets of inlets to create a net circulatory motion. (Courtesy: Sarah Perry [Perry *et al.*, 2009]).

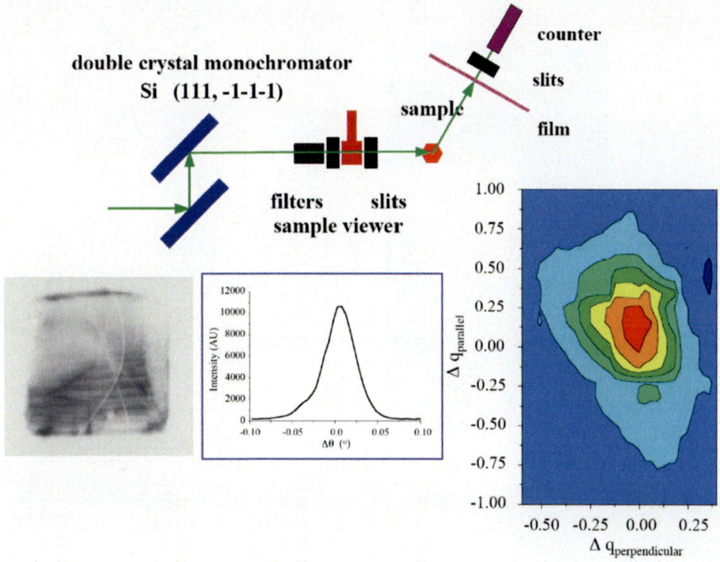

Figure 14.6 High-resolution diffraction imaging. Top: Scheme of the experimental layout at ID19 at the European Synchrotron Radiation Facility; after the identification of the reflection spot by the oscillation method, the 2D detector is removed and a film is set at the proper angle to record the X-ray topography image. The film is then removed to allow the measurement of the rocking curve and reciprocal space mapping (courtesy of J. Härtwig). Bottom: High-resolution diffraction image of reflection (0 14 −22) from a thaumatin crystal grown by the batch method in gel media at room temperature. From left to right: X-ray diffraction topography, dislocations (vertical lines at the bottom) and growth sectors are present in the image; rocking curve showing extended shoulders for angles smaller than the Bragg peak; the rather small angular misorientations (mosaicity) is also observed in the reciprocal space map ($\Delta q_{perpendicular}$) the extensive variation observed along the scattering vector ($\Delta q_{parallel}$) are a reflection of the strain field, and consequent interplanar (d-spacing) spacing variations result of defect formations such as the dislocations observed at the bottom of the X-ray topography image.

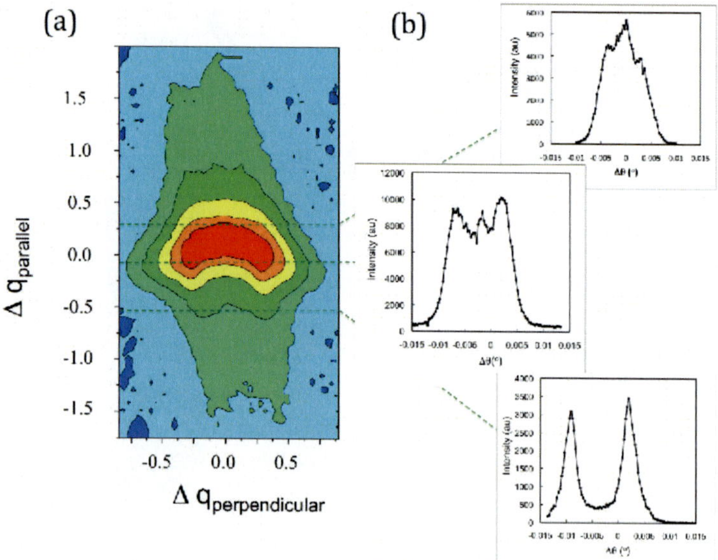

Figure 14.7 High-resolution diffraction images of reflection (15 15 0) of a hen egg white lysozyme crystal grown by the batch method in gel media at room temperature in supersaturation conditions. (a) Reciprocal space map, both interplanar spacing variations ($\Delta q_{\text{parallel}}$, i.e., d-spacing variations), and orientational changes $\Delta q_{\text{perpendicular}}$, i.e., $\Delta\theta$) are observed. (b) The rocking curves were collected at different positions along the scattering vector showing the differentiated angular misorientation in the crystal volume.

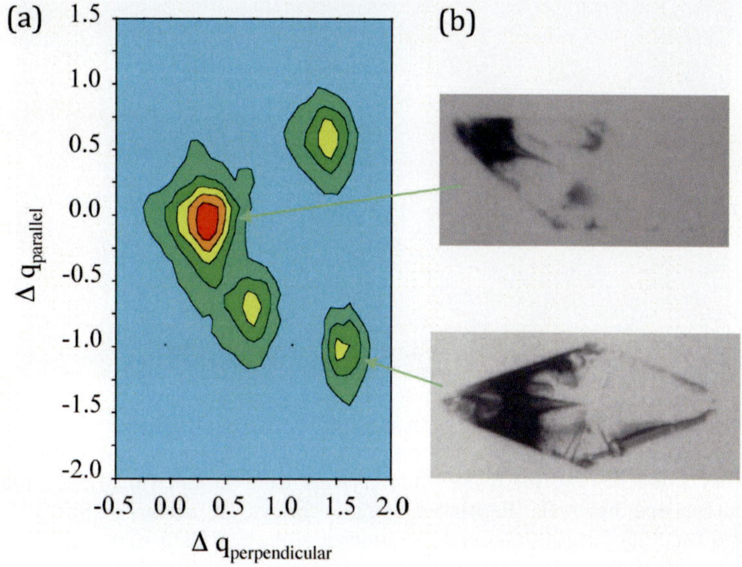

Figure 14.8 High-resolution diffraction image from a thaumatin crystal grown by the batch method at room temperature in the presence of agarose gel media. The *c*-axis is horizontal and parallel to the plane of the figure so is the *a*-axis. (a) Reciprocal space map of reflection (17–1 0), the four distinct peaks are result of differentiated lattice spacing ($\Delta q_{\text{parallel}}$) and orientational, angular ($\Delta q_{\text{perpendicular}}$) variations found in the crystal. (b) The rhombohedral outline of the diffraction topography images correspond to the projection of the crystal form in the scattering direction. The two images were recorded at different positions in reciprocal space. It is worth noting the straight lines at the bottom of the images moving outwards from the centre of the crystal toward the crystal faces (Stojanoff *et al.*, 2003).

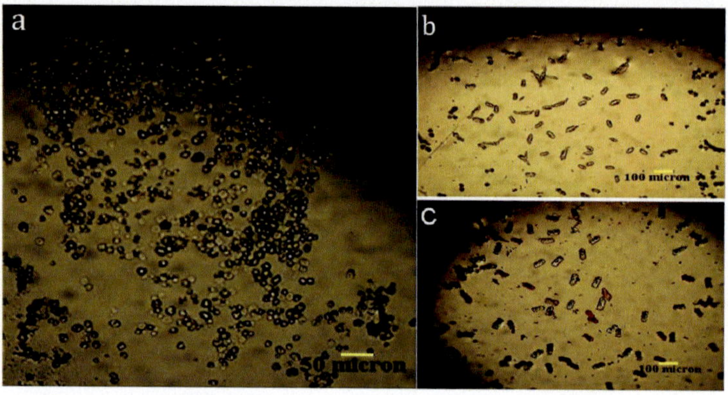

Figure 15.1 (a) aIFα microcrystals from archaebacteria Sulfolobus solfataricus obtained by LB nanotemplate; (b) aIF2β microcrystals from same source; (c) Groel microcrystals from *E. Coli* (polarized light microscope images). (Reprinted from Pechkova, E., Vasile, F., Spera, R. and Nicolini, C. (2008a). Crystallization of alpha and beta subunits of IF2 translation initiation factor from archabacteria *Sulfolobus solfataricus*. *Journal of Crystal Growth*, 310, pp. 3767–3770, © 2008 with the permission from Elsevier.)

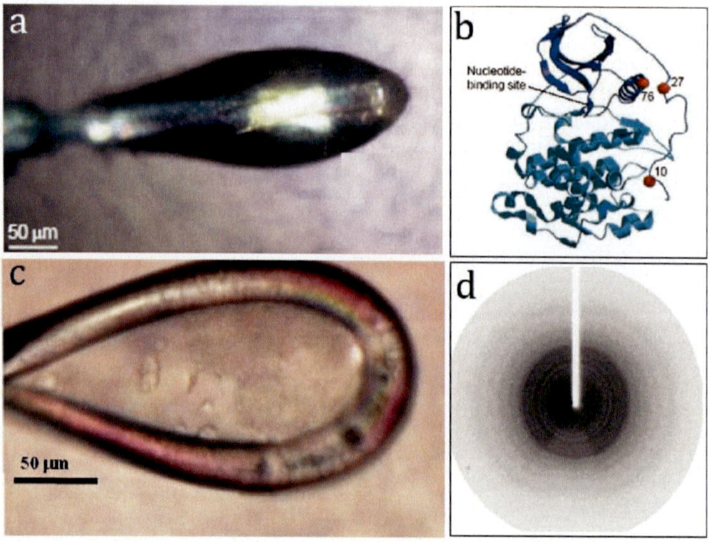

Figure 15.2 (a) Single crystal of human kinase CK2α in cryoloop. (b) Atomic structure of human kinase CK2α. (c) Microcrystals of P450scc protein. (d) P450scc powder pattern.

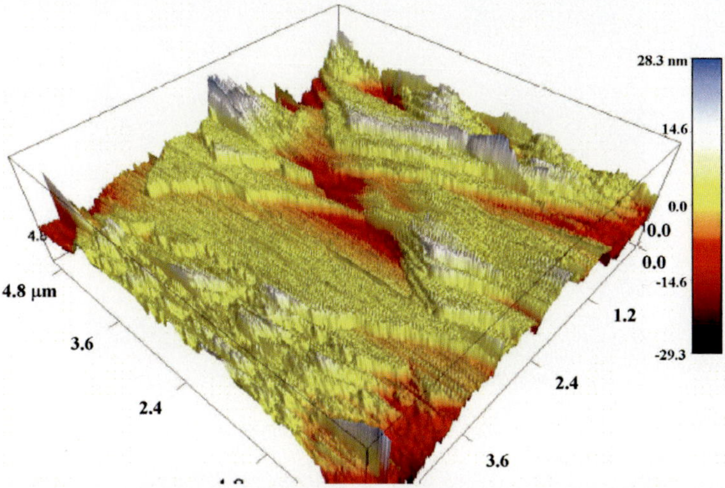

Figure 15.4 AFM image of lysozyme crystal in solution prepared according to standard crystallization techniques at the 5 μm scale (Nicolini and Pechkova, 2010).

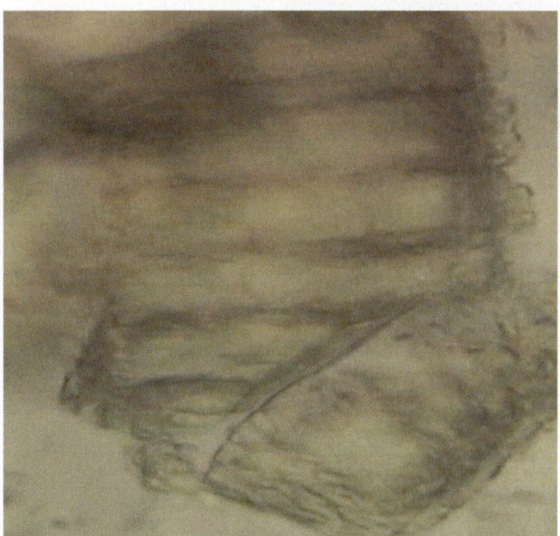

Figure 15.5 Domains in LB crystals induced by laser-microdissection (Nicolini *et al.*, unpublished).

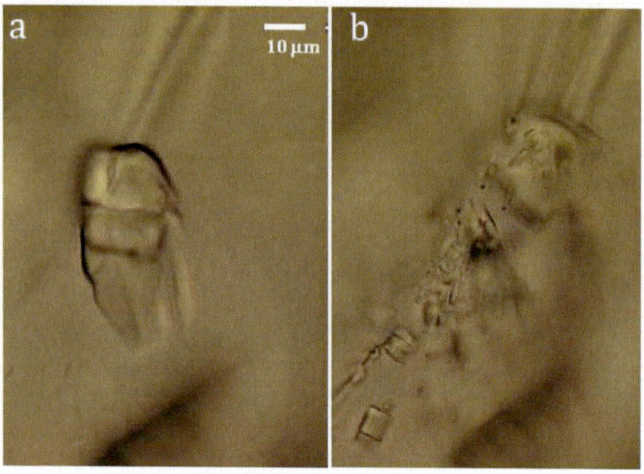

Figure 15.6 LB lysozyme crystal in a nylon cryoloop before (a) and after (b) laser irradiation.

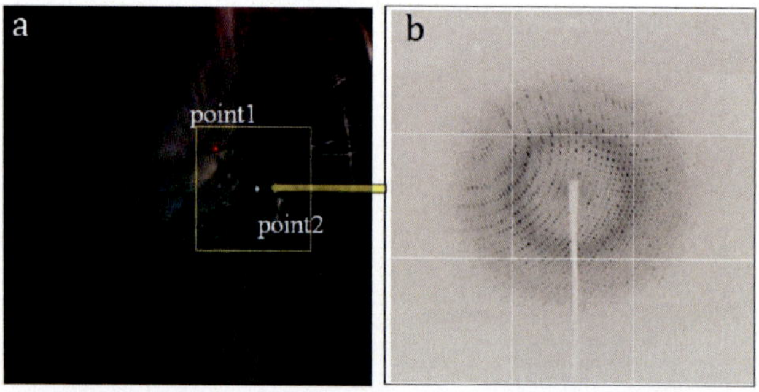

Figure 15.7 (a) Optical image of laser-microdissected lysozyme crystal shown in Fig. 15.6b after cryofreezing in nylon loop. A two-dimensional raster scan was performed in the yellow rectangle with a 0.5 μm beam around point 2. (b) Typical diffraction pattern recorded at point 1.

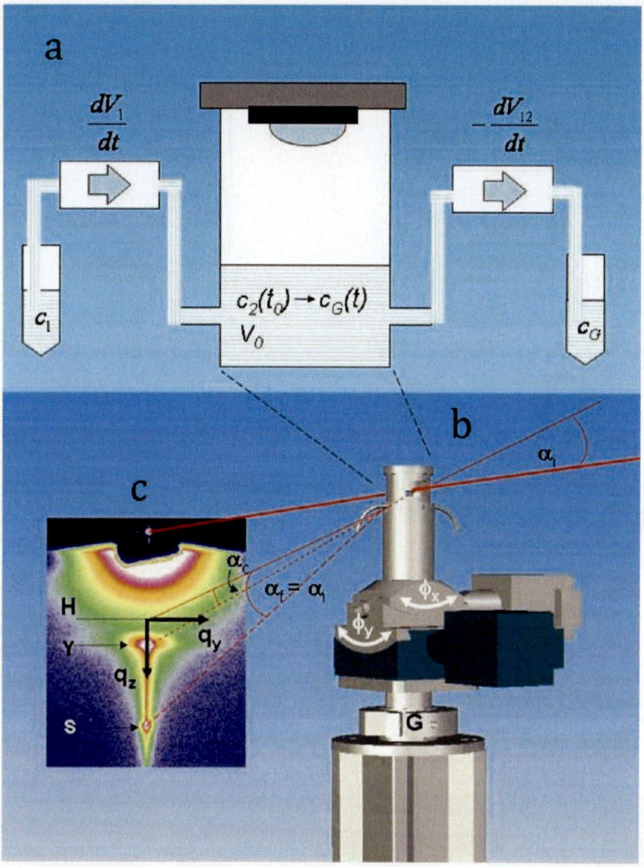

Figure 15.8 (a) Schematic design of in situ cell for μGISAXS experiments (Pechkova and Nicolini, 2010). The initial amount of salt in the reservoir with volume V_0 is m_0 at $t = 0$. dV_1/dt denotes for the incoming volume flux with the density of salt ρ_1. The absolute value of the outgoing flux is also dV_1/dt. (b) Schematic design of the installation of the μGISAXS cell on the two orthogonal *x*/*y* inclination stages which are fixed to the supporting element via a clamp G. An *x*/*y*/*z* stage attached to the supporting element (not shown) allows positioning of a specific point of the cell in the beam or provides raster-capabilities (Pechkova and Nicolini, 2002). (c) Typical μGISAXS pattern; *H*: height of the direct beam; *Y*: Yoneda peak at the critical angle $\alpha_c = 0.11°$; *S*: specular peak at $\alpha_f = \alpha_i$ with $\alpha_i = 0.71°$. (Reprinted with the permission from Pechkova *et al.* In situ GISAXS: I. Experimental setup for submicron study of protein nucleation and growth, *Biophysical Journal*, in press, doi:10.1016/j.bpj.2010.03.069, © 2010 Cell Press.)

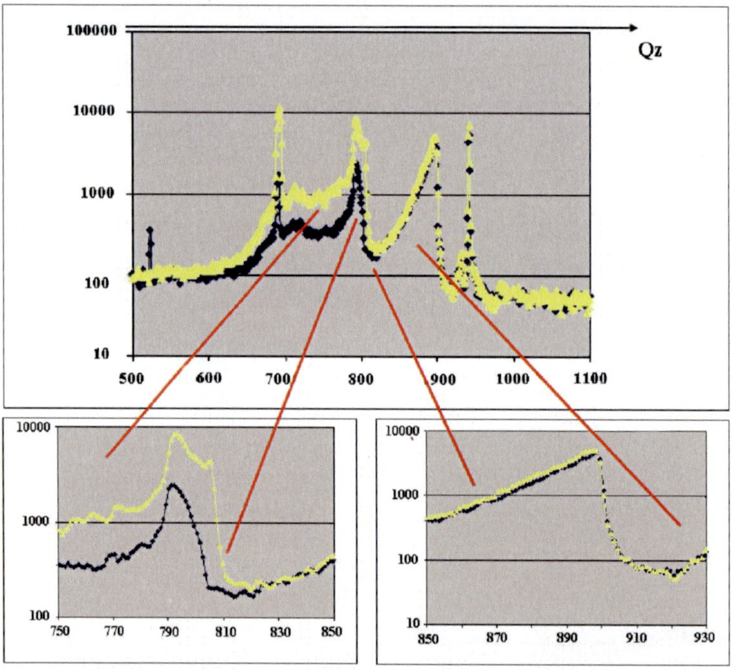

Figure 15.9 Detector-cut of GISAXS of thaumatin LB crystal.

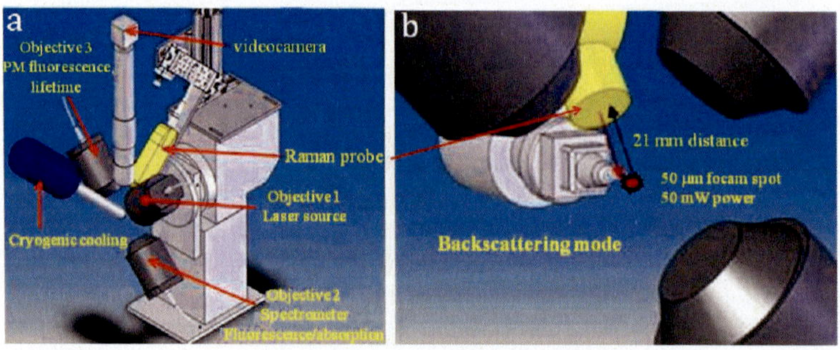

Figure 15.11 (a) Schematic design of IBS cryobench setup at ESRF for microspectroscopy and Raman spectroscopy. (b) Zoom of sample environment showing space available for lysozyme crystal nucleation and growth with flow-through cell.